Statistical Rethinking

A Bayesian Course with
Examples in R and Stan

CHAPMAN & HALL/CRC
Texts in Statistical Science Series

Series Editors

Francesca Dominici, *Harvard School of Public Health, USA*
Julian J. Faraway, *University of Bath, UK*
Martin Tanner, *Northwestern University, USA*
Jim Zidek, *University of British Columbia, Canada*

Texts in Statistical Science

Statistical Rethinking

A Bayesian Course with Examples in R and Stan

Richard McElreath

Max Planck Institute for Evolutionary Anthropology

Leipzig, Germany

CRC Press
Taylor & Francis Group
Boca Raton London New York

CRC Press is an imprint of the
Taylor & Francis Group an **informa** business

A CHAPMAN & HALL BOOK

CRC Press
Taylor & Francis Group
6000 Broken Sound Parkway NW, Suite 300
Boca Raton, FL 33487-2742

Printed on acid-free paper
Version Date: 20150910

International Standard Book Number-13: 978-1-4822-5344-3 (Hardback)

Library of Congress Cataloging-in-Publication Data

McElreath, Richard, 1973-
  Statistical rethinking : a Bayesian course with examples in R and Stan / Richard McElreath.
    pages cm. -- (Chapman & Hall/CRC texts in statistical science series ; 122)
  "A CRC title."
  Includes bibliographical references and index.
  ISBN 978-1-4822-5344-3 (alk. paper)
  1. Bayesian statistical decision theory. 2. R (Computer program language) I. Title.

QA279.5 .M3975
519.5'42--dc23   20150320089

Visit the Taylor & Francis Web site at
http://www.taylorandfrancis.com

and the CRC Press Web site at
http://www.crcpress.com

Printed in Canada

Contents

Preface

Masons, when they start upon a building,
Are careful to test out the scaffolding;

Make sure that planks won't slip at busy points,
Secure all ladders, tighten bolted joints.

And yet all this comes down when the job's done
Showing off walls of sure and solid stone.

So if, my dear, there sometimes seem to be
Old bridges breaking between you and me

Never fear. We may let the scaffolds fall
Confident that we have built our wall.

("Scaffolding" by Seamus Heaney, 1939–2013)

This book means to help you raise your knowledge of and confidence in statistical modeling. It is meant as a scaffold, one that will allow you to construct the wall that you need, even though you will discard it afterwards. As a result, this book teaches the material in often inconvenient fashion, forcing you to perform step-by-step calculations that are usually automated. The reason for all the algorithmic fuss is to ensure that you understand enough of the details to make reasonable choices and interpretations in your own modeling work. So although you will move on to use more automation, it's important to take things slow at first. Put up your wall, and then let the scaffolding fall.

Audience

The principle audience is researchers in the natural and social sciences, whether new PhD students or seasoned professionals, who have had a basic course on regression but nevertheless remain uneasy about statistical modeling. This audience accepts that there is something vaguely wrong about typical statistical practice in the early 21st century, dominated as it is by p-values and a confusing menagerie of testing procedures. They see alternative methods in journals and books. But these people are not sure where to go to learn about these methods.

As a consequence, this book doesn't really argue against p-values and the like. The problem in my opinion isn't so much p-values as the set of odd rituals that have evolved around them, in the wilds of the sciences, as well as the exclusion of so many other useful tools. So the book assumes the reader is ready to try doing statistical inference without p-values. This isn't the ideal situation. It would be better to have material that helps you spot common mistakes and misunderstandings of p-values and tests in general, as all of us have to understand such things, even if we don't use them. So I've tried to sneak in a little material of that kind, but unfortunately cannot devote much space to it. The book would be too long, and it would disrupt the teaching flow of the material.

It's important to realize, however, that the disregard paid to p-values is not a uniquely Bayesian attitude. Indeed, significance testing can be—and has been—formulated as a Bayesian procedure as well. So the choice to avoid significance testing is stimulated instead by epistemological concerns, some of which are briefly discussed in the first chapter.

Teaching strategy

The book uses much more computer code than formal mathematics. Even excellent mathematicians can have trouble understanding an approach, until they see a working algorithm. This is because implementation in code form removes all ambiguities. So material of this sort is easier to learn, if you also learn how to implement it.

In addition to any pedagogical value of presenting code, so much of statistics is now computational that a purely mathematical approach is anyways insufficient. As you'll see in later parts of this book, the same mathematical statistical model can sometimes be implemented in different ways, and the differences matter. So when you move beyond this book to more advanced or specialized statistical modeling, the computational emphasis here will help you recognize and cope with all manner of practical troubles.

Every section of the book is really just the tip of an iceberg. I've made no attempt to be exhaustive. Rather I've tried to explain something well. In this attempt, I've woven a lot of concepts and material into data analysis examples. So instead of having traditional units on, for example, centering predictor variables, I've developed those concepts in the context of a narrative about data analysis. This is certainly not a style that works for all readers. But it has worked for a lot of my students. I suspect it fails dramatically for those who are being forced to learn this information. For the internally motivated, it reflects how we really learn these skills in the context of our research.

How to use this book

This book is not a reference, but a course. It doesn't try to support random access. Rather, it expects sequential access. This has immense pedagogical advantages, but it has the disadvantage of violating how most scientists actually read books.

This book has a lot of code in it, integrated fully into the main text. The reason for this is that doing model-based statistics in the 21st century really requires programming, of at least a minor sort. The code is not optional. Everyplace, I have erred on the side of including too much code, rather than too little. In my experience teaching scientific programming, novices learn more quickly when they have working code to modify, rather than needing to write an algorithm from scratch. My generation was probably the last to have to learn some programming to use a computer, and so coding has gotten harder and harder to teach as time goes on. My students are very computer literate, but they have no idea what computer code looks like.

What the book assumes. This book does not try to teach the reader to program, in the most basic sense. It assumes that you have made a basic effort to learn how to install and process data in R. In most cases, a short introduction to R programming will be enough. I know many people have found Emmanuel Paradis' *R for Beginners* helpful. You can find it and many other beginner guides here:

http://cran.r-project.org/other-docs.html

To make use of this book, you should know already that y<-7 stores the value 7 in the symbol y. You should know that symbols which end in parentheses are functions. You should recognize a loop and understand that commands can be embedded inside other commands (recursion). Knowing that R *vectorizes* a lot of code, instead of using loops, is important. But you don't have to yet be confident with R programming.

Inevitably you will come across elements of the code in this book that you haven't seen before. I have made an effort to explain any particularly important or unusual programming tricks in my own code. In fact, this book spends a lot of time explaining code. I do this because students really need it. Unless they can connect each command to the recipe and the goal, when things go wrong, they won't know whether it is because of a minor or major error. The same issue arises when I teach mathematical evolutionary theory—students and colleagues often suffer from rusty algebra skills, so when they can't get the right answer, they often don't know whether it's because of some small mathematical misstep or instead some problem in strategy. The protracted explanations of code in this book aim to build a level of understanding that allows the reader to diagnose and fix problems.

Using the code. Code examples in the book are marked by a shaded box, and output from example code is often printed just beneath a shaded box, but marked by a fixed-width typeface. For example:

```
print( "All models are wrong, but some are useful." )
```
R code
0.1

```
[1] "All models are wrong, but some are useful."
```

Next to each snippet of code, you'll find a number that you can search for in the accompanying code snippet file, available from the book's website. The intention is that the reader follow along, executing the code in the shaded boxes and comparing their own output to that printed in the book. I really want you to execute the code, because just as one cannot learn martial arts by watching Bruce Lee movies, you can't learn to program statistical models by only reading a book. You have to get in there and throw some punches and, likewise, take some hits.

If you ever get confused, remember that you can execute each line independently and inspect the intermediate calculations. That's how you learn as well as solve problems. For example, here's a confusing way to multiply the numbers 10 and 20:

```
x <- 1:2
x <- x*10
x <- log(x)
x <- sum(x)
x <- exp(x)
x
```
R code
0.2

If you don't understand any particular step, you can always print out the contents of the symbol x immediately after that step. For the code examples, this is how you come to understand them. For your own code, this is how you find the source of any problems and then fix them.

Optional sections. Reflecting realism in how books like this are actually read, there are two kinds of optional sections: (1) Rethinking and (2) Overthinking. The Rethinking sections look like this:

> **Rethinking: Think again.** The point of these Rethinking boxes is to provide broader context for the material. They allude to connections to other approaches, provide historical background, or call out common misunderstandings. These boxes are meant to be optional, but they round out the material and invite deeper thought.

The Overthinking sections look like this:

Overthinking: Getting your hands dirty. These sections, set in smaller type, provide more detailed explanations of code or mathematics. This material isn't essential for understanding the main text. But it does have a lot of value, especially on a second reading. For example, sometimes it matters how you perform a calculation. Mathematics tells that these two expressions are equivalent:

$$p_1 = \log(0.01^{200})$$
$$p_2 = 200 \times \log(0.01)$$

But when you use R to compute them, they yield different answers:

```
( log( 0.01^200 ) )
( 200 * log(0.01) )
```

```
[1] -Inf
[1] -921.034
```

The second line is the right answer. This problem arises because of rounding error, when the computer rounds very small decimal values to zero. This loses *precision* and can introduce substantial errors in inference. As a result, we nearly always do statistical calculations using the logarithm of a probability, rather than the probability itself.

You can ignore most of these Overthinking sections on a first read.

The command line is the best tool. Programming at the level needed to perform 21st century statistical inference is not that complicated, but it is unfamiliar at first. Why not just teach the reader how to do all of this with a point-and-click program? There are big advantages to doing statistics with text commands, rather than pointing and clicking on menus.

Everyone knows that the command line is more powerful. But it also saves you time and fulfills ethical obligations. With a command script, each analysis documents itself, so that years from now you can come back to your analysis and replicate it exactly. You can re-use your old files and send them to colleagues. Pointing and clicking, however, leaves no trail of breadcrumbs. A file with your R commands inside it does. Once you get in the habit of planning, running, and preserving your statistical analyses in this way, it pays for itself many times over. With point-and-click, you pay down the road, rather than only up front. It is also a basic ethical requirement of science that our analyses be fully documented and repeatable. The integrity of peer review and the cumulative progress of research depend upon it. A command line statistical program makes this documentation natural. A point-and-click interface does not. Be ethical.

So we don't use the command line because we are hardcore or elitist (although we might be). We use the command line because it is better. It is harder at first. Unlike the point-and-click interface, you do have to learn a basic set of commands to get started with a command line interface. However, the ethical and cost saving advantages are worth the inconvenience.

How you should work. But I would be cruel, if I just told the reader to use a command-line tool, without also explaining something about how to do it. You do have to relearn some habits, but it isn't a major change. For readers who have only used menu-driven statistics software before, there will be some significant readjustment. But after a few days, it will seem natural to you. For readers who have used command-driven statistics software like Stata and SAS, there is still some readjustment ahead. I'll explain the overall approach first. Then I'll say why even Stata and SAS users are in for a change.

First, the sane approach to scripting statistical analyses is to work back and forth between two applications: (1) a *plain text editor* of your choice and (2) the R program itself. A plain text editor is a program that creates and edits simple formatting-free text files. Common examples include Notepad (in Windows) and TextEdit (in Mac OS X) and Emacs (in most *NIX distributions, including Mac OS X). There is also a wide selection of fancy text editors specialized for programmers. You might investigate for example RStudio and the Atom text editor, both of which are free. Note that MSWord files are not plain text. Do not use them.

You will use a plain text editor to keep a running log of the commands you feed into the R application for processing. You absolutely do not want to just type out commands directly into R itself. Instead, you want to either copy and paste lines of code from your plain text editor into R, or instead read entire script files directly into R. You might enter commands directly into R as you explore data or debug or merely play. But your serious work should be implemented through the plain text editor, for the reasons explained in the previous section.

You can add comments to your R scripts to help you plan the code and remember later what the code is doing. To make a comment, just begin a line with the # symbol. To help clarify the approach, below I provide a very short complete script for running a linear regression on one of R's built-in sets of data. Even if you don't know what the code does yet, hopefully you will see it as a basic model of clarity of formatting and use of comments.

R code
0.4

```
# Load the data:
# car braking distances in feet paired with speeds in km/h
# see ?cars for details
data(cars)

# fit a linear regression of distance on speed
m <- lm( dist ~ speed , data=cars )

# estimated coefficients from the model
coef(m)

# plot residuals against speed
plot( resid(m) ~ speed , data=cars )
```

Finally, even those who are familiar with scripting Stata or SAS will be in for some readjustment. Programs like Stata and SAS have a different paradigm for how information is processed. In those applications, procedural commands like PROC GLM are issued in imitation of menu commands. These procedures produce a mass of default output that the user then sifts through. R does not behave this way. Instead, R forces the user to decide which bits of information she wants. One fits a statistical model in R and then must issue later commands to ask questions about it. This more interrogative paradigm will become familiar through the examples in the text. But be aware that you are going to take a more active role in deciding what questions to ask about your models.

Installing the `rethinking` R package

The code examples require that you have installed the `rethinking` R package. This package contains the data examples and many of the modeling tools that the text uses. The `rethinking` package itself relies upon another package, `rstan`, for fitting the more advanced models in the second half of the book.

You should install `rstan` first. Navigate your internet browser to `mc-stan.org` and follow the instructions for your platform. You will need to install both a C++ compiler (also called the "tool chain") and the `rstan` package. Instructions for doing both are at `mc-stan.org`.

Then from within R, you can install `rethinking` and its dependencies with this code:

R code
0.5
```
install.packages(c("coda","mvtnorm","devtools"))
library(devtools)
devtools::install_github("rmcelreath/rethinking")
```

Note that `rethinking` is not on the CRAN package archive, at least not yet. There's no real benefit to having a package on CRAN. You'll always be able to perform a simple internet search and figure out the current installation instructions for the most recent version of the `rethinking` package. If you encounter any bugs while using the package, you can check `github.com/rmcelreath/rethinking` to see if a solution is already posted. If not, you can leave a bug report and be notified when a solution becomes available. In addition, all of the source code for the package is found there, in case you aspire to do some tinkering of your own. Feel free to "fork" the package and bend it to your will.

Acknowledgments

Many people have contributed advice, ideas, and complaints to this book. Most important among them have been the graduate students who have taken statistics courses from me over the last decade, as well as the colleagues who have come to me for advice. These people taught me how to teach them this material, and in some cases I learned the material only because they needed it. A large number of individuals donated their time to comment on sections of the book or accompanying computer code. These include: Rasmus Bååth, Ryan Baldini, Bret Beheim, Maciek Chudek, John Durand, Andrew Gelman, Ben Goodrich, Mark Grote, Dave Harris, Chris Howerton, James Holland Jones, Jeremy Koster, Andrew Marshall, Sarah Mathew, Karthik Panchanathan, Pete Richerson, Alan Rogers, Cody Ross, Noam Ross, Aviva Rossi, Kari Schroeder, Paul Smaldino, Rob Trangucci, Shravan Vasishth, Annika Wallin, and a score of anonymous reviewers. Bret Beheim and Dave Harris were brave enough to provide extensive comments on an early draft. Caitlin DeRango and Kotrina Kajokaite invested their time in improving several chapters and problem sets. Mary Brooke McEachern provided crucial opinions on content and presentation, as well as calm support and tolerance. A number of anonymous reviewers provided detailed feedback on individual chapters. None of these people agree with all of the choices I have made, and all mistakes and deficiencies remain my responsibility. But especially when we haven't agreed, their opinions have made the book stronger.

The book is dedicated to Dr. Parry M. R. Clarke (1977–2012), who asked me to write it. Parry's inquisition of statistical and mathematical and computational methods helped everyone around him. He made us better.

1 The Golem of Prague

In the 16th century, the House of Habsburg controlled much of Central Europe, the Netherlands, and Spain, as well as Spain's colonies in the Americas. The House was maybe the first true world power, with the Sun always shining on some portion of it. Its ruler was also Holy Roman Emperor, and his seat of power was Prague. The Emperor in the late 16th century, Rudolph II, loved intellectual life. He invested in the arts, the sciences (including astrology and alchemy), and mathematics, making Prague into a world center of learning and scholarship. It is appropriate then that in this learned atmosphere arose an early robot, the Golem of Prague.

A golem (GOH-lem) is a clay robot known in Jewish folklore, constructed from dust and fire and water. It is brought to life by inscribing *emet*, Hebrew for "truth," on its brow. Animated by truth, but lacking free will, a golem always does exactly what it is told. This is lucky, because the golem is incredibly powerful, able to withstand and accomplish more than its creators could. However, its obedience also brings danger, as careless instructions or unexpected events can turn a golem against its makers. Its abundance of power is matched by its lack of wisdom.

In some versions of the golem legend, Rabbi Judah Loew ben Bezalel sought a way to defend the Jews of Prague. As in many parts of 16th century Central Europe, the Jews of Prague were persecuted. Using secret techniques from the *Kabbalah*, Rabbi Judah was able to build a golem, animate it with "truth," and order it to defend the Jewish people of Prague. Not everyone agreed with Judah's action, fearing unintended consequences of toying with the power of life. Ultimately Judah was forced to destroy the golem, as its combination of extraordinary power with clumsiness eventually led to innocent deaths. Wiping away one letter from the inscription *emet* to spell instead *met*, "death," Rabbi Judah decommissioned the robot.

1.1. Statistical golems

Scientists also make golems.[1] Our golems rarely have physical form, but they too are often made of clay, living in silicon as computer code. These golems are scientific models. But these golems have real effects on the world, through the predictions they make and the intuitions they challenge or inspire. A concern with "truth" enlivens these models, but just like a golem or a modern robot, scientific models are neither true nor false, neither prophets nor charlatans. Rather they are constructs engineered for some purpose. These constructs are incredibly powerful, dutifully conducting their programmed calculations.

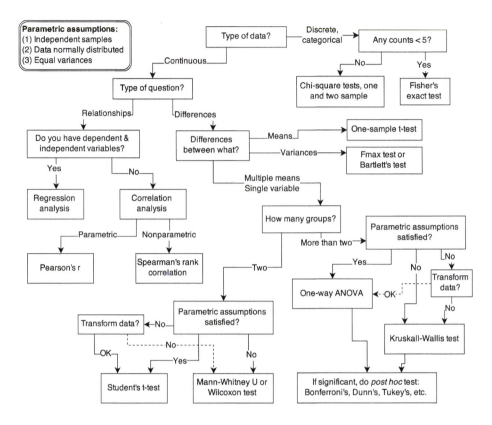

FIGURE 1.1. Example decision tree, or flowchart, for selecting an appropriate statistical procedure. Beginning at the top, the user answers a series of questions about measurement and intent, arriving eventually at the name of a procedure. Many such decision trees are possible.

Sometimes their unyielding logic reveals implications previously hidden to their designers. These implications can be priceless discoveries. Or they may produce silly and dangerous behavior. Rather than idealized angels of reason, scientific models are powerful clay robots without intent of their own, bumbling along according to the myopic instructions they embody. Like with Rabbi Judah's golem, the golems of science are wisely regarded with both awe and apprehension. We absolutely have to use them, but doing so always entails some risk.

There are many kinds of statistical models. Whenever someone deploys even a simple statistical procedure, like a classical t-test, she is deploying a small golem that will obediently carry out an exact calculation, performing it the same way (nearly[2]) every time, without complaint. Nearly every branch of science relies upon the senses of statistical golems. In many cases, it is no longer possible to even measure phenomena of interest, without making use of a model. To measure the strength of natural selection or the speed of a neutrino or the number of species in the Amazon, we must use models. The golem is a prosthesis, doing the measuring for us, performing impressive calculations, finding patterns where none are obvious.

However, there is no wisdom in the golem. It doesn't discern when the context is inappropriate for its answers. It just knows its own procedure, nothing else. It just does as it's told.

And so it remains a triumph of statistical science that there are now so many diverse golems, each useful in a particular context. Viewed this way, statistics is neither mathematics nor a science, but rather a branch of engineering. And like engineering, a common set of design principles and constraints produces a great diversity of specialized applications.

This diversity of applications helps to explain why introductory statistics courses are so often confusing to the initiates. Instead of a single method for building, refining, and critiquing statistical models, students are offered a zoo of pre-constructed golems known as "tests." Each test has a particular purpose. Decision trees, like the one in FIGURE 1.1, are common. By answering a series of sequential questions, users choose the "correct" procedure for their research circumstances.

Unfortunately, while experienced statisticians grasp the unity of these procedures, students and researchers rarely do. Advanced courses in statistics do emphasize engineering principles, but most scientists never get that far. Teaching statistics this way is somewhat like teaching engineering backwards, starting with bridge building and ending with basic physics. So students and many scientists tend to use charts like FIGURE 1.1 without much thought to their underlying structure, without much awareness of the models that each procedure embodies, and without any framework to help them make the inevitable compromises required by real research. It's not their fault.

For some, the toolbox of pre-manufactured golems is all they will ever need. Provided they stay within well-tested contexts, using only a few different procedures in appropriate tasks, a lot of good science can be completed. This is similar to how plumbers can do a lot of useful work without knowing much about fluid dynamics. Serious trouble begins when scholars move on to conducting innovative research, pushing the boundaries of their specialties. It's as if we got our hydraulic engineers by promoting plumbers.

Why aren't the tests enough for innovative research? The classical procedures of introductory statistics tend to be inflexible and fragile. By inflexible, I mean that they have very limited ways to adapt to unique research contexts. By fragile, I mean that they fail in unpredictable ways when applied to new contexts. This matters, because at the boundaries of most sciences, it is hardly ever clear which procedure is appropriate. None of the traditional golems has been evaluated in novel research settings, and so it can be hard to choose one and then to understand how it behaves. A good example is *Fisher's exact test*, which applies (exactly) to an extremely narrow empirical context, but is regularly used whenever cell counts are small. I have personally read hundreds of uses of Fisher's exact test in scientific journals, but aside from Fisher's original use of it, I have never seen it used appropriately. Even a procedure like ordinary linear regression, which is quite flexible in many ways, being able to encode a large diversity of interesting hypotheses, is sometimes fragile. For example, if there is substantial measurement error on prediction variables, then the procedure can fail in spectacular ways. But more importantly, it is nearly always possible to do better than ordinary linear regression, largely because of a phenomenon known as OVERFITTING (Chapter 6).

The point isn't that statistical tools are specialized. Of course they are. The point is that classical tools are not diverse enough to handle many common research questions. Every active area of science contends with unique difficulties of measurement and interpretation, converses with idiosyncratic theories in a dialect barely understood by other scientists from other tribes. Statistical experts outside the discipline can help, but they are limited by lack of fluency in the empirical and theoretical concerns of the discipline. In such settings, pre-manufactured golems may do nothing useful at all. Worse, they might wreck Prague. And if we keep adding new types of tools, soon there will be far too many to keep track of.

Instead, what researchers need is some unified theory of golem engineering, a set of principles for designing, building, and refining special-purpose statistical procedures. Every major branch of statistical philosophy possesses such a unified theory. But the theory is never taught in introductory—and often not even in advanced—courses. So there are benefits in rethinking statistical inference as a set of strategies, instead of a set of pre-made tools.

1.2. Statistical rethinking

A lot can go wrong with statistical inference, and this is one reason that beginners are so anxious about it. When the framework is to choose a pre-made test from a flowchart, then the anxiety can mount as one worries about choosing the "correct" test. Statisticians, for their part, can derive pleasure from scolding scientists, which just makes the psychological battle worse.

But anxiety can be cultivated into wisdom. That is the reason that this book insists on working with the computational nuts and bolts of each golem. If you don't understand how the golem processes information, then you can't interpret the golem's output. This requires knowing the statistical model in greater detail than is customary, and it requires doing the computations the hard way, at least until you are wise enough to use the push-button solutions.

There are conceptual obstacles as well, obstacles with how scholars define statistical objectives and interpret statistical results. Understanding any individual golem is not enough, in these cases. Instead, we need some statistical epistemology, an appreciation of how statistical models relate to hypotheses and the natural mechanisms of interest. What are we supposed to be doing with these little computational machines, anyway?

The greatest obstacle that I encounter among students and colleagues is the tacit belief that the proper objective of statistical inference is to test null hypotheses.[3] This is the proper objective, the thinking goes, because Karl Popper argued that science advances by falsifying hypotheses. Karl Popper (1902–1994) is possibly the most influential philosopher of science, at least among scientists. He did persuasively argue that science works better by developing hypotheses that are, in principle, falsifiable. Seeking out evidence that might embarrass our ideas is a normative standard, and one that most scholars—whether they describe themselves as scientists or not—subscribe to. So maybe statistical procedures should falsify hypotheses, if we wish to be good statistical scientists.

But the above is a kind of folk Popperism, an informal philosophy of science common among scientists but not among philosophers of science. Science is not described by the falsification standard, as Popper recognized and argued.[4] In fact, deductive falsification is impossible in nearly every scientific context. In this section, I review two reasons for this impossibility.

(1) Hypotheses are not models. The relations among hypotheses and different kinds of models are complex. Many models correspond to the same hypothesis, and many hypotheses correspond to a single model. This makes strict falsification impossible.

(2) Measurement matters. Even when we think the data falsify a model, another observer will debate our methods and measures. They don't trust the data. Sometimes they are right.

For both of these reasons, deductive falsification never works. The scientific method cannot be reduced to a statistical procedure, and so our statistical methods should not pretend. Statistical evidence is part of the hot mess that is science, with all of its combat and egotism and mutual coercion. If you believe, as I do, that science does very often work, then learning that it doesn't work via falsification shouldn't change your mind. But it might help you do better science, because it will open your eyes to the many legitimately useful functions of statistical golems.

> **Rethinking: Is NHST falsificationist?** Null hypothesis significance testing, NHST, is often identified with the falsificationist, or Popperian, philosophy of science. However, usually NHST is used to falsify a null hypothesis, not the actual research hypothesis. So the falsification is being done to something other than the explanatory model. This seems the reverse from Karl Popper's philosophy.[5]

1.2.1. Hypotheses are not models. When we attempt to falsify a hypothesis, we must work with a model of some kind. Even when the attempt is not explicitly statistical, there is always a tacit model of measurement, of evidence, that operationalizes the hypothesis. All models are false,[6] so what does it mean to falsify a model? One consequence of the requirement to work with models is that it's no longer possible to deduce that a hypothesis is false, just because we reject a model derived from it.

Let's explore this consequence in the context of an example from population biology (FIGURE 1.2). Beginning in the 1960s, many evolutionary biologists became interested in the proposal that the vast majority of evolution—changes in gene frequency—are caused not by natural selection, by rather by mutation and drift. No one really doubted that natural selection is responsible for functional design. This was a debate about genetic sequences. So began several productive decades of scholarly combat over "neutral" models of molecular evolution.[7] This combat is most strongly associated with Motoo Kimura (1924–1994), who was perhaps the strongest advocate of neutral models. But many other population geneticists participated. As time has passed, related disciplines such as community ecology[8] and anthropology[9] have experienced (or are currently experiencing) their own versions of the neutrality debate.

Let's use the schematic in FIGURE 1.2 to explore connections between motivating hypotheses and different models, in the context of the neutral evolution debate. On the left, there are two stereotyped, informal hypotheses: Either evolution is "neutral" (H_0) or natural selection matters somehow (H_1). These hypotheses have vague boundaries, because they begin as verbal conjectures, not precise models. There are thousands of possible detailed processes that can be described as "neutral," depending upon choices about, for example, population structure, number of sites, number of alleles at each site, mutation rates, and recombination.

Once we have made these choices, we have the middle column in FIGURE 1.2, detailed process models of evolution. P_{0A} and P_{0B} differ in that one assumes the population size and structure have been constant long enough for the distribution of alleles to reach a steady state. The other imagines instead that population size fluctuates through time, which can be true even when there is no selective difference among alleles. The "selection matters" hypothesis H_1 likewise corresponds to many different process models. I've shown two big players: a model in which selection always favors certain alleles and another in which selection fluctuates through time, favoring different alleles.[10]

In order to challenge these process models with evidence, they have to be made into statistical models. This usually means deriving the expected frequency distribution of some quantity—a "statistic"—in the model. For example, a very common statistic in this context is the frequency distribution (histogram) of the frequency of different genetic variants (alleles).

Hypotheses   Process models   Statistical models

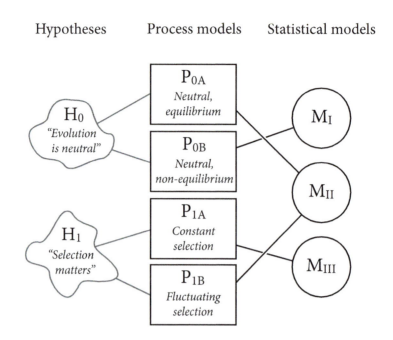

FIGURE 1.2. Relations among hypotheses (left), detailed process models (middle), and statistical models (right), illustrated by the example of "neutral" models of evolution. Hypotheses (H) are typically vague, and so correspond to more than one process model (P). Statistical evaluations of hypotheses rarely address process models directly. Instead, they rely upon statistical models (M), all of which reflect only some aspects of the process models. As a result, relations are multiple in both directions: Hypotheses do not imply unique models, and models do not imply unique hypotheses. This fact greatly complicates statistical inference.

Some alleles are rare, appearing in only a few individuals. Others are very common, appearing in very many individuals in the population. A famous result in population genetics is that a model like P_{0A} produces a *power law* distribution of allele frequencies. And so this fact yields a statistical model, M_{II}, that predicts a power law in the data. In contrast the constant selection process model P_{1A} predicts something quite different, M_{III}.

Unfortunately, other selection models (P_{1B}) imply the same statistical model, M_{II}, as the neutral model. They also produce power laws. So we've reached the uncomfortable lesson:

(1) Any given statistical model (M) may correspond to more than one process model (P).

(2) Any given hypothesis (H) may correspond to more than one process model (P).

(3) Any given statistical model (M) may correspond to more than one hypothesis (H).

Now look what happens when we compare the statistical models to data. The classical approach is to take the "neutral" model as a null hypothesis. If the data are not sufficiently similar to the expectation under the null, then we say that we "reject" the null hypothesis. Suppose we follow the history of this subject and take P_{0A} as our null hypothesis. This implies data corresponding to M_{II}. But since the same statistical model corresponds to a selection model P_{1B}, it's not at all clear what we are to make of either rejecting or accepting the null. The null model is not unique to any process model nor hypothesis. If we reject the null, we can't really conclude that selection matters, because there are other neutral models that predict different distributions of alleles. And if we fail to reject the null, we can't really conclude that evolution is neutral, because some selection models expect the same frequency distribution.

This is a huge bother. Once we have the diagram in FIGURE 1.2, it's easy to see the problem. But few of us are so lucky. While population genetics has recognized this issue, scholars in other disciplines continue to test frequency distributions against power law expectations, arguing even that there is only one neutral model.[11] Even if there were only one neutral model, there are so many non-neutral models that mimic the predictions of neutrality, that neither rejecting nor failing to reject the null model carries much inferential power.

And while you might think that more routine statistical models, like linear regressions (Chapter 4), don't carry such risk, think again. A typical "null" in these contexts is just that there is zero *average* difference between groups. But there are usually many different ways for this average to be close to or consistent with zero, just as there are many different ways to get a power law. This recognition lies behind many common practices in statistical inference, such as consideration of unobserved variables and sampling bias.

So what can be done? Well, if you have multiple process models, a lot can be done. If it turns out that all of the process models of interest make very similar predictions, then you know to search for a different description of the evidence, a description under which the processes look different. For example, while P_{0A} and P_{1B} make very similar power law predictions for the frequency distribution of alleles, they make very dissimilar predictions for the distribution of changes in allele frequency over time. In other words, explicitly compare predictions of more than one model, and you can save yourself from some ordinary kinds of folly.

Rethinking: Entropy and model identification. One reason that statistical models routinely correspond to many different detailed process models is because they rely upon distributions like the normal, binomial, Poisson, and others. These distributions are members of a family, the EXPONENTIAL FAMILY. Nature loves the members of this family. Nature loves them because nature loves entropy, and all of the exponential family distributions are MAXIMUM ENTROPY distributions. Taking the natural personification out of that explanation will wait until Chapter 9. The practical implication is that one can no more infer evolutionary process from a power law than one can infer developmental process from the fact that height is normally distributed. This fact should make us humble about what typical regression models—the meat of this book—can teach us about mechanistic process. On the other hand, the maximum entropy nature of these distributions means we can use them to do useful statistical work, even when we can't identify the underlying process. Not only can we not identify it, but we don't have to.

1.2.2. Measurement matters.

The logic of falsification is very simple. We have a hypothesis H, and we show that it entails some observation D. Then we look for D. If we don't find it, we must conclude that H is false. Logicians call this kind of reasoning *modus tollens*, which is Latin shorthand for "the method of destruction." In contrast, finding D tells us nothing certain about H, because other hypotheses might also predict D.

A compelling scientific fable that employs *modus tollens* concerns the color of swans. Before discovering Australia, all swans that any European had ever seen had white feathers. This led to the belief that all swans are white. Let's call this a formal hypothesis:

H_0: All swans are white.

When Europeans reached Australia, however, they encountered swans with black feathers. This evidence seemed to instantly prove H_0 to be false. Indeed, not all swans are white. Some are certainly black, according to all observers. The key insight here is that, before voyaging to Australia, no number of observations of white swans could prove H_0 to be true. However it required only one observation of a black swan to prove it false.

This is a seductive story. If we can believe that important scientific hypotheses can be stated in this form, then we have a powerful method for improving the accuracy of our theories: look for evidence that disconfirms our hypotheses. Whenever we find a black swan, H_0 must be false. Progress!

Seeking disconfirming evidence is important, but it cannot be as powerful as the swan story makes it appear. In addition to the correspondence problems among hypotheses and models, discussed in the previous section, most of the problems scientists confront are not so logically discrete. Instead, we most often face two simultaneous problems that make the swan fable misrepresentative. First, observations are prone to error, especially at the boundaries of scientific knowledge. Second, most hypotheses are quantitative, concerning degrees of existence, rather than discrete, concerning total presence or absence. Let's briefly consider each of these problems.

1.2.2.1. *Observation error.* All observers will agree under most conditions that a swan is either black or white. There are few intermediate shades, and most observers' eyes work similarly enough that there will be little, if any, disagreement about which swans are white and which are black. But this kind of example is hardly commonplace in science, at least in mature fields. Instead, we routinely confront contexts in which we are not sure if we have detected a disconfirming result. At the edges of scientific knowledge, the ability to measure a hypothetical phenomenon is often in question as much as the phenomenon itself.

Here are two examples.

In 2005, a team of ornithologists from Cornell claimed to have evidence of an individual Ivory-billed Woodpecker (*Campephilus principalis*), a species thought extinct. The hypothesis implied here is:

H_0: The Ivory-billed Woodpecker is extinct.

It would only take one observation to falsify this hypothesis. However, many doubted the evidence. Despite extensive search efforts and a $50,000 cash reward for information leading to a live specimen, no evidence satisfying all parties has yet (by 2015) emerged. Even if good physical evidence does eventually arise, this episode should serve as a counterpoint to the swan story. Finding disconfirming cases is complicated by the difficulties of observation. Black swans are not always really black swans, and sometimes white swans are really black swans. There are mistaken confirmations (false positives) and mistaken disconfirmations (false negatives). Against this background of measurement difficulties, scientists who already believe that the Ivory-billed Woodpecker is extinct will always be suspicious of a claimed falsification. Those who believe it is still alive will tend to count the vaguest evidence as falsification.

Another example, this one from physics, focuses on the detection of faster-than-light (FTL) neutrinos.[12] In September 2011, a large and respected team of physicists announced detection of neutrinos—small, neutral sub-atomic particles able to pass easily and harmlessly through most matter—that arrived from Switzerland to Italy in slightly faster-than-lightspeed time. According to Einstein, neutrinos cannot travel faster than the speed of light. So this seems to be a falsification of special relativity. If so, it would turn physics on its head.

The dominant reaction from the physics community was not "Einstein was wrong!" but instead "How did the team mess up the measurement?" The team that made the measurement had the same reaction, and asked others to check their calculations and attempt to replicate the result.

What could go wrong in the measurement? You might think measuring speed is a simple matter of dividing distance by time. It is, at the scale and energy you live at. But with a fundamental particle like a neutrino, if you measure when it starts its journey, you stop the journey. The particle is consumed by the measurement. So more subtle approaches are needed. The detected difference from light-speed, furthermore, is quite small, and so even the latency of the time it takes a signal to travel from a detector to a control room can be orders of magnitude larger. And since the "measurement" in this case is really an estimate from a statistical model, all of the assumptions of the model are now suspect. By 2013, the physics community was unanimous that the FTL neutrino result was measurement error. They found the technical error, which involved a poorly attached cable, among other things.[13] Furthermore, neutrinos clocked from supernova events are consistent with Einstein, and those distances are much larger and so would reveal differences in speed much better.

In both the woodpecker and neutrino dramas, the key dilemma is whether the falsification is real or spurious. Measurement is complicated in both cases, but in quite different ways, rendering both true-detection and false-detection plausible. Popper himself was aware of this limitation inherent in measurement, and it may be one reason that Popper himself saw science as being broader than falsification. But the probabilistic nature of evidence rarely appears when practicing scientists discuss the philosophy and practice of falsification.[14] My reading of the history of science is that these sorts of measurement problems are the norm, not the exception.[15]

1.2.2.2. *Continuous hypotheses.* Another problem for the swan story is that most interesting scientific hypotheses are not of the kind "all swans are white" but rather of the kind:

$$H_0\text{: 80\% of swans are white.}$$

Or maybe:

$$H_0\text{: Black swans are rare.}$$

Now what are we to conclude, after observing a black swan? The null hypothesis doesn't say black swans do not exist, but rather that they have some frequency. The task here is not to disprove or prove a hypothesis of this kind, but rather to estimate and explain the distribution of swan coloration as accurately as we can. Even when there is no measurement error of any kind, this problem will prevent us from applying the *modus tollens* swan story to our science.[16]

You might object that the hypothesis above is just not a good scientific hypothesis, because it isn't easy to disprove. But if that's the case, then most of the important questions about the world are not good scientific hypotheses. In that case, we should conclude that the definition of a "good hypothesis" isn't doing us much good. Now, nearly everyone agrees that it is a good practice to design experiments and observations that can differentiate competing hypotheses. But in many cases, the comparison must be probabilistic, a matter of degree, not kind.[17]

1.2.3. Falsification is consensual. The scientific community does come to regard some hypotheses as false. The caloric theory of heat and the geocentric model of the universe are no longer taught in science courses, unless it's to teach how they were falsified. And evidence often—but not always—has something to do with such falsification.

But falsification is always *consensual*, not *logical*. In light of the real problems of measurement error and the continuous nature of natural phenomena, scientific communities argue towards consensus about the meaning of evidence. These arguments can be messy. After the fact, some textbooks misrepresent the history so it appears like logical falsification.[18] Such historical revisionism may hurt everyone. It may hurt scientists, by rendering it impossible for their own work to live up to the legends that precede them. It may make science an easy target, by promoting an easily attacked model of scientific epistemology. And it may hurt the public, by exaggerating the definitiveness of scientific knowledge.[19]

1.3. Three tools for golem engineering

So if attempting to mimic falsification is not a generally useful approach to statistical methods, what are we to do? We are to model. Models can be made into testing procedures— all statistical tests are also models[20]—but they can also be used to measure, forecast, and argue. Doing research benefits from the ability to produce and manipulate statistical models, both because scientific problems are more general than "testing" and because the pre-made golems you maybe met in introductory statistics courses are ill-fit to many research contexts. If you want to reduce your chances of wrecking Prague, then some golem engineering know-how is needed. Make no mistake: You will wreck Prague eventually. But if you are a good golem engineer, at least you'll notice the destruction. And since you'll know a lot about how your golem works, you stand a good chance to figure out what went wrong. Then your next golem won't be as bad. Without the engineering training, you're always at someone else's mercy.

It can be hard to get a good education in statistical model building and criticism, though. Applied statistical modeling in the early 21st century is marked by the heavy use of several engineering tools that are almost always absent from introductory, and even many advanced, statistics courses. These tools aren't really new, but they are newly popular. And many of the recent advances in statistical inference depend upon computational innovations that feel more like computer science than classical statistics, so it's not clear who is responsible for teaching them, if anyone.

There are many tools worth learning. In this book I've chosen to focus on three broad ones that are in demand in both the social and biological sciences. These tools are:

(1) Bayesian data analysis
(2) Multilevel models
(3) Model comparison using information criteria

These tools are deeply related to one another, so it makes sense to teach them together. Understanding of these tools comes, as always, only with implementation—you can't comprehend golem engineering until you do it. And so this book focuses mostly on code, how to do things. But in the rest of this section, I provide brief introductions to the three tools.

1.3.1. Bayesian data analysis. For the classical Greeks and Romans, wisdom and chance were enemies. Minerva (Athena), symbolized by the owl, was the personification of wisdom. Fortuna (Tyche), symbolized by the wheel of fortune, was the personification of luck, both good and bad. Minerva was mindful and measuring, while Fortuna was fickle and unreliable. Only a fool would rely on Fortuna, while all wise folk appealed to Minerva.[21]

The rise of probability theory changed that. Statistical inference compels us instead to rely on Fortuna as a servant of Minerva, to use chance and uncertainty to discover reliable knowledge. All flavors of statistical inference have this motivation. But Bayesian data analysis embraces it most fully, by using the language of chance to describe the plausibility of different possibilities.

There are many ways to use the term "Bayesian." But mainly it denotes a particular interpretation of probability. In modest terms, Bayesian inference is no more than counting the numbers of ways things can happen, according to our assumptions. Things that can happen more ways are more plausible. And since probability theory is just a calculus for counting, this means that we can use probability theory as a general way to represent plausibility, whether in reference to countable events in the world or rather theoretical constructs like parameters. Once you accept this gambit, the rest follows logically. Once we have defined our assumptions, Bayesian inference forces a purely logical way of processing that information to produce inference.

Chapter 2 explains this in depth. For now, it will help to have another probability concept to compare. Bayesian probability is a very general approach to probability, and it includes as a special case another important approach, the FREQUENTIST approach. The frequentist approach requires that all probabilities be defined by connection to countable events and their frequencies in very large samples.[22] This leads to frequentist uncertainty being premised on imaginary resampling of data—if we were to repeat the measurement many many times, we would end up collecting a list of values that will have some pattern to it. It means also that parameters and models cannot have probability distributions, only measurements can. The distribution of these measurements is called a SAMPLING DISTRIBUTION. This resampling is never done, and in general it doesn't even make sense—it is absurd to consider repeat sampling of the diversification of song birds in the Andes. As Sir Ronald Fisher, one of the most important frequentist statisticians of the 20th century, put it:

> [...] the only populations that can be referred to in a test of significance have no objective reality, being exclusively the product of the statistician's imagination [...][23]

But in many contexts, like controlled greenhouse experiments, it's a useful device for describing uncertainty. Whatever the context, it's just part of the model, an assumption about what the data would look like under resampling. It's just as fantastical as the Bayesian gambit of using probability to describe all types of uncertainty, whether empirical or epistemological.[24]

But these different attitudes towards probability do enforce different trade-offs. It will help to encounter a simple example where the difference between Bayesian and frequentist probability matters. In the year 1610, Galileo turned a primitive telescope to the night sky and became the first human to see Saturn's rings. Well, he probably saw a blob, with some smaller blobs attached to it (FIGURE 1.3). Since the telescope was primitive, it couldn't really focus the image very well. Saturn always appeared blurred. This is a statistical problem, of a sort. There's uncertainty about the planet's shape, but notice that none of the uncertainty is a result of variation in repeat measurements. We could look through the telescope a thousand times, and it will always give the same blurred image (for any given position of the Earth and Saturn). So the sampling distribution of any measurement is constant, because the measurement is deterministic—there's nothing "random" about it. Frequentist statistical inference has a lot of trouble getting started here. In contrast, Bayesian inference proceeds as usual, because the deterministic "noise" can still be modeled using probability, as long as we don't

FIGURE 1.3. Saturn, much like Galileo must have seen it. The true shape is uncertain, but not because of any sampling variation. Probability theory can still help.

identify probability with frequency. As a result, the field of image reconstruction and processing is dominated by Bayesian algorithms.[25]

In more routine statistical procedures, like linear regression, this difference in probability concepts has less of an effect. However, it is important to realize that even when a Bayesian procedure and frequentist procedure give exactly the same answer, our Bayesian golems aren't justifying their inferences with imagined repeat sampling. More generally, Bayesian golems treat "randomness" as a property of information, not of the world. Nothing in the real world—excepting controversial interpretations of quantum physics—is actually random. Presumably, if we had more information, we could exactly predict everything. We just use randomness to describe our uncertainty in the face of incomplete knowledge. From the perspective of our golem, the coin toss is "random," but it's really the golem that is random, not the coin.

Note that the preceding description of Bayesian analysis doesn't invoke anyone's "beliefs" or subjective opinions. Bayesian data analysis is just a logical procedure for processing information. There is a tradition of using this procedure as a normative description of rational belief, a tradition called BAYESIANISM.[26] But this book neither describes nor advocates it.

Rethinking: Probability is not unitary. It will make some readers uncomfortable to suggest that there is more than one way to define "probability." Aren't mathematical concepts uniquely correct? They are not. Once you adopt some set of premises, or axioms, everything does follow logically in mathematical systems. But the axioms are open to debate and interpretation. So not only is there "Bayesian" and "frequentist" probability, but there are different versions of Bayesian probability even, relying upon different arguments to justify the approach. In more advanced Bayesian texts, you'll come across names like Bruno de Finetti, Richard T. Cox, and Leonard "Jimmie" Savage. Each of these figures is associated with a somewhat different conception of Bayesian probability. There are others. This book mainly follows the "logical" Cox (or Laplace-Jeffreys-Cox-Jaynes) interpretation. This interpretation is presented beginning in the next chapter, but unfolds fully only in Chapter 9.

How can different interpretations of probability theory thrive? By themselves, mathematical entities don't necessarily "mean" anything, in the sense of real world implication. What does it mean to take the square root of a negative number? What does mean to take a limit as something approaches infinity? These are essential and routine concepts, but their meanings depend upon context and analyst, upon beliefs about how well abstraction represents reality. Mathematics doesn't access the real world directly. So answering such questions remains a contentious and entertaining project, in all branches of applied mathematics. So while everyone subscribes to the same axioms of probability, not everyone agrees in all contexts about how to interpret probability.

Before moving on to describe the next two tools, it's worth emphasizing an advantage of Bayesian data analysis, at least when scholars are learning statistical modeling. This entire book could be rewritten to remove any mention of "Bayesian." In places, it would become easier. In others, it would become much harder. But having taught applied statistics both ways, I have found that the Bayesian framework presents a distinct pedagogical advantage: many people find it more intuitive. Perhaps best evidence for this is that very many scientists interpret non-Bayesian results in Bayesian terms, for example interpreting ordinary p-values as Bayesian posterior probabilities and non-Bayesian confidence intervals as Bayesian ones (you'll learn posterior probability and confidence intervals in Chapters 2 and 3). Even statistics instructors make these mistakes.[27] In this sense then, Bayesian models lead to more intuitive interpretations, the ones scientists tend to project onto statistical results. The opposite pattern of mistake—interpreting a posterior probability as a p-value—seems to happen only rarely, if ever.

None of this ensures that Bayesian models will be more correct than non-Bayesian models. It just means that the scientist's intuitions will less commonly be at odds with the actual logic of the framework. This simplifies some of the aspects of teaching statistical modeling.

Rethinking: A little history. Bayesian statistical inference is much older than the typical tools of introductory statistics, most of which were developed in the early 20th century. Versions of the Bayesian approach were applied to scientific work in the late 1700s and repeatedly in the 19th century. But after World War I, anti-Bayesian statisticians, like Sir Ronald Fisher, succeeded in marginalizing the approach. All Fisher said about Bayesian analysis (then called *inverse probability*) in his influential 1925 handbook was:

> [...] the theory of inverse probability is founded upon an error, and must be wholly rejected.[28]

Bayesian data analysis became increasingly accepted within statistics during the second half of the 20th century, because it proved not to be founded upon an error. All philosophy aside, it worked. Beginning in the 1990s, new computational approaches led to a rapid rise in application of Bayesian methods.[29] Bayesian methods remain computationally expensive, however. And so as data sets have increased in scale—millions of rows is common in genomic analysis, for example—alternatives to or approximations to Baycsian inference remain important, and probably always will.

1.3.2. Multilevel models.

In an apocryphal telling of Hindu cosmology, it is said that the Earth rests on the back of a great elephant, who in turn stands on the back of a massive turtle. When asked upon what the turtle stands, a guru is said to reply, "it's turtles all the way down."

Statistical models don't contain turtles, but they do contain parameters. And parameters support inference. Upon what do parameters themselves stand? Sometimes, in some of the most powerful models, it's parameters all the way down. What this means is that any particular parameter can be usefully regarded as a placeholder for a missing model. Given some model of how the parameter gets its value, it is simple enough to embed the new model inside the old one. This results in a model with multiple levels of uncertainty, each feeding into the next—a MULTILEVEL MODEL.

Multilevel models—also known as hierarchical, random effects, varying effects, or mixed effects models—are becoming *de rigueur* in the biological and social sciences. Fields as diverse as educational testing and bacterial phylogenetics now depend upon routine multilevel models to process data. Like Bayesian data analysis, multilevel modeling is not particularly new, but it has only been available on desktop computers for a few decades. And since such models have a natural Bayesian representation, they have grown hand-in-hand with Bayesian data analysis.

There are four typical and complementary reasons to use multilevel models:

(1) *To adjust estimates for repeat sampling.* When more than one observation arises from the same individual, location, or time, then traditional, single-level models may mislead us.

(2) *To adjust estimates for imbalance in sampling.* When some individuals, locations, or times are sampled more than others, we may also be misled by single-level models.

(3) *To study variation.* If our research questions include variation among individuals or other groups within the data, then multilevel models are a big help, because they model variation explicitly.

(4) *To avoid averaging.* Frequently, scholars pre-average some data to construct variables for a regression analysis. This can be dangerous, because averaging removes variation. It therefore manufactures false confidence. Multilevel models allow us to preserve the uncertainty in the original, pre-averaged values, while still using the average to make predictions.

All four apply to contexts in which the researcher recognizes clusters or groups of measurements that may differ from one another. These clusters or groups may be individuals such as different students, locations such as different cities, or times such as different years. Since each cluster may well have a different average tendency or respond differently to any treatment, clustered data often benefit from being modeled by a golem that expects such variation.

But the scope of multilevel modeling is much greater than these examples. Diverse model types turn out to be multilevel: models for missing data (imputation), measurement error, factor analysis, some time series models, types of spatial and network regression, and phylogenetic regressions all are special applications of the multilevel strategy. This is why grasping the concept of multilevel modeling may lead to a perspective shift. Suddenly single-level models end up looking like mere components of multilevel models. The multilevel strategy provides an engineering principle to help us to introduce these components into a particular analysis, exactly where we think we need them.

I want to convince the reader of something that appears unreasonable: *multilevel regression deserves to be the default form of regression.* Papers that do not use multilevel models should have to justify not using a multilevel approach. Certainly some data and contexts do not need the multilevel treatment. But most contemporary studies in the social and natural sciences, whether experimental or not, would benefit from it. Perhaps the most important reason is that even well-controlled treatments interact with unmeasured aspects of the individuals, groups, or populations studied. This leads to variation in treatment effects, in which individuals or groups vary in how they respond to the same circumstance. Multilevel models attempt to quantify the extent of this variation, as well as identify which units in the data responded in which ways.

These benefits don't come for free, however. Fitting and interpreting multilevel models can be considerably harder than fitting and interpreting a traditional regression model. In practice, many researchers simply trust their black-box software and interpret multilevel regression exactly like single-level regression. In time, this will change. There was a time in applied statistics when even ordinary multiple regression was considered cutting edge, something for only experts to fiddle with. Instead, scientists used many simple procedures, like *t*-tests. Now, almost everyone uses the better multivariate tools. The same will eventually be true of multilevel models. But scholarly culture and curriculum still have a little catching up to do.

Rethinking: Multilevel election forecasting. One of the older applications of multilevel modeling is to forecast the outcomes of democratic elections. In the early 1960s, John Tukey (1915–2000) began working for the National Broadcasting Company (NBC) in the United States, developing real-time election prediction models that could exploit diverse types of data: polls, past elections, partial results, and complete results from related districts. The models used a multilevel framework similar to the models presented in Chapters 12 and 13. Tukey developed and used such models for NBC through 1978.[30] Contemporary election prediction and poll aggregation remains an active topic for multilevel modeling.[31]

1.3.3. Model comparison and information criteria.

Beginning seriously in the 1960s and 1970s, statisticians began to develop a peculiar family of metrics for comparing structurally different models: INFORMATION CRITERIA. All of these criteria aim to let us compare models based upon future predictive accuracy. But they do so in more and less general ways and by using different approximations for different model types. So the number of unique information criteria in the statistical literature has grown quite large. Still, they all share this common enterprise.

The most famous information criterion is AIC, the Akaike (ah-kah-ee-kay) information criterion. AIC and related metrics—we'll discuss DIC and WAIC as well—explicitly build a model of the prediction task and use that model to estimate performance of each model you might wish to compare. Because the prediction is modeled, it depends upon assumptions. So information criteria do not in fact achieve the impossible, by seeing the future. They are still golems.

AIC and its kin are known as "information" criteria, because they develop their measure of model accuracy from INFORMATION THEORY. Information theory has a scope far beyond comparing statistical models. But it will be necessary to understand a little bit of general information theory, in order to really comprehend information criteria. So in Chapter 6, you'll also find a conceptual crash course in information theory.

What AIC and its kin actually do for a researcher is help with two common difficulties in model comparison.

(1) The most important statistical phenomenon that you may have never heard of is OVERFITTING.[32] Overfitting is the subject of Chapter 6. For now, you can understand overfitting with this mantra: *fitting is easy; prediction is hard*. Future data will not be exactly like past data, and so any model that is unaware of this fact tends to make worse predictions than it could. So if we wish to make good predictions, we

cannot judge our models simply on how well they fit our data. Information criteria provide estimates of predictive accuracy, rather than merely fit. So they compare models where it matters.

(2) A major benefit of using AIC and its kin is that they allow for comparison of multiple non-null models to the same data. Frequently, several plausible models of a phenomenon are known to us. The neutral evolution debate (page 6) is one example. In some empirical contexts, like social networks and evolutionary phylogenies, there are no reasonable or uniquely "null" models. This was also true of the neutral evolution example. In such cases, it's not only a good idea to explicitly compare models. It's also mandatory. Information criteria aren't the only way to conduct the comparison. But they are an accessible and widely used way.

Multilevel modeling and Bayesian data analysis have been worked on for decades and centuries, respectively. Information criteria are comparatively very young. Many statisticians have never used information criteria in an applied problem, and there is no consensus about which metrics are best and how best to use them. Still, information criteria are already in frequent use in the sciences—appearing in prominent publications and featuring in prominent debates[33]—and a great deal is known about them, both from analysis and experience.

> **Rethinking: The Neanderthal in you.** Even simple models need alternatives. In 2010, a draft genome of a Neanderthal demonstrated more DNA sequences in common with non-African contemporary humans than with African ones. This finding is consistent with interbreeding between Neanderthals and modern humans, as the latter dispersed from Africa. However, just finding DNA in common between modern Europeans and Neanderthals is not enough to demonstrate interbreeding. It is also consistent with ancient structure in the African continent.[34] In short, if ancient north-east Africans had unique DNA sequences, then both Neanderthals and modern Europeans could possess these sequences from a common ancestor, rather than from direct interbreeding. So even in the seemingly simple case of estimating whether Neanderthals and modern humans share unique DNA, there is more than one process-based explanation. Model comparison is necessary.

1.4. Summary

This first chapter has argued for a rethinking of popular statistical and scientific philosophy. Instead of choosing among various black-box tools for testing null hypotheses, we should learn to build and analyze multiple non-null models of natural phenomena. To support this goal, the chapter introduced Bayesian inference, multilevel models, and information theoretic model comparison.

The remainder of the book is organized into four interdependent parts.

(1) Chapters 2 and 3 are foundational. They introduce Bayesian inference and the basic tools for performing Bayesian calculations. They move quite slowly and emphasize a purely logical interpretation of probability theory.

(2) The next four chapters, 4 through 7, build multiple linear regression as a Bayesian tool. These chapters also move rather slowly, largely because of the emphasis on plotting results, including interaction effects. Problems of model complexity—overfitting—also feature prominently. So you'll also get an introduction to information theory in Chapter 6.

(3) The third part of the book, Chapters 8 through 11, presents generalized linear models of several types. Chapter 8 is something of a divider, as it introduces Markov

chain Monte Carlo, used to fit the non-linear models in Chapters 10 through 14. Chapter 9 introduces maximum entropy as an explicit procedure to help us design these models. Then Chapters 10 and 11 detail the models themselves.

(4) The last part, Chapters 12 through 14, gets around to multilevel models, both linear and generalized linear, as well as specialized types that address measurement error, missing data, and spatial correlation modeled through Gaussian processes. This material is fairly advanced, but it proceeds in the same mechanistic way as earlier material.

The final chapter, Chapter 15, returns to some of the issues raised in this first one.

At the end of each chapter, there are practice problems ranging from easy to hard. These problems help you test your comprehension. The harder ones expand on the material, introducing new examples and obstacles. The solutions to these problems are available online.

2 Small Worlds and Large Worlds

When Cristoforo Colombo (Christopher Columbus) infamously sailed west in the year 1492, he believed that the Earth was spherical. In this, he was like most educated people of his day. He was unlike most people, though, in that he also believed the planet was much smaller than it actually is—only 30,000 km around its middle instead of the actual 40,000 km (FIGURE 2.1).[35] This was one of the most consequential mistakes in European history. If Colombo had believed instead that the Earth was 40,000 km around, he would have correctly reasoned that his fleet could not carry enough food and potable water to complete a journey all the way westward to Asia. But at 30,000 km around, Asia would lie a bit west of the coast of California. It was possible to carry enough supplies to make it that far. Emboldened in part by his unconventional estimate, Colombo set sail, eventually making landfall in the Bahamas.

Colombo made a prediction based upon his view that the world was small. But since he lived in a large world, aspects of the prediction were wrong. In his case, the error was lucky. His small world model was wrong in an unanticipated way: There was a lot of land in the way. If he had been wrong in the expected way, with nothing but ocean between Europe and Asia, he and his entire expedition would have run out of supplies long before reaching the East Indies.

Colombo's small and large worlds provide a contrast between model and reality. All statistical modeling has these same two frames: the *small world* of the model itself and the *large world* we hope to deploy the model in.[36] Navigating between these two worlds remains a central challenge of statistical modeling. The challenge is aggravated by forgetting the distinction.

The SMALL WORLD is the self-contained logical world of the model. Within the small world, all possibilities are nominated. There are no pure surprises, like the existence of a huge continent between Europe and Asia. Within the small world of the model, it is important to be able to verify the model's logic, making sure that it performs as expected under favorable assumptions. Bayesian models have some advantages in this regard, as they have reasonable claims to optimality: No alternative model could make better use of the information in the data and support better decisions, assuming the small world is an accurate description of the real world.[37]

The LARGE WORLD is the broader context in which one deploys a model. In the large world, there may be events that were not imagined in the small world. Moreover, the model is always an incomplete representation of the large world, and so will make mistakes, even if all kinds of events have been properly nominated. The logical consistency of a model in the small world is no guarantee that it will be optimal in the large world. But it is certainly a warm comfort.

FIGURE 2.1. Illustration of Martin Behaim's 1492 globe, showing the small world that Colombo anticipated. Europe lies on the right-hand side. Asia lies on the left. The big island labeled "Cipangu" is Japan.

In this chapter, you will begin to build Bayesian models. The way that Bayesian models learn from evidence is arguably optimal in the small world. When their assumptions approximate reality, they also perform well in the large world. But large world performance has to be demonstrated rather than logically deduced. Passing back and forth between these two worlds allows both formal methods, like Bayesian inference, and informal methods, like peer review, to play an indispensable role.

This chapter focuses on the small world. It explains probability theory in its essential form: counting the ways things can happen. Bayesian inference arises automatically from this perspective. Then the chapter presents the stylized components of a Bayesian statistical model, a model for learning from data. Then it shows you how to animate the model, to produce estimates.

All this work provides a foundation for the next chapter, in which you'll learn to summarize Bayesian estimates, as well as begin to consider large world obligations.

Rethinking: Fast and frugal in the large world. The natural world is complex, as trying to do science serves to remind us. Yet everything from the humble tick to the industrious squirrel to the idle sloth manages to frequently make adaptive decisions. But it's a good bet that most animals are not Bayesian, if only because being Bayesian is expensive and depends upon having a good model. Instead, animals use various heuristics that are fit to their environments, past or present. These heuristics take adaptive shortcuts and so may outperform a rigorous Bayesian analysis, once costs of information gathering and processing (and overfitting, Chapter 6) are taken into account.[38] Once you already know which information to ignore or attend to, being fully Bayesian is a waste. It's neither necessary nor sufficient for making good decisions, as real animals demonstrate. But for human animals, Bayesian analysis provides a general way to discover relevant information and process it logically. Just don't think that it is the only way.

2.1. The garden of forking data

Our goal in this section will be to build Bayesian inference up from humble beginnings, so there is no superstition about it. Bayesian inference is really just counting and comparing of possibilities. Consider by analogy Jorge Luis Borges' short story "The Garden of Forking Paths." The story is about a man who encounters a book filled with contradictions. In most books, characters arrive at plot points and must decide among alternative paths. A protagonist may arrive at a man's home. She might kill the man, or rather take a cup of tea. Only one of these paths is taken—murder or tea. But the book within Borges' story explores all paths, with each decision branching outward into an expanding garden of forking paths.

This is the same device that Bayesian inference offers. In order to make good inference about what actually happened, it helps to consider everything that could have happened. A Bayesian analysis is a garden of forking data, in which alternative sequences of events are cultivated. As we learn about what did happen, some of these alternative sequences are pruned. In the end, what remains is only what is logically consistent with our knowledge.

This approach provides a quantitative ranking of hypotheses, a ranking that is maximally conservative, given the assumptions and data that go into it. The approach cannot guarantee a correct answer, on large world terms. But it can guarantee the best possible answer, on small world terms, that could be derived from the information fed into it.

Consider the following toy example.

2.1.1. Counting possibilities. Suppose there's a bag, and it contains four marbles. These marbles come in two colors: blue and white. We know there are four marbles in the bag, but we don't know how many are of each color. We do know that there are five possibilities: (1) [○○○○], (2) [●○○○], (3) [●●○○], (4) [●●●○], (5) [●●●●]. These are the only possibilities consistent with what we know about the contents of the bag. Call these five possibilities the *conjectures*.

Our goal is to figure out which of these conjectures is most plausible, given some evidence about the contents of the bag. We do have some evidence: A sequence of three marbles is pulled from the bag, one at a time, replacing the marble each time and shaking the bag before drawing another marble. The sequence that emerges is: ●○●, in that order. These are the data.

So now let's plant the garden and see how to use the data to infer what's in the bag. Let's begin by considering just the single conjecture, [●○○○], that the bag contains one blue and three white marbles. On the first draw from the bag, one of four things could happen, corresponding to one of four marbles in the bag. So we can visualize the possibilities branching outward:

Notice that even though the three white marbles look the same from a data perspective—we just record the color of the marbles, after all—they are really different events. This is important, because it means that there are three more ways to see ○ than to see ●.

Now consider the garden as we get another draw from the bag. It expands the garden out one layer:

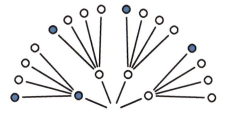

Now there are 16 possible paths through the garden, one for each pair of draws. On the second draw from the bag, each of the paths above again forks into four possible paths. Why?

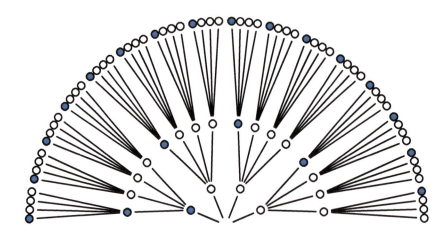

FIGURE 2.2. The 64 possible paths generated by assuming the bag contains one blue and three white marbles.

Because we believe that our shaking of the bag gives each marble a fair chance at being drawn, regardless of which marble was drawn previously. The third layer is built in the same way, and the full garden is shown in FIGURE 2.2. There are $4^3 = 64$ possible paths in total.

As we consider each draw from the bag, some of these paths are logically eliminated. The first draw tuned out to be ●, recall, so the three white paths at the bottom of the garden are eliminated right away. If you imagine the real data tracing out a path through the garden, it must have passed through the one blue path near the origin. The second draw from the bag produces ○, so three of the paths forking out of the first blue marble remain. As the data trace out a path, we know it must have passed through one of those three white paths (after the first blue path), but we don't know which one, because we recorded only the color of each marble. Finally, the third draw is ●. Each of the remaining three paths in the middle layer sustain one blue path, leaving a total of three ways for the sequence ●○● to appear, assuming the bag contains [●○○○]. FIGURE 2.3 shows the garden again, now with logically eliminated paths grayed out. We can't be sure which of those three paths the actual data took. But as long as we're considering only the possibility that the bag contains one blue and three white marbles, we can be sure that the data took one of those three paths. Those are the only paths consistent with both our knowledge of the bag's contents (four marbles, white or blue) and the data (●○●).

This demonstrates that there are three (out of 64) ways for a bag containing [●○○○] to produce the data ●○●. We have no way to decide among these three ways. The inferential power comes from comparing this count to the numbers of ways each of the other conjectures of the bag's contents could produce the same data. For example, consider the conjecture [○○○○]. There are zero ways for this conjecture to produce the observed data, because even one ● is logically incompatible with it. The conjecture [●●●●] is likewise logically incompatible with the data. So we can eliminate these two conjectures, because neither provides even a single path that is consistent with the data.

FIGURE 2.4 displays the full garden now, for the remaining three conjectures: [●○○○], [●●○○], and [●●●○]. The upper-left wedge displays the same garden as FIGURE 2.3. The upper-right shows the analogous garden for the conjecture that the bag contains three blue marbles and one white marble. And the bottom wedge shows the garden for two blue

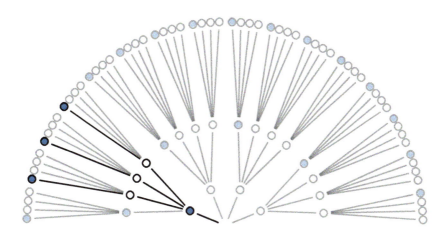

FIGURE 2.3. After eliminating paths inconsistent with the observed sequence, only 3 of the 64 paths remain.

and two white marbles. Now we count up all of the ways each conjecture could produce the observed data. For one blue and three white, there are three ways, as we counted already. For two blue and two white, there are eight paths forking through the garden that are logically consistent with the observed sequence. For three blue and one white, there are nine paths that survive.

To summarize, we've considered five different conjectures about the contents of the bag, ranging from zero blue marbles to four blue marbles. For each of these conjectures, we've counted up how many sequences, paths through the garden of forking data, could potentially produce the observed data, ●○●:

Conjecture	Ways to produce ●○●
[○○○○]	$0 \times 4 \times 0 = 0$
[●○○○]	$1 \times 3 \times 1 = 3$
[●●○○]	$2 \times 2 \times 2 = 8$
[●●●○]	$3 \times 1 \times 3 = 9$
[●●●●]	$4 \times 0 \times 4 = 0$

Notice that the number of ways to produce the data, for each conjecture, can be computed by first counting the number of paths in each "ring" of the garden and then by multiplying these counts together. This is just a computational device. It tells us the same thing as FIGURE 2.4, but without having to draw the garden. The fact that numbers are multiplied during calculation doesn't change the fact that this is still just counting of logically possible paths. This point will come up again, when you meet the more formal representation of Bayesian inference.

So what good are these counts? By comparing these counts, we have part of a solution for a way to rate the relative plausibility of each conjectured bag composition. But it's only a part of a solution, because in order to compare these counts we first have to decide how many ways each conjecture could itself be realized. We might argue that when we have no reason to assume otherwise, we can just consider each conjecture equally plausible and compare the counts directly. But often we do have reason to assume otherwise.

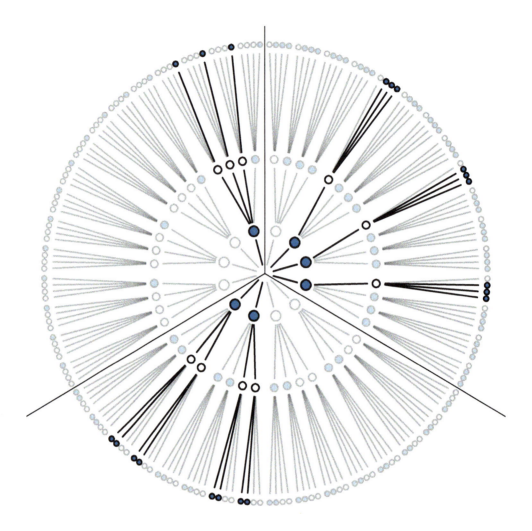

FIGURE 2.4. The garden of forking data, showing for each possible composition of the bag the forking paths that are logically compatible with the data.

Rethinking: Justification. Using these counts of paths through the garden as measures of relative plausibility can be justified in several ways. The justification here is logical: If we wish to reason about plausibility and remain consistent with ordinary logic—statements about *true* and *false*—then we should obey this procedure.[39] There are several other justifications that lead to the same mathematical procedure. Regardless of how you choose to philosophically justify it, notice that it actually works. Justifications and philosophy motivate procedures, but it is the results that matter. The many successful real world applications of Bayesian inference may be all the justification you need. Twentieth century opponents of Bayesian data analysis argued that Bayesian inference was easy to justify, but hard to apply.[40] That is luckily no longer true. Just be careful not to assume that because Bayesian inference is justified that no other approach can also be justified. Golems come in many types, and some of all types are useful.

2.1.2. Using prior information. We may have prior information about the relative plausibility of each conjecture. This prior information could arise from knowledge of how the contents of the bag were generated. It could also arise from previous data. Or we might want to act as if we had prior information, so we can build conservatism into the analysis. Whatever the source, it would help to have a way to use prior information. Luckily there is a natural solution: Just multiply the prior count by the new count.

To grasp this solution, suppose we're willing to say each conjecture is equally plausible at the start. So we just compare the counts of ways in which each conjecture is compatible with the observed data. This comparison suggests that [●●●○] is slightly more plausible than [●●○○], and both are about three times more plausible than [●○○○].

Now suppose we draw another marble from the bag to get another observation: ●. Now you have two choices. You could start all over again, making a garden with four layers to trace out the paths compatible with the data sequence ●○●●. Or you could take the previous counts over conjectures (0, 3, 8, 9, 0) and just update them in light of the new observation. It turns out that these two methods are mathematically identical, as long as the new observation is logically independent of the previous observations.

Here's how to do it. First we count the numbers of ways each conjecture could produce the new observation, ●. Then we multiply each of these new counts by the previous numbers of ways for each conjecture. In table form:

Conjecture	Ways to produce ●	Previous counts	New count
[○○○○]	0	0	$0 \times 0 = 0$
[●○○○]	1	3	$3 \times 1 = 3$
[●●○○]	2	8	$8 \times 2 = 16$
[●●●○]	3	9	$9 \times 3 = 27$
[●●●●]	4	0	$0 \times 4 = 0$

The new counts in the right-hand column above summarize all the evidence for each conjecture. As new data arrive, and provided those data are independent of previous observations, then the number of logically possible ways for a conjecture to produce all the data up to that point can be computed just by multiplying the new count by the old count.

This updating approach amounts to nothing more than asserting that (1) when we have previous information suggesting there are W_{prior} ways for a conjecture to produce a previous observation D_{prior} and (2) we acquire new observations D_{new} that the same conjecture can produce in W_{new} ways, then (3) the number of ways the conjecture can account for both D_{prior} as well as D_{new} is just the product $W_{prior} \times W_{new}$. For example, in the table above the conjecture [●●○○] has $W_{prior} = 8$ ways to produce $D_{prior} = $ ●○●. It also has $W_{new} = 2$ ways to produce the new observation $D_{new} = $ ●. So there are $8 \times 2 = 16$ ways for the conjecture to produce both D_{prior} and D_{new}. Why multiply? Multiplication is just a shortcut to enumerating and counting up all of the paths through the garden that could produce all the observations.

In this example, the prior data and new data are of the same type: marbles drawn from the bag. But in general, the prior data and new data can be of different types. Suppose for example that someone from the marble factory tells you that blue marbles are rare. So for every bag containing [●●●○], they made two bags containing [●●○○] and three bags containing [●○○○]. They also ensured that every bag contained at least one blue and one white marble. We can update our counts again:

Conjecture	Prior count	Factory count	New count
[○○○○]	0	0	$0 \times 0 = 0$
[●○○○]	3	3	$3 \times 3 = 9$
[●●○○]	16	2	$16 \times 2 = 32$
[●●●○]	27	1	$27 \times 1 = 27$
[●●●●]	0	0	$0 \times 0 = 0$

Now the conjecture [●●○○] is most plausible, but barely better than [●●●○]. Is there a threshold difference in these counts at which we can safely decide that one of the conjectures is the correct one? You'll spend the next chapter exploring that question.

> **Rethinking: Original ignorance.** Which assumption should we use, when there is no previous information about the conjectures? The most common solution is to assign an equal number of ways that each conjecture could be correct, before seeing any data. This is sometimes known as the PRINCIPLE OF INDIFFERENCE: When there is no reason to say that one conjecture is more plausible than another, weigh all of the conjectures equally. The issue of choosing a representation of "ignorance" is surprisingly complicated. The issue will arise again in later chapters. For the sort of problems we examine in this book, the principle of indifference results in inferences very comparable to mainstream non-Bayesian approaches, most of which contain implicit equal weighting of possibilities. For example a typical non-Bayesian confidence interval weighs equally all of the possible values a parameter could take, regardless of how implausible some of them are. Many non-Bayesian procedures have moved away from this, through the use of penalized likelihood and other methods. We'll discuss these in Chapter 6.

2.1.3. From counts to probability.
It is helpful to think of this strategy as adhering to a principle of honest ignorance: *When we don't know what caused the data, potential causes that may produce the data in more ways are more plausible.* This leads us to count paths through the garden of forking data.

It's hard to use these counts though, so we almost always standardize them in a way that transforms them into probabilities. Why is it hard to work with the counts? First, since relative value is all that matters, the size of the counts 3, 8, and 9 contain no information of value. They could just as easily be 30, 80, and 90. The meaning would be the same. It's just the relative values that matter. Second, as the amount of data grows, the counts will very quickly grow very large and become difficult to manipulate. By the time we have 10 data points, there are already more than one million possible sequences. We'll want to analyze data sets with thousands of observations, so explicitly counting these things isn't practical.

Luckily, there's a mathematical way to compress all of this. Specifically, we define the updated plausibility of each possible composition of the bag, after seeing the data, as:

$$\text{plausibility of } [●○○○] \text{ after seeing } ●○●$$
$$\propto$$
$$\text{ways } [●○○○] \text{ can produce } ●○●$$
$$\times$$
$$\text{prior plausibility } [●○○○]$$

That little $\propto$ means *proportional to*. We want to compare the plausibility of each possible bag composition. So it'll be helpful to define p as the proportion of marbles that are blue. For [

●○○○], $p = 1/4 = 0.25$. Also let $D_{new} = $ ●○●. And now we can write:

plausibility of p after D_{new} ∝ ways p can produce D_{new} × prior plausibility of p

The above just means that for any value p can take, we judge the plausibility of that value p as proportional to the number of ways it can get through the garden of forking data. This expression just summarizes the calculations you did in the tables of the previous section.

Finally, we construct probabilities by standardizing the plausibility so that the sum of the plausibilities for all possible conjectures will be one. All you need to do in order to standardize is to add up all of the products, one for each value p can take, and then divide each product by the sum of products:

$$\text{plausibility of } p \text{ after } D_{new} = \frac{\text{ways } p \text{ can produce } D_{new} \times \text{prior plausibility } p}{\text{sum of products}}$$

A worked example is needed for this to really make sense. So consider again the table from before, now updated using our definitions of p and "plausibility":

Possible composition	p	Ways to produce data	Plausibility
[○○○○]	0	0	0
[●○○○]	0.25	3	0.15
[●●○○]	0.5	8	0.40
[●●●○]	0.75	9	0.45
[●●●●]	1	0	0

You can quickly compute these plausibilities in R:

```
ways <- c( 0 , 3 , 8 , 9 , 0 )
ways/sum(ways)
```

R code
2.1

```
[1] 0.00 0.15 0.40 0.45 0.00
```

The values in ways are the products mentioned before. And sum(ways) is the denominator "sum of products" in the expression near the top of the page.

These plausibilities are also *probabilities*—they are non-negative (zero or positive) real numbers that sum to one. And all of the mathematical things you can do with probabilities you can also do with these values. Specifically, each piece of the calculation has a direct partner in applied probability theory. These partners have stereotyped names, so it's worth learning them, as you'll see them again and again.

- A conjectured proportion of blue marbles, p, is usually called a PARAMETER value. It's just a way of indexing possible explanations of the data.
- The relative number of ways that a value p can produce the data is usually called a LIKELIHOOD. It is derived by enumerating all the possible data sequences that could have happened and then eliminating those sequences inconsistent with the data.
- The prior plausibility of any specific p is usually called the PRIOR PROBABILITY.
- The new, updated plausibility of any specific p is usually called the POSTERIOR PROBABILITY.

In the next major section, you'll meet the more formal notation for these objects and see how they compose a simple statistical model.

Rethinking: Randomization. When you shuffle a deck of cards or assign subjects to treatments by flipping a coin, it is common to say that the resulting deck and treatment assignments are *randomized*. What does it mean to randomize something? It just means that we have processed the thing so that we know almost nothing about its arrangement. Shuffling a deck of cards changes our state of knowledge, so that we no longer have any specific information about the ordering of cards. However, the bonus that arises from this is that, if we really have shuffled enough to erase any prior knowledge of the ordering, then the order the cards end up in is very likely to be one of the many orderings with high INFORMATION ENTROPY. The concept of information entropy will be increasingly important as we progress, and will be unpacked in Chapters 6 and 9.

2.2. Building a model

By working with probabilities instead of raw counts, Bayesian inference is made much easier, but it looks much harder. So in this section, we follow up on the garden of forking data by presenting the conventional form of a Bayesian statistical model. The toy example we'll use here has the anatomy of a typical statistical analysis, so it's the style that you'll grow accustomed to. But every piece of it can be mapped onto the garden of forking data. The logic is the same.

Suppose you have a globe representing our planet, the Earth. This version of the world is small enough to hold in your hands. You are curious how much of the surface is covered in water. You adopt the following strategy: You will toss the globe up in the air. When you catch it, you will record whether or not the surface under your right index finger is water or land. Then you toss the globe up in the air again and repeat the procedure.[41] This strategy generates a sequence of surface samples from the globe. The first nine samples might look like:

W L W W W L W L W

where W indicates water and L indicates land. So in this example you observe six W (water) observations and three L (land) observations. Call this sequence of observations the *data*.

To get the logic moving, we need to make assumptions, and these assumptions constitute the model. Designing a simple Bayesian model benefits from a design loop with three steps.

(1) Data story: Motivate the model by narrating how the data might arise.
(2) Update: Educate your model by feeding it the data.
(3) Evaluate: All statistical models require supervision, leading possibly to model revision.

The next sections walk through these steps, in the context of the globe tossing evidence.

2.2.1. A data story. Bayesian data analysis usually means producing a story for how the data came to be. This story may be *descriptive*, specifying associations that can be used to predict outcomes, given observations. Or it may be *causal*, a theory of how some events produce other events. Typically, any story you intend to be causal may also be descriptive. But many descriptive stories are hard to interpret causally. But all data stories are complete, in the sense that they are sufficient for specifying an algorithm for simulating new data. In the next chapter, you'll see examples of doing just that, as simulating new data is a useful part of model criticism.

You can motivate your data story by trying to explain how each piece of data is born. This usually means describing aspects of the underlying reality as well as the sampling process. The data story in this case is simply a restatement of the sampling process:

(1) The true proportion of water covering the globe is p.
(2) A single toss of the globe has a probability p of producing a water (W) observation. It has a probability $1 - p$ of producing a land (L) observation.
(3) Each toss of the globe is independent of the others.

The data story is then translated into a formal probability model. This probability model is easy to build, because the construction process can be usefully broken down into a series of component decisions. Before meeting these components, however, it'll be useful to visualize how a Bayesian model behaves. After you've become acquainted with how such a model learns from data, we'll pop the machine open and investigate its engineering.

> **Rethinking: The value of storytelling.** The data story has value, even if you quickly abandon it and never use it to build a model or simulate new observations. Indeed, it is important to eventually discard the story, because many different stories always correspond to the same model. As a result, showing that a model does a good job does not in turn uniquely support our data story. Still, the story has value because in trying to outline the story, often one realizes that additional questions must be answered. Most data stories are much more specific than are the verbal hypotheses that inspire data collection. Hypotheses can be vague, such as "it's more likely to rain on warm days." When you are forced to consider sampling and measurement and make a precise statement of how temperature predicts rain, many stories and resulting models will be consistent with the same vague hypothesis. Resolving that ambiguity often leads to important realizations and model revisions, before any model is fit to data.

2.2.2. Bayesian updating.

Our problem is one of using the evidence—the sequence of globe tosses—to decide among different possible proportions of water on the globe. These proportions are like the conjectured marbles inside the bag, from earlier in the chapter. Each possible proportion may be more or less plausible, given the evidence. A Bayesian model begins with one set of plausibilities assigned to each of these possibilities. These are the prior plausibilities. Then it updates them in light of the data, to produce the posterior plausibilities. This updating process is a kind of learning, called BAYESIAN UPDATING. The details of this updating—how it is mechanically achieved—can wait until later in the chapter. For now, let's look only at how such a machine behaves.

For the sake of the example only, let's program our Bayesian machine to initially assign the same plausibility to every proportion of water, every value of p. Now look at the top-left plot in FIGURE 2.5. The dashed horizontal line represents this initial plausibility of each possible value of p. After seeing the first toss, which is a "W," the model updates the plausibilities to the solid line. The plausibility of $p = 0$ has now fallen to exactly zero—the equivalent of "impossible." Why? Because we observed at least one speck of water on the globe, so now we know there is *some* water. The model executes this logic automatically. You don't have it instruct it to account for this consequence. Probability theory takes care of it for you, because it is essentially counting paths through the garden of forking data, as in the previous section.

Likewise, the plausibility of $p > 0.5$ has increased. This is because there is not yet any evidence that there is land on the globe, so the initial plausibilities are modified to be consistent with this. Note however that the relative plausibilities are what matter, and there isn't

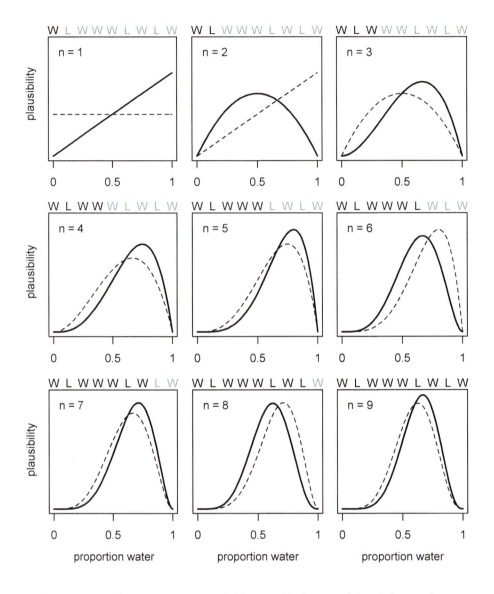

FIGURE 2.5. How a Bayesian model learns. Each toss of the globe produces an observation of water (W) or land (L). The model's estimate of the proportion of water on the globe is a plausibility for every possible value. The lines and curves in this figure are these collections of plausibilities. In each plot, previous plausibilities (dashed curve) are updated in light of the latest observation to produce a new set of plausibilities (solid curve).

yet much evidence. So the differences in plausibility are not yet very large. In this way, the amount of evidence seen so far is embodied in the plausibilities of each value of p.

In the remaining plots in FIGURE 2.5, the additional samples from the globe are introduced to the model, one at a time. Each dashed curve is just the solid curve from the previous plot, moving left to right and top to bottom. Every time a "W" is seen, the peak of the plausibility curve moves to the right, towards larger values of p. Every time an "L" is seen, it moves the other direction. The maximum height of the curve increases with each sample, meaning that fewer values of p amass more plausibility as the amount of evidence increases. As each new observation is added, the curve is updated consistent with all previous observations.

Notice that every updated set of plausibilities becomes the initial plausibilities for the next observation. Every conclusion is the starting point for future inference. However, this updating process works backwards, as well as forwards. Given the final set of plausibilities in the bottom-right plot of FIGURE 2.5, and knowing the final observation (W), it is possible to mathematically divide out the observation, to infer the previous plausibility curve. So the data could be presented to your model in any order, or all at once even. In most cases, you will present the data all at once, for the sake of convenience. But it's important to realize that this merely represents abbreviation of an iterated learning process.

Rethinking: Sample size and reliable inference. It is common to hear that there is a minimum number of observations for a useful statistical estimate. For example, there is a widespread superstition that 30 observations are needed before one can use a Gaussian distribution. Why? In non-Bayesian statistical inference, procedures are often justified by the method's behavior at very large sample sizes, so-called *asymptotic* behavior. As a result, performance at small samples sizes is questionable.

In contrast, Bayesian estimates are valid for any sample size. This does not mean that more data isn't helpful—it certainly is. Rather, the estimates have a clear and valid interpretation, no matter the sample size. But the price for this power is dependency upon the initial estimates, the prior. If the prior is a bad one, then the resulting inference will be misleading. There's no free lunch,[42] when it comes to learning about the world. A Bayesian golem must choose an initial plausibility, and a non-Bayesian golem must choose an estimator. Both golems pay for lunch with their assumptions.

2.2.3. Evaluate.

The Bayesian model learns in a way that is demonstrably optimal, provided that the real, large world is accurately described by the model. This is to say that your Bayesian machine guarantees perfect inference, within the small world. No other way of using the available information, and beginning with the same state of information, could do better.

Don't get too excited about this logical virtue, however. The calculations may malfunction, so results always have to be checked. And if there are important differences between the model and reality, then there is no logical guarantee of large world performance. And even if the two worlds did match, any particular sample of data could still be misleading. So it's worth keeping in mind at least two cautious principles.

First, the model's certainty is no guarantee that the model is a good one. As the amount of data increases, the globe tossing model will grow increasingly sure of the proportion of water. This means that the curves in FIGURE 2.5 will become increasingly narrow and tall, restricting plausible values within a very narrow range. But models of all sorts—Bayesian or not—can be very confident about an estimate, even when the model is seriously misleading. This is because the estimates are conditional on the model. What your model is telling you is that, given a commitment to this particular model, it can be very sure that the plausible values are in a narrow range. Under a different model, things might look differently.

Second, it is important to supervise and critique your model's work. Consider again the fact that the updating in the previous section works in any order of data arrival. We could shuffle the order of the observations, as long as six W's and three L's remain, and still end up with the same final plausibility curve. That is only true, however, because the model assumes that order is irrelevant to inference. When something is irrelevant to the machine, it won't affect the inference directly. But it may affect it indirectly, because the data will depend upon order. So it is important to check the model's inferences in light of aspects of the data it does not know about. Such checks are an inherently creative enterprise, left to the analyst and the scientific community. Golems are very bad at it.

In Chapter 3, you'll see some examples of such checks. For now, note that the goal is not to test the truth value of the model's assumptions. We know the model's assumptions are never exactly right, in the sense of matching the true data generating process. Therefore there's no point in checking if the model is true. Failure to conclude that a model is false must be a failure of our imagination, not a success of the model. Moreover, models do not need to be exactly true in order to produce highly precise and useful inferences. All manner of small world assumptions about error distributions and the like can be violated in the large world, but a model may still produce a perfectly useful estimate. This is because models are essentially information processing machines, and there are some surprising aspects of information that cannot be easily captured by framing the problem in terms of the truth of assumptions.[43]

Instead, the objective is to check the model's adequacy for some purpose. This usually means asking and answering additional questions, beyond those that originally constructed the model. Both the questions and answers will depend upon the scientific context. So it's hard to provide general advice. There will be many examples, throughout the book, and of course the scientific literature is replete with evaluations of the suitability of models for different jobs—prediction, comprehension, measurement, and persuasion.

> **Rethinking: Deflationary statistics.** It may be that Bayesian inference is the best general purpose method of inference known. However, Bayesian inference is much less powerful than we'd like it to be. There is no approach to inference that provides universal guarantees. No branch of applied mathematics has unfettered access to reality, because math is not discovered, like the proton. Instead it is invented, like the shovel.[44]

2.3. Components of the model

Now that you've seen how the Bayesian model behaves, it's time to open up the machine and learn how it works. Consider three different kinds of things we counted in the previous sections.

(1) The number of ways each conjecture could produce an observation
(2) The accumulated number of ways each conjecture could produce the entire data
(3) The initial plausibility of each conjectured cause of the data

Each of these things has a direct analog in conventional probability theory. And so the usual way we build a statistical model involves choosing distributions and devices for each that represent the relative numbers of ways things can happen.

These distributions and devices are (1) a likelihood function, (2) one or more parameters, and (3) a prior. These components are usually chosen in that order. In this section, you'll meet these components in some detail and see how each relates to the counting you did earlier in the chapter.

2.3.1. Likelihood. The first and most influential component of a Bayesian model is the LIKE-LIHOOD. The likelihood is a mathematical formula that specifies the plausibility of the data.

What this means is that the likelihood maps each conjecture—such as a proportion of water on the globe—onto the relative number of ways the data could occur, given that possibility.

You can build your own likelihood formula from basic assumptions of your story for how the data arise. That's what we did in the globe tossing example earlier. Or you can use one of several off-the-shelf likelihoods that are common in the sciences. Later in the book, you'll see how information theory justifies many of the conventional choices of likelihood. But however you get the likelihood function, the likelihood needs to tell you the probability of any possible observation, for any possible state of the (small) world, such as a proportion of water on a globe.

In the case of the globe tossing model, the likelihood can be derived from the data story. Begin by nominating all of the possible events. There are two: *water* (W) and *land* (L). There are no other events. The globe never gets stuck to the ceiling, for example. When we observe a sample of W's and L's of length N (nine in the actual sample), we need to say how likely that exact sample is, out of the universe of potential samples of the same length. That might sound challenging, but it's the kind of thing you get good at very quickly, once you start practicing.

In this case, once we add our assumptions that (1) every toss is independent of the other tosses and (2) the probability of W is the same on every toss, probability theory provides a unique answer, known as the *binomial distribution*. This is the common "coin tossing" distribution. And so the probability of observing w W's in n tosses, with a probability p of W, is:

$$\Pr(w|n, p) = \frac{n!}{w!(n-w)!}p^w(1-p)^{n-w}$$

Read the above as:

> The count of "water" observations w is distributed binomially, with proba- bility p of "water" on each toss and n tosses in total.

And the binomial distribution formula is built into R, so you can easily compute the likeli- hood of the data—six W's in nine tosses—under any value of p with:

```
dbinom( 6 , size=9 , prob=0.5 )
```
R code
2.2

```
[1] 0.1640625
```

That number is the relative number of ways to get six W's, holding p at 0.5 and n at nine. So it does the job of counting relative number of paths through the garden. Change the 0.5 to any other value, to see how the value changes.

Sometimes, likelihoods are written $L(p|w, n)$: the likelihood of p, conditional on w and n. Note however that this notation reverses what is on the left side of the | symbol. Just keep in mind that the job of the likelihood is to tell us the relative number of ways to see the data w, given values for p and n.

Overthinking: Names and probability distributions. The "d" in dbinom stands for *density*. Func- tions named in this way almost always have corresponding partners that begin with "r" for random samples and that begin with "p" for cumulative probabilities. See for example the help ?dbinom.

Rethinking: A central role for likelihood. A great deal of ink has been spilled focusing on how Bayesian and non-Bayesian data analyses differ. Focusing on differences is useful, but sometimes it distracts us from fundamental similarities. Notably, the most influential assumptions in both Bayesian and many non-Bayesian models are the likelihood functions and their relations to the parameters. The assumptions about the likelihood influence inference for every piece of data, and as sample size increases, the likelihood matters more and more. This helps to explain why Bayesian and non-Bayesian inferences are often so similar.

2.3.2. Parameters. For most likelihood functions, there are adjustable inputs. In the binomial likelihood, these inputs are p (the probability of seeing a W), n (the sample size), and w (the number of W's). One or all of these may also be quantities that we wish to estimate from data, PARAMETERS, and they represent the different conjectures for causes or explanations of the data. In our globe tossing example, both n and w are data—we believe that we have observed their values without error. That leaves p as an unknown parameter, and our Bayesian machine's job is to describe what the data tell us about it.

But in other analyses, different inputs of the likelihood may be the targets of our analysis. It's common in field biology, for example, to estimate both p and n. Such an analysis may be called mark-recapture, capture-recapture, or sight-resight. The goal is usually to estimate population size, the n input of the binomial distribution. In other cases, we might know p and n, but not w. Your Bayesian machine can tell you what the data say about any parameter, once you tell it the likelihood and which bits have been observed. In some cases, the golem will just tell you that not much can be learned—Bayesian data analysis isn't magic, after all. But it is usually an advance to learn that our data don't discriminate among the possibilities.

In future chapters, there will be more parameters in your models, in excess of the small number that may appear in the likelihood. In statistical modeling, many of the most common questions we ask about data are answered directly by parameters:

- What is the average difference between treatment groups?
- How strong is the association between a treatment and an outcome?
- Does the effect of the treatment depend upon a covariate?
- How much variation is there among groups?

You'll see how these questions become extra parameters inside the likelihood.

Rethinking: Datum or parameter? It is typical to conceive of data and parameters as completely different kinds of entities. Data are measured and known; parameters are unknown and must be estimated from data. Usefully, in the Bayesian framework the distinction between a datum and a parameter is fuzzy. A datum can be recast as a very narrow probability density for a parameter, and a parameter as a datum with uncertainty. Much later in the book (Chapter 14), you'll see how to exploit this continuity between certainty (data) and uncertainty (parameters) to incorporate measurement error and missing data into your modeling.

2.3.3. Prior. For every parameter you intend your Bayesian machine to estimate, you must provide to the machine a PRIOR. A Bayesian machine must have an initial plausibility assignment for each possible value of the parameter. The prior is this initial set of plausibilities. When you have a previous estimate to provide to the machine, that can become the prior, as in the steps in FIGURE 2.5. Back in FIGURE 2.5, the machine did its learning one piece of data at a time. As a result, each estimate becomes the prior for the next step. But this doesn't resolve the problem of providing a prior, because at the dawn of time, when $n = 0$, the machine still had an initial state of information for the parameter p: a flat line specifying equal plausibility for every possible value.

Overthinking: Prior as probability distribution. You could write the prior in the example here as:

$$\Pr(p) = \frac{1}{1 - 0} = 1.$$

The prior is a probability distribution for the parameter. In general, for a uniform prior from a to b, the probability of any point in the interval is $1/(b - a)$. If you're bothered by the fact that the probability of every value of p is 1, remember that every probability distribution must sum (integrate) to 1. The expression $1/(b-a)$ ensures that the area under the flat line from a to b is equal to 1. There will be more to say about this in Chapter 4.

So where do priors come from? They are engineering assumptions, chosen to help the machine learn. The flat prior in FIGURE 2.5 is very common, but it is hardly ever the best prior. You'll see later in the book that priors that gently nudge the machine usually improve inference. Such priors are sometimes called REGULARIZING or WEAKLY INFORMATIVE priors. They are so useful that non-Bayesian statistical procedures have adopted a mathematically equivalent approach, PENALIZED LIKELIHOOD. These priors are conservative, in that they tend to guard against inferring strong associations between variables.

More generally, priors are useful for constraining parameters to reasonable ranges, as well as for expressing any knowledge we have about the parameter, before any data are observed. For example, in the globe tossing case, you know before the globe is tossed even once that the values $p = 0$ and $p = 1$ are completely implausible. You also know that any values of p very close to 0 and 1 are less plausible than values near $p = 0.5$. Even vague knowledge like this can be useful, when evidence is scarce.

There is a school of Bayesian inference that emphasizes choosing priors based upon the personal beliefs of the analyst.[45] While this SUBJECTIVE BAYESIAN approach thrives in some statistics and philosophy and economics programs, it is rare in the sciences. Within Bayesian data analysis in the natural and social sciences, the prior is considered to be just part of the model. As such it should be chosen, evaluated, and revised just like all of the other components of the model. In practice, the subjectivist and the non-subjectivist will often analyze data in nearly the same way.

None of this should be understood to mean that any statistical analysis is not inherently subjective, because of course it is—lots of little subjective decisions are involved in all parts of science. It's just that priors and Bayesian data analysis are no more inherently subjective than are likelihoods and the repeat sampling assumptions required for significance testing.[46] Anyone who has visited a statistics help desk at a university has probably experienced this subjectivity—statisticians do not in general exactly agree on how to analyze anything but the simplest of problems. The fact that statistical inference uses mathematics does not imply that there is only one reasonable or useful way to conduct an analysis. Engineering uses math as well, but there are many ways to build a bridge.

Beyond all of the above, there's no law mandating we use only one prior. If you don't have a strong argument for any particular prior, then try different ones. Because the prior is an assumption, it should be interrogated like other assumptions: by altering it and checking how sensitive inference is to the assumption. No one is required to swear an oath to the assumptions of a model, and no set of assumptions deserves our obedience.

Rethinking: Prior, prior pants on fire. Historically, some opponents of Bayesian inference objected to the arbitrariness of priors. It's true that priors are very flexible, being able to encode many different states of information. If the prior can be anything, isn't it possible to get any answer you want? Indeed it is. Regardless, after a couple hundred years of Bayesian calculation, it hasn't turned out that people use priors to lie. If your goal is to lie with statistics, you'd be a fool to do it with priors, because such a lie would be easily uncovered. Better to use the more opaque machinery of the likelihood. Or better yet—don't actually take this advice!—massage the data, drop some "outliers," and otherwise engage in motivated data transformation.

It is true though that choice of the likelihood is much more conventionalized than choice of prior. But conventional choices are often poor ones, smuggling in influences that can be hard to discover. In this regard, both Bayesian and non-Bayesian models are equally harried, because both traditions depend heavily upon likelihood functions and conventionalized model forms. And the fact that the non-Bayesian procedure doesn't have to make an assumption about the prior is of little comfort. This is because non-Bayesian procedures need to make choices that Bayesian ones do not, such as choice of estimator or likelihood penalty. Often, such choices can be shown to be equivalent to some Bayesian choice of prior or rather choice of loss function. (You'll meet loss functions later in Chapter 3.)

2.3.4. Posterior. Once you have chosen a likelihood, which parameters are to be estimated, and a prior for each parameter, a Bayesian model treats the estimates as a purely logical consequence of those assumptions. For every unique combination of data, likelihood, parameters, and prior, there is a unique set of estimates. The resulting estimates—the relative plausibility of different parameter values, conditional on the data—are known as the POSTERIOR DISTRIBUTION. The posterior distribution takes the form of the probability of the parameters, conditional on the data: $\Pr(p|n, w)$.

The mathematical procedure that defines the logic of the posterior distribution is BAYES' THEOREM. This is the theorem that gives Bayesian data analysis its name. But the theorem itself is a trivial implication of probability theory. Here's a quick derivation of it, in the context of the globe tossing example. Really this will just be a re-expression of the garden of forking data derivation from earlier in the chapter. What makes it look different is that it will use the rules of probability theory to coax out the updating rule. But it is still just counting.

For simplicity, I'm going to omit n from the notation. The joint probability of the data w and any particular value of p is:

$$\Pr(w, p) = \Pr(w|p)\Pr(p)$$

This just says that the probability of w and p is the product of likelihood $\Pr(w|p)$ and the prior probability $\Pr(p)$. This is like saying that the probability of rain and cold on the same day is equal to the probability of rain, when it's cold, times the probability that it's cold. This much is just definition. But it's just as true that:

$$\Pr(w, p) = \Pr(p|w)\Pr(w)$$

All I've done is reverse which probability is conditional, on the right-hand side. It is still a true definition. It's like saying that the probability of rain and cold on the same day is equal to the probability that it's cold, when it's raining, times the probability of rain. Compare this statement to the one in the previous paragraph.

Now since both right-hand sides are equal to the same thing, we can set them equal to one another and solve for the posterior probability, $\Pr(p|w)$:

$$\Pr(p|w) = \frac{\Pr(w|p)\,\Pr(p)}{\Pr(w)}$$

And this is Bayes' theorem. It says that the probability of any particular value of p, considering the data, is equal to the product of the likelihood and prior, divided by this thing $\Pr(w)$, which I'll call the *average likelihood*. In word form:

$$\text{Posterior} = \frac{\text{Likelihood} \times \text{Prior}}{\text{Average Likelihood}}$$

The average likelihood, $\Pr(w)$, can be confusing. It is commonly called the "evidence" or the "probability of the data," neither of which is a transparent name. The probability $\Pr(w)$ is merely the *average likelihood* of the data. Averaged over what? Averaged over the prior. It's job is to standardize the posterior, to ensure it sums (integrates) to one. In mathematical form:

$$\Pr(w) = \mathrm{E}\big(\Pr(w|p)\big) = \int \Pr(w|p)\,\Pr(p)dp$$

The operator E means to take an *expectation*. Such averages are commonly called *marginals* in mathematical statistics, and so you may also see this same probability called a *marginal likelihood*. And the integral above just defines the proper way to compute the average over a continuous distribution of values, like the infinite possible values of p.

The key lesson is that the posterior is proportional to the product of the prior and the likelihood. Why? Because the number of paths through the garden of forking data is the product of the prior number of paths and the new number of paths. The likelihood indicates the number of paths, and the prior indicates the prior number. Multiplication is just compressed counting. The average likelihood on the bottom just standardizes the counts so they sum to one. So while Bayes' theorem looks complicated, because the relationship with counting paths is obscured, it just expresses the counting that logic demands.

FIGURE 2.6 illustrates the multiplicative interaction of a prior and a likelihood. On each row, a prior on the left is multiplied by the likelihood in the middle to produce a posterior on the right. The likelihood in each case is the same, the likelihood for the globe toss data. The priors however vary. As a result, the posterior distributions vary.

Rethinking: Bayesian data analysis isn't about Bayes' theorem. A common notion about Bayesian data analysis, and Bayesian inference more generally, is that it is distinguished by the use of Bayes' theorem. This is a mistake. Inference under any probability concept will eventually make use of Bayes' theorem. Common introductory examples of "Bayesian" analysis using HIV and DNA testing are not uniquely Bayesian. Since all of the elements of the calculation are frequencies of observations, a non-Bayesian analysis would do exactly the same thing. Instead, Bayesian approaches get to use Bayes' theorem more generally, to quantify uncertainty about theoretical entities that cannot be observed, like parameters and models. Powerful inferences can be produced under both Bayesian and non-Bayesian probability concepts, but different justifications and sacrifices are necessary.

2.4. Making the model go

Recall that your Bayesian model is a machine, a figurative golem. It has built-in definitions for the likelihood, the parameters, and the prior. And then at its heart lies a motor that

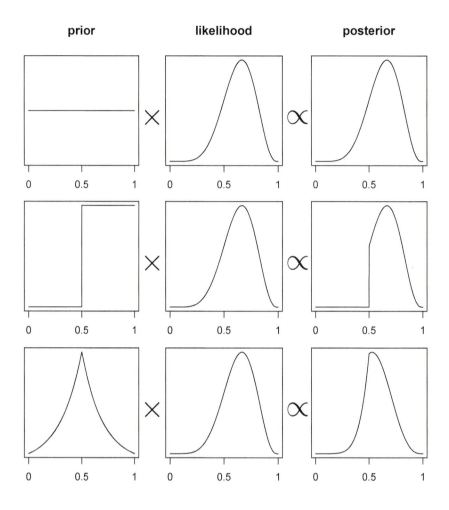

FIGURE 2.6. The posterior distribution, as a product of the prior distribution and likelihood. Top row: A flat prior constructs a posterior that is simply proportional to the likelihood. Middle row: A step prior, assigning zero probability to all values less than 0.5, resulting in a truncated posterior. Bottom row: A peaked prior that shifts and skews the posterior, relative to the likelihood.

processes data, producing a posterior distribution. The action of this motor can be thought of as *conditioning* the prior on the data. As explained in the previous section, this conditioning is governed by the rules of probability theory, which defines a uniquely logical posterior for every prior, likelihood, and data.

However, knowing the mathematical rule is often of little help, because many of the interesting models in contemporary science cannot be conditioned formally, no matter your skill in mathematics. And while some broadly useful models like linear regression can be conditioned formally, this is only possible if you constrain your choice of prior to special forms that are easy to do mathematics with. We'd like to avoid forced modeling choices of this kind, instead favoring conditioning engines that can accommodate whichever prior is most useful for inference.

What this means is that various numerical techniques are needed to approximate the mathematics that follows from the definition of Bayes' theorem. In this book, you'll meet three different conditioning engines, numerical techniques for computing posterior distributions:

(1) Grid approximation
(2) Quadratic approximation
(3) Markov chain Monte Carlo (MCMC)

There are many other engines, and new ones are being invented all the time. But the three you'll get to know here are common and widely useful. In addition, as you learn them, you'll also learn principles that will help you understand other techniques.

> **Rethinking: How you fit the model is part of the model.** Earlier in this chapter, I implicitly defined the model as a composite of a prior and a likelihood. That definition is typical. But in practical terms, we should also consider how the model is fit to data as part of the model. In very simple problems, like the globe tossing example that consumes this chapter, calculation of the posterior density is trivial and foolproof. In even moderately complex problems, however, the details of fitting the model to data force us to recognize that our numerical technique influences our inferences. This is because different mistakes and compromises arise under different techniques. The same model fit to the same data using different techniques may produce different answers. When something goes wrong, every piece of the machine may be suspect. And so our golems carry with them their updating engines, as much slaves to their engineering as they are to the priors and likelihoods we program into them.

2.4.1. Grid approximation. One of the simplest conditioning techniques is grid approximation. While most parameters are *continuous*, capable of taking on an infinite number of values, it turns out that we can achieve an excellent approximation of the continuous posterior distribution by considering only a finite grid of parameter values. At any particular value of a parameter, p', it's a simple matter to compute the posterior probability: just multiply the prior probability of p' by the likelihood at p'. Repeating this procedure for each value in the grid generates an approximate picture of the exact posterior distribution. This procedure is called GRID APPROXIMATION. In this section, you'll see how to perform a grid approximation, using simple bits of R code.

Grid approximation will mainly be useful as a pedagogical tool, as learning it forces the user to really understand the nature of Bayesian updating. But in most of your real modeling, grid approximation isn't practical. The reason is that it scales very poorly, as the number of parameters increases. So in later chapters, grid approximation will fade away, to be replaced by other, more efficient techniques. Still, the conceptual value of this exercise will carry forward, as you graduate to other techniques.

In the context of the globe tossing problem, grid approximation works extremely well. So let's build a grid approximation for the model we've constructed so far. Here is the recipe:

(1) Define the grid. This means you decide how many points to use in estimating the posterior, and then you make a list of the parameter values on the grid.
(2) Compute the value of the prior at each parameter value on the grid.
(3) Compute the likelihood at each parameter value.
(4) Compute the unstandardized posterior at each parameter value, by multiplying the prior by the likelihood.

(5) Finally, standardize the posterior, by dividing each value by the sum of all values.

In the globe tossing context, here's the code to complete all five of these steps:

R code
2.3

```
# define grid
p_grid <- seq( from=0 , to=1 , length.out=20 )

# define prior
prior <- rep( 1 , 20 )

# compute likelihood at each value in grid
likelihood <- dbinom( 6 , size=9 , prob=p_grid )

# compute product of likelihood and prior
unstd.posterior <- likelihood * prior

# standardize the posterior, so it sums to 1
posterior <- unstd.posterior / sum(unstd.posterior)
```

The above code makes a grid of only 20 points. To display the posterior distribution now:

R code
2.4

```
plot( p_grid , posterior , type="b" ,
    xlab="probability of water" , ylab="posterior probability" )
mtext( "20 points" )
```

You'll get the right-hand plot in FIGURE 2.7. Try sparser grids (5 points) and denser grids (100 or 1000 points). The correct density for your grid is determined by how accurate you want your approximation to be. More points means more precision. In this simple example, you can go crazy and use 100,000 points, but there won't be much change in inference after the first 100.

Now to replicate the different priors in FIGURE 2.5, try these lines of code—one at a time—for the prior grid:

R code
2.5

```
prior <- ifelse( p_grid < 0.5 , 0 , 1 )
prior <- exp( -5*abs( p_grid - 0.5 ) )
```

The rest of the code remains the same.

Overthinking: Vectorization. One of R's useful features is that it makes working with lists of numbers almost as easy as working with single values. So even though both lines of code above say nothing about how dense your grid is, whatever length you chose for the vector p_grid will determine the length of the vector prior. In R jargon, the calculations above are *vectorized*, because they work on lists of values, *vectors*. In a vectorized calculation, the calculation is performed on each element of the input vector—p_grid in this case—and the resulting output therefore has the same length. In other computing environments, the same calculation would require a *loop*. R can also use loops, but vectorized calculations are typically faster. They can however be much harder to read, when you are starting out with R. Be patient, and you'll soon grow accustomed to vectorized calculations.

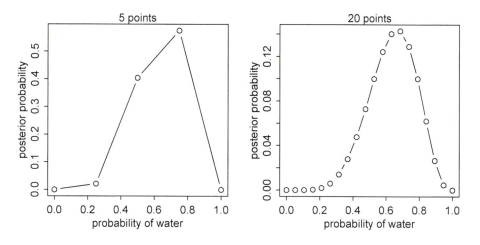

FIGURE 2.7. Computing posterior distribution by grid approximation. In each plot, the posterior distribution for the globe toss data and model is approximated with a finite number of evenly spaced points. With only 5 points (left), the approximation is terrible. But with 20 points (right), the approximation is already quite good. Compare to the analytically solved, exact posterior distribution in FIGURE 2.5 (page 30).

2.4.2. Quadratic approximation. We'll stick with the grid approximation to the globe tossing posterior, for the rest of this chapter and the next. But before long you'll have to resort to another approximation, one that makes stronger assumptions. The reason is that the number of unique values to consider in the grid grows rapidly as the number of parameters in your model increases. For the single-parameter globe tossing model, it's no problem to compute a grid of 100 or 1000 values. But for two parameters approximated by 100 values each, that's already $100^2 = 10000$ values to compute. For 10 parameters, the grid becomes many billions of values. These days, it's routine to have models with hundreds or thousands of parameters. The grid approximation strategy scales very poorly with model complexity, so it won't get us very far.

A useful approach is QUADRATIC APPROXIMATION. Under quite general conditions, the region near the peak of the posterior distribution will be nearly Gaussian—or "normal"—in shape. This means the posterior distribution can be usefully approximated by a Gaussian distribution. A Gaussian distribution is convenient, because it can be completely described by only two numbers: the location of its center (mean) and its spread (variance).

A Gaussian approximation is called "quadratic approximation" because the logarithm of a Gaussian distribution forms a parabola. And a parabola is a quadratic function. So this approximation essentially represents any log-posterior with a parabola.

We'll use quadratic approximation for much of the first half of this book. For many of the most common procedures in applied statistics—linear regression, for example—the approximation works very well. Often, it is even exactly correct, not actually an approximation at all. Computationally, quadratic approximation is very inexpensive, at least compared to grid approximation and MCMC (discussed next). The procedure, which R will happily conduct at your command, contains two steps.

(1) Find the posterior mode. This is usually accomplished by some optimization algorithm, a procedure that virtually "climbs" the posterior distribution, as if it were a
mountain. The golem doesn't know where the peak is, but it does know the slope
under its feet. There are many well-developed optimization procedures, most of
them more clever than simple hill climbing. But all of them try to find peaks.

(2) Once you find the peak of the posterior, you must estimate the curvature near the
peak. This curvature is sufficient to compute a quadratic approximation of the
entire posterior distribution. In some cases, these calculations can be done analytically, but usually your computer uses some numerical technique instead.

To compute the quadratic approximation for the globe tossing data, we'll use a tool in
the rethinking package: map. The abbreviation MAP stands for MAXIMUM A POSTERIORI,
which is just a fancy Latin name for the mode of the posterior distribution. We're going to
be using map a lot in the book. It's a flexible model fitting tool that will allow us to specify a
large number of different "regression" models. So it'll be worth trying it out right now. You'll
get a more thorough understanding of it later.

To compute the quadratic approximation to the globe tossing data:

R code
2.6
```
library(rethinking)
globe.qa <- map(
    alist(
        w ~ dbinom(9,p) ,   # binomial likelihood
        p ~ dunif(0,1)      # uniform prior
    ) ,
    data=list(w=6) )

# display summary of quadratic approximation
precis( globe.qa )
```

To use map, you provide a *formula*, a list of *data*, and a list of *start* values for the parameters.
The formula defines the likelihood and prior. I'll say much more about these formulas in
Chapter 4. The data list is just the count of water (6). Now let's see the output:

```
 Mean StdDev 5.5% 94.5%
p 0.67   0.16 0.42  0.92
```

The function precis presents a brief summary of the quadratic approximation. In this case,
it shows a MAP value of $p = 0.67$, which it calls the "Mean." The curvature is labeled "Std-
Dev." This stands for *standard deviation*. This value is the standard deviation of the posterior
distribution, while the mean value is its peak. Finally, the last two values in the precis output show the 89% percentile interval, which you'll learn more about in the next chapter. You
can read this kind of approximation like: *Assuming the posterior is Gaussian, it is maximized
at 0.67, and its standard deviation is 0.16.*

Since we already know the posterior, let's compare to see how good the approximation
is. I'll use the analytical approach here, which uses dbeta. I won't explain this calculation,
but it ensures that we have exactly the right answer, with no approximations. You can find
an explanation and derivation of it in just about any mathematical textbook on Bayesian
inference.

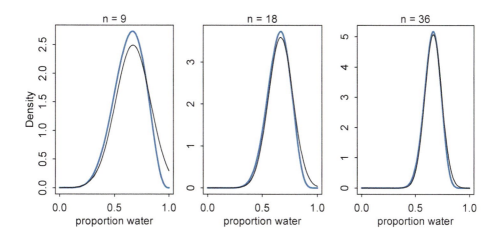

FIGURE 2.8. Accuracy of the quadratic approximation. In each plot, the exact posterior distribution is plotted in blue, and the quadratic approximation is plotted as the black curve. Left: The globe tossing data with $n = 9$ tosses and $w = 6$ waters. Middle: Double the amount of data, with the same fraction of water, $n = 18$ and $w = 12$. Right: Four times as much data, $n = 36$ and $w = 24$.

```
# analytical calculation
w <- 6
n <- 9
curve( dbeta( x , w+1 , n-w+1 ) , from=0 , to=1 )
# quadratic approximation
curve( dnorm( x , 0.67 , 0.16 ) , lty=2 , add=TRUE )
```

R code
2.7

You can see this plot (with a little extra formatting) on the left in FIGURE 2.8. The blue curve is the analytical posterior and the black curve is the quadratic approximation. The black curve does alright on its left side, but looks pretty bad on its right side. It even assigns positive probability to $p = 1$, which we know is impossible, since we saw at least one land sample.

As the amount of data increases, however, the quadratic approximation gets better. In the middle of FIGURE 2.8, the sample size is doubled to $n = 18$ tosses, but with the same fraction of water, so that the mode of the posterior is in the same place. The quadratic approximation looks better now, although still not great. At quadruple the data, on the right side of the figure, the two curves are nearly the same now.

This phenomenon, where the quadratic approximation improves with the amount of data, is very common. It's one of the reasons that so many classical statistical procedures are nervous about small samples: Those procedures use quadratic (or other) approximations that are only known to be safe with infinite data. Often, these approximations are useful with less than infinite data, obviously. But the rate of improvement as sample size increases varies greatly depending upon the details. In some model types, the quadratic approximation can remain terrible even with thousands of samples.

Using the quadratic approximation in a Bayesian context brings with it all the same concerns. But you can always lean on some algorithm other than quadratic approximation, if you have doubts. Indeed, grid approximation works very well with small samples, because in such cases the model must be simple and the computations will be quite fast. You can also use MCMC, which is introduced next.

> **Rethinking: Maximum likelihood estimation.** The quadratic approximation, either with a uniform prior or with a lot of data, is usually equivalent to a MAXIMUM LIKELIHOOD ESTIMATE (MLE) and its STANDARD ERROR. The MLE is a very common non-Bayesian parameter estimate. This equivalence between a Bayesian approximation and a common non-Bayesian estimator is both a blessing and a curse. It is a blessing, because it allows us to re-interpret a wide range of published non-Bayesian model fits in Bayesian terms. It is a curse, because maximum likelihood estimates have some curious drawbacks, which we'll explore in later chapters.

Overthinking: The Hessians are coming. Sometimes it helps to know more about how the quadratic approximation is computed. In particular, the approximation sometimes fails. When it does, chances are you'll get a confusing error message that says something about the "Hessian." Students of world history may know that the Hessians were German mercenaries hired by the British in the 18th century to do various things, including fight against the American revolutionary George Washington. These mercenaries are named after a region of what is now central Germany, Hesse.

The Hessian that concerns us here has little to do with mercenaries. It is named after mathematician Ludwig Otto Hesse (1811–1874). A *Hessian* is a square matrix of second derivatives. It is used for many purposes in mathematics, but in the quadratic approximation it is second derivatives of the log of posterior probability with respect to the parameters. It turns out that these derivatives are sufficient to describe a Gaussian distribution, because the logarithm of a Gaussian distribution is just a parabola. Parabolas have no derivatives beyond the second, so once we know the center of the parabola (the posterior mode) and its second derivative, we know everything about it. And indeed the second derivative (with respect to the outcome) of the logarithm of a Gaussian distribution is proportional to its inverse squared standard deviation (its "precision": page 76). So knowing the standard deviation tells us everything about its shape.

The standard deviation is typically computed from the Hessian, so computing the Hessian is nearly always a necessary step. But sometimes the computation goes wrong, and your golem will choke while trying to compute the Hessian. In those cases, you have several options. Not all hope is lost. But for now it's enough to recognize the term and associate it with an attempt to find the standard deviation for a quadratic approximation.

2.4.3. Markov chain Monte Carlo.

There are lots of important model types, like multilevel (mixed-effects) models, for which neither grid approximation nor quadratic approximation is always satisfactory. Such models can easily have hundreds or thousands or tens-of-thousands of parameters. Grid approximation obviously fails here, because it just takes too long—the Sun will go dark before your computer finishes the grid. Special forms of quadratic approximation might work, if everything is just right. But commonly, something is not just right. Furthermore, multilevel models do not always allow us to write down a single, unified function for the posterior distribution. This means that the function to maximize (when finding the MAP) is not known, but must be computed in pieces.

As a result, various counterintuitive model fitting techniques have arisen. The most popular of these is MARKOV CHAIN MONTE CARLO (MCMC), which is a family of conditioning engines capable of handling highly complex models. It is fair to say that MCMC is largely responsible for the insurgence of Bayesian data analysis that began in the 1990s. While MCMC is older than the 1990s, affordable computer power is not, so we must also thank the engineers. Much later in the book (Chapter 8), you'll meet simple and precise examples of MCMC model fitting, aimed at helping you understand the technique.

The conceptual challenge with MCMC lies in its highly non-obvious strategy. Instead of attempting to compute or approximate the posterior distribution directly, MCMC techniques merely draw samples from the posterior. You end up with a collection of parameter values, and the frequencies of these values correspond to the posterior plausibilities. You can then build a picture of the posterior from the histogram of these samples.

We nearly always work directly with these samples, rather than first constructing some mathematical estimate from them. And the samples are in many ways more convenient than having the posterior, because they are easier to think with. And so that's where we turn in the next chapter, to thinking with samples.

2.5. Summary

This chapter introduced the conceptual mechanics of Bayesian data analysis. The target of inference in Bayesian inference is a posterior probability distribution. Posterior probabilities state the relative numbers of ways each conjectured cause of the data could have produced the data. These relative numbers indicate plausibilities of the different conjectures. These plausibilities are updated in light of observations, a process known as Bayesian updating.

More mechanically, a Bayesian model is a composite of a likelihood, a choice of parameters, and a prior. The likelihood provides the plausibility of an observation (data), given a fixed value for the parameters. The prior provides the plausibility of each possible value of the parameters, before accounting for the data. The rules of probability tell us that the logical way to compute the plausibilities, after accounting for the data, is to use Bayes' theorem. This results in the posterior distribution.

In practice, Bayesian models are fit to data using numerical techniques, like grid approximation, quadratic approximation, and Markov chain Monte Carlo. Each method imposes different trade-offs.

2.6. Practice

Easy.

2E1. Which of the expressions below correspond to the statement: *the probability of rain on Monday*?

   (1) $Pr(rain)$
   (2) $Pr(rain|Monday)$
   (3) $Pr(Monday|rain)$
   (4) $Pr(rain, Monday)/Pr(Monday)$

2E2. Which of the following statements corresponds to the expression: $Pr(Monday|rain)$?

   (1) The probability of rain on Monday.
   (2) The probability of rain, given that it is Monday.
   (3) The probability that it is Monday, given that it is raining.
   (4) The probability that it is Monday and that it is raining.

2E3. Which of the expressions below correspond to the statement: *the probability that it is Monday, given that it is raining*?

    (1) $\Pr(\text{Monday}|\text{rain})$
    (2) $\Pr(\text{rain}|\text{Monday})$
    (3) $\Pr(\text{rain}|\text{Monday})\Pr(\text{Monday})$
    (4) $\Pr(\text{rain}|\text{Monday})\Pr(\text{Monday})/\Pr(\text{rain})$
    (5) $\Pr(\text{Monday}|\text{rain})\Pr(\text{rain})/\Pr(\text{Monday})$

2E4. The Bayesian statistician Bruno de Finetti (1906–1985) began his book on probability theory with the declaration: "PROBABILITY DOES NOT EXIST." The capitals appeared in the original, so I imagine de Finetti wanted us to shout this statement. What he meant is that probability is a device for describing uncertainty from the perspective of an observer with limited knowledge; it has no objective reality. Discuss the globe tossing example from the chapter, in light of this statement. What does it mean to say "the probability of water is 0.7"?

Medium.

2M1. Recall the globe tossing model from the chapter. Compute and plot the grid approximate posterior distribution for each of the following sets of observations. In each case, assume a uniform prior for p.

    (1) W, W, W
    (2) W, W, W, L
    (3) L, W, W, L, W, W, W

2M2. Now assume a prior for p that is equal to zero when $p < 0.5$ and is a positive constant when $p \geq 0.5$. Again compute and plot the grid approximate posterior distribution for each of the sets of observations in the problem just above.

2M3. Suppose there are two globes, one for Earth and one for Mars. The Earth globe is 70% covered in water. The Mars globe is 100% land. Further suppose that one of these globes—you don't know which—was tossed in the air and produced a "land" observation. Assume that each globe was equally likely to be tossed. Show that the posterior probability that the globe was the Earth, conditional on seeing "land" ($\Pr(\text{Earth}|\text{land})$), is 0.23.

2M4. Suppose you have a deck with only three cards. Each card has two sides, and each side is either black or white. One card has two black sides. The second card has one black and one white side. The third card has two white sides. Now suppose all three cards are placed in a bag and shuffled. Someone reaches into the bag and pulls out a card and places it flat on a table. A black side is shown facing up, but you don't know the color of the side facing down. Show that the probability that the other side is also black is 2/3. Use the counting method (Section 2 of the chapter) to approach this problem. This means counting up the ways that each card could produce the observed data (a black side facing up on the table).

2M5. Now suppose there are four cards: B/B, B/W, W/W, and another B/B. Again suppose a card is drawn from the bag and a black side appears face up. Again calculate the probability that the other side is black.

2M6. Imagine that black ink is heavy, and so cards with black sides are heavier than cards with white sides. As a result, it's less likely that a card with black sides is pulled from the bag. So again assume there are three cards: B/B, B/W, and W/W. After experimenting a number of times, you conclude that for every way to pull the B/B card from the bag, there are 2 ways to pull the B/W card and 3 ways to pull the W/W card. Again suppose that a card is pulled and a black side appears face up. Show that the probability the other side is black is now 0.5. Use the counting method, as before.

2M7. Assume again the original card problem, with a single card showing a black side face up. Before looking at the other side, we draw another card from the bag and lay it face up on the table. The face that is shown on the new card is white. Show that the probability that the first card, the one showing a black side, has black on its other side is now 0.75. Use the counting method, if you can. Hint: Treat this like the sequence of globe tosses, counting all the ways to see each observation, for each possible first card.

Hard.

2H1. Suppose there are two species of panda bear. Both are equally common in the wild and live in the same places. They look exactly alike and eat the same food, and there is yet no genetic assay capable of telling them apart. They differ however in their family sizes. Species A gives birth to twins 10% of the time, otherwise birthing a single infant. Species B births twins 20% of the time, otherwise birthing singleton infants. Assume these numbers are known with certainty, from many years of field research.

Now suppose you are managing a captive panda breeding program. You have a new female panda of unknown species, and she has just given birth to twins. What is the probability that her next birth will also be twins?

2H2. Recall all the facts from the problem above. Now compute the probability that the panda we have is from species A, assuming we have observed only the first birth and that it was twins.

2H3. Continuing on from the previous problem, suppose the same panda mother has a second birth and that it is not twins, but a singleton infant. Compute the posterior probability that this panda is species A.

2H4. A common boast of Bayesian statisticians is that Bayesian inference makes it easy to use all of the data, even if the data are of different types.

So suppose now that a veterinarian comes along who has a new genetic test that she claims can identify the species of our mother panda. But the test, like all tests, is imperfect. This is the information you have about the test:

- The probability it correctly identifies a species A panda is 0.8.
- The probability it correctly identifies a species B panda is 0.65.

The vet administers the test to your panda and tells you that the test is positive for species A. First ignore your previous information from the births and compute the posterior probability that your panda is species A. Then redo your calculation, now using the birth data as well.

3   Sampling the Imaginary

Lots of books on Bayesian statistics introduce posterior inference by using a medical testing scenario. To repeat the structure of common examples, suppose there is a blood test that correctly detects vampirism 95% of the time. This implies $\Pr(\text{positive}|\text{vampire}) = 0.95$. It's a very accurate test. It does make mistakes, though, in the form of false positives. One percent of the time, it incorrectly diagnoses normal people as vampires, implying $\Pr(\text{positive}|\text{mortal}) = 0.01$. The final bit of information we are told is that vampires are rather rare, being only 0.1% of the population, implying $\Pr(\text{vampire}) = 0.001$. Suppose now that someone tests positive for vampirism. What's the probability that he or she is a bloodsucking immortal?

The correct approach is just to use Bayes' theorem to invert the probability, to compute $\Pr(\text{vampire}|\text{positive})$. The calculation can be presented as:

$$\Pr(\text{vampire}|\text{positive}) = \frac{\Pr(\text{positive}|\text{vampire})\,\Pr(\text{vampire})}{\Pr(\text{positive})}$$

where $\Pr(\text{positive})$ is the average probability of a positive test result, that is,

$$\Pr(\text{positive}) = \Pr(\text{positive}|\text{vampire})\,\Pr(\text{vampire})$$
$$+ \Pr(\text{positive}|\text{mortal})\left(1 - \Pr(\text{vampire})\right)$$

Performing the calculation in R:

R code
3.1

```
PrPV <- 0.95
PrPM <- 0.01
PrV <- 0.001
PrP <- PrPV*PrV + PrPM*(1-PrV)
( PrVP <- PrPV*PrV / PrP )
```

```
[1] 0.08683729
```

That corresponds to an 8.7% chance that the suspect is actually a vampire.

Most people find this result counterintuitive. And it's a very important problem, because it mimics the structure of many realistic testing contexts, such as HIV and DNA testing, criminal profiling, and even statistical significance testing (see the rethinking box at the end of this section). Whenever the condition of interest is very rare, having a test that finds all the true cases is still no guarantee that a positive result carries much information at all. The reason is that most positive results are false positives, even when all the true positives are detected correctly.

But I don't like these examples, for two reasons. First, there's nothing really "Bayesian" about them. Remember: Bayesian inference is distinguished by a broad view of probability, not by the use of Bayes' theorem. Since all of the probabilities I provided above reference frequencies of events, rather than theoretical parameters, all major statistical philosophies would agree to use Bayes' theorem in this case. Second, and more important to our work in this chapter, these examples make Bayesian inference seem much harder than it has to be. Few people find it easy to remember which number goes where, probably because they never grasp the logic of the procedure. It's just a formula that descends from the sky.

There is a way to present the same problem that does make it more intuitive, however. Suppose that instead of reporting probabilities, as before, I tell you the following:

   (1) In a population of 100,000 people, 100 of them are vampires.
   (2) Of the 100 who are vampires, 95 of them will test positive for vampirism.
   (3) Of the 99,900 mortals, 999 of them will test positive for vampirism.

Now tell me, if we test all 100,000 people, what proportion of those who test positive for vampirism actually are vampires? Many people, although certainly not all people, find this presentation a lot easier.[47] Now we can just count up the number of people who test positive: $95 + 999 = 1094$. Out of these 1094 positive tests, 95 of them are real vampires, so that implies:

$$\Pr(\text{vampire}|\text{positive}) = \frac{95}{1094} \approx 0.087$$

It's exactly the same answer as before, but without a seemingly arbitrary rule.

The second presentation of the problem, using counts rather than probabilities, is often called the *frequency format* or *natural frequencies*. Why a frequency format helps people intuit the correct approach remains contentious. Some people think that human psychology naturally works better when it receives information in the form a person in a natural environment would receive it. In the real world, we encounter counts only. No one has ever seen a probability, the thinking goes. But everyone sees counts ("frequencies") in their daily lives. Maybe so.

> **Rethinking: The natural frequency phenomenon is not unique.** Changing the representation of a problem often makes it easier to address or inspires new ideas that were not available in an old representation.[48] In physics, switching between Newtonian and Lagrangian mechanics can make problems much easier. In evolutionary biology, switching between inclusive fitness and multilevel selection sheds new light on old models. And in statistics, switching between Bayesian and non-Bayesian representations often teaches us new things about both approaches.

Regardless of the explanation for this phenomenon, we can exploit it. And in this chapter we exploit it by taking the probability distributions from the previous chapter and sampling from them to produce counts. The posterior distribution is a probability distribution. And like all probability distributions, we can imagine drawing *samples* from it. The sampled events in this case are parameter values. Most parameters have no exact empirical realization. The Bayesian formalism treats parameter distributions as relative plausibility, not as any physical random process. In any event, randomness is always a property of information, never of the real world. But inside the computer, parameters are just as empirical as the outcome of a coin flip or a die toss or an agricultural experiment. The posterior defines the expected frequency that different parameter values will appear, once we start plucking parameters out of it.

This chapter teaches you basic skills for working with samples from the posterior distribution. It will seem a little silly to work with samples at this point, because the posterior distribution for the globe tossing model is very simple. It's so simple that it's no problem to work directly with the grid approximation or even the exact mathematical form.[49] But there are two reasons to adopt the sampling approach early on, before it's really necessary.

First, many scientists are quite shaky about integral calculus, even though they have strong and valid intuitions about how to summarize data. Working with samples transforms a problem in calculus into a problem in data summary, into a frequency format problem. An integral in a typical Bayesian context is just the total probability in some interval. That can be a challenging calculus problem. But once you have samples from the probability distribution, it's just a matter of counting values in the interval. Even seemingly simple calculations, like confidence intervals, are made difficult once a model has many parameters. In those cases, one must average over the uncertainty in all other parameters, when describing the uncertainty in a focal parameter. This requires a complicated integral, but only a very simple data summary. An empirical attack on the posterior allows the scientist to ask and answer more questions about the model, without relying upon a captive mathematician. For this reason, it is often easier and more intuitive to work with samples from the posterior, than to work with probabilities and integrals directly.

Second, some of the most capable methods of computing the posterior produce nothing but samples. Many of these methods are variants of Markov chain Monte Carlo techniques (MCMC). So if you learn early on how to conceptualize and process samples from the posterior, when you inevitably must fit a model to data using MCMC—and chances are you will—you will already know how to make sense of the output. Beginning with Chapter 8 of this book, you will use MCMC to open up the types and complexity of models you can practically fit to data. MCMC is no longer a technique only for experts, but rather part of the standard toolkit of quantitative science. So it's worth planning ahead.

So in this chapter we'll begin to use samples to summarize and simulate model output. The skills you learn here will apply to every problem in the remainder of the book, even though the details of the models, how they are fit to data, and how the samples are produced will vary.

Rethinking: Why statistics can't save bad science. The vampirism example at the start of this chapter has the same logical structure as many different *signal detection* problems: (1) There is some binary state that is hidden from us; (2) we observe an imperfect cue of the hidden state; (3) we (should) use Bayes' theorem to logically deduce the impact of the cue on our uncertainty.

Scientific inference is often framed in similar terms: (1) An hypothesis is either true or false; (2) we use a statistical procedure and get an imperfect cue of the hypothesis' falsity; (3) we (should) use Bayes' theorem to logically deduce the impact of the cue on the status of the hypothesis. It's the third step that is hardly ever done. But let's do it, for a toy example, so you can see how little statistical procedures—Bayesian or not—may do for us.

Suppose the probability of a positive finding, when an hypothesis is true, is $\Pr(\text{sig}|\text{true}) = 0.95$. That's the *power* of the test. Suppose that the probability of a positive finding, when an hypothesis is false, is $\Pr(\text{sig}|\text{false}) = 0.05$. That's the false-positive rate, like the 5% of conventional significance testing. Finally, we have to state the *base rate* at which hypotheses are true. Suppose for example that 1 in every 100 hypotheses turns out to be true. Then $\Pr(\text{true}) = 0.01$. No one knows this value, but the history of science suggests it's small. See Chapter 15 for more discussion. Now use Bayes' to compute the posterior:

$$\Pr(\text{true}|\text{pos}) = \frac{\Pr(\text{pos}|\text{true})\,\Pr(\text{true})}{\Pr(\text{pos})} = \frac{\Pr(\text{pos}|\text{true})\,\Pr(\text{true})}{\Pr(\text{pos}|\text{true})\,\Pr(\text{true}) + \Pr(\text{pos}|\text{false})\,\Pr(\text{false})}$$

Plug in the appropriate values, and the answer is approximately $\Pr(\text{true}|\text{pos}) = 0.16$. So a positive finding corresponds to a 16% chance that the hypothesis is true. This is the same low base-rate phenomenon that applies in medical (and vampire) testing. You can shrink the false-positive rate to 1% and get this posterior probability up to around 0.5, only as good as a coin flip. The most important thing to do is to improve the base rate, $\Pr(\text{true})$, and that requires thinking, not testing.[50]

3.1. Sampling from a grid-approximate posterior

Before beginning to work with samples, we need to generate them. Here's a reminder for how to compute the posterior for the globe tossing model, using grid approximation:

R code
3.2
```
p_grid <- seq( from=0 , to=1 , length.out=1000 )
prior <- rep( 1 , 1000 )
likelihood <- dbinom( 6 , size=9 , prob=p_grid )
posterior <- likelihood * prior
posterior <- posterior / sum(posterior)
```

Now we wish to draw 10,000 samples from this posterior. Imagine the posterior is a bucket full of parameter values, numbers such as 0.1, 0.7, 0.5, 1, etc. Within the bucket, each value exists in proportion to its posterior probability, such that values near the peak are much more common than those in the tails. We're going to scoop out 10,000 values from the bucket. Provided the bucket is well mixed, the resulting samples will have the same proportions as the exact posterior density. Therefore the individual values of *p* will appear in our samples in proportion to the posterior plausibility of each value.

Here's how you can do this in R, with one line of code:

R code
3.3
```
samples <- sample( p_grid , prob=posterior , size=1e4 , replace=TRUE )
```

The workhorse here is `sample`, which randomly pulls values from a vector. The vector in this case is `p_grid`, the grid of parameter values. The probability of each value is given by `posterior`, which you computed just above.

The resulting samples are displayed in FIGURE 3.1. On the left, all 10,000 (`1e4`) random samples are shown sequentially.

R code
3.4
```
plot( samples )
```

In this plot, it's as if you are flying over the posterior distribution, looking down on it. There are many more samples from the dense region near 0.6 and very few samples below 0.25. On the right, the plot shows the *density estimate* computed from these samples.

R code
3.5
```
library(rethinking)
dens( samples )
```

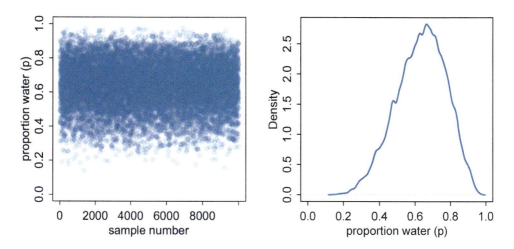

FIGURE 3.1. Sampling parameter values from the posterior distribution. Left: 10,000 samples from the posterior implied by the globe tossing data and model. Right: The density of samples (vertical) at each parameter value (horizontal).

You can see that the estimated density is very similar to ideal posterior you computed via grid approximation. If you draw even more samples, maybe 1e5 or 1e6, the density estimate will get more and more similar to the ideal.

All you've done so far is crudely replicate the posterior density you had already computed. So next it is time to use these samples to describe and understand the posterior.

3.2. Sampling to summarize

Once your model produces a posterior distribution, the model's work is done. But your work has just begun. It is necessary to summarize and interpret the posterior distribution. Exactly how it is summarized depends upon your purpose. But common questions include:

- How much posterior probability lies below some parameter value?
- How much posterior probability lies between two parameter values?
- Which parameter value marks the lower 5% of the posterior probability?
- Which range of parameter values contains 90% of the posterior probability?
- Which parameter value has highest posterior probability?

These simple questions can be usefully divided into questions about (1) intervals of *defined boundaries*, (2) questions about intervals of *defined probability mass*, and (3) questions about *point estimates*. We'll see how to approach these questions using samples from the posterior.

3.2.1. Intervals of defined boundaries.
Suppose I ask you for the posterior probability that the proportion of water is less than 0.5. Using the grid-approximate posterior, you can just add up all of the probabilities, where the corresponding parameter value is less than 0.5:

```
# add up posterior probability where p < 0.5
sum( posterior[ p_grid < 0.5 ] )
```

R code
3.6

```
[1] 0.1718746
```

So about 17% of the posterior probability is below 0.5. Couldn't be easier. But since grid approximation isn't practical in general, it won't always be so easy. Once there is more than one parameter in the posterior distribution (wait until the next chapter for that complication), even this simple sum is no longer very simple.

So let's see how to perform the same calculation, using samples from the posterior. This approach does generalize to complex models with many parameters, and so you can use it everywhere. All you have to do is similarly add up all of the samples below 0.5, but also divide the resulting count by the total number of samples. In other words, find the frequency of parameter values below 0.5:

R code
3.7
```
sum( samples < 0.5 ) / 1e4
```

```
[1] 0.1726
```

And that's nearly the same answer as the grid approximation provided, although your answer will not be exactly the same, because the exact samples you drew from the posterior will be different. This region is shown in the upper-left plot in FIGURE 3.2. Using the same approach, you can ask how much posterior probability lies between 0.5 and 0.75:

R code
3.8
```
sum( samples > 0.5 & samples < 0.75 ) / 1e4
```

```
[1] 0.6059
```

So about 61% of the posterior probability lies between 0.5 and 0.75. This region is shown in the upper-right plot of FIGURE 3.2.

Overthinking: Counting with sum. In the R code examples just above, I used the function sum to effectively count up how many samples fulfill a logical criterion. Why does this work? It works because R internally converts a logical expression, like samples < 0.5, to a vector of TRUE and FALSE results, one for each element of samples, saying whether or not each element matches the criterion. Go ahead and enter samples < 0.5 on the R prompt, to see this for yourself. Then when you sum this vector of TRUE and FALSE, R counts each TRUE as 1 and each FALSE as 0. So it ends up counting how many TRUE values are in the vector, which is the same as the number of elements in samples that match the logical criterion.

3.2.2. Intervals of defined mass. It is more common to see scientific journals reporting an interval of defined mass, usually known as a CONFIDENCE INTERVAL. An interval of posterior probability, such as the ones we are working with, may instead be called a CREDIBLE INTERVAL, although the terms may also be used interchangeably, in the usual polysemy that arises when commonplace words are used in technical definitions. It's easy to keep track of what's being summarized, however, as long as you pay attention to how the model is defined.

These posterior intervals report two parameter values that contain between them a specified amount of posterior probability, a probability mass. For this type of interval, it is easier to find the answer by using samples from the posterior than by using a grid approximation. Suppose for example you want to know the boundaries of the lower 80% posterior probability. You know this interval starts at $p = 0$. To find out where it stops, think of the samples as data and ask where the 80th percentile lies:

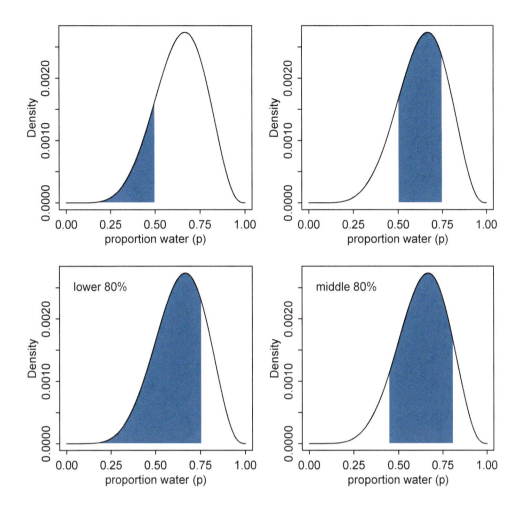

FIGURE 3.2. Two kinds of posterior interval. Top row: Intervals of defined boundaries. Top-left: The blue area is the posterior probability below a parameter value of 0.5. Top-right: The posterior probability between 0.5 and 0.75. Bottom row: Intervals of defined mass. Bottom-left: Lower 80% posterior probability exists below a parameter value of about 0.75. Bottom-right: Middle 80% posterior probability lies between the 10% and 90% quantiles.

```
quantile( samples , 0.8 )
```

R code
3.9

```
      80%
0.7607608
```

This region is shown in the bottom-left plot in FIGURE 3.2. Similarly, the middle 80% interval lies between the 10th percentile and the 90th percentile. These boundaries are found using the same approach:

R code
3.10
```
quantile( samples , c( 0.1 , 0.9 ) )
```

```
     10%       90%
0.4464464 0.8118118
```

This region is shown in the bottom-right plot in FIGURE 3.2.

Intervals of this sort, which assign equal probability mass to each tail, are very common in the scientific literature. We'll call them PERCENTILE INTERVALS (PI). These intervals do a good job of communicating the shape of a distribution, as long as the distribution isn't too asymmetrical. But in terms of supporting inferences about which parameters are consistent with the data, they are not perfect. Consider the posterior distribution and different intervals in FIGURE 3.3. This posterior is consistent with observing three waters in three tosses and a uniform (flat) prior. It is highly skewed, having its maximum value at the boundary, $p = 1$. You can compute it, via grid approximation, with:

R code
3.11
```
p_grid <- seq( from=0 , to=1 , length.out=1000 )
prior <- rep(1,1000)
likelihood <- dbinom( 3 , size=3 , prob=p_grid )
posterior <- likelihood * prior
posterior <- posterior / sum(posterior)
samples <- sample( p_grid , size=1e4 , replace=TRUE , prob=posterior )
```

This code also goes ahead to sample from the posterior. Now, on the left of FIGURE 3.3, the 50% percentile confidence interval is shaded. You can conveniently compute this from the samples with PI (part of rethinking):

R code
3.12
```
PI( samples , prob=0.5 )
```

```
     25%       75%
0.7037037 0.9329329
```

This interval assigns 25% of the probability mass above and below the interval. So it provides the central 50% probability. But in this example, it ends up excluding the most probable parameter values, near $p = 1$. So in terms of describing the shape of the posterior distribution—which is really all these intervals are asked to do—the percentile interval can be misleading.

In contrast, the right-hand plot in FIGURE 3.3 displays the 50% HIGHEST POSTERIOR DENSITY INTERVAL (HPDI).[51] The HPDI is the narrowest interval containing the specified probability mass. If you think about it, there must be an infinite number of posterior intervals with the same mass. But if you want an interval that best represents the parameter values most consistent with the data, then you want the densest of these intervals. That's what the HPDI is. Compute it from the samples with HPDI (also part of rethinking):

R code
3.13
```
HPDI( samples , prob=0.5 )
```

```
    |0.5       0.5|
0.8408408 1.0000000
```

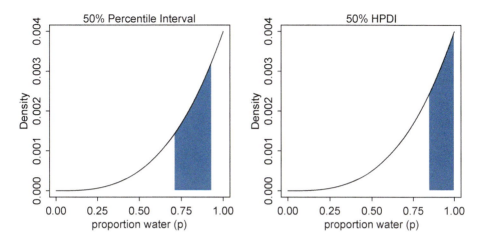

FIGURE 3.3. The difference between percentile and highest posterior density confidence intervals. The posterior density here corresponds to a flat prior and observing three water samples in three total tosses of the globe. Left: 50% percentile interval. This interval assigns equal mass (25%) to both the left and right tail. As a result, it omits the most probable parameter value, $p = 1$. Right: 50% highest posterior density interval, HPDI. This interval finds the narrowest region with 50% of the posterior probability. Such a region always includes the most probable parameter value.

This interval captures the parameters with highest posterior probability, as well as being noticeably narrower: 0.16 in width rather than 0.23 for the percentile interval.

So the HPDI has some advantages over the PI. But in most cases, these two types of interval are very similar.[52] They only look so different in this case because the posterior distribution is highly skewed. If we instead used samples from the posterior distribution for six waters in nine tosses, these intervals would be nearly identical. Try it for yourself, using different probability masses, such as prob=0.8 and prob=0.95. When the posterior is bell shaped, it hardly matters which type of interval you use. Remember, we're not launching rockets or calibrating atom smashers, so fetishizing precision to the 5th decimal place will not improve your science.

The HPDI also has some disadvantages. HPDI is more computationally intensive than PI and suffers from greater *simulation variance*, which is a fancy way of saying that it is sensitive to how many samples you draw from the posterior. It is also harder to understand and many scientific audiences will not appreciate its features, while they will immediately understand a percentile interval, as ordinary non-Bayesian intervals are nearly always percentile intervals (although of sampling distributions, not posterior distributions).

Overall, if the choice of interval type makes a big difference, then you shouldn't be using intervals to summarize the posterior. Remember, the entire posterior distribution is the Bayesian estimate. It summarizes the relative plausibilities of each possible value of the parameter. Intervals of the distribution are just helpful for summarizing it. If choice of interval leads to different inferences, then you'd be better off just plotting the entire posterior distribution.

Rethinking: Why 95%? The most common interval mass in the natural and social sciences is the 95% interval. This interval leaves 5% of the probability outside, corresponding to a 5% chance of the parameter not lying within the interval (although see below). This customary interval also reflects the customary threshold for *statistical significance*, which is 5% or $p < 0.05$. It is not easy to defend the choice of 95% (5%), outside of pleas to convention. Ronald Fisher is sometimes blamed for this choice, but his widely cited 1925 invocation of it was not enthusiastic:

> "The [number of standard deviations] for which $P = .05$, or 1 in 20, is 1.96 or nearly 2; it is convenient to take this point as a limit in judging whether a deviation is to be considered significant or not."[53]

Most people don't think of convenience as a serious criterion. Later in his career, Fisher actively advised against always using the same threshold for significance.[54]

So what are you supposed to do then? There is no consensus, but thinking is always a good idea. If you are trying to say that an interval doesn't include some value, then you might use the widest interval that excludes the value. Often, all confidence intervals do is communicate the shape of a distribution. In that case, a series of nested intervals may be more useful than any one interval. For example, why not present 67%, 89%, and 97% intervals, along with the median? Why these values? No reason. They are prime numbers, which makes them easy to remember. But all that matters is they be spaced enough to illustrate the shape of the posterior. And these values avoid 95%, since conventional 95% intervals encourage many readers to conduct unconscious hypothesis tests.

Rethinking: What do confidence intervals mean? It is common to hear that a 95% confidence interval means that there is a probability 0.95 that the true parameter value lies within the interval. In strict non-Bayesian statistical inference, such a statement is never correct, because strict non-Bayesian inference forbids using probability to measure uncertainty about parameters. Instead, one should say that if we repeated the study and analysis a very large number of times, then 95% of the computed intervals would contain the true parameter value. If the distinction is not entirely clear to you, then you are in good company. Most scientists find the definition of a confidence interval to be bewildering, and many of them slip unconsciously into a Bayesian interpretation.

But whether you use a Bayesian interpretation or not, a 95% interval does not contain the true value 95% of the time. The history of science teaches us that confidence intervals exhibit chronic overconfidence.[55] The word *true* should set off alarms that something is wrong with a statement like "contains the true value." The 95% is a *small world* number (see the introduction to Chapter 2), only true in the model's logical world. So it will never apply exactly to the real or *large world*. It is what the golem believes, but you are free to believe something else. Regardless, the width of the interval, and the values it covers, can provide valuable advice.

3.2.3. Point estimates. The third and final common summary task for the posterior is to produce point estimates of some kind. Given the entire posterior distribution, what value should you report? This seems like an innocent question, but it is difficult to answer. The Bayesian parameter estimate is precisely the entire posterior distribution, which is not a single number, but instead a function that maps each unique parameter value onto a plausibility value. So really the most important thing to note is that you don't have to choose a point estimate. It's hardly ever necessary.

But if you must produce a point estimate from the posterior, you'll have to ask and answer more questions. Consider the following example. Suppose again the globe tossing experiment in which we observe 3 waters out of 3 tosses, as in FIGURE 3.3. Let's consider three alternative point estimates. First, it is very common for scientists to report the parameter value with highest posterior probability, a *maximum a posteriori* (MAP) estimate. You can easily compute the MAP in this example:

```
p_grid[ which.max(posterior) ]
```
R code
3.14

```
[1] 1
```

Or if you instead have samples from the posterior, you can still approximate the same point:

```
chainmode( samples , adj=0.01 )
```
R code
3.15

```
[1] 0.9985486
```

But why is this point, the mode, interesting? Why not report the posterior mean or median?

```
mean( samples )
median( samples )
```
R code
3.16

```
[1] 0.8005558
[1] 0.8408408
```

These are also point estimates, and they also summarize the posterior. But all three—the mode (MAP), mean, and median—are different in this case. How can we choose among them? FIGURE 3.4 shows this posterior distribution and the locations of these point summaries.

One principled way to go beyond using the entire posterior as the estimate is to choose a LOSS FUNCTION. A loss function is a rule that tells you the cost associated with using any particular point estimate. While statisticians and game theorists have long been interested in loss functions, and how Bayesian inference supports them, scientists hardly ever use them explicitly. The key insight is that *different loss functions imply different point estimates.*

Here's an example to help us work through the procedure. Suppose I offer you a bet. Tell me which value of p, the proportion of water on the Earth, you think is correct. I will pay you $100, if you get it exactly right. But I will subtract money from your gain, proportional to the distance of your decision from the correct value. Precisely, your loss is proportional to the absolute value of $d - p$, where d is your decision and p is the correct answer. We could change the precise dollar values involved, without changing the important aspects of this problem. What matters is that the loss is proportional to the distance of your decision from the true value.

Now once you have the posterior distribution in hand, how should you use it to maximize your expected winnings? It turns out that the parameter value that maximizes expected winnings (minimizes expected loss) is the median of the posterior distribution. Let's calculate that fact, without using a mathematical proof. Those interested in the proof should follow the endnote.[56]

Calculating expected loss for any given decision means using the posterior to average over our uncertainty in the true value. Of course we don't know the true value, in most cases. But if we are going to use our model's information about the parameter, that means using the entire posterior distribution. So suppose we decide $p = 0.5$ will be our decision. Then the expected loss will be:

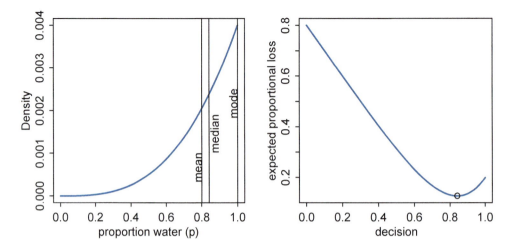

FIGURE 3.4. Point estimates and loss functions. Left: Posterior distribution
(blue) after observing 3 water in 3 tosses of the globe. Vertical lines show
the locations of the mode, median, and mean. Each point implies a different
loss function. Right: Expected loss under the rule that loss is proportional
to absolute distance of decision (horizontal axis) from the true value. The
point marks the value of p that minimizes the expected loss, the posterior
median.

R code
3.17

```
sum( posterior*abs( 0.5 - p_grid ) )
```

```
[1] 0.3128752
```

The symbols `posterior` and `p_grid` are the same ones we've been using throughout this
chapter, containing the posterior probabilities and the parameter values, respectively. All
the code above does is compute the weighted average loss, where each loss is weighted by its
corresponding posterior probability. There's a trick for repeating this calculation for every
possible decision, using the function `sapply`.

R code
3.18

```
loss <- sapply( p_grid , function(d) sum( posterior*abs( d - p_grid ) ) )
```

Now the symbol `loss` contains a list of loss values, one for each possible decision, corresponding the values in `p_grid`. From here, it's easy to find the parameter value that minimizes the loss:

R code
3.19

```
p_grid[ which.min(loss) ]
```

```
[1] 0.8408408
```

And this is actually the posterior median, the parameter value that splits the posterior density
such that half of the mass is above it and half below it. Try `median(samples)` for comparison.
It may not be exactly the same value, due to sampling variation, but it will be close.

So what are we to learn from all of this? In order to decide upon a *point estimate*, a single-value summary of the posterior distribution, we need to pick a loss function. Different loss functions nominate different point estimates. The two most common examples are the absolute loss as above, which leads to the median as the point estimate, and the quadratic loss $(d - p)^2$, which leads to the posterior mean (`mean(samples)`) as the point estimate. When the posterior distribution is symmetrical and normal-looking, then the median and mean converge to the same point, which relaxes some anxiety we might have about choosing a loss function. For the original globe tossing data (6 waters in 9 tosses), for example, the mean and median are barely different.

In principle, though, the details of the applied context may demand a rather unique loss function. Consider a practical example like deciding whether or not to order an evacuation, based upon an estimate of hurricane wind speed. Damage to life and property increases very rapidly as wind speed increases. There are also costs to ordering an evacuation when none is needed, but these are much smaller. Therefore the implied loss function is highly asymmetric, rising sharply as true wind speed exceeds our guess, but rising only slowly as true wind speed falls below our guess. In this context, the optimal point estimate would tend to be larger than posterior mean or median. Moreover, the real issue is whether or not to order an evacuation, and so producing a point estimate of wind speed may not be necessary at all.

Usually, research scientists don't think about loss functions. And so any point estimate like the mean or MAP that they may report isn't intended to support any particular decision, but rather to describe the shape of the posterior. You might argue that the decision to make is whether or not to accept an hypothesis. But the challenge then is to say what the relevant costs and benefits would be, in terms of the knowledge gained or lost.[57] Usually it's better to communicate as much as you can about the posterior distribution, as well as the data and the model itself, so that others can build upon your work. Premature decisions to accept or reject hypotheses can cost lives.[58]

It's healthy to keep these issues in mind, if only because they remind us that many of the routine questions in statistical inference can only be answered under consideration of a particular empirical context and applied purpose. Statisticians can provide general outlines and standard answers, but a motivated and attentive scientist will always be able to improve upon such general advice.

3.3. Sampling to simulate prediction

Another common job for samples from the posterior is to ease SIMULATION of the model's implied observations. Generating implied observations from a model is useful for at least four distinct reasons.

(1) *Model checking.* After a model is fit to real data, it is worth simulating implied observations, to check both whether the fit worked correctly and to investigate model behavior.

(2) *Software validation.* In order to be sure that our model fitting software is working, it helps to simulate observations under a known model and then attempt to recover the values of the parameters the data were simulated under.

(3) *Research design.* If you can simulate observations from your hypothesis, then you can evaluate whether the research design can be effective. In a narrow sense, this means doing *power analysis*, but the possibilities are much broader.

(4) *Forecasting.* Estimates can be used to simulate new predictions, for new cases and future observations. These forecasts can be useful as applied prediction, but also for model criticism and revision.

In this final section of the chapter, we'll look at how to produce simulated observations and how to perform some simple model checks.

3.3.1. Dummy data. Let's summarize the globe tossing model that you've been working with for two chapters now. A fixed true proportion of water p exists, and that is the target of our inference. Tossing the globe in the air and catching it produces observations of "water" and "land" that appear in proportion to p and $1 - p$, respectively.

Now note that these assumptions not only allow us to infer the plausibility of each possible value of p, after observation. That's what you did in the previous chapter. These assumptions also allow us to simulate the observations that the model implies. They allow this, because likelihood functions work in both directions. Given a realized observation, the likelihood function says how plausible the observation is. And given only the parameters, the likelihood defines a distribution of possible observations that we can sample from, to simulate observation. In this way, Bayesian models are always *generative*, capable of simulating predictions. Many non-Bayesian models are also generative, but many are not.

We will call such simulated data DUMMY DATA, to indicate that it is a stand-in for actual data. With the globe tossing model, the dummy data arises from a binomial likelihood:

$$\Pr(w|n, p) = \frac{n!}{w!(n-w)!} p^w (1-p)^{n-w}$$

where w is an observed count of "water" and n is the number of tosses. Suppose $n = 2$, two tosses of the globe. Then there are only three possible observations: 0 water, 1 water, 2 water. You can quickly compute the likelihood of each, for any given value of p. Let's use $p = 0.7$, which is just about the true proportion of water on the Earth:

R code
3.20
```
dbinom( 0:2 , size=2 , prob=0.7 )
```

```
[1] 0.09 0.42 0.49
```
This means that there's a 9% chance of observing $w = 0$, a 42% chance of $w = 1$, and a 49% chance of $w = 2$. If you change the value of p, you'll get a different distribution of implied observations.

Now we're going to simulate observations, using these likelihoods. This is done by sampling from the distribution just described above. You could use `sample` to do this, but R provides convenient sampling functions for all the ordinary probability distributions, like the binomial. So a single dummy data observation of w can be sampled with:

R code
3.21
```
rbinom( 1 , size=2 , prob=0.7 )
```

```
[1] 1
```
The "r" in rbinom stands for "random." It can also generate more than one simulation at a time. A set of 10 simulations can be made by:

R code
3.22
```
rbinom( 10 , size=2 , prob=0.7 )
```

```
[1] 2 2 2 1 2 1 1 1 0 2
```

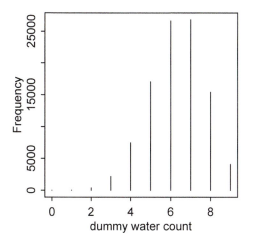

FIGURE 3.5. Distribution of simulated sample observations from 9 tosses of the globe. These samples assume the proportion of water is 0.7.

Let's generate 100,000 dummy observations, just to verify that each value (0, 1, or 2) appears in proportion to its likelihood:

```
dummy_w <- rbinom( 1e5 , size=2 , prob=0.7 )
table(dummy_w)/1e5
```
R code
3.23

```
dummy_w
      0       1       2
0.08904 0.41948 0.49148
```

And those values are very close to the analytically calculated likelihoods further up. You will see slightly different values, due to simulation variance. Execute the code above multiple times, to see how the exact realized frequencies fluctuate from simulation to simulation.

Only two tosses of the globe isn't much of a sample, though. So now let's simulate the same sample size as before, 9 tosses.

```
dummy_w <- rbinom( 1e5 , size=9 , prob=0.7 )
simplehist( dummy_w , xlab="dummy water count" )
```
R code
3.24

The resulting plot is shown in FIGURE 3.5. Notice that most of the time the expected observation does not contain water in its true proportion, 0.7. That's the nature of observation: There is a one-to-many relationship between data and data-generating processes. You should experiment with sample size, the size input in the code above, as well as the prob, to see how the distribution of simulated samples changes shape and location.

So that's how to perform a basic simulation of observations. What good is this? There are many useful jobs for these samples. In this chapter, we'll put them to use in examining the implied predictions of a model. But to do that, we'll have to combine them with samples from the posterior distribution. That's next.

Rethinking: Sampling distributions. Many readers will already have seen simulated observations. SAMPLING DISTRIBUTIONS are the foundation of common non-Bayesian statistical traditions. In those approaches, inference about parameters is made through the sampling distribution. In this book, inference about parameters is never done directly through a sampling distribution. The posterior distribution is not sampled, but deduced logically. Then samples can be drawn from the posterior, as earlier in this chapter, to aid in inference. In neither case is "sampling" a physical act. In both cases, it's just a mathematical device and produces only *small world* (Chapter 2) numbers.

3.3.2. Model checking. MODEL CHECKING means (1) ensuring the model fitting worked correctly and (2) evaluating the adequacy of a model for some purpose. Since Bayesian models are always *generative*, able to simulate observations as well as estimate parameters from observations, once you condition a model on data, you can simulate to examine the model's empirical expectations.

3.3.2.1. Did the software work? In the simplest case, we can check whether the software worked by checking for correspondence between implied predictions and the data used to fit the model. You might also call these implied predictions *retrodictions*, as they ask how well the model reproduces the data used to educate it. An exact match is neither expected nor desired. But when there is no correspondence at all, it probably means the software did something wrong.

There is no way to really be sure that software works correctly. Even when the retrodictions correspond to the observed data, there may be subtle mistakes. And when you start working with multilevel models, you'll have to expect a certain pattern of lack of correspondence between retrodictions and observations. Despite there being no perfect way to ensure software has worked, the simple check I'm encouraging here often catches silly mistakes, mistakes of the kind everyone makes from time to time.

In the case of the globe tossing analysis, the software implementation is simple enough that it can be checked against analytical results. So instead let's move directly to considering the model's adequacy.

3.3.2.2. Is the model adequate? After assessing whether the posterior distribution is the correct one, because the software worked correctly, it's useful to also look for aspects of the data that are not well described by the model's expectations. The goal is not to test whether the model's assumptions are "true," because all models are false. Rather, the goal is to assess exactly how the model fails to describe the data, as a path towards model comprehension, revision, and improvement.

All models fail in some respect, so you have to use your judgment—as well as the judgments of your colleagues—to decide whether any particular failure is or is not important. Few scientists want to produce models that do nothing more than re-describe existing samples. So imperfect prediction (retrodiction) is not a bad thing. Typically we hope to either predict future observations or understand enough that we might usefully tinker with the world. We'll consider these problems again, in Chapter 6.

For now, we need to learn how to combine sampling of simulated observations, as in the previous section, with sampling parameters from the posterior distribution. We expect to do better when we use the entire posterior distribution, not just some point estimate derived from it. Why? Because there is a lot of information about uncertainty in the entire posterior distribution. We lose this information when we pluck out a single parameter value and then perform calculations with it. This loss of information leads to overconfidence.

Let's do some basic model checks, using simulated observations for the globe tossing model. The observations in our example case are counts of water, over tosses of the globe.

The implied predictions of the model are uncertain in two ways, and it's important to be aware of both.

First, there is observation uncertainty. For any unique value of the parameter p, there is a unique implied pattern of observations that the model expects. These patterns of observations are the same gardens of forking data that you explored in the previous chapter. These patterns are also what you sampled in the previous section. There is uncertainty in the predicted observations, because even if you know p with certainty, you won't know the next globe toss with certainty (unless $p = 0$ or $p = 1$).

Second, there is uncertainty about p. The posterior distribution over p embodies this uncertainty. And since there is uncertainty about p, there is uncertainty about everything that depends upon p. The uncertainty in p will interact with the sampling variation, when we try to assess what the model tells us about outcomes.

We'd like to *propagate* the parameter uncertainty—carry it forward—as we evaluate the implied predictions. All that is required is averaging over the posterior density for p, while computing the predictions. For each possible value of the parameter p, there is an implied distribution of outcomes. So if you were to compute the sampling distribution of outcomes at each value of p, then you could average all of these prediction distributions together, using the posterior probabilities of each value of p, to get a POSTERIOR PREDICTIVE DISTRIBUTION.

FIGURE 3.6 illustrates this averaging. At the top, the posterior distribution is shown, with 10 unique parameter values highlighted by the vertical lines. The implied distribution of observations specific to each of these parameter values is shown in the middle row of plots. Observations are never certain for any value of p, but they do shift around in response to it. Finally, at the bottom, the sampling distributions for all values of p are combined, using the posterior probabilities to compute the weighted average frequency of each possible observation, zero to nine water samples.

The resulting distribution is for predictions, but it incorporates all of the uncertainty embodied in the posterior distribution for the parameter p. As a result, it is honest. While the model does a good job of predicting the data—the most likely observation is indeed the observed data—predictions are still quite spread out. If instead you were to use only a single parameter value to compute implied predictions, say the most probable value at the peak of posterior distribution, you'd produce an overconfident distribution of predictions, narrower than the posterior predictive distribution in FIGURE 3.6 and more like the sampling distribution shown for $p = 0.6$ in the middle row. The usual effect of this overconfidence will be to lead you to believe that the model is more consistent with the data than it really is— the predictions will cluster around the observations more tightly. This illusion arises from tossing away uncertainty about the parameters.

So how do you actually do the calculations? To simulate predicted observations for a single value of p, say $p = 0.6$, you can use rbinom to generate random binomial samples:

```
w <- rbinom( 1e4 , size=9 , prob=0.6 )
```

R code
3.25

This generates 10,000 (1e4) simulated predictions of 9 globe tosses (size=9), assuming $p = 0.6$. The predictions are stored as counts of water, so the theoretical minimum is zero and the theoretical maximum is nine. You can use simplehist(w) (in the rethinking package) to get a clean histogram of your simulated outcomes.

All you need to propagate parameter uncertainty into these predictions is replace the value 0.6 with samples from the posterior:

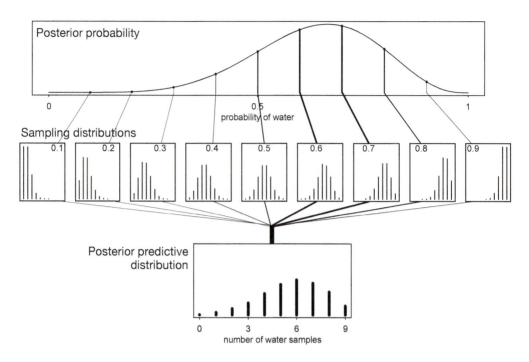

FIGURE 3.6. Simulating predictions from the total posterior. Top: The familiar posterior distribution for the globe tossing data. Ten example parameter values are marked by the vertical lines. Values with greater posterior probability indicated by thicker lines. Middle row: Each of the ten parameter values implies a unique sampling distribution of predictions. Bottom: Combining simulated observation distributions for all parameter values (not just the ten shown), each weighted by its posterior probability, produces the posterior predictive distribution. This distribution propagates uncertainty about parameter to uncertainty about prediction.

R code
3.26
```
w <- rbinom( 1e4 , size=9 , prob=samples )
```

The symbol samples above is the same list of random samples from the posterior distribution that you've used in previous sections. For each sampled value, a random binomial observation is generated. Since the sampled values appear in proportion to their posterior probabilities, the resulting simulated observations are averaged over the posterior. You can manipulate these simulated observations just like you manipulate samples from the posterior—you can compute intervals and point statistics using the same procedures. If you plot these samples, you'll see the distribution shown in the right-hand plot in FIGURE 3.6.

The simulated model predictions are quite consistent with the observed data in this case—the actual count of 6 lies right in the middle of the simulated distribution. There is quite a lot of spread to the predictions, but a lot of this spread arises from the binomial process itself, not uncertainty about p. Still, it'd be premature to conclude that the model is perfect. So far, we've only viewed the data just as the model views it: Each toss of the globe is completely independent of the others. This assumption is questionable. Unless the person

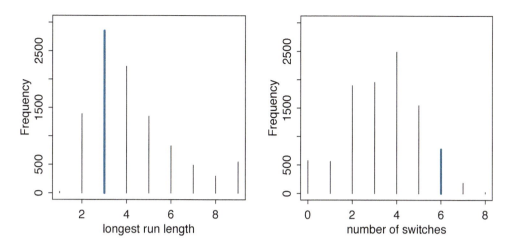

FIGURE 3.7. Alternative views of the same posterior predictive distribution (see FIGURE 3.6). Instead of considering the data as the model saw it, as a sum of water samples, now we view the data as both the length of the maximum run of water or land (left) and the number of switches between water and land samples (right). Observed values highlighted in blue. While the simulated predictions are consistent with the run length (3 water in a row), they are much less consistent with the frequent switches (6 switches in 9 tosses).

tossing the globe is careful, it is easy to induce correlations and therefore patterns among the sequential tosses. Consider for example that about half of the globe (and planet) is covered by the Pacific Ocean. As a result, water and land are not uniformly distributed on the globe, and therefore unless the globe spins and rotates enough while in the air, the position when tossed could easily influence the sample once it lands. The same problem arises in coin tosses, and indeed skilled individuals can influence the outcome of a coin toss, by exploiting the physics of it.[59]

So with the goal of seeking out aspects of prediction in which the model fails, let's look at the data in two different ways. Recall that the sequence of nine tosses was W L W W W L W L W. First, consider the length of the longest run of either water or land. This will provide a crude measure of correlation between tosses. So in the observed data, the longest run is 3 W's. Second, consider the number of times in the data that the sample switches from water to land or from land to water. This is another measure of correlation between samples. In the observed data, the number of switches is 6. There is nothing special about these two new ways of describing the data. They just serve to inspect the data in new ways. In your own modeling, you'll have to imagine aspects of the data that are relevant in your context, for your purposes.

FIGURE 3.7 shows the simulated predictions, viewed in these two new ways. On the left, the length of the longest run of water or land is plotted, with the observed value of 3 highlighted by the bold line. Again, the true observation is the most common simulated observation, but with a lot of spread around it. On the right, the number of switches from water to land and land to water is shown, with the observed value of 6 highlighted in bold. Now the simulated predictions appear less consistent with the data, as the majority of simulated observations have fewer switches than were observed in the actual sample. This is consistent with lack of independence between tosses of the globe, in which each toss is negatively correlated with the last.

Does this mean that the model is bad? That depends. The model will always be wrong in some sense, be *mis-specified*. But whether or not the mis-specification should lead us to try other models will depend upon our specific interests. In this case, if tosses do tend to switch from W to L and L to W, then each toss will provide less information about the true coverage of water on the globe. In the long run, even the wrong model we've used throughout the chapter will converge on the correct proportion. But it will do so more slowly than the posterior distribution may lead us to believe.

> **Rethinking: What does more extreme mean?** A common way of measuring deviation of observation from model is to count up the tail area that includes the observed data and any more extreme data. Ordinary *p*-values are an example of such a tail-area probability. When comparing observations to distributions of simulated predictions, as in FIGURE 3.6 and FIGURE 3.7, we might wonder how far out in the tail the observed data must be before we conclude that the model is a poor one. Because statistical contexts vary so much, it's impossible to give a universally useful answer.
>
> But more importantly, there are usually very many ways to view data and define "extreme." Ordinary *p*-values view the data in just the way the model expects it, and so provide a very weak form of model checking. For example, the far-right plot in FIGURE 3.6 evaluates model fit in the best way for the model. Alternative ways of defining "extreme" may provide a more serious challenge to a model. The different definitions of extreme in FIGURE 3.7 can more easily embarrass it.
>
> Model fitting remains an objective procedure—everyone and every golem conducts Bayesian updating in a way that doesn't depend upon personal preferences. But model checking is inherently subjective, and this actually allows it to be quite powerful, since subjective knowledge of an empirical domain provides expertise. Expertise in turn allows for imaginative checks of model performance. Since golems have terrible imaginations, we need the freedom to engage our own imaginations. In this way, the objective and subjective work together.[60]

3.4. Summary

This chapter introduced the basic procedures for manipulating posterior distributions. Our fundamental tool is samples of parameter values drawn from the posterior distribution. Working with samples transforms a problem of integral calculus into a problem of data summary. These samples can be used to produce intervals, point estimates, posterior predictive checks, as well as other kinds of simulations.

Posterior predictive checks combine uncertainty about parameters, as described by the posterior distribution, with uncertainty about outcomes, as described by the assumed likelihood function. These checks are useful for verifying that your software worked correctly. They are also useful for prospecting for ways in which your models are inadequate.

Once models become more complex, posterior predictive simulations will be used for a broader range of applications. Even understanding a model often requires simulating implied observations. We'll keep working with samples from the posterior, to make these tasks as easy and customizable as possible.

3.5. Practice

Easy. These problems use the samples from the posterior distribution for the globe tossing example. This code will give you a specific set of samples, so that you can check your answers exactly.

R code
3.27

```
p_grid <- seq( from=0 , to=1 , length.out=1000 )
prior <- rep( 1 , 1000 )
likelihood <- dbinom( 6 , size=9 , prob=p_grid )
posterior <- likelihood * prior
posterior <- posterior / sum(posterior)
set.seed(100)
samples <- sample( p_grid , prob=posterior , size=1e4 , replace=TRUE )
```

Use the values in `samples` to answer the questions that follow.

3E1. How much posterior probability lies below $p = 0.2$?

3E2. How much posterior probability lies above $p = 0.8$?

3E3. How much posterior probability lies between $p = 0.2$ and $p = 0.8$?

3E4. 20% of the posterior probability lies below which value of p?

3E5. 20% of the posterior probability lies above which value of p?

3E6. Which values of p contain the narrowest interval equal to 66% of the posterior probability?

3E7. Which values of p contain 66% of the posterior probability, assuming equal posterior probability both below and above the interval?

Medium.

3M1. Suppose the globe tossing data had turned out to be 8 water in 15 tosses. Construct the posterior distribution, using grid approximation. Use the same flat prior as before.

3M2. Draw 10,000 samples from the grid approximation from above. Then use the samples to calculate the 90% HPDI for p.

3M3. Construct a posterior predictive check for this model and data. This means simulate the distribution of samples, averaging over the posterior uncertainty in p. What is the probability of observing 8 water in 15 tosses?

3M4. Using the posterior distribution constructed from the new (8/15) data, now calculate the probability of observing 6 water in 9 tosses.

3M5. Start over at 3M1, but now use a prior that is zero below $p = 0.5$ and a constant above $p = 0.5$. This corresponds to prior information that a majority of the Earth's surface is water. Repeat each problem above and compare the inferences. What difference does the better prior make? If it helps, compare inferences (using both priors) to the true value $p = 0.7$.

Hard.

Introduction. The practice problems here all use the data below. These data indicate the gender (male=1, female=0) of officially reported first and second born children in 100 two-child families.

R code
3.28
```
birth1 <- c(1,0,0,0,1,1,0,1,0,1,0,0,1,1,0,1,1,0,0,0,1,0,0,0,1,0,
0,0,0,1,1,1,0,1,0,1,1,1,0,1,0,1,1,0,1,0,0,1,1,0,1,0,0,0,0,0,0,0,
1,1,0,1,0,0,1,0,0,0,1,0,0,1,1,1,1,0,1,0,1,1,1,1,1,0,0,1,0,1,1,0,
1,0,1,1,1,0,1,1,1,1)
birth2 <- c(0,1,0,1,0,1,1,1,0,0,1,1,1,1,1,0,0,1,1,1,0,0,1,1,1,0,
1,1,1,0,1,1,1,0,1,0,0,1,1,1,1,0,0,1,0,1,1,1,1,1,1,1,1,1,1,1,1,1,
1,1,1,0,1,1,0,1,1,0,1,1,1,0,0,0,0,0,0,1,0,0,0,1,1,0,0,1,0,0,1,1,
0,0,0,1,1,1,0,0,0,0)
```

So for example, the first family in the data reported a boy (1) and then a girl (0). The second family reported a girl (0) and then a boy (1). The third family reported two girls. You can load these two vectors into R's memory by typing:

R code
3.29
```
library(rethinking)
data(homeworkch3)
```

Use these vectors as data. So for example to compute the total number of boys born across all of these births, you could use:

R code
3.30
```
sum(birth1) + sum(birth2)
```

```
[1] 111
```

3H1. Using grid approximation, compute the posterior distribution for the probability of a birth being a boy. Assume a uniform prior probability. Which parameter value maximizes the posterior probability?

3H2. Using the `sample` function, draw 10,000 random parameter values from the posterior distribution you calculated above. Use these samples to estimate the 50%, 89%, and 97% highest posterior density intervals.

3H3. Use `rbinom` to simulate 10,000 replicates of 200 births. You should end up with 10,000 numbers, each one a count of boys out of 200 births. Compare the distribution of predicted numbers of boys to the actual count in the data (111 boys out of 200 births). There are many good ways to visualize the simulations, but the `dens` command (part of the `rethinking` package) is probably the easiest way in this case. Does it look like the model fits the data well? That is, does the distribution of predictions include the actual observation as a central, likely outcome?

3H4. Now compare 10,000 counts of boys from 100 simulated first borns only to the number of boys in the first births, `birth1`. How does the model look in this light?

3H5. The model assumes that sex of first and second births are independent. To check this assumption, focus now on second births that followed female first borns. Compare 10,000 simulated counts of boys to only those second births that followed girls. To do this correctly, you need to count the number of first borns who were girls and simulate that many births, 10,000 times. Compare the counts of boys in your simulations to the actual observed count of boys following girls. How does the model look in this light? Any guesses what is going on in these data?

4 Linear Models

History has been unkind to Ptolemy. Claudius Ptolemy (born 90 CE, died 168 CE) was an Egyptian mathematician and astronomer, most famous for his geocentric model of the solar system. These days, when scientists wish to mock someone, they might compare him to a supporter of the geocentric model. But Ptolemy was a genius. His mathematical model of the motions of the planets (FIGURE 4.1) was extremely accurate. To achieve its accuracy, it employed a device known as an *epicycle*, a circle on a circle. It is even possible to have epi-epicycles, circles on circles on circles. With enough epicycles in the right places, Ptolemy's model could predict with accuracy greater than anyone had achieved before him. And so the model was utilized for over a thousand years. And Ptolemy and people like him, toiling over centuries, worked it all out without the aid of a computer. Anyone should be flattered to be compared to Ptolemy.

The trouble of course is that the geocentric model is wrong, in many respects. If you used it to plot the path of your Mars probe, you'd miss the red planet by quite a distance. But for spotting Mars in the night sky, it remains an excellent model. It would have to be re-calibrated every century or so, depending upon which heavenly body you wish to locate. But the geocentric model continues to make useful predictions, provided those predictions remain within a narrow domain of questioning.

The strategy of using epicycles might seem crazy, once you know the correct structure of the solar system. But it turns out that the ancients had hit upon a generalized system of approximation. Given enough circles embedded in enough places, the Ptolemaic strategy is the same as a *Fourier series*, a way of decomposing a periodic function (like an orbit) into a series of sine and cosine functions. So no matter the actual arrangement of planets and moons, a geocentric model can be built to describe their paths against the night sky.

LINEAR REGRESSION is the geocentric model of applied statistics. By "linear regression," we will mean a family of simple statistical golems that attempt to learn about the mean and variance of some measurement, using an additive combination of other measurements. Like geocentrism, linear regression can usefully describe a very large variety of natural phenomena. Like geocentrism, linear regression is a descriptive model that corresponds to many different process models. If we read its structure too literally, we're likely to make mistakes. But used wisely, these little linear golems continue to be useful.

This chapter introduces linear regression as a Bayesian procedure. Under a probability interpretation, which is necessary for Bayesian work, linear regression uses a Gaussian (normal) distribution to describe our golem's uncertainty about some measurement of interest. This type of model is simple, flexible, and commonplace. Like all statistical models, it is not universally useful. But linear regression has a strong claim to being foundational, in the

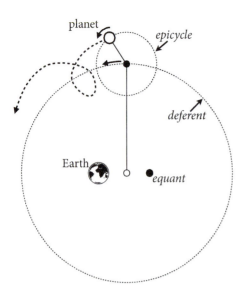

FIGURE 4.1. The Ptolemaic Universe, in which complex motion of the planets in the night sky was explained by orbits within orbits, called *epicycles*. The model is incredibly wrong, yet makes quite good predictions.

sense that once you learn to build and interpret linear regression models, you can quickly move on to other types of regression which are less normal.

4.1. Why normal distributions are normal

Suppose you and a thousand of your closest friends line up on the halfway line of a soccer field (football pitch). Each of you has a coin in your hand. At the sound of the whistle, you begin flipping the coins. Each time a coin comes up heads, that person moves one step towards the left-hand goal. Each time a coin comes up tails, that person moves one step towards the right-hand goal. Each person flips the coin 16 times, follows the implied moves, and then stands still. Now we measure the distance of each person from the halfway line. Can you predict what proportion of the thousand people who are standing on the halfway line? How about the proportion 5 yards left of the line?

It's hard to say where any individual person will end up, but you can say with great confidence what the collection of positions will be. The distances will be distributed in approximately normal, or Gaussian, fashion. This is true even though the underlying distribution is binomial. It does this because there are so many more possible ways to realize a sequence of left-right steps that sums to zero. There are slightly fewer ways to realize a sequence that ends up one step left or right of zero, and so on, with the number of possible sequences declining in the characteristic bell curve of the normal distribution.

4.1.1. Normal by addition. Let's see this result, by simulating this experiment in R. To show that there's nothing special about the underlying coin flip, assume instead that each step is different from all the others, a random distance between zero and one yard. Thus a coin is flipped, a distance between zero and one yard is taken in the indicated direction, and the process repeats. To simulate this, we generate for each person a list of 16 random numbers between -1 and 1. These are the individual steps. Then we add these steps together to get the position after 16 steps. Then we need to replicate this procedure 1000 times. This is the sort of task that would be harrowing in a point-and-click interface, but it is made trivial by the command line. Here's a single line to do the whole thing:

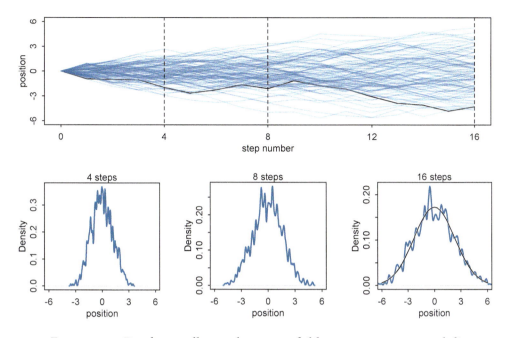

FIGURE 4.2. Random walks on the soccer field converge to a normal distribution. The more steps are taken, the closer the match between the real empirical distribution of positions and the ideal normal distribution, superimposed in the last plot in the bottom panel.

R code
4.1

```
pos <- replicate( 1000 , sum( runif(16,-1,1) ) )
```

You can plot the distribution of final positions in a number of different ways, including hist(pos) and plot(density(pos)). In FIGURE 4.2, I show the result of these random walks and how their distribution evolves as the number of steps increases. The top panel plots 100 different, independent random walks, with one highlighted in black. The vertical dashes indicate the locations corresponding to the distribution plots underneath, measured after 4, 8, and 16 steps. Although the distribution of positions starts off seemingly idiosyncratic, after 16 steps, it has already taken on a familiar outline. The familiar "bell" curve of the Gaussian distribution is emerging from the randomness. Go ahead and experiment with even larger numbers of steps to verify for yourself that the distribution of positions is stabilizing on the Gaussian. You can square the step sizes and transform them in a number of arbitrary ways, without changing the result: Normality emerges. Where does it come from?

Any process that adds together random values from the same distribution converges to a normal. But it's not easy to grasp why addition should result in a bell curve of sums.[61] Here's a conceptual way to think of the process. Whatever the average value of the source distribution, each sample from it can be thought of as a fluctuation from that average value. When we begin to add these fluctuations together, they also begin to cancel one another out. A large positive fluctuation will cancel a large negative one. The more terms in the sum, the more chances for each fluctuation to be canceled by another, or by a series of smaller ones in the opposite direction. So eventually the most likely sum, in the sense that there are the most ways to realize it, will be a sum in which every fluctuation is canceled by another, a sum of zero (relative to the mean).[62]

It doesn't matter what shape the underlying distribution possesses. It could be uniform, like in our example above, or it could be (nearly) anything else.[63] Depending upon the underlying distribution, the convergence might be slow, but it will be inevitable. Often, as in this example, convergence is rapid.

4.1.2. Normal by multiplication. Here's another way to get a normal distribution. Suppose the growth rate of an organism is influenced by a dozen loci, each with several alleles that code for more growth. Suppose also that all of these loci interact with one another, such that each increase growth by a percentage. This means that their effects multiply, rather than add. For example, we can sample a random growth rate for this example with this line of code:

R code
4.2
```
prod( 1 + runif(12,0,0.1) )
```

This code just samples 12 random numbers between 1.0 and 1.1, each representing a proportional increase in growth. Thus 1.0 means no additional growth and 1.1 means a 10% increase. The product of all 12 is computed and returned as output. Now what distribution do you think these random products will take? Let's generate 10,000 of them and see:

R code
4.3
```
growth <- replicate( 10000 , prod( 1 + runif(12,0,0.1) ) )
dens( growth , norm.comp=TRUE )
```

The reader should execute this code in R and see that the distribution is approximately normal again. I said normal distributions arise from summing random fluctuations, which is true. But the effect at each locus was multiplied by the effects at all the others, not added. So what's going on here?

We again get convergence towards a normal distribution, because the effect at each locus is quite small. Multiplying small numbers is approximately the same as addition. For example, if there are two loci with alleles increasing growth by 10% each, the product is:

$$1.1 \times 1.1 = 1.21$$

We could also approximate this product by just adding the increases, and be off by only 0.01:

$$1.1 \times 1.1 = (1+0.1)(1+0.1) = 1 + 0.2 + 0.01 \approx 1.2$$

The smaller the effect of each locus, the better this additive approximation will be. In this way, small effects that multiply together are approximately additive, and so they also tend to stabilize on Gaussian distributions. Verify this for yourself by comparing:

R code
4.4
```
big <- replicate( 10000 , prod( 1 + runif(12,0,0.5) ) )
small <- replicate( 10000 , prod( 1 + runif(12,0,0.01) ) )
```

The interacting growth deviations, as long as they are sufficiently small, converge to a Gaussian distribution. In this way, the range of casual forces that tend towards Gaussian distributions extends well beyond purely additive interactions.

4.1.3. Normal by log-multiplication. But wait, there's more. Large deviates that are multiplied together do not produce Gaussian distributions, but they do tend to produce Gaussian distributions on the log scale. For example:

```
log.big <- replicate( 10000 , log(prod(1 + runif(12,0,0.5))) )
```

R code
4.5

Yet another Gaussian distribution. We get the Gaussian distribution back, because adding logs is equivalent to multiplying the original numbers. So even multiplicative interactions of large deviations can produce Gaussian distributions, once we measure the outcomes on the log scale. Since measurement scales are arbitrary, there's nothing suspicious about this transformation. After all, it's natural to measure sound and earthquakes and even information (Chapter 6) on a log scale.

4.1.4. Using Gaussian distributions. We're going to spend the rest of this chapter using the Gaussian distribution as a skeleton for our hypotheses, building up models of measurements as aggregations of normal distributions. The justifications for using the Gaussian distribution fall into two broad categories: (1) ontological and (2) epistemological.

4.1.4.1. *Ontological justification.* The world is full of Gaussian distributions, approximately. As a mathematical idealization, we're never going to experience a perfect Gaussian distribution. But it is a widespread pattern, appearing again and again at different scales and in different domains. Measurement errors, variations in growth, and the velocities of molecules all tend towards Gaussian distributions. These processes do this because at their heart, these processes add together fluctuations. And repeatedly adding finite fluctuations results in a distribution of sums that have shed all information about the underlying process, aside from mean and spread.

One consequence of this is that statistical models based on Gaussian distributions cannot reliably identify micro-process. This recalls the modeling philosophy from Chapter 1 (page 6). But it also means that these models can do useful work, even when they cannot identify process. If we had to know the development biology of height before we could build a statistical model of height, human biology would be sunk.

There are many other patterns in nature, so make no mistake in assuming that the Gaussian pattern is universal. In later chapters, we'll see how other useful and common patterns, like the exponential and gamma and Poisson, also arise from natural processes. The Gaussian is a member of a family of fundamental natural distributions known as the EXPONENTIAL FAMILY. All of the members of this family are important for working science, because they populate our world.

4.1.4.2. *Epistemological justification.* But the natural occurrence of the Gaussian distribution is only one reason to build models around it. Another route to justifying the Gaussian as our choice of skeleton, and a route that will help us appreciate later why it is often a poor choice, is that it represents a particular state of ignorance. When all we know or are willing to say about a distribution of measures (measures are continuous values on the real number line) is their mean and variance, then the Gaussian distribution arises as the most consistent with our assumptions.

That is to say that the Gaussian distribution is the most natural expression of our state of ignorance, because if all we are willing to assume is that a measure has finite variance, the Gaussian distribution is the shape that can be realized in the largest number of ways and does not introduce any new assumptions. It is the least surprising and least informative assumption to make. In this way, the Gaussian is the distribution most consistent with our assumptions. Or rather, it is the most consistent with our golem's assumptions. If you don't think the distribution should be Gaussian, then that implies that you know something else that you should tell your golem about, something that would improve inference.

This epistemological justification is premised on INFORMATION THEORY and MAXIMUM ENTROPY. We'll dwell on information theory in Chapter 6 and maximum entropy in Chapter 9. Then in later chapters, other common and useful distributions will be used to build *generalized linear models* (GLMs). When these other distributions are introduced, you'll learn the constraints that make them the uniquely most appropriate (consistent with our assumptions) distributions.

For now, let's take the ontological and epistemological justifications of just the Gaussian distribution as reasons to start building models of measures around it. Throughout all of this modeling, keep in mind that using a model is not equivalent to swearing an oath to it. The golem is your servant, not the other way around. And so the golem's beliefs are not your own. There is no contract that forces us to believe the assumptions of our models. They just need to be useful robots.

Overthinking: Gaussian distribution. You don't have to memorize the Gaussian probability distribution formula to make good use of it. You're computer already knows it. But a little knowledge of its form can help demystify it. The probability *density* (see below) of some value y, given a Gaussian (normal) distribution with mean μ and standard deviation σ, is:

$$p(y|\mu, \sigma) = \frac{1}{\sqrt{2\pi\sigma^2}} \exp\left(-\frac{(y - \mu)^2}{2\sigma^2}\right)$$

This looks monstrous. But the important bit is just the $(y - \mu)^2$ bit. This is the part that gives the normal distribution its fundamental shape, a quadratic shape. Once you exponentiate the quadratic shape, you get the classic bell curve. The rest of it just scales and standardizes the distribution so that it sums to one, as all probability distributions must. But an expression as simple as $\exp(-y^2)$ yields the Gaussian prototype.

The Gaussian is also a continuous distribution, unlike the binomial probabilities of earlier chapters. This means that the value y in the Gaussian distribution can be any continuous value. The binomial, in contrast, requires integers. Probability distributions with only discrete outcomes, like the binomial, are usually called *probability mass* functions and denoted Pr. Continuous ones like the Gaussian are called *probability density* functions, denoted with p or just plain old f, depending upon author and tradition. For mathematical reasons, probability densities, but not masses, can be greater than 1. Try dnorm(0,0,0.1), for example, which is the way to make R calculate $p(0|0, 0.1)$. The answer, about 4, is no mistake. Probability *density* is the rate of change in cumulative probability. So where cumulative probability is increasing rapidly, density can easily exceed 1. But if we calculate the area under the density function, it will never exceed 1. Such areas are also called *probability mass*.

You can usually ignore all these density/mass details while doing computational work. In the conceptual parts of this book, I'll treat them identically. But it's good to be aware of the distinction. Sometimes the difference matters.

The Gaussian distribution is routinely seen without σ but with another parameter, τ. The parameter τ in this context is usually called *precision* and defined as $\tau = 1/\sigma^2$. This change of parameters gives us the equivalent formula (just substitute $\sigma = 1/\sqrt{\tau}$):

$$p(y|\mu, \tau) = \sqrt{\frac{\tau}{2\pi}} \exp\left(-\tfrac{1}{2}\tau(y - \mu)^2\right)$$

This form is common in Bayesian data analysis, and Bayesian model fitting software, such as BUGS or JAGS, sometimes requires using τ rather than σ.

4.2. A language for describing models

This book adopts a standard language for describing and coding statistical models. You find this language in many statistical texts and in nearly all statistical journals, as it is general to both Bayesian and non-Bayesian modeling. Scientists increasingly use this same language to describe their statistical methods, as well. So learning this language is an investment, no matter where you are headed next.

Here's the approach, in abstract. There will be many examples later. These numbered items describe the choices that will be encoded in your model description.

(1) First, we recognize a set of measurements that we hope to predict or understand, the *outcome* variable or variables.
(2) For each of these outcome variables, we define a likelihood distribution that defines the plausibility of individual observations. In linear regression, this distribution is always Gaussian.
(3) Then we recognize a set of other measurements that we hope to use to predict or understand the outcome. Call these *predictor* variables.
(4) We relate the exact shape of the likelihood distribution—its precise location and variance and other aspects of its shape, if it has them—to the predictor variables. In choosing a way to relate the predictors to the outcomes, we are forced to name and define all of the parameters of the model.
(5) Finally, we choose priors for all of the parameters in the model. These priors define the initial information state of the model, before seeing the data.

After all these decisions are made—and most of them will come to seem automatic to you before long—we summarize the model with something mathy like:

$$\text{outcome}_i \sim \text{Normal}(\mu_i, \sigma)$$
$$\mu_i = \beta \times \text{predictor}_i$$
$$\beta \sim \text{Normal}(0, 10)$$
$$\sigma \sim \text{HalfCauchy}(0, 1)$$

If that doesn't make much sense, good. That indicates that you are holding the right textbook, since this book teaches you how to read and write these mathematical model descriptions. We won't do any mathematical manipulation of them. Instead, they provide an unambiguous way to define and communicate our models. Once you get comfortable with their grammar, when you start reading these mathematical descriptions in other books or in scientific journals, you'll find them less obtuse.

The approach above surely isn't the only way to describe statistical modeling, but it is a widespread and productive language. Once a scientist learns this language, it becomes easier to communicate the assumptions of our models. We no longer have to remember seemingly arbitrary lists of bizarre conditions like *homoscedasticity* (constant variance), because we can just read these conditions from the model definitions. We will also be able to see natural ways to change these assumptions, instead of feeling trapped within some procrustean model type, like regression or multiple regression or ANOVA or ANCOVA or such. These are all the same kind of model, and that fact becomes obvious once we know how to talk about models as mappings of one set of variables through a probability distribution onto another set of variables. Fundamentally, these models define the ways values of some variables can arise, given values of other variables (Chapter 2).

4.2.1. Re-describing the globe tossing model. It's good to work with examples. Recall the proportion of water problem from previous chapters. The model in that case was always:

$$w \sim \text{Binomial}(n, p)$$
$$p \sim \text{Uniform}(0, 1)$$

where w was the observed count of water samples, n was the total number of samples, and p was the proportion of water on the actual globe. Read the above statement as:

> The count w is distributed binomially with sample size n and probability p.
> The prior for p is assumed to be uniform between zero and one.

Once we know the model in this way, we automatically know all of its assumptions. We know the binomial distribution assumes that each sample (globe toss) is independent of the others, and so we also know that the model assumes that sample points are independent of one another.

For now, we'll focus on simple models like the above. In these models, the first line defines the likelihood function used in Bayes' theorem. The other lines define priors. Both of the lines in this model are STOCHASTIC, as indicated by the $\sim$ symbol. A stochastic relationship is just a mapping of a variable or parameter onto a distribution. It is *stochastic* because no single instance of the variable on the left is known with certainty. Instead, the mapping is probabilistic: Some values are more plausible than others, but very many different values are plausible under any model. Later, we'll have models with deterministic definitions in them as well.

Overthinking: From model definition to Bayes' theorem. To relate the mathematical format above to Bayes' theorem, you could use the model definition to define the posterior distribution:

$$\Pr(p|w, n) = \frac{\text{Binomial}(w|n, p)\text{Uniform}(p|0, 1)}{\int \text{Binomial}(w|n, p)\text{Uniform}(p|0, 1)dp}$$

That monstrous denominator is just the average likelihood again. It standardizes the posterior to sum to 1. The action is in the numerator, where the posterior probability of any particular value of p is seen again to be proportional to the product of the likelihood and prior. In R code form, this is the same grid approximation calculation you've been using all along. In a form recognizable as the above expression:

R code
4.6
```
w <- 6; n <- 9;
p_grid <- seq(from=0,to=1,length.out=100)
posterior <- dbinom(w,n,p_grid)*dunif(p_grid,0,1)
posterior <- posterior/sum(posterior)
```

Compare to the calculations in earlier chapters.

4.3. A Gaussian model of height

Let's build a linear regression model now. Well, it'll be a "regression" once we have a predictor variable in it. For now, we'll get the scaffold in place and construct the predictor variable in the next section.

We'll work through this material by using real sets of data. In this case, we want a single measurement variable to model as a Gaussian distribution. There will be two parameters describing the distribution's shape, the mean μ and the standard deviation σ. Bayesian updating will allow us to consider every possible combination of values for μ and σ and to score each combination by its relative plausibility, in light of the data. These relative plausibilities are the posterior probabilities of each combination of values μ, σ.

Another way to say the above is this. There are an infinite number of possible Gaussian distributions. Some have small means. Others have large means. Some are wide, with a large σ. Others are narrow. We want our Bayesian machine to consider every possible distribution, each defined by a combination of μ and σ, and rank them by posterior plausibility. Posterior plausibility provides a measure of the logical compatibility of each possible distribution with the data and model.

In practice we'll use approximations to the formal analysis. So we won't really consider every possible value of μ and σ. But that won't cost us anything in most cases. Instead the thing to worry about is keeping in mind that the "estimate" here will be the entire posterior distribution, not any point within it. And as a result, the posterior distribution will be a distribution of Gaussian distributions. Yes, a distribution of distributions. If that doesn't make sense yet, then that just means you are being honest with yourself. Hold on, work hard, and it will make plenty of sense before long.

4.3.1. The data. The data contained in `data(Howell1)` are partial census data for the Dobe area !Kung San, compiled from interviews conducted by Nancy Howell in the late 1960s.[64] For the non-anthropologists reading along, the !Kung San are the most famous foraging population of the 20th century, largely because of detailed quantitative studies by people like Howell.

Load the data and place them into a convenient object with:

```
library(rethinking)
data(Howell1)
d <- Howell1
```

R code
4.7

What you have now is a *data frame* named simply d. I use the name d over and over again in this book to refer to the data frame we are working with at the moment. I keep its name short to save you typing. A *data frame* is a special kind of object in R. It is a table with named columns, corresponding to variables, and numbered rows, corresponding to individual cases. In this example, the cases are individuals. Inspect the structure of the data frame, the same way you'd inspect the structure of any symbol in R:

```
str( d )
```

R code
4.8

```
'data.frame': 544 obs. of  4 variables:
 $ height: num  152 140 137 157 145 ...
 $ weight: num  47.8 36.5 31.9 53 41.3 ...
 $ age   : num  63 63 65 41 51 35 32 27 19 54 ...
 $ male  : int  1 0 0 1 0 1 0 1 0 1 ...
```

This data frame contains four columns. Each column has 544 entries, so there are 544 individuals in these data. Each individual has a recorded height (centimeters), weight (kilograms), age (years), and "maleness" (0 indicating female and 1 indicating male).

We're going to work with just the `height` column, for the moment. The column containing the heights is really just a regular old R *vector*, the kind of list we have been working with in many of the code examples. You can access this vector by using its name:

```
d$height
```

Read the symbol $ as *extract*, as in *extract the column named* `height` *from the data frame* d. All it does is give you the column that follows it.

Overthinking: Data frames. It might seem like this whole data frame thing is kind of annoying, right now. If we're working with only one column here, why bother with this d thing at all? You don't have to use a data frame, as you can just pass raw vectors to every command we'll use in this book. But keeping related variables in the same data frame is a huge convenience. Once we have more than one variable, and we wish to model one as a function of the others, you'll better see the value of the data frame. You won't have to wait long.

More technically, a data frame is a special kind of `list` in R. So you access the individual variables with the usual list "double bracket" notation, like `d[[1]]` for the first variable or `d[['x']]` for the variable named x. Unlike regular lists, however, data frames force all variables to have the same length. That isn't always a good thing. And that's why some statistical packages, like the powerful Stan Markov chain sampler (mc-stan.org), accept plain lists of data, rather than proper data frames.

All we want for now are heights of adults in the sample. The reason to filter out non-adults for now is that height is strongly correlated with age, before adulthood. Later in the chapter, I'll ask you to tackle the age problem. But for now, better to postpone it. You can filter the data frame down to individuals of age 18 or greater with:

```
d2 <- d[ d$age >= 18 , ]
```

We'll be working with the data frame d2 now. It should have 352 rows (individuals) in it.

Overthinking: Index magic. The square bracket notation used in the code above is *index* notation. It is very powerful, but also quite compact and confusing. The data frame d is a matrix, a rectangular grid of values. You can access any value in the matrix with `d[row,col]`, replacing row and col with row and column numbers. If row or col are lists of numbers, then you get more than one row or column. If you leave the spot for row or col blank, then you get all of whatever you leave blank. For example, `d[ 3 , ]` gives all columns at row 3. Typing `d[,]` just gives you the entire matrix, because it returns all rows and all columns.

So what `d[ d$age >= 18 , ]` does is give you all of the rows in which d$age is greater-than-or-equal-to 18. It also gives you all of the columns, because the spot after the comma is blank. The result is stored in d2, the new data frame containing only adults. With a little practice, you can use this square bracket index notion to perform custom searches of your data, much like performing a database query.

4.3.2. The model. Our goal is to model these values using a Gaussian distribution. First, go ahead and plot the distribution of heights, with `dens(d2$height)`. These data look rather Gaussian in shape, as is typical of height data. This may be because height is a sum of many small growth factors. As you saw at the start of the chapter, a distribution of sums tends to converge to a Gaussian distribution. Whatever the reason, adult heights are nearly always approximately normal.

So it's reasonable for the moment to adopt the stance that the model's likelihood should be Gaussian. But be careful about choosing the Gaussian distribution only when the plotted outcome variable looks Gaussian to you. Gawking at the raw data, to try to decide how to model them, is usually not a good idea. The data could be a mixture of different Gaussian distributions, for example, and in that case you won't be able to detect the underlying normality just by eyeballing the outcome distribution. Furthermore, as mentioned earlier in this chapter, the empirical distribution needn't be actually Gaussian in order to justify using a Gaussian likelihood.

So which Gaussian distribution? There are an infinite number of them, with an infinite number of different means and standard deviations. We're ready to write down the general model and compute the plausibility of each combination of μ and σ. To define the heights as normally distributed with a mean μ and standard deviation σ, we write:

$$h_i \sim \text{Normal}(\mu, \sigma)$$

In many books you'll see the same model written as $h_i \sim \mathcal{N}(\mu, \sigma)$, which means the same thing. The symbol h refers to the list of heights, and the subscript i means *each individual element of this list*. It is conventional to use i because it stands for *index*. The index i takes on row numbers, and so in this example can take any value from 1 to 352 (the number of heights in d2$height). As such, the model above is saying that all the golem knows about each height measurement is defined by the same normal distribution, with mean μ and standard deviation σ. Before long, those little i's are going to show up on the right-hand side of the model definition, and you'll be able to see why we must bother with them. So don't ignore the i, even if it seems like useless ornamentation right now.

Rethinking: Independent and identically distributed. The short model above is sometimes described as assuming that the values h_i are *independent and identically distributed*, which may be abbreviated i.i.d., iid, or IID. You might even see the same model written:

$$h_i \overset{\text{iid}}{\sim} \text{Normal}(\mu, \sigma).$$

"iid" indicates that each value h_i has the same probability function, independent of the other h values and using the same parameters. A moment's reflection tells us that this is hardly ever true, in a physical sense. Whether measuring the same distance repeatedly or studying a population of heights, it is hard to argue that every measurement is independent of the others. For example, heights within families are correlated because of alleles shared through recent shared ancestry.

The i.i.d. assumption doesn't have to seem awkward, however, as long as you remember that probability is inside the golem, not outside in the world. The i.i.d. assumption is about how the golem represents its uncertainty. It is an *epistemological* assumption. It is not a physical assumption about the world, an *ontological* one, unless you insist that it is. E. T. Jaynes (1922–1998) called this the *mind projection fallacy*, the mistake of confusing epistemological claims with ontological claims.[65]

The point isn't to say epistemology trumps reality, but rather that in ignorance of such correlations the most conservative distribution to use is i.i.d.[66] This issue will return in Chapter 9. Furthermore, there is a mathematical result known as *de Finetti's theorem* that tells us that values which are EXCHANGEABLE can be approximated by mixtures of i.i.d. distributions. Colloquially, exchangeable values can be reordered. The practical impact of this is that "i.i.d." as an assumption cannot be read too literally, as different process models again correspond to the same statistical model (as argued in

Chapter 1). Even furthermore, there are many types of correlation that do little or nothing to the overall shape of a distribution, but only affect the precise sequence in which values appear. For example, pairs of sisters have highly correlated heights. But the overall distribution of female height remains almost perfectly normal. In such cases, i.i.d. remains perfectly useful, despite ignoring the correlations. Consider for example that Markov chain Monte Carlo (Chapter 8) can use highly correlated sequential samples to estimate most any iid distribution we like.

To complete the model, we're going to need some priors. The parameters to be estimated are both μ and σ, so we need a prior $\Pr(\mu, \sigma)$, the joint prior probability for all parameters. In most cases, priors are specified independently for each parameter, which amounts to assuming $\Pr(\mu, \sigma) = \Pr(\mu)\Pr(\sigma)$. Then we can write:

$$h_i \sim \text{Normal}(\mu, \sigma) \qquad \text{[likelihood]}$$
$$\mu \sim \text{Normal}(178, 20) \qquad \text{[}\mu \text{ prior]}$$
$$\sigma \sim \text{Uniform}(0, 50) \qquad \text{[}\sigma \text{ prior]}$$

The labels on the right are not part of the model, but instead just notes to help you keep track of the purpose of each line. The prior for μ is a broad Gaussian prior, centered on 178cm, with 95% of probability between 178 ± 40.

Why 178 cm? Your author is 178 cm tall. And the range from 138 cm to 218 cm encompasses a huge range of plausible mean heights for human populations. So domain-specific information has gone into this prior. Everyone knows something about human height and can set a reasonable and vague prior of this kind. But in many regression problems, as you'll see later, using prior information is more subtle, because parameters don't always have such clear physical meaning.

Whatever the prior, it's a very good idea to plot your priors, so you have a sense of the assumption they build into the model. In this case:

R code
4.11
```
curve( dnorm( x , 178 , 20 ) , from=100 , to=250 )
```

Execute that code yourself, to see that the golem is assuming that the average height (not each individual height) is almost certainly between 140 cm and 220 cm. So this prior carries a little information, but not a lot. The σ prior is a truly flat prior, a uniform one, that functions just to constrain σ to have positive probability between zero and 50cm. View it with:

R code
4.12
```
curve( dunif( x , 0 , 50 ) , from=-10 , to=60 )
```

A standard deviation like σ must be positive, so bounding it at zero makes sense. How should we pick the upper bound? In this case, a standard deviation of 50cm would imply that 95% of individual heights lie within 100cm of the average height. That's a very large range.

All this talk is nice, but it'll help to really see what these priors imply about the distribution of individual heights. You didn't specify a prior probability distribution of heights directly, but once you've chosen priors for μ and σ, these imply a prior distribution of individual heights. You can quickly simulate heights by sampling from the prior, like you sampled from the posterior back in Chapter 3. Remember, every posterior is also potentially a prior for a subsequent analysis, so you can process priors just like posteriors.

R code
4.13

```
sample_mu <- rnorm( 1e4 , 178 , 20 )
sample_sigma <- runif( 1e4 , 0 , 50 )
prior_h <- rnorm( 1e4 , sample_mu , sample_sigma )
dens( prior_h )
```

The density plot you get shows a vaguely bell-shaped density with thick tails. It is the expected distribution of heights, averaged over the prior. Notice that the prior probability distribution of height is not itself Gaussian. This is okay. The distribution you see is not an empirical expectation, but rather the distribution of relative plausibilities of different heights, before seeing the data.

Play around with the numbers in the priors above, to explore their effects on the prior probability density of heights.

Rethinking: A farewell to epsilon. Some readers will have already met an alternative notation for a Gaussian linear model:

$$h_i = \mu + \epsilon_i$$

$$\epsilon_i \sim \text{Normal}(0, \sigma)$$

This is equivalent to the $h_i \sim \text{Normal}(\mu, \sigma)$ form, with the ϵ standing in for the Gaussian density. But this ϵ form is poor form. The reason is that it does not usually generalize to other types of models. This means it won't be possible to express non-Gaussian models using tricks like ϵ. Better to learn one system that does generalize.

Overthinking: Model definition to Bayes' theorem again. It can help to see how the model definition on the previous page allows us to build up the posterior distribution. The height model, with its priors for μ and σ, defines this posterior distribution:

$$\Pr(\mu, \sigma | h) = \frac{\prod_i \text{Normal}(h_i | \mu, \sigma) \text{Normal}(\mu | 178, 20) \text{Uniform}(\sigma | 0, 50)}{\int \int \prod_i \text{Normal}(h_i | \mu, \sigma) \text{Normal}(\mu | 178, 20) \text{Uniform}(\sigma | 0, 50) d\mu d\sigma}$$

This looks monstrous, but it's the same creature as before. There are two new things that make it seem complicated. The first is that there is more than one observation in h, so to get the joint likelihood across all the data, we have to compute the probability for each h_i and then multiply all these likelihoods together. The product on the right-hand side takes care of that. The second complication is the two priors, one for μ and one for σ. But these just stack up. In the grid approximation code in the section to follow, you'll see the implications of this definition in the R code. Everything will be calculated on the log scale, so multiplication will become addition. But otherwise it's just a matter of executing Bayes' theorem.

4.3.3. Grid approximation of the posterior distribution.
Since this is the first Gaussian model in the book, and indeed the first model with more than one parameter, it's worth quickly mapping out the posterior distribution through brute force calculations. This isn't the approach I encourage in any other place, because it is laborious and computationally expensive. Indeed, it is usually so impractical as to be essentially impossible. But as always, it is worth knowing what the target actually looks like, before you start accepting approximations of it. A little later in this chapter, you'll use quadratic approximation to estimate the posterior distribution, and that's the approach you'll use for several chapters more. Once you

have the samples you'll produce in this subsection, you can compare them to the quadratic approximation in the next.

Unfortunately, doing the calculations here requires some technical tricks that add little, if any, conceptual insight. So I'm going to present the code here without explanation. You can execute it and keep going for now, but later return and follow the endnote for an explanation of the algorithm.[67] For now, here are the guts of the golem:

R code
4.14
```
mu.list <- seq( from=140, to=160 , length.out=200 )
sigma.list <- seq( from=4 , to=9 , length.out=200 )
post <- expand.grid( mu=mu.list , sigma=sigma.list )
post$LL <- sapply( 1:nrow(post) , function(i) sum( dnorm(
                d2$height ,
                mean=post$mu[i] ,
                sd=post$sigma[i] ,
                log=TRUE ) ) )
post$prod <- post$LL + dnorm( post$mu , 178 , 20 , TRUE ) +
    dunif( post$sigma , 0 , 50 , TRUE )
post$prob <- exp( post$prod - max(post$prod) )
```

You can inspect this posterior distribution, now residing in post$prob, using a variety of plotting commands. You can get a simple contour plot with:

R code
4.15
```
contour_xyz( post$mu , post$sigma , post$prob )
```

Or you can plot a simple heat map with:

R code
4.16
```
image_xyz( post$mu , post$sigma , post$prob )
```

The functions contour_xyz and image_xyz are both in the rethinking package.

4.3.4. Sampling from the posterior. To study this posterior distribution in more detail, again I'll push the flexible approach of sampling parameter values from it. This works just like it did in Chapter 3, when you sampled values of p from the posterior distribution for the globe tossing example. The only new trick is that since there are two parameters, and we want to sample combinations of them, we first randomly sample row numbers in post in proportion to the values in post$prob. Then we pull out the parameter values on those randomly sampled rows. This code will do it:

R code
4.17
```
sample.rows <- sample( 1:nrow(post) , size=1e4 , replace=TRUE ,
    prob=post$prob )
sample.mu <- post$mu[ sample.rows ]
sample.sigma <- post$sigma[ sample.rows ]
```

You end up with 10,000 samples, with replacement, from the posterior for the height data. Take a look at these samples:

R code
4.18
```
plot( sample.mu , sample.sigma , cex=0.5 , pch=16 , col=col.alpha(rangi2,0.1) )
```

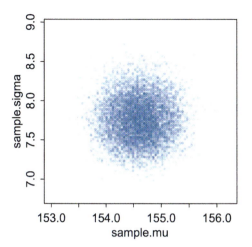

FIGURE 4.3. Samples from the posterior distribution for the heights data. The density of points is highest in the center, reflecting the most plausible combinations of μ and σ. There are many more ways for these parameter values to produce the data, conditional on the model.

I reproduce this plot in FIGURE 4.3. Note that the function col.alpha is part of the rethinking R package. All it does is make colors transparent, which helps the plot in FIGURE 4.3 more easily show density, where samples overlap. Adjust the plot to your tastes by playing around with cex (character expansion, the size of the points), pch (plot character), and the 0.1 transparency value.

Now that you have these samples, you can describe the distribution of confidence in each combination of μ and σ by summarizing the samples. Think of them like data and describe them, just like in Chapter 3. For example, to characterize the shapes of the *marginal* posterior densities of μ and σ, all we need to do is:

```
dens( sample.mu )
dens( sample.sigma )
```

R code
4.19

The jargon "marginal" here means "averaging over the other parameters." Execute the above code and inspect the plots. These densities are very close to being normal distributions. And this is quite typical. As sample size increases, posterior densities approach the normal distribution. If you look closely, though, you'll notice that the density for σ has a longer right-hand tail. I'll exaggerate this tendency a bit later, to show you that this condition is very common for standard deviation parameters.

To summarize the widths of these densities with highest posterior density intervals, just like in Chapter 3:

```
HPDI( sample.mu )
HPDI( sample.sigma )
```

R code
4.20

Since these samples are just vectors of numbers, you can compute any statistic from them that you could from ordinary data. If you want the mean or median, just use the corresponding R functions.

Overthinking: Sample size and the normality of σ's posterior. Before moving on to using quadratic approximation (map) as shortcut to all of this inference, it is worth repeating the analysis of the height data above, but now with only a fraction of the original data. The reason to do this is to demonstrate that, in principle, the posterior is not always so Gaussian in shape. There's no trouble with the mean, μ. For a Gaussian likelihood and a Gaussian prior on μ, the posterior distribution is always Gaussian as well, regardless of sample size. It is the standard deviation σ that causes problems. So if you care about σ—often people do not—you do need to be careful of abusing the quadratic approximation.

The deep reasons for the posterior of σ tending to have a long right-hand tail are complex. But a useful way to conceive of the problem is that variances must be positive. As a result, there must be more uncertainty about how big the variance (or standard deviation) is than about how small it is. For example, if the variance is estimated to be near zero, then you know for sure that it can't be much smaller. But it could be a lot bigger.

Let's quickly analyze only 20 of the heights from the height data to reveal this issue. To sample 20 random heights from the original list:

R code
4.21

```
d3 <- sample( d2$height , size=20 )
```

Now I'll repeat all the code from the previous subsection, modified to focus on the 20 heights in d3 rather than the original data. I'll compress all of the code together, but it's just what you've already seen above.

R code
4.22

```
mu.list <- seq( from=150, to=170 , length.out=200 )
sigma.list <- seq( from=4 , to=20 , length.out=200 )
post2 <- expand.grid( mu=mu.list , sigma=sigma.list )
post2$LL <- sapply( 1:nrow(post2) , function(i)
    sum( dnorm( d3 , mean=post2$mu[i] , sd=post2$sigma[i] ,
    log=TRUE ) ) )
post2$prod <- post2$LL + dnorm( post2$mu , 178 , 20 , TRUE ) +
    dunif( post2$sigma , 0 , 50 , TRUE )
post2$prob <- exp( post2$prod - max(post2$prod) )
sample2.rows <- sample( 1:nrow(post2) , size=1e4 , replace=TRUE ,
    prob=post2$prob )
sample2.mu <- post2$mu[ sample2.rows ]
sample2.sigma <- post2$sigma[ sample2.rows ]
plot( sample2.mu , sample2.sigma , cex=0.5 ,
    col=col.alpha(rangi2,0.1) ,
    xlab="mu" , ylab="sigma" , pch=16 )
```

After executing the code above, you'll see another scatter plot of the samples from the posterior density, but this time you'll notice a distinctly longer tail at the top of the cloud of points. You should also inspect the marginal posterior density for σ, averaging over μ, produced with:

R code
4.23

```
dens( sample2.sigma , norm.comp=TRUE )
```

This code will also show a normal approximation with the same mean and variance. Now you can see that the posterior for σ is not Gaussian, but rather has a long tail of uncertainty towards higher values.

4.3.5. Fitting the model with map.
Now we leave grid approximation behind and move on to one of the great engines of applied statistics, the quadratic approximation. Our interest in quadratic approximation, recall, is as a handy way to quickly make inferences about the shape of the posterior. The posterior's peak will lie at the *maximum a posteriori* estimate (MAP), and we can get a useful image of the posterior's shape by using the quadratic approximation of the posterior distribution at this peak.

To find the values of μ and σ that maximize the posterior probability, we'll use map, a command in the rethinking package. The way that map works is by using the model definition you were introduced to earlier in this chapter. Each line in the definition has a corresponding definition in the form of R code. The engine inside map then uses these definitions to define the posterior probability at each combination of parameter values. Then it can climb the posterior distribution and find the peak, its MAP.

Let's begin by repeating the code to load the data and select out the adults:

```
library(rethinking)
data(Howell1)
d <- Howell1
d2 <- d[ d$age >= 18 , ]
```

R code
4.24

Now we're ready to define the model, using R's formula syntax. The model definition in this case is just as before, but now we'll repeat it with each corresponding line of R code shown on the right-hand margin:

$$h_i \sim \text{Normal}(\mu, \sigma) \qquad\qquad \text{height} \sim \text{dnorm(mu,sigma)}$$
$$\mu \sim \text{Normal}(178, 20) \qquad\qquad \text{mu} \sim \text{dnorm(156,10)}$$
$$\sigma \sim \text{Uniform}(0, 50) \qquad\qquad \text{sigma} \sim \text{dunif(0,50)}$$

Now place the R code equivalents into an alist. Here's an alist of the formulas above:

```
flist <- alist(
    height ~ dnorm( mu , sigma ) ,
    mu ~ dnorm( 178 , 20 ) ,
    sigma ~ dunif( 0 , 50 )
)
```

R code
4.25

Note the commas at the end of each line, except the last. These commas separate each line of the model definition.

Fit the model to the data in the data frame d2 with:

```
m4.1 <- map( flist , data=d2 )
```

R code
4.26

After executing this code, you'll have a fit model stored in the symbol m4.1. Now take a look at the fit *maximum a posteriori* model:

```
precis( m4.1 )
```

R code
4.27

```
      Mean StdDev   5.5%  94.5%
mu  154.61   0.41 153.95 155.27
sigma  7.73   0.29   7.27   8.20
```

These numbers provide Gaussian approximations for each parameter's *marginal* distribution. This means the plausibility of each value of μ, after averaging over the plausibilities of each value of σ, is given by a Gaussian distribution with mean 154.6 and standard deviation 0.4.

The 5.5% and 94.5% quantiles are percentile interval boundaries, corresponding to an 89% interval. Why 89%? It's just the default. It displays a quite wide interval, so it shows a high-probability range of parameter values. If you want another interval, such as the conventional and mindless 95%, you can use `precis(m4.1,prob=0.95)`. But I don't recommend 95% intervals, because readers will have a hard time not viewing them as significance tests. 89 is also a prime number, so if someone asks you to justify it, you can stare at them meaningfully and incant, "Because it is prime." That's no worse justification than the conventional justification for 95%.

I encourage you to compare these 89% boundaries to the HPDIs from the grid approximation earlier. You'll find that they are almost identical. When the posterior is approximately Gaussian, then this is what you should expect.

Overthinking: Start values for `map`. `map` estimates the posterior by climbing it like a hill. To do this, it has to start climbing someplace, at some combination of parameter values. Unless you tell it otherwise, `map` starts at random values sampled from the prior. But it's also possible to specify a starting value for any parameter in the model. In the example in the previous section, that means the parameters μ and σ. Here's a good list of starting values in this case:

R code
4.28

```
start <- list(
    mu=mean(d2$height),
    sigma=sd(d2$height)
)
```

These start values are good guesses of the rough location of the MAP values.

Note that the list of start values is a regular `list`, not an `alist` like the formula list is. The two functions `alist` and `list` do the same basic thing: allow you to make a collection of arbitrary R objects. They differ in one important respect: `list` evaluates the code you embed inside it, while `alist` does not. So when you define a list of formulas, you should use `alist`, so the code isn't executed. But when you define a list of start values for parameters, you should use `list`, so that code like `mean(d2$height)` will be evaluated to a numeric value.

The priors we used before are very weak, both because they are nearly flat and because there is so much data. So I'll splice in a more informative prior for μ, so you can see the effect. All I'm going to do is change the standard deviation of the prior to 0.1, so it's a very narrow prior. I'll also build the formula right into the call to `map`, so you can see how to build it all at once.

R code
4.29

```
m4.2 <- map(
        alist(
            height ~ dnorm( mu , sigma ) ,
            mu ~ dnorm( 178 , 0.1 ) ,
            sigma ~ dunif( 0 , 50 )
        ) ,
        data=d2 )
precis( m4.2 )
```

```
        Mean StdDev   5.5%  94.5%
mu    177.86   0.10 177.70 178.02
sigma  24.52   0.93  23.03  26.00
```

Notice that the estimate for μ has hardly moved off the prior. The prior was very concentrated around 178. So this is not surprising. But also notice that the estimate for σ has changed quite a lot, even though we didn't change its prior at all. Once the golem is certain that the mean is near 178—as the prior insists—then the golem has to estimate σ conditional on that fact. This results in a different posterior for σ, even though all we changed is prior information about the other parameter.

At this point, it pays to play around with different priors, to get a sense of their effect on inference. There's so much data here that you'll have to use pretty extreme priors to have any effect on inference. Also keep in mind that most non-Bayesian estimates implicitly use flat priors. Before long, you're going to see that perfectly flat priors are hardly ever the best priors. But with as much data as we have in this example, and with such a simple model, it hardly matters which priors you use.

Overthinking: How strong is a prior? A prior can usually be interpreted as a former posterior inference, as previous data. So it's sometimes useful to talk about the strength of a prior in terms of which data would lead to the same posterior distribution, beginning with a flat prior. In the very narrow prior used just above, the $\mu \sim \text{Normal}(178, 0.1)$ prior, we can compute the implied amount of data easily, because there is a simple formula for the standard deviation of a Gaussian posterior for μ:

$$\sigma_{\text{post}} = 1/\sqrt{n}$$

(This is by no coincidence the same as the formula for the standard error of the sampling distribution of the mean, in non-Bayesian inference.) This formula implies that the implied amount of data (with mean 178) is $n = 1/\sigma_{\text{post}}^2$. So in this case the implied amount of data corresponding to the prior with standard deviation 0.1 is $n = 1/0.01 = 100$. So the $\mu \sim \text{Normal}(178, 0.1)$ is equivalent to having previously observed 100 heights with mean value 178. That's a pretty strong prior. In contrast, the former Normal(178, 20) prior implies $n = 1/20^2 = 0.0025$ of an observation. This is an extremely weak prior. But of course exactly how strong or weak either prior is will depend upon how much data is used to update it.

4.3.6. Sampling from a `map` fit. The above explains how to get a MAP quadratic approximation of the posterior, using `map`. But how do you then get samples from the quadratic approximate posterior distribution? The answer is rather simple, but non-obvious, and it requires recognizing that a quadratic approximation to a posterior distribution with more than one parameter dimension—μ and σ each contribute one dimension—is just a multi-dimensional Gaussian distribution.

As a consequence, when R constructs a quadratic approximation, it calculates not only standard deviations for all parameters, but also the covariances among all pairs of parameters. Just like a mean and standard deviation (or its square, a variance) are sufficient to describe a one-dimensional Gaussian distribution, a list of means and a matrix of variances and covariances are sufficient to describe a multi-dimensional Gaussian distribution. To see this matrix of variances and covariances, for model `m4.1`, use:

R code
4.30

```
vcov( m4.1 )
```

```
              mu           sigma
mu     0.1697395865  0.0002180593
sigma  0.0002180593  0.0849057933
```

The above is a **VARIANCE-COVARIANCE** matrix. It is the multi-dimensional glue of a quadratic approximation, because it tells us how each parameter relates to every other parameter in the posterior distribution. A variance-covariance matrix can be factored into two elements: (1) a vector of variances for the parameters and (2) a correlation matrix that tells us how changes in any parameter lead to correlated changes in the others. This decomposition is usually easier to understand. So let's do that now:

```
diag( vcov( m4.1 ) )
cov2cor( vcov( m4.1 ) )
```

```
          mu       sigma
0.16973959 0.08490579
```

```
                mu        sigma
mu     1.000000000 0.001816412
sigma 0.001816412 1.000000000
```

The two-element vector in the output is the list of variances. If you take the square root of this vector, you get the standard deviations that are shown in precis output. The two-by-two matrix in the output is the correlation matrix. Each entry shows the correlation, bounded between −1 and +1, for each pair of parameters. The 1's indicate a parameter's correlation with itself. If these values were anything except 1, we would be worried. The other entries are typically closer to zero, and they are very close to zero in this example. This indicates that learning μ tells us nothing about σ and likewise that learning σ tells us nothing about μ. This is typical of simple Gaussian models of this kind. But it is quite rare more generally, as you'll see in later chapters.

Okay, so how do we get samples from this multi-dimensional posterior? Now instead of sampling single values from a simple Gaussian distribution, we sample vectors of values from a multi-dimensional Gaussian distribution. The rethinking package provides a convenience function to do exactly that:

```
library(rethinking)
post <- extract.samples( m4.1 , n=1e4 )
head(post)
```

```
        mu      sigma
1 155.0031 7.443893
2 154.0347 7.771255
3 154.9157 7.822178
4 154.4252 7.530331
5 154.5307 7.655490
6 155.1772 7.974603
```

You end up with a data frame, post, with 10,000 (1e4) rows and two columns, one column for μ and one for σ. Each value is a sample from the posterior, so the mean and standard deviation of each column will be very close to the MAP values from before. You can confirm this by summarizing the samples:

```
precis(post)
```

```
        Mean StdDev |0.89  0.89|
mu    154.61   0.42 153.95 155.27
sigma   7.73   0.29   7.26   8.19
```

The 89% HPDI is displayed now, because we used samples. |0.89 means the lower boundary, and 0.89| means the upper boundary. Compare these values to the output from pre-cis(m4.1). And you can use plot(post) to see how much they resemble the samples from the grid approximation in FIGURE 4.3 (page 85).

These samples also preserve the covariance between μ and σ. This hardly matters right now, because μ and σ don't covary at all in this model. But once you add a predictor variable to your model, covariance will matter a lot.

Overthinking: Under the hood with multivariate sampling. The function extract.samples is for convenience. It is just running a simple simulation of the sort you conducted near the end of Chapter 3. Here's a peak at the motor. The work is done by a multi-dimensional version of rnorm, mvrnorm. The function rnorm simulates random Gaussian values, while mvrnorm simulates random vectors of multivariate Gaussian values. Here's how to use it directly to do what extract.samples does:

```
library(MASS)
post <- mvrnorm( n=1e4 , mu=coef(m4.1) , Sigma=vcov(m4.1) )
```
R code
4.34

You don't usually need to use mvrnorm directly like this, but sometimes you want to simulate multivariate Gaussian outcomes. In that case, you'll need to access mvrnorm directly. And of course it's always good to know a little about how the machine operates.

Overthinking: Getting σ right. The quadratic assumption for σ can be problematic, as seen on page 85. A conventional way to improve the situation is the estimate $\log(\sigma)$ instead. Why does this help? While the posterior distribution of σ will often not be Gaussian, the distribution of its logarithm can be much closer to Gaussian. So if we impose the quadratic approximation on the logarithm, rather than the standard deviation itself, we can often get a better approximation of the uncertainty. Here's how you can do this, using map.

```
m4.1_logsigma <- map(
        alist(
            height ~ dnorm( mu , exp(log_sigma) ) ,
            mu ~ dnorm( 178 , 20 ) ,
            log_sigma ~ dnorm( 2 , 10 )
        ) , data=d2 )
```
R code
4.35

Notice the exp inside the likelihood. That converts a continuous parameter, log_sigma, to be strictly positive, because $\exp(x) > 0$ for any real value x. Notice also the prior for log_sigma. Since log_sigma is continuous now, it can have a Gaussian prior.

When you extract samples, it is log_sigma that has a Gaussian distribution. To get the distribution of sigma, you just need to use the same exp as in the model definition, to get back on the natural scale:

```
post <- extract.samples( m4.1_logsigma )
sigma <- exp( post$log_sigma )
```
R code
4.36

When you have a lot of data, this won't make any noticeable difference. But the use of exp to effectively constrain a parameter to be positive is a robust and useful one. And it relates to *link functions*, which will be very important when we arrive at generalized linear models in Chapter 9.

4.4. Adding a predictor

What we've done above is a Gaussian model of height in a population of adults. But it doesn't really have the usual feel of "regression" to it. Typically, we are interested in modeling how an outcome is related to some predictor variable. And by including a predictor variable in a particular way, we'll have linear regression.

So now let's look at how height in these Kalahari foragers covaries with weight. This isn't the most thrilling scientific question, I know. But it is an easy relationship to start with, and it only seems dull because you don't have a theory about growth and life history in mind. If you did, it would be thrilling. Go ahead and plot height and weight against one another to get an idea of how strongly they covary:

R code
4.37
```
plot( d2$height ~ d2$weight )
```

The resulting plot is not shown here. You really should do it yourself. Once you can see the plot, you'll see that there's obviously a relationship: Knowing a person's weight helps you predict height.

To make this vague observation into a more precise quantitative model that relates values of weight to plausible values of height, we need some more technology. How do we take our Gaussian model from the previous section and incorporate predictor variables?

> **Rethinking: What is "regression"?** Many diverse types of models are called "regression." The term has come to mean using one or more predictor variables to model the distribution of one or more outcome variables. The original use of term, however, arose from anthropologist Francis Galton's (1822–1911) observation that the sons of tall and short men tended to be more similar to the population mean, hence *regression to the mean*.[68]
>
> This phenomenon arises statistically whenever individual measurements are assigned a common distribution, leading to *shrinkage* as each measurement informs the others. In the context of Galton's height data, attempting to predict each son's height on the basis of only his father's height is folly. Better to use the population of fathers. This leads to a prediction for each son which is similar to each father but "shrunk" towards the overall mean. Such predictions are routinely better. This same regression/shrinkage phenomenon applies at higher levels of abstraction and forms one basis of multilevel modeling (Chapter 12).

4.4.1. The linear model strategy. The strategy is to make the parameter for the mean of a Gaussian distribution, μ, into a linear function of the predictor variable and other, new parameters that we invent. This strategy is often simply called the LINEAR MODEL. The linear model strategy instructs the golem to assume that the predictor variable has a perfectly constant and additive relationship to the mean of the outcome. The golem then computes the posterior distribution of this constant relationship.

What this means, recall, is that the machine considers every possible combination of the parameter values. With a linear model, some of the parameters now stand for the strength of association between the mean of the outcome and the value of the predictor. For each combination of values, the machine computes the posterior probability, which is a measure of relative plausibility, given the model and data. So the posterior distribution ranks the infinite possible combinations of parameter values by their logical plausibility. As a result, the posterior distribution provides relative plausibilities of the different possible strengths of association, given the assumptions you programmed into the model.

Here's how it works, in the simplest case of only one predictor variable. We'll wait until the next chapter to confront more than one predictor. Recall the basic Gaussian model:

$$h_i \sim \text{Normal}(\mu, \sigma) \qquad \text{[likelihood]}$$
$$\mu \sim \text{Normal}(178, 20) \qquad \text{[μ prior]}$$
$$\sigma \sim \text{Uniform}(0, 50) \qquad \text{[σ prior]}$$

Now how do we get weight into a Gaussian model of height? Let x be the mathematical name for the column of weight measurements, d2$weight. Now we have a predictor variable x, which is a list of measures of the same length as h. We'd like to say how knowing the values in x can help us describe or predict the values in h. To get weight into the model in this way, we define the mean μ as a function of the values in x. This is what it looks like, with explanation to follow:

$$h_i \sim \text{Normal}(\mu_i, \sigma) \qquad \text{[likelihood]}$$
$$\mu_i = \alpha + \beta x_i \qquad \text{[linear model]}$$
$$\alpha \sim \text{Normal}(178, 100) \qquad \text{[α prior]}$$
$$\beta \sim \text{Normal}(0, 10) \qquad \text{[β prior]}$$
$$\sigma \sim \text{Uniform}(0, 50) \qquad \text{[σ prior]}$$

Again, I've labeled each line on the right-hand side, by the type of definition it encodes. We'll discuss each in turn.

4.4.1.1. *Likelihood.* To decode all of this, let's begin with just the likelihood, the first line of the model. This is nearly identical to before, except now there is a little index i on the μ, as well as the h. This is necessary now, because the mean μ now depends upon unique predictor values on each row i. So the little i on μ_i indicates that *the mean depends upon the row.*

4.4.1.2. *Linear model.* The mean μ is no longer a parameter to be estimated. Rather, as seen in the second line of the model, μ_i is constructed from other parameters, α and β, and the predictor variable x. This line is not a stochastic relationship—there is no $\sim$ in it, but rather an $=$ in it—because the definition of μ_i is deterministic, not probabilistic. That is to say that, once we know α and β and x_i, we also know μ_i.

The value x_i is just the weight value on row i. It refers to the same individual as the height value, h_i, on the same row. The parameters α and β are more mysterious. Where did they come from? We made them up. The parameters μ and σ are necessary and sufficient to describe a Gaussian distribution. But α and β are instead devices we invent for manipulating μ, allowing it to vary systematically across cases in the data.

You'll be making up all manner of parameters as your skills improve. One way to understand these made-up parameters is to think of them as targets of learning. Each parameter is something that must be described in the posterior density. So when you want to know something about the data, you ask your golem by inventing a parameter for it. This will make more and more sense as you progress. Here's how it works in this context. The second line of the model definition is just:

$$\mu_i = \alpha + \beta x_i$$

What this tells the regression golem is that you are asking two questions about the mean of the outcome.

(1) What is the expected height, when $x_i = 0$? The parameter α answers this question. For this reason, α is often called the *intercept*.
(2) What is the change in expected height, when x_i changes by 1 unit? The parameter β answers this question.

Jointly these two parameters, together with x, ask the golem to find a line that relates x to h, a line that passes through α when $x_i = 0$ and has slope β. That is a task that golems are very good at. It's up to you, though, to be sure it's a good question.

Rethinking: Nothing special or natural about linear models. Note that there's nothing special about the linear model, really. You can choose a different relationship between α and β and μ. For example, the following is a perfectly legitimate definition for μ_i:

$$\mu_i = \alpha \exp(-\beta x_i)$$

This does not define a linear regression, but it does define a regression model. The linear relationship we are using instead is conventional, but nothing requires that you use it. It is very common in some fields, like ecology and demography, to use functional forms for μ that come from theory, rather than the geocentrism of linear models. Models built out of substantive theory can dramatically outperform linear models of the same phenomena.[69]

Overthinking: Units and regression models. Readers who had a traditional training in physical sciences will know how to carry units through equations of this kind. For their benefit, here's the model again (omitting priors for brevity), now with units of each symbol added.

$$h_i \text{cm} \sim \text{Normal}(\mu_i \text{cm}, \sigma \text{cm})$$

$$\mu_i \text{cm} = \alpha \text{cm} + \beta \frac{\text{cm}}{\text{kg}} x_i \text{kg}$$

So you can see that β must have units of cm/kg in order for the mean μ_i to have units of cm. One of the facts that labeling with units clears up is that a parameter like β is a kind of rate. There's also a tradition called *dimensional analysis* that advocates constructing variables so that they are unit-less ratios. In this context, for example, we might divide height by a reference height, removing its units. Measurement scales are arbitrary human constructions, and sometimes the unit-less analysis is more natural.

4.4.1.3. *Priors.* The remaining lines in the model define priors for the parameters to be estimated: α, β, and σ. All of these are weak priors, leading to inferences that will echo non-Bayesian methods of model fitting, such as maximum likelihood. But as always, you should play around with the priors—plotting them, changing them, and refitting the model—to get a sense of their influence.

You've seen priors for α and σ before, although α was called μ back then. I've widened the prior for α, since as you'll see it is common for the intercept in a linear model to swing a long way from the mean of the outcome variable. The flat prior here with a huge standard deviation will allow it to move wherever it needs to. This won't make sense right now, but once you see the posterior distribution, it will.

The prior for β deserves explanation. Why have a Gaussian prior with mean zero? This prior places just as much probability below zero as it does above zero, and when $\beta = 0$, weight has no relationship to height. So many people see it as conservative assumption. And such a prior will pull probability mass towards zero, leading to more conservative estimates than a perfectly flat prior will. But note that a Gaussian prior with standard deviation of 10 is still very weak, so the amount of conservatism it induces will be very small. As you make the standard deviation in this prior smaller, the amount of shrinkage towards zero increases and your model produces more and more conservative estimates about the relationship between height and weight. In Chapter 6, you'll see why such conservative priors are useful for improving inference.

Before fitting this model, though, consider whether this zero-centered prior really makes sense. Do you think there's just as much chance that the relationship between height and weight is negative as that it is positive? Of course you don't. In this context, such a silly prior is harmless, because there is a lot of data. But in other contexts, your golem may need a little nudge in the right direction.

Rethinking: What's the correct prior? People commonly ask what the correct prior is for a given analysis. The question sometimes implies that for any given set of data, there is a uniquely correct prior that must be used, or else the analysis will be invalid. This is a mistake. There is no more a uniquely correct prior than there is a uniquely correct likelihood. Statistical models are machines for inference. Many machines will work, but some work better than others. Priors can be wrong, but only in the same sense that a kind of hammer can be wrong for building a table.

In choosing priors, there are simple guidelines to get you started. Priors encode states of information before seeing data. So priors allow us to explore the consequences of beginning with different information. In cases in which we have good prior information that discounts the plausibility of some parameter values, like negative associations between height and weight, we can encode that information directly into priors. When we don't have such information, we still usually know enough about the plausible range of values. And you can vary the priors and repeat the analysis in order to study how different states of initial information influence inference. Frequently, there are many reasonable choices for a prior, and all of them produce the same inference. And conventional Bayesian priors are *conservative*, relative to conventional non-Bayesian approaches.

Making choices tends to make novices nervous. There's an illusion sometimes that default procedures are more objective than procedures that require user choice, such as choosing priors. If that's true, then all "objective" means is that everyone does the same thing. It carries no guarantees of realism or accuracy.

4.4.2. Fitting the model. The code needed to fit this model via quadratic approximation is a straightforward modification of the kind of code you've already seen. All we have to do is incorporate our new model for the mean into the model specification inside map and be sure to add our new parameters to the start list. Let's repeat the model definition, now with the corresponding R code on the right-hand side:

$$h_i \sim \text{Normal}(\mu_i, \sigma) \qquad \text{height} \sim \text{dnorm(mu,sigma)}$$
$$\mu_i = \alpha + \beta x_i \qquad \text{mu} \leftarrow \text{a + b*weight}$$
$$\alpha \sim \text{Normal}(178, 100) \qquad \text{a} \sim \text{dnorm(156,100)}$$
$$\beta \sim \text{Normal}(0, 10) \qquad \text{b} \sim \text{dnorm(0,10)}$$
$$\sigma \sim \text{Uniform}(0, 50) \qquad \text{sigma} \sim \text{dunif(0,50)}$$

Notice that the linear model, in the R code on the right-hand side, uses the R assignment operator, <-, even though the mathematical definition uses the symbol =. This is a code convention shared by several Bayesian model fitting engines, so it's worth getting used to the switch. You just have to remember to use <- instead of = when defining a linear model. That's it.

And the above allows us to build the MAP model fit:

```
# load data again, since it's a long way back
library(rethinking)
data(Howell1)
d <- Howell1
d2 <- d[ d$age >= 18 , ]

# fit model
m4.3 <- map(
    alist(
        height ~ dnorm( mu , sigma ) ,
        mu <- a + b*weight ,
        a ~ dnorm( 156 , 100 ) ,
        b ~ dnorm( 0 , 10 ) ,
        sigma ~ dunif( 0 , 50 )
    ) ,
    data=d2 )
```

The parameter mu is no longer really a parameter here, because it has been replaced by the linear model, a+b*weight, where a is α and b is β and weight is of course our x in this instance. So there is a prior for the parameter a now, but not one for mu, since mu is defined by the linear model instead.

In the start list, mu is replaced by a, which starts at the overall mean, just like mu used to. We also have to add the new parameter b to this list, and as is usually a conservative first guess, we start the slope out at zero (0), which is equivalent to no relationship between the outcome and predictor.

Note that starting b at zero is not the same as having β's prior with mean zero. The values in the start list don't alter the posterior probabilities, while priors definitely do. But you do have to think a little about your choice of start values, because if you start the MAP search far outside the high density region of the posterior, R may give up looking before it finds the MAP values.

Rethinking: Everything that depends upon parameters has a posterior distribution. In the model introduced above, the parameter μ is no longer a parameter, since it has become a function of the parameters α and β. But since the parameters α and β have a joint posterior, so too does μ. Later in the chapter, you'll work directly the posterior distribution of μ, even though it's not a parameter anymore. Since parameters are uncertain, everything that depends upon them is also uncertain. This includes statistics like μ, as well as model-based predictions, measures of fit, and everything else that uses parameters. By working with samples from the posterior, all you have to do to account for posterior uncertainty in any quantity is to compute that quantity for each sample from the posterior. The resulting quantities, one for each posterior sample, will approximate the quantity's posterior distribution.

Overthinking: Embedding linear models. It may help to understand what the linear model is doing, by seeing another way to fit the same model, but without a separate line for the linear model. You can just merge the linear model into the likelihood definition, like this:

```
m4.3 <- map(
    alist(
        height ~ dnorm( a + b*weight , sigma ) ,
        a ~ dnorm( 178 , 100 ) ,
        b ~ dnorm( 0 , 10 ) ,
        sigma ~ dunif( 0 , 50 )
    ) ,
    data=d2 )
```

This is exactly the same model, but with the linear definition for μ embedded in the first line of code. The form with the explicit linear model, mu <- a + b*weight, is easier to read. But the form just above better reflects the actual computation. Some of the helper functions you'll use later depend upon the explicit linear model however, so I recommend using it. It'll make life a little easier.

4.4.3. Interpreting the model fit.
One trouble with statistical models is that they are hard to understand. Once you've fit the model, it can only report posterior probabilities. These are the right answer to the question that is this combination of model and data. But it's your responsibility to process the answer and make sense of it.

There are two broad categories of processing: (1) reading tables and (2) plotting. For some simple questions, it's possible to learn a lot just from tables of MAP values, their standard deviations, and intervals. But most models are very hard to understand from tables of estimates alone. A major difficulty with tables alone is their apparent simplicity compared to the complexity of the model and data that generated them. Once you have more than a couple of parameters in a model, it is very hard to figure out from numbers alone how all of them act to influence prediction. Once you begin adding interaction terms (Chapter 7) or polynomials (later in this chapter), it may not even be possible to guess the direction of influence a predictor variable has on an outcome.

So throughout this book, I emphasize plotting posterior distributions and posterior predictions, instead of attempting to understand a table. Once you become experienced with a particular type of model and kind of data, you'll be able to confidently read tables of estimates. But that will be because you will have won valuable contextual knowledge about the problem at hand, knowledge that will transfer, but incompletely, to other problems.

To win such knowledge, most of us must do a lot of plotting. Plotting the implications of your estimates will allow you to inquire about several things that are sometimes hard to read from tables:

(1) Whether or not the model fitting procedure worked correctly
(2) The *absolute* magnitude, rather than merely *relative* magnitude, of a relationship between outcome and predictor
(3) The uncertainty surrounding an average relationship
(4) The uncertainty surrounding the implied predictions of the model, as these are distinct from mere parameter uncertainty

In addition, once you get the hang of processing estimates into plots, you can ask any question you can think of, for any model type. And readers of your results will appreciate a figure much more than they will a table of estimates.

So in the remainder of this section, I first spend a little time talking about tables of estimates. I use this opportunity to briefly introduce a data transformation called CENTERING that aids in interpreting estimates. Then I move on to show how to plot estimates that always incorporate information from the full posterior distribution, including correlations among parameters.

> **Rethinking: What do parameters mean?** A basic issue with interpreting model-based estimates is in knowing the meaning of parameters. There is no consensus about what a parameter means, however, because different people take different philosophical stances towards models, probability, and prediction. The perspective in this book is a common Bayesian perspective: *Posterior probabilities of parameter values describe the relative compatibility of different states of the world with the data, according to the model.* These are small world (Chapter 2) numbers. So reasonable people may disagree about the large world meaning, and the details of those disagreements depend strongly upon context. Such disagreements are productive, because they lead to model criticism and revision, something that golems cannot do for themselves.

4.4.3.1. *Tables of estimates.* Before looking closely at the new table of estimates, it's important to realize that models cannot in general be understood by tables of estimates. In this simple model, a lot can be learned from the summary output. But this is not a general property of models, Bayesian or not, because of the covariation among parameters.

With the new linear regression fit to the Kalahari data, we inspect the estimates:

R code
4.40
```
precis( m4.3 )
```

```
          Mean StdDev   5.5%   94.5%
a       113.90   1.91 110.85 116.94
b         0.90   0.04   0.84    0.97
sigma     5.07   0.19   4.77    5.38
```

The first row gives the quadratic approximation for α, the second the approximation for β, and the third approximation for σ. Let's try to make some sense of them in this very simple model.

Best to begin with b (β), because it's the new parameter. Since β is a slope, the value 0.90 can be read as *a person 1 kg heavier is expected to be 0.90 cm taller.* 89% of the posterior probability lies between 0.84 and 0.97. That suggests that β values close to zero or greatly above one are highly incompatible with these data and this model. If you were thinking that perhaps there was no relationship at all between height and weight, then this estimate indicates strong evidence of a positive relationship instead. But maybe you just wanted as precise a measurement as possible of the relationship between height and weight. This estimate embodies that measurement, conditional on the model. For a different model, the measure of the relationship might be different.

The estimate of α, a in the precis table, indicates that a person of weight 0 should be 114cm tall. This is nonsense, since real people always have positive weight, yet it is also true. Parameters like α are "intercepts" that tell us the value of μ when all of the predictor variables have value zero. As a consequence, the value of the intercept is frequently uninterpretable without also studying any β parameters. This is why we need very weak priors for intercepts, in many cases.

Finally, the estimate for σ, sigma, informs us of the width of the distribution of heights around the mean. A quick way to interpret it is to recall that about 95% of the probability in a Gaussian distribution lies between two standard deviations. So in this case, the estimate tells us that 95% of plausible heights lie within 10cm (2σ) of the mean height. But there is also uncertainty about this, as indicated by the 89% percentile interval.

As I mentioned at the start of this section, the numbers in the default precis output aren't sufficient to describe the quadratic posterior completely. For that, we also require the variance-covariance matrix. We're interested in correlations among parameters—we already have their variance in the table above—so let's go straight to the correlation matrix:

```
precis( m4.3 , corr=TRUE )
```
R code
4.41

	Mean	StdDev	5.5%	94.5%	a	b	sigma
a	113.90	1.91	110.85	116.94	1.00	-0.99	0
b	0.90	0.04	0.84	0.97	-0.99	1.00	0
sigma	5.07	0.19	4.77	5.38	0.00	0.00	1

The new columns on the far right show the correlations among the parameters. This is the same information you'd get by using cov2cor(vcov(m4.3)). Notice that α and β are almost perfectly negatively correlated. Right now, this is harmless. It just means that these two parameters carry the same information—as you change the slope of the line, the best intercept changes to match it. But in more complex models, strong correlations like this can make it difficult to fit the model to the data. So we'll want to use some golem engineering tricks to avoid it, when possible.

The first trick is CENTERING. Centering is the procedure of subtracting the mean of a variable from each value. To create a centered version of the weight variable:

```
d2$weight.c <- d2$weight - mean(d2$weight)
```
R code
4.42

You can confirm that the average value of weight.c is zero: mean(d2$weight.c). Now let's refit the model and see what this gains us:

```
m4.4 <- map(
    alist(
        height ~ dnorm( mu , sigma ) ,
        mu <- a + b*weight.c ,
        a ~ dnorm( 178 , 100 ) ,
        b ~ dnorm( 0 , 10 ) ,
        sigma ~ dunif( 0 , 50 )
    ) ,
    data=d2 )
```
R code
4.43

The above code just replaces weight with weight.c, the new variable. Now for the new estimates:

```
precis( m4.4 , corr=TRUE )
```
R code
4.44

	Mean	StdDev	5.5%	94.5%	a	b	sigma
a	154.60	0.27	154.17	155.03	1	0	0
b	0.91	0.04	0.84	0.97	0	1	0
sigma	5.07	0.19	4.77	5.38	0	0	1

The estimates for β and σ are unchanged (within rounding error), but the estimate for α (a) is now the same as the average height value in the raw data. Try it yourself: mean(d2$height). And the correlations among parameters are now all zero. What has happened here?

The estimate for the intercept, α, still means the same thing it did before: the expected value of the outcome variable, when the predictor variable is equal to zero. But now the mean value of the predictor is also zero. So the intercept also means: the expected value of the outcome, when the predictor is at its average value. This makes interpreting the intercept a lot easier.

4.4.3.2. *Plotting posterior inference against the data.* In truth, tables of estimates are usually insufficient for understanding the information contained in the posterior distribution. It's almost always much more useful to plot the posterior inference against the data. Not only does plotting help in interpreting the posterior, but it also provides an informal check on model assumptions. When the model's predictions don't come close to key observations or patterns in the plotted data, then you might suspect the model either did not fit correctly or is rather badly specified.

But even if you only treat plots as a way to help in interpreting the posterior, they are invaluable. For simple models like this one, it is easy to just read the table of numbers and understand what the model says. But for even slightly more complex models, especially those that include interaction effects (Chapter 7), interpreting posterior distributions is hard. Combine with this the problem of incorporating the information in vcov into your interpretations, and the plots are irreplaceable.

We're going to start with a simple version of that task, superimposing just the MAP values over the height and weight data. Then we'll slowly add more and more information to the prediction plots, until we've used the entire posterior distribution.

To superimpose the MAP values for mean height over the actual data:

R code
4.45
```
plot( height ~ weight , data=d2 )
abline( a=coef(m4.3)["a"] , b=coef(m4.3)["b"] )
```

You can see the resulting plot in FIGURE 4.4. Each point in this plot is a single individual. The black line is defined by the MAP slope β and MAP intercept α. Notice that in the code above, I pulled the numbers straight from the fit model. The function coef returns a vector of MAP values, and the names of the parameters extract the intercept and slope.

4.4.3.3. *Adding uncertainty around the mean.* The MAP line is just the posterior mean, the most plausible line in the infinite universe of lines the posterior distribution has considered. Plots of the MAP line, like FIGURE 4.4, are useful for getting an impression of the magnitude of the estimated influence of a variable, like weight, on an outcome, like height.

But they do a poor job of communicating uncertainty. Remember, the posterior distribution considers every possible regression line connecting height to weight. It assigns a relative plausibility to each. This means that each combination of α and β has a posterior probability. It could be that there are many lines with nearly the same posterior probability

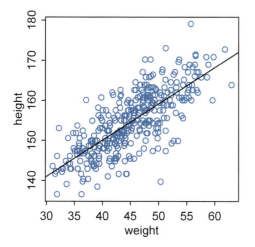

FIGURE 4.4. Height in centimeters (vertical) plotted against weight in kilograms (horizontal), with the *maximum a posteriori* line for the mean height at each weight plotted in black.

as the MAP line. Or it could be instead that the posterior distribution is rather narrow near the MAP line.

So how can we get that uncertainty onto the plot? Together, a combination of α and β define a line. And so we could sample a bunch of lines from the posterior distribution. Then we could display those lines on the plot, to visualize the uncertainty in the regression relationship.

To better appreciate how the posterior distribution contains lines, extract some samples from the model:

```
post <- extract.samples( m4.3 )
```

R code
4.46

Then inspect the first 5 rows of the samples:

```
post[1:5,]
```

R code
4.47

```
          a         b     sigma
1 114.7880 0.8822921 5.121102
2 112.7115 0.9230855 4.907987
3 114.4557 0.9018482 5.276036
4 114.7696 0.8831561 5.021958
5 112.6333 0.9383632 4.898554
```

Each row is a correlated random sample from the joint posterior of all three parameters, using the covariances provided by vcov(m4.3). The paired values of a and b on each row define a line. The average of very many of these lines is the MAP line. But the scatter around that average is meaningful, because it alters our confidence in the relationship between the predictor and the outcome.

So now let's display a bunch of these lines, so you can see the scatter. This lesson will be easier to appreciate, if we use only some of the data to begin. Then you can see how adding in more data changes the scatter of the lines. So we'll begin with just the first 10 cases in d2. The following code extracts the first 10 cases and re-estimates the model:

R code
4.48

```
N <- 10
dN <- d2[ 1:N , ]
mN <- map(
    alist(
        height ~ dnorm( mu , sigma ) ,
        mu <- a + b*weight ,
        a ~ dnorm( 178 , 100 ) ,
        b ~ dnorm( 0 , 10 ) ,
        sigma ~ dunif( 0 , 50 )
    ) , data=dN )
```

Now let's plot 20 of these lines, to see what the uncertainty looks like.

R code
4.49

```
# extract 20 samples from the posterior
post <- extract.samples( mN , n=20 )

# display raw data and sample size
plot( dN$weight , dN$height ,
    xlim=range(d2$weight) , ylim=range(d2$height) ,
    col=rangi2 , xlab="weight" , ylab="height" )
mtext(concat("N = ",N))

# plot the lines, with transparency
for ( i in 1:20 )
    abline( a=post$a[i] , b=post$b[i] , col=col.alpha("black",0.3) )
```

The last line loops over all 20 lines, using abline to display each.

The result is shown in the upper-left plot in FIGURE 4.5. By plotting multiple regression lines, sampled from the posterior, it is easy to see both the highly confident aspects of the relationship and the less confident aspects. The cloud of regression lines displays greater uncertainty at extreme values for weight. This is very common.

The other plots in FIGURE 4.5 show the same relationships, but for increasing amounts of data. Just re-use the code from before, but change N <- 10 to some other value. Notice that the cloud of regression lines grows more compact as the sample size increases. This is a result of the model growing more confident about the location of the mean.

4.4.3.4. *Plotting regression intervals and contours.* The cloud of regression lines in FIGURE 4.5 is an appealing display, because it communicates uncertainty about the relationship in a way that many people find intuitive. But it's much more common to see the uncertainty displayed by plotting an interval or contour around the MAP regression line. In this section, I'll walk you through how to compute any arbitrary interval you like, using the underlying cloud of regression lines embodied in the posterior distribution. Then we'll plot a shaded region around the MAP line, to display the interval.

Here's how to plot an interval around the regression line. This interval incorporates uncertainty in both the slope β and intercept α at the same time. To understand how it works, focus for the moment on a single weight value, say 50 kilograms. You can quickly make a list of 10,000 values of μ for an individual who weighs 50 kilograms, by using your samples from the posterior:

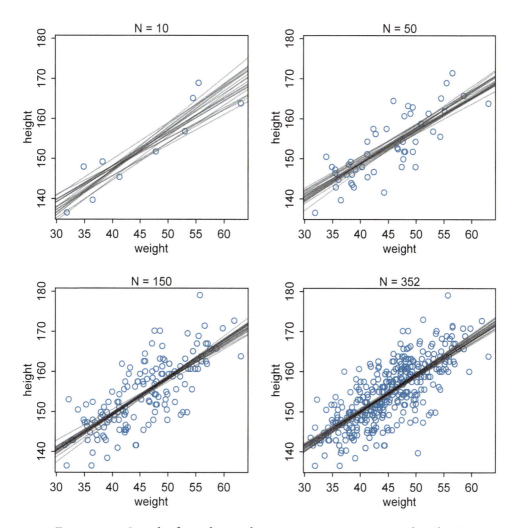

FIGURE 4.5. Samples from the quadratic approximate posterior distribution for the height/weight model, m4.3, with increasing amounts of data. In each plot, 20 lines sampled from the posterior distribution, showing the uncertainty in the regression relationship.

R code
4.50

```
mu_at_50 <- post$a + post$b * 50
```

The code to the right of the <- above takes its form from the equation for μ_i:

$$\mu_i = \alpha + \beta x_i$$

The value of x_i in this case is 50. Go ahead and take a look inside the result, mu_at_50. It's a vector of predicted means, one for each random sample from the posterior. Since joint a and b went into computing each, the variation across those means incorporates the uncertainty in and correlation between both parameters. It might be helpful at this point to actually plot the density for this vector of means:

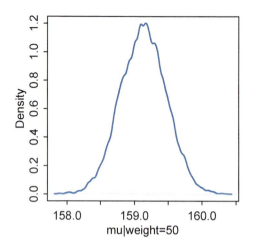

FIGURE 4.6. The quadratic approximate poste-rior distribution of the mean height, μ, when weight is 50 kg. This distribution represents the relative plausibility of different values of the mean.

R code
4.51
```
dens( mu_at_50 , col=rangi2 , lwd=2 , xlab="mu|weight=50" )
```

I reproduce this plot in FIGURE 4.6. Since the components of μ have distributions, so too does μ. And since the distributions of α and β are Gaussian, so to is the distribution of μ (adding Gaussian distributions always produces a Gaussian distribution).

Since the posterior for μ is a distribution, you can find intervals for it, just like for any posterior distribution. To find the 89% highest posterior density interval of μ at 50 kg, just use the HPDI command as usual:

R code
4.52
```
HPDI( mu_at_50 , prob=0.89 )
```

```
 |0.89      0.89|
158.5642 159.6616
```

What these numbers mean is that the central 89% of the ways for the model to produce the data place the average height between about 159 cm and 160 cm (conditional on the model and data), assuming the weight is 50 kg.

That's good so far, but we need to repeat the above calculation for every weight value on the horizontal axis, not just when it is 50 kg. We want to draw 89% HPDIs around the MAP slope in Figure 4.4.

This is made simple by strategic use of the link function, a part of the rethinking pack-age. What link will do is take your map model fit, sample from the posterior distribution, and then compute μ for each case in the data and sample from the posterior distribution. Here's what it looks like for the data you used to fit the model:

R code
4.53
```
mu <- link( m4.3 )
str(mu)
```

```
 num [1:1000, 1:352] 157 157 157 157 157 ...
```

You end up with a big matrix of values of μ. Each row is a sample from the posterior distribu-tion. The default is 1000 samples, but you can use as many or as few as you like. Each column is a case (row) in the data. There are 352 rows in d2, corresponding to 352 individuals. So there are 352 columns in the matrix mu above.

Now what can we do with this big matrix? Lots of things. The function link provides a posterior distribution of μ for each case we feed it. So above we have a distribution of μ for each individual in the original data. We actually want something slightly different: a distribution of μ for each unique weight value on the horizontal axis. It's only slightly harder to compute that, by just passing link some new data:

```
# define sequence of weights to compute predictions for
# these values will be on the horizontal axis
weight.seq <- seq( from=25 , to=70 , by=1 )

# use link to compute mu
# for each sample from posterior
# and for each weight in weight.seq
mu <- link( m4.3 , data=data.frame(weight=weight.seq) )
str(mu)
```

R code
4.54

```
num [1:1000, 1:46] 137 136 137 137 136 ...
```

And now there are only 46 columns in mu, because we fed it 46 different values for weight. To visualize what you've got here, let's plot the distribution of μ values at each height, on the plot.

```
# use type="n" to hide raw data
plot( height ~ weight , d2 , type="n" )

# loop over samples and plot each mu value
for ( i in 1:100 )
    points( weight.seq , mu[i,] , pch=16 , col=col.alpha(rangi2,0.1) )
```

R code
4.55

The result is shown on the left-hand side of FIGURE 4.7. At each weight value in weight.seq, a pile of computed μ values are shown. Each of these piles is a Gaussian distribution, like that in FIGURE 4.6. You can see now that the amount of uncertainty in μ depends upon the value of weight. And this is the same fact you saw in the right-hand plot in FIGURE 4.5.

The final step is to summarize the distribution for each weight value. We'll use apply, which applies a function of your choice to a matrix.

```
# summarize the distribution of mu
mu.mean <- apply( mu , 2 , mean )
mu.HPDI <- apply( mu , 2 , HPDI , prob=0.89 )
```

R code
4.56

Read apply(mu,2,mean) as *compute the mean of each column (dimension "2") of the matrix* mu. Now mu.mean contains the average μ at each weight value, and mu.HPDI contains 89% lower and upper bounds for each weight value. Be sure to take a look inside mu.mean and mu.HPDI, to demystify them. They are just different kinds of summaries of the distributions in mu, with each column being for a different weight value.

You can plot these summaries on top of the data with a few lines of R code:

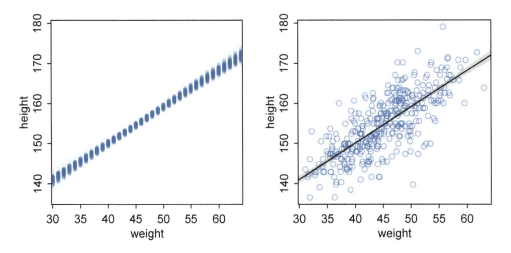

FIGURE 4.7. Left: The first 100 values in the distribution of μ at each weight value. Right: The !Kung height data again, now with 89% HPDI of the mean indicated by the shaded region. Compare this region to the distributions of blue points on the left.

R code
4.57

```
# plot raw data
# fading out points to make line and interval more visible
plot( height ~ weight , data=d2 , col=col.alpha(rangi2,0.5) )

# plot the MAP line, aka the mean mu for each weight
lines( weight.seq , mu.mean )

# plot a shaded region for 89% HPDI
shade( mu.HPDI , weight.seq )
```

You can see the results in the right-hand plot in FIGURE 4.7.

Using this approach, you can derive and plot posterior prediction means and intervals for quite complicated models, for any data you choose. It's true that it is possible to use analytical formulas to compute intervals like this. I have tried teaching such an analytical approach before, and it has always been disaster. Part of the reason is probably my own failure as a teacher, but another part is that most social and natural scientists have never had much training in probability theory and tend to get very nervous around $\int$'s. I'm sure with enough effort, every one of them could learn to do the mathematics. But all of them can quickly learn to generate and summarize samples derived from the posterior distribution. So while the mathematics would be a more elegant approach, and there is some additional insight that comes from knowing the mathematics, the pseudo-empirical approach presented here is very flexible and allows a much broader audience of scientists to pull insight from their statistical modeling. And again, when you start estimating models with MCMC (Chapter 8), this is really the only approach available. So it's worth learning now.

To summarize, here's the recipe for generating predictions and intervals from the posterior of a fit model.

(1) Use link to generate distributions of posterior values for μ. The default behavior of link is to use the original data, so you have to pass it a list of new horizontal axis values you want to plot posterior predictions across.

(2) Use summary functions like mean or HPDI or PI to find averages and lower and upper bounds of μ for each value of the predictor variable.

(3) Finally, use plotting functions like lines and shade to draw the lines and intervals. Or you might plot the distributions of the predictions, or do further numerical calculations with them. It's really up to you.

This recipe works for every model we fit in the book. As long as you know how parameters relate to the data, you can use samples from the posterior to describe any aspect of the model's behavior.

Rethinking: Overconfident confidence intervals. The confidence interval for the regression line in FIGURE 4.7 clings tightly to the MAP line. Thus there is very little uncertainty about the average height as a function of average weight. But you have to keep in mind that these inferences are always conditional on the model. Even a very bad model can have very tight confidence intervals. It may help if you think of the regression line in FIGURE 4.7 as saying: *Conditional on the assumption that height and weight are related by a straight line, then this is the most plausible line, and these are its plausible bounds.*

Overthinking: How link works. The function link is not really very sophisticated. All it is doing is using the formula you provided when you fit the model to compute the value of the linear model. It does this for each sample from the posterior distribution, for each case in the data. You could accomplish the same thing for any model, fit by any means, by performing these steps yourself. This is how it'd look for m4.3.

R code
4.58

```
post <- extract.samples(m4.3)
mu.link <- function(weight) post$a + post$b*weight
weight.seq <- seq( from=25 , to=70 , by=1 )
mu <- sapply( weight.seq , mu.link )
mu.mean <- apply( mu , 2 , mean )
mu.HPDI <- apply( mu , 2 , HPDI , prob=0.89 )
```

And the values in mu.mean and mu.HPDI should be very similar (allowing for simulation variance) to what you got the automated way, using link.

Knowing this manual method is useful both for (1) understanding and (2) sheer power. Whatever the model you find yourself with, this approach can be used to generate posterior predictions for any component of it. Automated tools like link save effort, but they are never as flexible as the code you can write yourself.

4.4.3.5. *Prediction intervals.* Now let's walk through generating an 89% prediction interval for actual heights, not just the average height, μ. This means we'll incorporate the standard deviation σ and its uncertainty as well. Remember, the statistical model here is (omitting priors for brevity):

$$h_i \sim \text{Normal}(\mu_i, \sigma)$$
$$\mu_i = \alpha + \beta x_i$$

What you've done so far is just use samples from the posterior to visualize the uncertainty in μ_i, the linear model of the mean. But actual predictions of heights depend also upon the stochastic definition in the first line. The Gaussian distribution on the first line tells us that the model expects observed heights to be distributed around μ, not right on top of it. And the spread around μ is governed by σ. All of this suggests we need to incorporate σ in the predictions somehow.

Here's how you do it. Imagine simulating heights. For any unique weight value, you sample from a Gaussian distribution with the correct mean μ for that weight, using the correct value of σ sampled from the same posterior distribution. If you do this for every sample from the posterior, for every weight value of interest, you end up with a collection of simulated heights that embody the uncertainty in the posterior *as well as* the uncertainty in the Gaussian likelihood.

R code
4.59

```
sim.height <- sim( m4.3 , data=list(weight=weight.seq) )
str(sim.height)
```

```
 num [1:1000, 1:46] 139 144 141 140 130 ...
```

This matrix is much like the earlier one, mu, but it contains simulated heights, not distributions of plausible average height, μ.

We can summarize these simulated heights in the same way we summarized the distributions of μ, by using apply:

R code
4.60

```
height.PI <- apply( sim.height , 2 , PI , prob=0.89 )
```

Now height.PI contains the 89% posterior prediction interval of observable (according to the model) heights, across the values of weight in weight.seq.

Let's plot everything we've built up: (1) the MAP line, (2) the shaded region of 89% plausible μ, and (3) the boundaries of the simulated heights the model expects.

R code
4.61

```
# plot raw data
plot( height ~ weight , d2 , col=col.alpha(rangi2,0.5) )

# draw MAP line
lines( weight.seq , mu.mean )

# draw HPDI region for line
shade( mu.HPDI , weight.seq )

# draw PI region for simulated heights
shade( height.PI , weight.seq )
```

The code above uses some objects computed in previous sections, so go back and execute that code, if you need to.

In FIGURE 4.8, I plot the result. The wide shaded region in the figure represents the area within which the model expects to find 89% of actual heights in the population, at each weight. There is nothing special about the value 89% here. You could plot the boundary for other percents, such as 67% and 97% (also both primes), and add those to the plot. Doing so would help you see more of the shape of the predicted distribution of heights. I leave that as

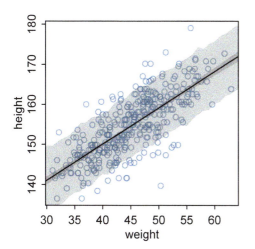

FIGURE 4.8. 89% prediction interval for height, as a function of weight. The solid line is the MAP estimate of the mean height at each weight. The two shaded regions show different 89% plausible regions. The narrow shaded interval around the line is the distribution of μ. The wider shaded region represents the region within which the model expects to find 89% of actual heights in the population, at each weight.

an exercise for the reader. Just go back to the code above and add `prob=0.67`, for example, to the call to `PI`. That will give you 67% intervals, instead of 89% ones.

Notice that the outline for the wide shaded interval is a little jagged. This is the simulation variance in the tails of the sampled Gaussian values. If it really bothers you, increase the number of samples you take from the posterior distribution. The optional n parameter for `sim.height` controls how many samples are used. Try for example:

```
sim.height <- sim( m4.3 , data=list(weight=weight.seq) , n=1e4 )
height.PI <- apply( sim.height , 2 , PI , prob=0.89 )
```

R code
4.62

Run the plotting code again, and you'll see the shaded boundary smooth out some. With extreme percentiles, it can be very hard to get out all of the jaggedness. Luckily, it hardly matters, except for aesthetics. Moreover, it serves to remind us that all statistical inference is approximate. The fact that we can compute an expected value to the 10th decimal place does not imply that our inferences are precise to the 10th decimal place.

Rethinking: Two kinds of uncertainty. In the procedure above, we encountered both uncertainty in parameter values and uncertainty in a sampling process. These are distinct concepts, even though they are processed much the same way and end up blended together in the posterior predictive simulation. The posterior distribution is a ranking of the relative plausibilities of every possible combination of parameter values. The distribution of simulated outcomes, like height, is instead a distribution that includes sampling variation from some process that generates Gaussian random variables. This sampling variation is still a model assumption. It's no more or less objective than the posterior distribution. Both kinds of uncertainty matter, at least sometimes. But it's important to keep them straight, because they depend upon different model assumptions. Furthermore, it's possible to view the Gaussian likelihood as a purely epistemological assumption (a device for estimating the mean and variance of a variable), rather than an ontological assumption about what future data will look like. In that case, it may not make complete sense to simulate outcomes.

Overthinking: Rolling your own `sim`. Just like with `link`, it's useful to know a little about how `sim` operates. For every distribution like `dnorm`, there is a companion simulation function. For the Gaussian distribution, the companion is `rnorm`, and it simulates sampling from a Gaussian distribution. What we want R to do is simulate a height for each set of samples, and to do this for each value of weight. The following will do it:

R code
4.63
```
post <- extract.samples(m4.3)
weight.seq <- 25:70
sim.height <- sapply( weight.seq , function(weight)
    rnorm(
        n=nrow(post) ,
        mean=post$a + post$b*weight ,
        sd=post$sigma ) )
height.PI <- apply( sim.height , 2 , PI , prob=0.89 )
```

The values in `height.PI` will be practically identical to the ones computed in the main text and displayed in FIGURE 4.8.

4.5. Polynomial regression

In the next chapter, you'll see how to use linear models to build regressions with more than one predictor variable. But before then, it helps to see how to model the outcome as a curved function of a single predictor. The models so far all assume that a straight line describes the relationship. But there's nothing special about straight lines, aside from their simplicity.

Let's work through an example, using the full !Kung data:

R code
4.64
```
library(rethinking)
data(Howell1)
d <- Howell1
str(d)
```

```
'data.frame':   544 obs. of  4 variables:
 $ height: num  152 140 137 157 145 ...
 $ weight: num  47.8 36.5 31.9 53 41.3 ...
 $ age   : num  63 63 65 41 51 35 32 27 19 54 ...
 $ male  : int  1 0 0 1 0 1 0 1 0 1 ...
```

Go ahead and plot height against weight. The relationship is visibly curved, now that we've included the non-adult individuals.

There are many ways to model a curved relationship between two variables. Here, I'll show you a very common one, POLYNOMIAL REGRESSION. In this context, "polynomial" means equations for μ_i that add additional terms with squares, cubes, and even higher powers of the predictor variable. There's still only one predictor variable in the model, so this is still a bivariate regression. But the definition of μ_i has more parameters now.

While this section teaches polynomial regression, in general it's a bad thing to do. Why? Because polynomials are very hard to interpret. Better would be to have a more mechanistic model of the data, one that builds the non-linear relationship up from a principled beginning. In the practice problems for this chapter, you'll see how to do this with the height data.

But it's still worth working through a polynomial example, both because it is very common and it will expose some general issues with modeling that we'll take up in Chapter 6. So here's the most common polynomial regression, a parabolic model of the mean:

$$\mu_i = \alpha + \beta_1 x_i + \beta_2 x_i^2$$

The above is a parabolic (second order) polynomial. The $\alpha + \beta_1 x_i$ part is the same linear function of x in a linear regression, just with a little "1" subscript added to the parameter name, so we can tell it apart from the new parameter. The additional term uses the square of x_i to construct a parabola, rather than a perfectly straight line. The new parameter β_2 measures the curvature of the relationship.

> **Rethinking: Linear, additive, funky.** The parabolic model of μ_i above is still called a "linear model" of the mean. This is so, even though the equation is clearly not of a straight line. Unfortunately, the word "linear" means different things in different contexts, and different people use it differently in the same context. What "linear" usually means in this context is that μ_i is a *linear function* of any single parameter. Such models have the advantage of being easier to fit to data. They are also often easier to interpret, because they assume that parameters act independently on the mean. They have the disadvantage of being strongly conventional. They are often used thoughtlessly. When you have real knowledge of your study system, it is often easy to do better than a linear model. These models are geocentric engines, devices for describing partial correlations among variables. We should feel embarrassed to use them, just so we don't become satisfied with the phenomenological explanations they provide.

Fitting these models to data is easy. Interpreting them can be hard. We'll begin with the easy part, fitting a parabolic model of height on weight. The first thing to do is to STANDARDIZE the predictor variable. This means to first center the variable and then divide it by its standard deviation. Why do this? You already have some sense of the value of centering. Going further to standardize leaves the mean at zero but also rescales the range of the data. This is helpful for two reasons:

(1) Interpretation might be easier. For a standardized variable, a change of one unit is equivalent to a change of one standard deviation. In many contexts, this is more interesting and more revealing than a one unit change on the natural scale. And once you start making regressions with more than one kind of predictor variable, standardizing all of them makes it easier to compare their relative influence on the outcome, using only estimates. On the other hand, you might want to interpret the data on the natural scale. So standardization can make interpretation harder, not easier. With a little practice, though, you can learn to quickly convert back and forth between natural and standard scales.

(2) More important though are the advantages for fitting the model to the data. When predictor variables have very large values in them, there are sometimes numerical glitches. Even well-known statistical software can suffer from these glitches, leading to mistaken estimates. These problems are very common for polynomial regression, because the square or cube of a large number can be truly massive. Standardizing largely resolves this issue.

To standardize `weight`, all you do is subtract the mean and then divide by the standard deviation. This will do it:

R code
4.65

```
d$weight.s <- ( d$weight - mean(d$weight) )/sd(d$weight)
```

This new variable weight.s has mean zero and standard deviation 1. No information has been lost in this procedure. Go ahead and plot height on weight.s to verify that. You'll see the same curved relationship as before, but now with a different range on the horizontal axis.

To fit the parabolic model, just modify the definition of μ_i. Here's the model (with very weak priors):

$$h_i \sim \text{Normal}(\mu_i, \sigma) \qquad\qquad \text{height} \sim \text{dnorm(mu,sigma)}$$
$$\mu_i = \alpha + \beta_1 x_i + \beta_2 x_i^2 \qquad \text{mu} \gets \text{a + b1*weight.s + b2*weight.s\textasciicircum2}$$
$$\alpha \sim \text{Normal}(178, 100) \qquad\qquad \text{a} \sim \text{dnorm(140,100)}$$
$$\beta_1 \sim \text{Normal}(0, 10) \qquad\qquad \text{b1} \sim \text{dnorm(0,10)}$$
$$\beta_2 \sim \text{Normal}(0, 10) \qquad\qquad \text{b2} \sim \text{dnorm(0,10)}$$
$$\sigma \sim \text{Uniform}(0, 50) \qquad\qquad \text{sigma} \sim \text{dunif(0,50)}$$

And fitting is straightforward, as well. Just modify the definition of mu so that it contains both the ordinary and quadratic terms. But in general it is better to pre-process any variable transformations. So I'll also build the square of weight.s as a separate variable:

R code
4.66

```
d$weight.s2 <- d$weight.s^2
m4.5 <- map(
    alist(
        height ~ dnorm( mu , sigma ) ,
        mu <- a + b1*weight.s + b2*weight.s2 ,
        a ~ dnorm( 178 , 100 ) ,
        b1 ~ dnorm( 0 , 10 ) ,
        b2 ~ dnorm( 0 , 10 ) ,
        sigma ~ dunif( 0 , 50 )
    ) ,
    data=d )
```

Interpreting the estimates from the summary table can be hard, though. Let's take a look:

R code
4.67

```
precis( m4.5 )
```

```
        Mean StdDev    5.5%   94.5%
a     146.66   0.37  146.07  147.26
b1     21.40   0.29   20.94   21.86
b2     -8.42   0.28   -8.87   -7.97
sigma   5.75   0.17    5.47    6.03
```

The parameter α (a) is still the intercept, so it tells us the expected value of height when weight.s is zero. But it is no longer equal to the mean height in the sample, since there is no guarantee it should in a polynomial regression.[70] And those β_1 and β_2 parameters are the linear and square components of the curve, respectively. But that doesn't make them transparent.

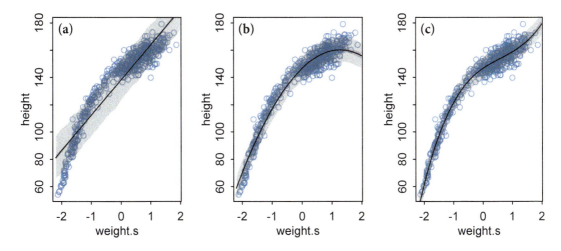

FIGURE 4.9. Polynomial regressions of height on weight (standardized), for the full !Kung data. In each plot, the raw data are shown by the circles. The solid curves show the path of μ in each model, and the shaded regions show the 89% interval of the mean (close to the solid curve) and the 89% interval of predictions (wider). (a) Linear regression. (b) A second order polynomial, a parabolic regression. (c) A third order polynomial, a cubic regression.

You have to plot these model fits to understand what they are saying. So let's do that. We'll calculate the mean relationship and the 89% intervals of the mean and the predictions, like in the previous section. Here's the working code:

R code
4.68

```
weight.seq <- seq( from=-2.2 , to=2 , length.out=30 )
pred_dat <- list( weight.s=weight.seq , weight.s2=weight.seq^2 )
mu <- link( m4.5 , data=pred_dat )
mu.mean <- apply( mu , 2 , mean )
mu.PI <- apply( mu , 2 , PI , prob=0.89 )
sim.height <- sim( m4.5 , data=pred_dat )
height.PI <- apply( sim.height , 2 , PI , prob=0.89 )
```

Plotting all of this is straightforward:

R code
4.69

```
plot( height ~ weight.s , d , col=col.alpha(rangi2,0.5) )
lines( weight.seq , mu.mean )
shade( mu.PI , weight.seq )
shade( height.PI , weight.seq )
```

The results are shown in FIGURE 4.9. Panel (a) of the figure shows the familiar linear regression from earlier in the chapter, but now with the standardized predictor and full data with both adults and non-adults. The linear model makes some spectularly poor predictions, at both very low and middle weights. Compare this to panel (b), our new parabolic regression. The curve does a much better job of finding a central path through the data.

Panel (c) in FIGURE 4.9 shows a higher-order polynomial regression, a cubic regression on weight. The model is (hiding the priors, but they are the same as before):

$$h_i \sim \text{Normal}(\mu_i, \sigma)$$

$$\mu_i = \alpha + \beta_1 x_i + \beta_2 x_i^2 + \beta_3 x_i^3$$

Fit the model with a slight modification of the parabolic model's code:

R code
4.70

```
d$weight.s3 <- d$weight.s^3
m4.6 <- map(
    alist(
        height ~ dnorm( mu , sigma ) ,
        mu <- a + b1*weight.s + b2*weight.s2 + b3*weight.s3 ,
        a ~ dnorm( 178 , 100 ) ,
        b1 ~ dnorm( 0 , 10 ) ,
        b2 ~ dnorm( 0 , 10 ) ,
        b3 ~ dnorm( 0 , 10 ) ,
        sigma ~ dunif( 0 , 50 )
    ) ,
    data=d )
```

Computing the curve and intervals is similarly a small modification of the previous code. This cubic curve is even more flexible than the parabola, so it fits the data even better.

But it's not clear that any of these models make a lot of sense. They are good geocentric descriptions of the sample, yes. But later in the book (Chapter 6), you'll see how a better fit might not actually be a better model. For the moment, consider that a pregnant promise. The same issue arises in regression models with more than one type of predictor. That's the subject of the next chapter.

Overthinking: Converting back to natural scale. The plots in FIGURE 4.9 have standard units on the horizontal axis. These units are sometimes called *z-scores*. But suppose you fit the model using standardized variables, but want to plot the estimates on the original scale. All that's really needed is first to turn off the horizontal axis when you plot the raw data:

R code
4.71

```
plot( height ~ weight.s , d , col=col.alpha(rangi2,0.5) , xaxt="n" )
```

The xaxt at the end there turns off the horizontal axis. Then you explicitly construct the axis, using the axis function.

R code
4.72

```
at <- c(-2,-1,0,1,2)
labels <- at*sd(d$weight) + mean(d$weight)
axis( side=1 , at=at , labels=round(labels,1) )
```

The first line above defines the location of the labels, in standardized units. The second line then takes those units and converts them back to the original scale. The third line draws the axis. Take a look at the help ?axis for more details.

4.6. Summary

This chapter introduced the simple linear regression model, a framework for estimating the association between a predictor variable and an outcome variable. The Gaussian distribution comprises the likelihood in such models, because it counts up the relative numbers of ways different combinations of means and standard deviations can produce an observation. To fit these models to data, the chapter introduced *maximum a prior* (MAP) estimation. It also introduced new procedures for visualizing posterior distributions and posterior predictions. The next chapter expands on these concepts by introducing regression models with more than one predictor variable.

4.7. Practice

Easy.

4E1. In the model definition below, which line is the likelihood?

$$y_i \sim \text{Normal}(\mu, \sigma)$$
$$\mu \sim \text{Normal}(0, 10)$$
$$\sigma \sim \text{Uniform}(0, 10)$$

4E2. In the model definition just above, how many parameters are in the posterior distribution?

4E3. Using the model definition above, write down the appropriate form of Bayes' theorem that includes the proper likelihood and priors.

4E4. In the model definition below, which line is the linear model?

$$y_i \sim \text{Normal}(\mu, \sigma)$$
$$\mu_i = \alpha + \beta x_i$$
$$\alpha \sim \text{Normal}(0, 10)$$
$$\beta \sim \text{Normal}(0, 1)$$
$$\sigma \sim \text{Uniform}(0, 10)$$

4E5. In the model definition just above, how many parameters are in the posterior distribution?

Medium.

4M1. For the model definition below, simulate observed heights from the prior (not the posterior).

$$y_i \sim \text{Normal}(\mu, \sigma)$$
$$\mu \sim \text{Normal}(0, 10)$$
$$\sigma \sim \text{Uniform}(0, 10)$$

4M2. Translate the model just above into a `map` formula.

4M3. Translate the `map` model formula below into a mathematical model definition.

```
flist <- alist(
    y ~ dnorm( mu , sigma ),
    mu <- a + b*x,
    a ~ dnorm( 0 , 50 ),
    b ~ dunif( 0 , 10 ),
    sigma ~ dunif( 0 , 50 )
)
```

4M4. A sample of students is measured for height each year for 3 years. After the third year, you want to fit a linear regression predicting height using year as a predictor. Write down the mathematical model definition for this regression, using any variable names and priors you choose. Be prepared to defend your choice of priors.

4M5. Now suppose I tell you that the average height in the first year was 120 cm and that every student got taller each year. Does this information lead you to change your choice of priors? How?

4M6. Now suppose I tell you that the variance among heights for students of the same age is never more than 64cm. How does this lead you to revise your priors?

Hard.

4H1. The weights listed below were recorded in the !Kung census, but heights were not recorded for these individuals. Provide predicted heights and 89% intervals (either HPDI or PI) for each of these individuals. That is, fill in the table below, using model-based predictions.

Individual	weight	expected height	89% interval
1	46.95		
2	43.72		
3	64.78		
4	32.59		
5	54.63		

4H2. Select out all the rows in the Howell1 data with ages below 18 years of age. If you do it right, you should end up with a new data frame with 192 rows in it.

   (a) Fit a linear regression to these data, using map. Present and interpret the estimates. For every 10 units of increase in weight, how much taller does the model predict a child gets?

   (b) Plot the raw data, with height on the vertical axis and weight on the horizontal axis. Superimpose the MAP regression line and 89% HPDI for the mean. Also superimpose the 89% HPDI for predicted heights.

   (c) What aspects of the model fit concern you? Describe the kinds of assumptions you would change, if any, to improve the model. You don't have to write any new code. Just explain what the model appears to be doing a bad job of, and what you hypothesize would be a better model.

4H3. Suppose a colleague of yours, who works on allometry, glances at the practice problems just above. Your colleague exclaims, "That's silly. Everyone knows that it's only the *logarithm* of body weight that scales with height!" Let's take your colleague's advice and see what happens.

   (a) Model the relationship between height (cm) and the natural logarithm of weight (log-kg). Use the entire Howell1 data frame, all 544 rows, adults and non-adults. Fit this model, using quadratic approximation:

$$h_i \sim \text{Normal}(\mu_i, \sigma)$$
$$\mu_i = \alpha + \beta \log(w_i)$$
$$\alpha \sim \text{Normal}(178, 100)$$
$$\beta \sim \text{Normal}(0, 100)$$
$$\sigma \sim \text{Uniform}(0, 50)$$

where h_i is the height of individual i and w_i is the weight (in kg) of individual i. The function for computing a natural log in R is just log. Can you interpret the resulting estimates?

(b) Begin with this plot:

R code
4.73

```
plot( height ~ weight , data=Howell1 ,
    col=col.alpha(rangi2,0.4) )
```

Then use samples from the quadratic approximate posterior of the model in (a) to superimpose on the plot: (1) the predicted mean height as a function of weight, (2) the 97% HPDI for the mean, and (3) the 97% HPDI for predicted heights.

5 Multivariate Linear Models

One of the most reliable sources of waffles in North America, if not the entire world, is a Waffle House diner. Waffle House is nearly always open, even just after a hurricane. Most diners invest in disaster preparedness, including having their own electrical generators. As a consequence, the United States' disaster relief agency (FEMA) informally uses Waffle House as an index of disaster severity.[71] If the Waffle House is closed, that's a serious event.

It is ironic then that steadfast Waffle House is associated with the nation's highest divorce rates (FIGURE 5.1). States with many Waffle Houses per person, like Georgia and Alabama, also have some of the highest divorce rates in the United States. The lowest divorce rates are found where there are zero Waffle Houses. Could always-available waffles and hash brown potatoes put marriage at risk?

Probably not. This is an example of a misleading correlation. No one thinks there is any plausible mechanism by which Waffle House diners make divorce more likely. Instead, when we see a correlation of this kind, we immediately start asking about other variables that are "really" driving the relationship between waffles and divorce. In this case, Waffle House began in Georgia in the year 1955. Over time, the diners spread across the Southern United States, remaining largely bounded within it. So Waffle House is associated with the South. Divorce is not a uniquely Southern institution, but is more common anyplace that people marry young, and many communities in the South still frown on young people "shacking up" and living together out of wedlock. So it's probably just an accident of history that Waffle House and high divorce rates both occur in the South.

The truth is that correlation is not rare in nature, but rather very common. In large data sets, every pair of variables has a statistically discernible non-zero correlation.[72] So we should never be surprised to find that two variables are correlated. But since most correlations do not indicate causal relationships, we need tools for distinguishing mere association from evidence of causation. This is why so much statistical effort is devoted to MULTIVARIATE REGRESSION, using more than one predictor variable to model an outcome. Reasons often given for multivariate models include:

(1) Statistical "control" for confounds. A *confound* is a variable that may be correlated with another variable of interest. The spurious waffles and divorce correlation is one possible type of confound, where the confound (Southernness) makes a variable with no real importance (Waffle House density) appear to be important. But confounds can hide real important variables just as easily as they can produce false ones. In a particularly important type of confound, known as SIMPSON'S PARADOX, the entire direction of an apparent association between a predictor and outcome can be reversed by considering a confound.[73]

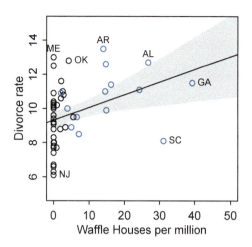

FIGURE 5.1. The number of Waffle House diners per million people is associated with divorce rate (in the year 2009) within the United States. Each point is a State. "Southern" (former Confederate) States shown in blue. Shaded region is 89% percentile interval of the mean. These data are in data(WaffleDivorce) in the rethinking package.

(2) Multiple causation. Even when confounds are absent, due for example to tight experimental control, a phenomenon may really arise from multiple causes. Measurement of each cause is useful, so when we can use the same data to estimate more than one type of influence, we should. Furthermore, when causation is multiple, one cause can hide another. Multivariate models can help in such settings.

(3) Interactions. Even when variables are completely uncorrelated, the importance of each may still depend upon the other. For example, plants benefit from both light and water. But in the absence of either, the other is no benefit at all. Such INTER-ACTIONS occur in a very large number of systems. So effective inference about one variable will usually depend upon consideration of other variables.

In this chapter, we begin to deal with the first of these two, using multivariate regression to deal with simple confounds and to take multiple measurements of influence. You'll see how to include any arbitrary number of *main effects* in your linear model of the Gaussian mean. These main effects are additive combinations of variables, the simplest type of multivariate model.

We'll focus on two valuable things multivariate models can help us with: (1) revealing *spurious* correlations like the Waffle House correlation with divorce and (2) revealing important correlations that may be *masked* by unrevealed correlations with other variables. But multiple predictor variables can hurt as much as they can help. So the chapter describes some dangers of multivariate models, notably *multicollinearity*. Along the way, you'll meet CATEGORICAL VARIABLES, which usually must be broken down into multiple predictor variables.

Rethinking: Causal inference. Despite its central importance, there is no unified approach to causal inference yet in the sciences or in statistics. There are even people who argue that cause does not really exist; it's just a psychological illusion.[74] And in complex dynamical systems, everything seems to cause everything else. "Cause" loses intuitive value. About one thing, however, there is general agreement: Causal inference always depends upon unverifiable assumptions. Another way to say this is that it's always possible to imagine some way in which your inference about cause is mistaken, no matter how careful the design or analysis. A lot can be accomplished, despite this ultimate barrier on inference.[75]

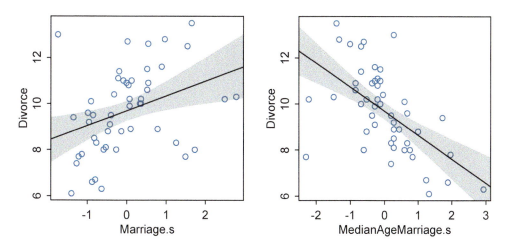

FIGURE 5.2. Divorce rate is associated with both marriage rate (left) and median age at marriage (right). Both predictor variables are standardized in this example. The average marriage rate across States is 20 per 1000 adults, and the average median age at marriage is 26 years.

5.1. Spurious association

Let's leave waffles behind, at least for the moment. An example that is easier to understand is the correlation between divorce rate and marriage rate (FIGURE 5.2). The rate at which adults marry is a great predictor of divorce rate, as seen in the left-hand plot in the figure. But does marriage *cause* divorce? In a trivial sense it obviously does: One cannot get a divorce without first getting married. But there's no reason high marriage rate must be correlated with divorce. It's easy to imagine high marriage rate indicating high cultural valuation of marriage and therefore being associated with *low* divorce rate. So something is suspicious here.

Another predictor associated with divorce is the median age at marriage, displayed in the right-hand plot in FIGURE 5.2. Age at marriage is also a good predictor of divorce rate—higher age at marriage predicts less divorce. You can replicate the right-hand plot in the figure by fitting this linear regression model:

$$D_i \sim \text{Normal}(\mu_i, \sigma)$$
$$\mu_i = \alpha + \beta_A A_i$$
$$\alpha \sim \text{Normal}(10, 10)$$
$$\beta_A \sim \text{Normal}(0, 1)$$
$$\sigma \sim \text{Uniform}(0, 10)$$

D_i is the divorce rate for State i, and A_i is State i's median age at marriage. There are no new code tricks or techniques here, but I'll add comments to help explain the mass of code. We're going to standardize the predictor here, because it's a good habit to get into.

R code
5.1

```
# load data
library(rethinking)
data(WaffleDivorce)
d <- WaffleDivorce

# standardize predictor
d$MedianAgeMarriage.s <- (d$MedianAgeMarriage-mean(d$MedianAgeMarriage))/
    sd(d$MedianAgeMarriage)

# fit model
m5.1 <- map(
    alist(
        Divorce ~ dnorm( mu , sigma ) ,
        mu <- a + bA * MedianAgeMarriage.s ,
        a ~ dnorm( 10 , 10 ) ,
        bA ~ dnorm( 0 , 1 ) ,
        sigma ~ dunif( 0 , 10 )
    ) , data = d )
```

And the following code will compute the shaded confidence region. The procedure is exactly like the examples from the previous chapter. Then it plots the raw data, draws the posterior mean regression line, and draws the shaded region.

R code
5.2

```
# compute percentile interval of mean
MAM.seq <- seq( from=-3 , to=3.5 , length.out=30 )
mu <- link( m5.1 , data=data.frame(MedianAgeMarriage.s=MAM.seq) )
mu.PI <- apply( mu , 2 , PI )

# plot it all
plot( Divorce ~ MedianAgeMarriage.s , data=d , col=rangi2 )
abline( m5.1 )
shade( mu.PI , MAM.seq )
```

If you inspect the precis output, you'll see that each additional standard deviation of delay in marriage (1.24 years) predicts a decrease of about one divorce per thousand adults, with an 89% interval from about -1.4 to -0.7. So it's reliably negative, even though the magnitude of the difference may vary quite a lot—the upper bound is half the lower bound. Of course there's nothing special about the interval boundaries, but the magnitude of the difference means that even though the association here (conditional on model and data) is implausibly positive, it could be both much stronger or weaker than the mean.

You can fit a similar regression for the relationship in the left-hand plot:

R code
5.3

```
d$Marriage.s <- (d$Marriage - mean(d$Marriage))/sd(d$Marriage)
m5.2 <- map(
    alist(
        Divorce ~ dnorm( mu , sigma ) ,
        mu <- a + bR * Marriage.s ,
        a ~ dnorm( 10 , 10 ) ,
```

```
        bR ~ dnorm( 0 , 1 ) ,
        sigma ~ dunif( 0 , 10 )
    ) , data = d )
```

And this shows an increase of 0.6 divorces for every additional standard deviation of marriage rate (3.8). As you may intuit and see in the figure, this relationship isn't as strong as the previous one.

But merely comparing parameter means between different bivariate regressions is no way to decide which predictor is better. Both of these predictors could provide independent value, or they could be redundant, or one could eliminate the value of the other. So we'll build a multivariate model with the goal of measuring the partial value of each predictor. The question we want answered is:

> What is the predictive value of a variable, once I already know all of the other predictor variables?

So for example once you fit a multivariate regression to predict divorce using both marriage rate and age at marriage, the model answers the questions:

(1) After I already know marriage rate, what additional value is there in also knowing age at marriage?
(2) After I already know age at marriage, what additional value is there in also knowing marriage rate?

The parameter estimates corresponding to each predictor are the (often opaque) answers to these questions. Next we'll fit the model that asks these questions.

Rethinking: "Control" is out of control. Very often, the question just above is spoken of as "statistical control," as in *controlling for* the effect of one variable while estimating the effect of another. But this is sloppy language, as it implies too much. It implies a causal interpretation ("effect"), and it implies an experimental disassociation of the predictor variables ("control"). You may be willing to make these assumptions, but they are not part of the model, so be wary.

The point here isn't to police language. Instead, the point is to observe the distinction between small world and large world interpretations. Since most people who use statistics are not statisticians, sloppy language like "control" can promote a sloppy culture of interpretation. Such cultures tend to overestimate the power of statistical methods, so resisting them can be difficult. Disciplining your own language may be enough. Disciplining another's language is hard to do, without seeming like a fastidious scold, as this very box must seem.

5.1.1. Multivariate notation. Multivariate regression formulas look a lot like the polynomial models at the end of the previous chapter—they add more parameters and variables to the definition of μ_i. The strategy is straightforward:

(1) Nominate the predictor variables you want in the linear model of the mean.
(2) For each predictor, make a parameter that will measure its association with the outcome.
(3) Multiply the parameter by the variable and add that term to the linear model.

Examples are always necessary, so here is the model that predicts divorce rate, using both marriage rate and age at marriage.

$$D_i \sim \text{Normal}(\mu_i, \sigma) \qquad \text{[likelihood]}$$
$$\mu_i = \alpha + \beta_R R_i + \beta_A A_i \qquad \text{[linear model]}$$
$$\alpha \sim \text{Normal}(10, 10) \qquad \text{[prior for } \alpha]$$
$$\beta_R \sim \text{Normal}(0, 1) \qquad \text{[prior for } \beta_R]$$
$$\beta_A \sim \text{Normal}(0, 1) \qquad \text{[prior for } \beta_A]$$
$$\sigma \sim \text{Uniform}(0, 10) \qquad \text{[prior for } \sigma]$$

You can use whatever symbols you like for the parameters and variables, but here I've chosen R for marriage rate and A for age at marriage, reusing these symbols as subscripts for the corresponding parameters. But feel free to use whichever symbols reduce the load on your own memory.

So what does it mean to assume $\mu_i = \alpha + \beta_R R_i + \beta_A A_i$? It means that the expected outcome for any State with marriage rate R_i and median age at marriage A_i is the sum of three independent terms. The first term is a constant, α. Every State gets this. The second term is the product of the marriage rate, R_i, and the coefficient, β_R, that measures the association between marriage rate and divorce rate. The third term is similar, but for the association with median age at marriage instead.

If you are like most people, this is still pretty mysterious. So it might help to read the $+$ symbols as "or" and then say: *A State's divorce rate can be a function of its marriage rate **or** its median age at marriage.* The "or" indicates independent associations, which may be purely statistical or rather causal.

Overthinking: Compact notation and the design matrix. Often, linear models are written using a compact form like:

$$\mu_i = \alpha + \sum_{j=1}^{n} \beta_j x_{ji}$$

where j is an index over predictor variables and n is the number of predictor variables. The same model may also be abbreviated:

$$\mu_i = \alpha + \beta_1 x_{1i} + \beta_2 x_{2i} + \dots + \beta_n x_{ni}$$

Both of these forms may be read as *the mean is a modeled as the sum of an intercept and an additive combination of the products of parameters and predictors.* Even more compactly, using matrix notation:

$$\mathbf{m} = \mathbf{Xb}$$

where $\mathbf{m}$ is a vector of predicted means, one for each row in the data, $\mathbf{b}$ is a (column) vector of parameters, one for each predictor variable, and $\mathbf{X}$ is a matrix. This matrix is called a *design matrix*. It has as many rows as the data, and as many columns as there are predictors plus one. So $\mathbf{X}$ is basically a data frame, but with an extra first column. The extra column is filled with 1s. These 1s are multiplied by the first parameter, which is the intercept, and so return the unmodified intercept. When $\mathbf{X}$ is matrix-multiplied by $\mathbf{b}$, you get the predicted means. In R notation, this operation is X %*% b.

We're not going to use the design matrix approach in this book. And in general you don't need to. But it's good to recognize it, and sometimes it can save you a lot of work. For example, for linear regressions, there is a nice matrix formula for the maximum likelihood (or *least squares*) estimates. Most statistical software exploits that formula, which requires using a design matrix.

5.1.2. Fitting the model. To fit this model to the divorce data, we just expand the linear model. Here's the model definition again, now with the code on the right-hand side:

$$D_i \sim \text{Normal}(\mu_i, \sigma) \qquad\qquad \texttt{Divorce ~ dnorm(mu,sigma)}$$
$$\mu_i = \alpha + \beta_R R_i + \beta_A A_i \qquad \texttt{mu <- a+bR*Marriage.s+bA*MedianAgeMarriage.s}$$
$$\alpha \sim \text{Normal}(10, 10) \qquad\qquad \texttt{a ~ dnorm(10,10)}$$
$$\beta_R \sim \text{Normal}(0, 1) \qquad\qquad \texttt{bR ~ dnorm(0,1)}$$
$$\beta_A \sim \text{Normal}(0, 1) \qquad\qquad \texttt{bA ~ dnorm(0,1)}$$
$$\sigma \sim \text{Uniform}(0, 10) \qquad\qquad \texttt{sigma ~ dunif(0,10)}$$

And here is the map fitting code:

```
m5.3 <- map(
    alist(
        Divorce ~ dnorm( mu , sigma ) ,
        mu <- a + bR*Marriage.s + bA*MedianAgeMarriage.s ,
        a ~ dnorm( 10 , 10 ) ,
        bR ~ dnorm( 0 , 1 ) ,
        bA ~ dnorm( 0 , 1 ) ,
        sigma ~ dunif( 0 , 10 )
    ) ,
    data = d )
precis( m5.3 )
```

R code
5.4

```
      Mean StdDev  5.5% 94.5%
a     9.69   0.20  9.36 10.01
bR   -0.13   0.28 -0.58  0.31
bA   -1.13   0.28 -1.58 -0.69
sigma 1.44   0.14  1.21  1.67
```

The posterior mean for marriage rate, bR, is now close to zero, with plenty of probability of both sides of zero. The posterior mean for age at marriage, ba, has actually gotten slightly farther from zero, but is essentially unchanged. It will help to visualize these posterior distribution estimates:

```
plot( precis(m5.3) )
```

R code
5.5

This is the result, with MAP values shown by the points and the percentile intervals by the solid horizontal lines:

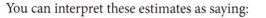

You can interpret these estimates as saying:

> *Once we know median age at marriage for a State, there is little or no addi-*
> *tional predictive power in also knowing the rate of marriage in that State.*

Note that this does not mean that there is no value in knowing marriage rate. If you didn't have access to age-at-marriage data, then you'd definitely find value in knowing the marriage rate. But how did the model achieve this result? To answer that question, we'll draw some pictures.

5.1.3. Plotting multivariate posteriors. Visualizing the posterior distribution in simple bivariate regressions, like those in the previous chapter, is easy. There's only one predictor variable, so a single scatterplot can convey a lot of information. And so in the previous chapter we used scatters of the data. Then we overlaid regression lines and intervals to both (1) visualize the size of the association between the predictor and outcome and (2) to get a crude sense of the ability of the model to predict the individual observations.

With multivariate regression, you'll need more plots. There is a huge literature detailing a variety of plotting techniques that all attempt to help one understand multiple linear regression. None of these techniques is suitable for all jobs, and most do not generalize beyond linear regression. So the approach I take here is to instead help you compute whatever you need from the model. I offer three types of interpretive plots:

(1) *Predictor residual plots.* These plots show the outcome against *residual* predictor values.
(2) *Counterfactual plots.* These show the implied predictions for imaginary experiments in which the different predictor variables can be changed independently of one another.
(3) *Posterior prediction plots.* These show model-based predictions against raw data, or otherwise display the error in prediction.

Each of these plot types has its advantages and deficiencies, depending upon the context and the question of interest. In the rest of this section, I show you how to manufacture each of these in the context of the divorce data.

5.1.3.1. *Predictor residual plots.* A predictor variable residual is the average prediction error when we use all of the other predictor variables to model a predictor of interest. That's a complicated concept, so we'll go straight to the example, where it will make sense. The benefit of computing these things is that, once plotted against the outcome, we have a bivariate regression of sorts that has already "controlled" for all of the other predictor variables. It just leaves in the variation that is not expected by the model of the mean, μ, as a function of the other predictors.

In our multivariate model of divorce rate, we have two predictors: (1) marriage rate (`Marriage.s`) and (2) median age at marriage (`MedianAgeMarriage.s`). To compute predictor residuals for either, we just use the other predictor to model it. So for marriage rate, this is the model we need:

$$R_i \sim \text{Normal}(\mu_i, \sigma)$$
$$\mu_i = \alpha + \beta A_i$$
$$\alpha \sim \text{Normal}(0, 10)$$
$$\beta \sim \text{Normal}(0, 1)$$
$$\sigma \sim \text{Uniform}(0, 10)$$

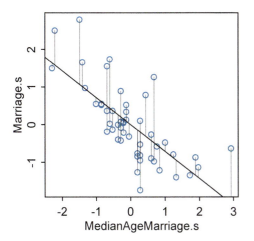

FIGURE 5.3. Residual marriage rate in each State, after accounting for the linear association with median age at marriage. Each gray line segment is a residual, the distance of each observed marriage rate from the expected value, attempting to predict marriage rate with median age at marriage alone. So States that lie above the black regression line have higher rates of marriage than expected, according to age at marriage. Those below the line have lower rates than expected.

As before, R is marriage rate and A is median age at marriage. Note that since we standardized both variables, we already expect the mean α to be around zero. So I've centered α's prior there, but it's still so flat that it hardly matters.

This code will fit the model:

```
m5.4 <- map(
    alist(
        Marriage.s ~ dnorm( mu , sigma ) ,
        mu <- a + b*MedianAgeMarriage.s ,
        a ~ dnorm( 0 , 10 ) ,
        b ~ dnorm( 0 , 1 ) ,
        sigma ~ dunif( 0 , 10 )
    ) ,
    data = d )
```

R code
5.6

And then we compute the *residuals* by subtracting the observed marriage rate in each State from the predicted rate, based upon using age at marriage:

```
# compute expected value at MAP, for each State
mu <- coef(m5.4)['a'] + coef(m5.4)['b']*d$MedianAgeMarriage.s
# compute residual for each State
m.resid <- d$Marriage.s - mu
```

R code
5.7

When a residual is positive, that means that the observed rate was in excess of what we'd expect, given the median age at marriage in that State. When a residual is negative, that means the observed rate was below what we'd expect. In simpler terms, States with positive residuals marry fast for their age of marriage, while States with negative residuals marry slow for their age of marriage. It'll help to plot the relationship between these two variables, and show the residuals as well. Here's some code to do just that, drawing a gray line segment for each residual for each State:

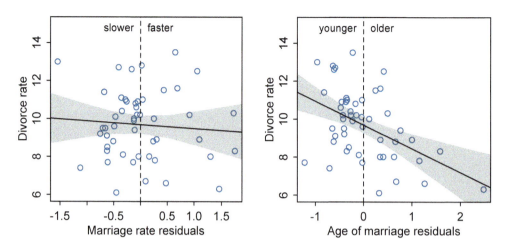

FIGURE 5.4. Predictor residual plots for the divorce data. Left: States with fast marriage rates for their median age of marriage have about the same divorce rates as do States with slow marriage rates. Right: States with old median age of marriage for their marriage rate have lower divorce rates, while States with young median age of marriage have higher divorce rates.

R code
5.8

```
plot( Marriage.s ~ MedianAgeMarriage.s , d , col=rangi2 )
abline( m5.4 )
# loop over States
for ( i in 1:length(m.resid) ) {
    x <- d$MedianAgeMarriage.s[i] # x location of line segment
    y <- d$Marriage.s[i] # observed endpoint of line segment
    # draw the line segment
    lines( c(x,x) , c(mu[i],y) , lwd=0.5 , col=col.alpha("black",0.7) )
}
```

The result is shown as FIGURE 5.3. Notice that the residuals are variation in marriage rate that is left over, after taking out the purely linear relationship between the two variables.

Now to use these residuals, let's put them on a horizontal axis and plot them against the actual outcome of interest, divorce rate. I plot these residuals against divorce rate in FIGURE 5.4 (left-hand plot), also overlaying the linear regression of the two variables. You can think of this plot as displaying the linear relationship between divorce and marriage rates, having statistically "controlled" for median age of marriage. The vertical dashed line indicates marriage rate that exactly matches the expectation from median age at marriage. So States to the right of the line marry faster than expected. States to the left of the line marry slower than expected. Average divorce rate on both sides of the line is about the same, and so the regression line demonstrates little relationship between divorce and marriage rates. The slope of the regression line is −0.13, exactly what we found in the multivariate model, m5.3.

The right-hand plot in FIGURE 5.4 displays the same kind of calculation, but now for median age at marriage, "controlling" for marriage rate. So States to the right of the vertical dashed line have older-than-expected median age at marriage, while those to the left have younger-than-expected median age at marriage. Now we find that the average divorce rate on the right is lower than the rate on the left, as indicated by the regression line. States in which people marry older than expected for a given rate of marriage tend to have less divorce. The slope of the regression line here is -1.13, again the same as in the multivariate model, m5.3.

So what's the point of all of this? There's direct value in seeing the model-based predictions displayed against the outcome, after subtracting out the influence of other predictors. The plots in FIGURE 5.4 do this. But this procedure also brings home the message that regression models answer with the remaining association of each predictor with the outcome, after already knowing the other predictors. In computing the predictor residual plots, you had to perform those calculations yourself. In the unified multivariate model, it all happens automatically.

Linear regression models do all of this with a very specific additive model of how the predictors relate to one another. But predictor variables can be related to one another in non-additive ways. The basic logic of statistical control does not change in those cases, but the details definitely do, and these residual plots do not in general work the same way. Luckily there are more general ways to plumb the mysteries of a model. That's where we turn next.

5.1.3.2. *Counterfactual plots.* A second sort of inferential plot displays the implied predictions of the model. I call these plots *counterfactual*, because they can be produced for any values of the predictor variables you like, even unobserved or impossible combinations like very high median age of marriage and very high marriage rate. There are no States with this combination, but in a counterfactual plot, you can ask the model for a prediction for such a State.

The simplest use of a counterfactual plot is to see how the predictions change as you change only one predictor at a time. This means holding the values of all predictors constant, except for a single predictor of interest. Such predictions will not necessarily look like your raw data—they are counterfactual after all—but they will help you understand the implications of the model. Since it's hard to interpret raw numbers in a table, plots that help you understand the model's implications are priceless.

Let's draw a pair of counterfactual plots for the divorce model. Beginning with a plot showing the impact of changes in Marriage.s on predictions:

R code
5.9

```
# prepare new counterfactual data
A.avg <- mean( d$MedianAgeMarriage.s )
R.seq <- seq( from=-3 , to=3 , length.out=30 )
pred.data <- data.frame(
    Marriage.s=R.seq,
    MedianAgeMarriage.s=A.avg
)

# compute counterfactual mean divorce (mu)
mu <- link( m5.3 , data=pred.data )
mu.mean <- apply( mu , 2 , mean )
mu.PI <- apply( mu , 2 , PI )

# simulate counterfactual divorce outcomes
R.sim <- sim( m5.3 , data=pred.data , n=1e4 )
```

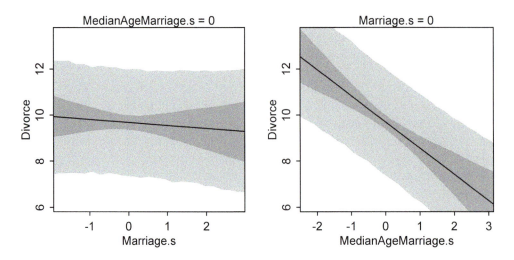

FIGURE 5.5. Counterfactual plots for the multivariate divorce model, m5.3. Each plot shows the change in predicted mean across values of a single predictor, holding the other predictor constant at its mean value (zero in both cases). Shaded regions show 89% percentile intervals of the mean (dark, narrow) and 89% prediction intervals (light, wide).

```
R.PI <- apply( R.sim , 2 , PI )

# display predictions, hiding raw data with type="n"
plot( Divorce ~ Marriage.s , data=d , type="n" )
mtext( "MedianAgeMarriage.s = 0" )
lines( R.seq , mu.mean )
shade( mu.PI , R.seq )
shade( R.PI , R.seq )
```

The strategy above is to build a new list of data that describe the counterfactual cases we wish to simulate predictions for. The list named pred.data holds these cases. Note that the observed values for MedianAgeMarriage.s are not used. Instead we compute the average value and then use this average inside the linear model. So Marriage.s changes across the values in MR.seq, while the other predictor is held constant at its mean, MAM.avg. Note that the average value of MedianAgeMarriage.s is of course zero, because it is a centered variable. This means you could omit it from the linear model and get the same predictions. But if you want to make a counterfactual plot for some other constant value of MedianAge-Marriage.s, or if you are working with un-centered variables, then you need the full linear model again.

The resulting plot is shown in FIGURE 5.5 (left side). The plot on the right side of the figure is constructed by the same strategy, but now with Marriage.s set to its average and MedianAgeMarriage.s allowed to vary:

R code
5.10

```
R.avg <- mean( d$Marriage.s )
A.seq <- seq( from=-3 , to=3.5 , length.out=30 )
pred.data2 <- data.frame(
    Marriage.s=R.avg,
    MedianAgeMarriage.s=A.seq
)

mu <- link( m5.3 , data=pred.data2 )
mu.mean <- apply( mu , 2 , mean )
mu.PI <- apply( mu , 2 , PI )

A.sim <- sim( m5.3 , data=pred.data2 , n=1e4 )
A.PI <- apply( A.sim , 2 , PI )

plot( Divorce ~ MedianAgeMarriage.s , data=d , type="n" )
mtext( "Marriage.s = 0" )
lines( A.seq , mu.mean )
shade( mu.PI , A.seq )
shade( A.PI , A.seq )
```

These plots have the same slopes as the residual plots in the previous section. But they don't display any data, raw or residual, because they are counterfactual. And they also show percentile intervals on the scale of the data, instead of on that weird residual scale. As a result, they are direct displays of the impact on prediction of a change in each variable.

A tension with such plots, however, lies in their counterfactual nature. In the small world of the model, it is possible to change median age of marriage without also changing the marriage rate. But is this also possible in the large world of reality? Probably not. Suppose for example that you pay young couples to postpone marriage until they are 35 years old. Surely this will also decrease the number of couples who ever get married—some people will die before turning 35, among other reasons—decreasing the overall marriage rate. An extraordinary and evil degree of control over people would be necessary to really hold marriage rate constant while forcing everyone to marry at a later age.

In this example, the difficulty of separately manipulating marriage rate and marriage age doesn't impede inference much, only because marriage rate has almost no effect on prediction, once median age of marriage is taken into account. But in many problems, including a later one in this chapter, more than one predictor variable has a sizable impact on the outcome. In that case, while these counterfactual plots always help in understanding the model, they may also mislead by displaying predictions for impossible combinations of predictor values. If our goal is to intervene in the world, there may not be any realistic way to manipulate each predictor without also manipulating the others. This is serious obstacle to applied science, whether you are an ecologist, an economist, or an epidemiologist.

5.1.3.3. *Posterior prediction plots.* In addition to understanding the estimates, it's important to check the model fit against the observed data. This is what you did in Chapter 3, when you simulated globe tosses, averaging over the posterior, and compared the simulated results to the observed. These kinds of checks are useful in many ways. For now, we'll focus on two uses for them.

(1) Did the model fit correctly? Golems do make mistakes, as do golem engineers. Many common software and user errors can be more easily diagnosed by comparing implied predictions to the raw data. Some caution is required, because not all models try to exactly match the sample. But even then, you'll know what to expect from a successful fit. You'll see some examples later.

(2) How does the model fail? All models are useful fictions, so they always fail in some way. Sometimes, the model fits correctly but is still so poor for our purposes that it must be discarded. More often, a model predicts well in some respects, but not in others. By inspecting the individual cases where the model makes poor predictions, you might get an idea of how to improve the model.

Let's begin by simulating predictions, averaging over the posterior.

R code
5.11
```
# call link without specifying new data
# so it uses original data
mu <- link( m5.3 )

# summarize samples across cases
mu.mean <- apply( mu , 2 , mean )
mu.PI <- apply( mu , 2 , PI )

# simulate observations
# again no new data, so uses original data
divorce.sim <- sim( m5.3 , n=1e4 )
divorce.PI <- apply( divorce.sim , 2 , PI )
```

This code is very similar to what you've seen before, but now using the original observed data.

For multivariate models, there are many different ways to display these simulations. So let's look at a few different ways. The simplest is to just plot predictions against observed. This code will do that, and then add a line to show perfect prediction and line segments for the confidence interval of each prediction:

R code
5.12
```
plot( mu.mean ~ d$Divorce , col=rangi2 , ylim=range(mu.PI) ,
    xlab="Observed divorce" , ylab="Predicted divorce" )
abline( a=0 , b=1 , lty=2 )
for ( i in 1:nrow(d) )
    lines( rep(d$Divorce[i],2) , c(mu.PI[1,i],mu.PI[2,i]) ,
        col=rangi2 )
```

The resulting plot appears in FIGURE 5.6(a). It's easy to see from this arrangement of the simulations that the model under-predicts for States with very high divorce rates while it over-predicts for States with very low divorce rates. Some States are very frustrating to the model, lying very far from the diagonal. I've labeled two points like this, Idaho (ID) and Utah (UT), both of which have much lower divorce rates than the model expects them to have. The easiest way to label a few select points is to use identify:

R code
5.13
```
identify( x=d$Divorce , y=mu.mean , labels=d$Loc , cex=0.8 )
```

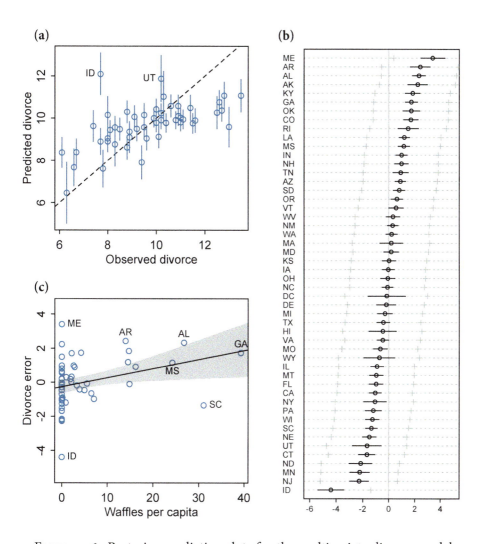

FIGURE 5.6. Posterior predictive plots for the multivariate divorce model, m5.3. (a) Predicted divorce rate against observed, with 89% confidence intervals of the average prediction. The dashed line shows perfect prediction. (b) Average prediction error for each State, with 89% interval of the mean (black line) and 89% prediction interval (gray +). (c) Average prediction error (residuals) against number of Waffle Houses per capita, with superimposed regression of the two variables.

After executing the line of code above, R will wait for you to click near a point in the active plot window. It'll then place a label near that point, on the side you choose. When you are done labeling points, press your right mouse button (or press ESC, on some platforms).

The plot in FIGURE 5.6(a) makes it hard to see the amount of prediction error, in many cases. For this reason, lots of people also use *residual* plots that show the mean prediction error for each row. I'm also going to use order to sort the States from lowest prediction error to highest. To compute residuals and display them:

R code
5.14

```
# compute residuals
divorce.resid <- d$Divorce - mu.mean
# get ordering by divorce rate
o <- order(divorce.resid)
# make the plot
dotchart( divorce.resid[o] , labels=d$Loc[o] , xlim=c(-6,5) , cex=0.6 )
abline( v=0 , col=col.alpha("black",0.2) )
for ( i in 1:nrow(d) ) {
    j <- o[i] # which State in order
    lines( d$Divorce[j]-c(mu.PI[1,j],mu.PI[2,j]) , rep(i,2) )
    points( d$Divorce[j]-c(divorce.PI[1,j],divorce.PI[2,j]) , rep(i,2),
        pch=3 , cex=0.6 , col="gray" )
}
```

The result is FIGURE 5.6(b). It's much easier now to see the large model failures, such as Idaho (ID) and Maine (ME). But if you have many different cases (rows) in the data, these plots can take up a lot of space.

Yet another common use for these simulations is to construct novel predictor residual plots. Once you've computed the divorce residuals, as you did just above, you can plot those residuals against new predictor variables. This is a quick way to see if remaining variation in the outcome is associated with another predictor. Let's consider Waffle House density again, for example. FIGURE 5.6(c) displays the divorce residuals (also known as "error") against Waffle Houses per capita (d$WaffleHouses/d$Population). A small positive correlation remains between divorce rate and Waffle House, despite already "controlling" for marriage rate and median age at marriage in each State. This does not mean that the correlation is real. No matter how many predictors you've already included in a regression, it's still possible to find spurious correlations with the remaining variation.

Rethinking: Stats, huh, yeah what is it good for? Often people want statistical modeling to do things that statistical modeling cannot do. For example, we'd like to know whether an effect is "real" or rather spurious. Unfortunately, modeling merely quantifies uncertainty in the precise way that the model understands the problem. Usually answers to large world questions about truth and causation depend upon information not included in the model. For example, any observed correlation between an outcome and predictor could be eliminated or reversed once another predictor is added to the model. But if we cannot think of another predictor, we might never notice this. Therefore all statistical models are vulnerable to and demand critique, regardless of the precision of their estimates and apparent accuracy of their predictions. Rounds of model criticism and revision embody the real tests of scientific hypotheses, while the statistical procedures often called "tests" are small components of the conversation.

Overthinking: Simulating spurious association. One way that spurious associations between a predictor and outcome can arise is when a truly causal predictor, call it x_{real}, influences both the outcome, y, and a spurious predictor, x_{spur}. This can be confusing, however, so it may help to simulate this scenario and see both how the spurious data arise and prove to yourself that multiple regression can reliably indicate the right predictor, x_{real}. So here's a very basic simulation:

```
N <- 100                    # number of cases
x_real <- rnorm( N )        # x_real as Gaussian with mean 0 and stddev 1
x_spur <- rnorm( N , x_real )    # x_spur as Gaussian with mean=x_real
y <- rnorm( N , x_real )    # y as Gaussian with mean=x_real
d <- data.frame(y,x_real,x_spur) # bind all together in data frame
```
R code
5.15

Now the data frame d has 100 simulated cases. Because x_real influences both y and x_spur, you can think of x_spur as another outcome of x_real, but one which we mistake as a potential predictor of y. As a result, both x_{real} and x_{spur} are correlated with y. You can see this in the scatterplots from pairs(d). But when you include both x variables in a linear regression predicting y, the posterior mean for the association between y and x_{spur} will be close to zero, while the comparable mean for x_{real} will be closer to 1.

5.2. Masked relationship

The divorce rate example demonstrates that multiple predictor variables are useful for knocking out spurious association. A second reason to use more than one predictor variable is to measure the direct influences of multiple factors on an outcome, when none of those influences is apparent from bivariate relationships. This kind of problem tends to arise when there are two predictor variables that are correlated with one another. However, one of these is positively correlated with the outcome and the other is negatively correlated with it.

You'll consider this kind of problem in a new data context, information about the composition of milk across primate species, as well as some facts about those species, like body mass and brain size.[76] Milk is a huge investment, being much more expensive than gestation. Such an expensive resource is likely adjusted in subtle ways, depending upon the physiological and development details of each mammal species. Let's load the data into R first:

```
library(rethinking)
data(milk)
d <- milk
str(d)
```
R code
5.16

You should see in the structure of the data frame that you have 29 rows (cases) for 8 variables (columns).

A popular hypothesis has it that primates with larger brains produce more energetic milk, so that brains can grow quickly. Answering questions of this sort consumes a lot of effort in evolutionary biology, because there are many subtle statistical issues that arise when comparing species. We'll start simple with these data, but by the end of the book we'll include a larger number of the subtle issues. The variables we'll consider for now are:

kcal.per.g: Kilocalories of energy per gram of milk.
mass: Average female body mass, in kilograms.
neocortex.perc: The percent of total brain mass that is neocortex mass.

The question here is to what extent energy content of milk, measured here by kilocalories, is related to the percent of the brain mass that is neocortex. Neocortex is the gray, outer part of the brain that is particularly elaborated in mammals and especially primates. We'll end up needing female body mass as well, to see the masking that hides the relationships among the variables.

The first model to consider is the simple bivariate regression between kilocalories and neocortex percent. You know how to set up this regression. Fit the model with:

R code
5.17
```
m5.5 <- map(
    alist(
        kcal.per.g ~ dnorm( mu , sigma ) ,
        mu <- a + bn*neocortex.perc ,
        a ~ dnorm( 0 , 100 ) ,
        bn ~ dnorm( 0 , 1 ) ,
        sigma ~ dunif( 0 , 1 )
    ) ,
    data=d )
```

When you execute this code, you'll get a confusing error message:

```
Error in map(alist(kcal.per.g ~ dnorm(mu, sigma), mu <- a + bn * neocortex.perc,
    initial value in 'vmmin' is not finite
The start values for the parameters were invalid. This could be caused by
missing values (NA) in the data or by start values outside the parameter
constraints. If there are no NA values in the data, try using explicit start
values.
```

What has gone wrong here? This particular error message means that the model didn't return a valid posterior probability for even the starting parameter values. In this case, the culprit is the missing values in the neocortex.perc column. Take a look inside that column and see for yourself:

R code
5.18
```
d$neocortex.perc
```

Each NA in the output is a missing value. If you pass a vector like this to a likelihood function like dnorm, it doesn't know what to do. After all, what's the probability of a missing value? Whatever the answer, it isn't a number, and so dnorm returns a NaN. Unable to even get started, map (or rather optim, which does the real work) gives up and barks about some weird thing called vmmin not being finite. This kind of opaque error message is unfortunately the norm in R.

This is easy to fix, though. What you need to do here is manually drop all the cases with missing values. More automated black-box commands, like lm and glm, will drop such cases for you. But this isn't always a good thing, if you aren't aware of it. In the next chapter, you'll see one reason why. So indulge me for now. It's worth learning how to do this yourself. To make a new data frame with only complete cases in it, just use:

R code
5.19
```
dcc <- d[ complete.cases(d) , ]
```

This makes a new data frame, dcc, that consists of the 17 rows from d that have values in all columns. Now let's work with the new data frame. All that is new in the code is using dcc instead of d:

R code
5.20
```
m5.5 <- map(
    alist(
        kcal.per.g ~ dnorm( mu , sigma ) ,
```

```
        mu <- a + bn*neocortex.perc ,
        a ~ dnorm( 0 , 100 ) ,
        bn ~ dnorm( 0 , 1 ) ,
        sigma ~ dunif( 0 , 1 )
    ) ,
  data=dcc )
```

You might get some poor random start values, but the model will now fit correctly. Take a look at the quadratic approximate posterior now:

```
precis( m5.5 , digits=3 )
```
R code
5.21

```
      Mean StdDev   5.5% 94.5%
a    0.353  0.471 -0.399 1.106
bn   0.005  0.007 -0.007 0.016
sigma 0.166 0.028  0.120 0.211
```

First, note that I added `digits=3` to the `precis` call. The reason is that the posterior mean for bn is very small. So we need more digits to see that it's not exactly zero—a change from the smallest neocortex percent in the data, 55%, to the largest, 76%, would result in an expected change of only:

```
coef(m5.5)["bn"] * ( 76 - 55 )
```
R code
5.22

```
0.09456654
```

That's less than 0.1 kilocalories. The kilocalories in the data range from less than 0.5 to more than 0.9 per gram, so this association isn't so impressive. More importantly, it isn't very precise. The 89% interval of the parameter extends a good distance on both sides of zero. You can plot the predicted mean and 89% interval for the mean to see this more easily:

```
np.seq <- 0:100
pred.data <- data.frame( neocortex.perc=np.seq )

mu <- link( m5.5 , data=pred.data , n=1e4 )
mu.mean <- apply( mu , 2 , mean )
mu.PI <- apply( mu , 2 , PI )

plot( kcal.per.g ~ neocortex.perc , data=dcc , col=rangi2 )
lines( np.seq , mu.mean )
lines( np.seq , mu.PI[1,] , lty=2 )
lines( np.seq , mu.PI[2,] , lty=2 )
```
R code
5.23

I display this plot in the upper-left of Figure 5.7. The MAP line is weakly positive, but it is highly imprecise. A lot of mildly positive and negative slopes are plausible, given this model and these data.

Now consider another predictor variable, adult female body mass, `mass` in the data frame. Let's use the logarithm of mass, `log(mass)`, as a predictor as well. Why the logarithm of mass instead of the raw mass in kilograms? It is often true that scaling measurements like body mass are related by magnitudes to other variables. Taking the log of a measure translates the

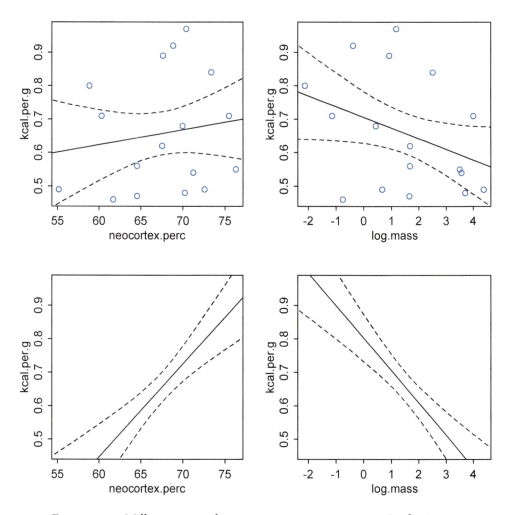

FIGURE 5.7. Milk energy and neocortex among primates. In the top two plots, simple bivariate regressions of kilocalories per gram of milk on (left) neocortex percent and (right) log female body mass show weak and uncertain associations. However, on the bottom, a single regression with both neocortex percent and log body mass suggests strong association with both variables. Both neocortex and body mass are associated with milk energy, but in opposite directions. This masks each variable's relationship with the outcome, unless both are considered simultaneously.

measure into magnitudes. So by using the logarithm of body mass here, we're saying that we suspect that the magnitude of a mother's body mass is related to milk energy, in a linear fashion.

To fit another bivariate regression, with the logarithm of body mass as the predictor variable now, I recommend transforming mass first to make a new column in the data:

```
dcc$log.mass <- log(dcc$mass)
```
R code
5.24

And this will fit the model:

```
m5.6 <- map(
    alist(
        kcal.per.g ~ dnorm( mu , sigma ) ,
        mu <- a + bm*log.mass ,
        a ~ dnorm( 0 , 100 ) ,
        bm ~ dnorm( 0 , 1 ) ,
        sigma ~ dunif( 0 , 1 )
    ) ,
    data=dcc )
precis(m5.6)
```
R code
5.25

```
        Mean StdDev  5.5% 94.5%
a       0.71   0.05  0.63  0.78
bm     -0.03   0.02 -0.06  0.00
sigma   0.16   0.03  0.11  0.20
```

We find that log-mass is negatively correlated with kilocalories. This influence does seem stronger than that of neocortex percent, although in the opposite direction. It is quite uncertain though, with a wide confidence interval that is consistent with a wide range of both weak and stronger relationships. This regression is shown in the upper-right of FIGURE 5.7. You should modify the code that plotted the left-hand plot in the same figure, to be sure you understand how to do this.

Now let's see what happens when we add both predictor variables at the same time to the regression. This is the multivariate model, in math form:

$$k_i \sim \text{Normal}(\mu_i, \sigma)$$
$$\mu_i = \alpha + \beta_n n_i + \beta_m \log(m_i)$$
$$\alpha \sim \text{Normal}(0, 100)$$
$$\beta_n \sim \text{Normal}(0, 1)$$
$$\beta_m \sim \text{Normal}(0, 1)$$
$$\sigma \sim \text{Uniform}(0, 10)$$

Above k is kcal.per.g, n is neocortex.perc, and m is mass. Fitting the joint model is just as you'd expect by now:

```
m5.7 <- map(
    alist(
        kcal.per.g ~ dnorm( mu , sigma ) ,
        mu <- a + bn*neocortex.perc + bm*log.mass ,
        a ~ dnorm( 0 , 100 ) ,
        bn ~ dnorm( 0 , 1 ) ,
        bm ~ dnorm( 0 , 1 ) ,
        sigma ~ dunif( 0 , 1 )
    ) ,
```
R code
5.26

```
     data=dcc )
precis(m5.7)
```

```
        Mean StdDev  5.5% 94.5%
a      -1.09   0.47 -1.83 -0.34
bn      0.03   0.01  0.02  0.04
bm     -0.10   0.02 -0.13 -0.06
sigma   0.11   0.02  0.08  0.15
```

By incorporating both predictor variables in the regression, the estimated association of both with the outcome has increased. The posterior mean for the association of neocortex percent has increased more than sixfold, and its 89% interval is now entirely above zero. The posterior mean for log body mass is more strongly negative.

Let's plot the intervals for the predicted mean kilocalories, for this new model. Here's the code for the relationship between kilocalories and neocortex percent. These are counterfactual plots, so we'll use the mean log body mass in this calculation, showing only how predicted energy varies as a function of neocortex percent:

R code
5.27
```
mean.log.mass <- mean( log(dcc$mass) )
np.seq <- 0:100
pred.data <- data.frame(
    neocortex.perc=np.seq,
    log.mass=mean.log.mass
)

mu <- link( m5.7 , data=pred.data , n=1e4 )
mu.mean <- apply( mu , 2 , mean )
mu.PI <- apply( mu , 2 , PI )

plot( kcal.per.g ~ neocortex.perc , data=dcc , type="n" )
lines( np.seq , mu.mean )
lines( np.seq , mu.PI[1,] , lty=2 )
lines( np.seq , mu.PI[2,] , lty=2 )
```

This plot is displayed in the lower-left of FIGURE 5.7. The analogous plot for log(mass) is shown in the lower-right. I leave it to the reader to modify the code above to replicate the plot in the lower-right. If you don't get in there and modify some code, make some mistakes, and fix them, you'll never grasp this stuff. Making mistakes is okay. It's not how much you know that matters; it's how fast you are learning.

Why did adding neocortex and body mass to the same model lead to larger estimated effects of both? This is a context in which there are two variables correlated with the outcome, but one is positively correlated with it and the other is negatively correlated with it. In addition, both of the explanatory variables are positively correlated with one another. As a result, they tend to cancel one another out. This is another case in which regression automatically finds the most revealing cases and uses them to produce estimates. What the regression model does is ask if species that have high neocortex percent *for their body mass* have higher milk energy. Likewise, the model asks if species with high body mass *for their neocortex percent* have higher milk energy. Bigger species, like apes, have milk with less energy. But species with more neocortex tend to have richer milk. The fact that these two variables, body size and neocortex, are correlated across species makes it hard to see these relationships, unless we statistically account for both.

Overthinking: Simulating a masking relationship. Just as with understanding spurious association (page 134), it may help to simulate data in which two meaningful predictors act to mask one another. Suppose again a single outcome, y, and two predictors, x_{pos} and x_{neg}. The predictor x_{pos} is positively associated with y, while x_{neg} is negatively associated with y. Furthermore, the two predictors are positively correlated with one another. Here's code to produce data meeting these criteria:

```
N <- 100                          # number of cases
rho <- 0.7                        # correlation btw x_pos and x_neg
x_pos <- rnorm( N )               # x_pos as Gaussian
x_neg <- rnorm( N , rho*x_pos ,   # x_neg correlated with x_pos
    sqrt(1-rho^2) )
y <- rnorm( N , x_pos - x_neg )   # y equally associated with x_pos, x_neg
d <- data.frame(y,x_pos,x_neg)    # bind all together in data frame
```

The first thing to do is enter `pairs(d)` to see the bivariate relationships among the variables. Now if you fit two bivariate regressions, predicting y using either `x_pos` or `x_neg`, you get posterior distributions that underestimate the true association (which should be about 1 or −1, respectively). But if you then fit a model predicting y using both predictors, you'll get a posterior distribution that better matches the underlying truth. If you move the value of `rho` closer to zero, this masking phenomenon will diminish. If you make `rho` closer to 1 or −1, it will magnify. But if `rho` gets very close to 1 or −1, then the two predictors contain exactly the same information, and there's no hope for any statistical model to tease out the true underlying association used in the simulation.

Why should two predictors be correlated in this way? They might both be influenced by another, unmeasured variable. Or one of them, say x_{neg}, is influenced partly by x_{pos}, but also by its own unique processes. They might both partly influence one another, in a case of reciprocal causation. In the primate milk example, it may be that the positive association between large body size and neocortex percent arises from a tradeoff between lifespan and learning. Large animals tend to live a long time. And in such animals, an investment in learning may be a better investment, because learning can be amortized over a longer lifespan. Both large body size and large neocortex then influence milk composition, but in different directions, for different reasons. Luckily, since lots of other factors influence both body size and brain structure, the correlation is far from perfect, which allows the statistical model to tease them apart. This is a speculative scenario, unlike the certainty of the simulated scenario above. But it may help in connecting the simulation to your own data analysis problems, whether or not it's true in the primate milk case.

5.3. When adding variables hurts

It is commonly true that there are many potential predictor variables to add to a regression model. In the case of the primate milk data frame, for example, there are 7 variables available to predict any column we choose as an outcome. Why not just fit a model that includes all 7? There are several good, purely statistical reasons to avoid doing this. We'll discuss three of them. The first is MULTICOLLINEARITY, a nasty word for a simple phenomenon. Once you know to watch for it, it's easy to spot and cope with. The second is POST-TREATMENT BIAS, which means statistically controlling for consequences of a causal factor. The third is OVERFITTING, which is a huge and perspective-changing problem that occupies us for much of the next chapter.

Multicollinearity means very strong correlation between two or more predictor variables. The consequence of it is that the posterior distribution will say that a very large range of parameter values are plausible, from tiny associations to massive ones, even if all of the variables are in reality strongly associated with the outcome variable. This frustrating phenomenon arises from the details of how statistical control works. So once you understand multicollinearity, you will better understand multivariate models in general.

To explore multicollinearity, let's begin with a simple simulation. Then we'll turn to the primate milk data again and see multicollinearity in a real data set.

5.3.1. Multicollinear legs.

The simulation example is predicting an individual's height using the length of his or her legs as predictor variables. Surely height is positively associated with leg length, or at least the simulation will assume it is. Nevertheless, once you put both leg lengths into the model, something vexing will happen.

The code below will simulate the heights and leg lengths of 100 individuals. For each, first a height is simulated from a Gaussian distribution. Then each individual gets a simulated proportion of height for their legs, ranging from 0.4 to 0.5. Finally, each leg is salted with a little measurement or developmental error, so the left and right legs are not exactly the same length, as is typical in real populations. At the end, the code puts height and the two leg lengths into a common data frame.

R code
5.29

```
N <- 100                        # number of individuals
height <- rnorm(N,10,2)         # sim total height of each
leg_prop <- runif(N,0.4,0.5)    # leg as proportion of height
leg_left <- leg_prop*height +   # sim left leg as proportion + error
    rnorm( N , 0 , 0.02 )
leg_right <- leg_prop*height +  # sim right leg as proportion + error
    rnorm( N , 0 , 0.02 )
                                # combine into data frame
d <- data.frame(height,leg_left,leg_right)
```

Now let's analyze these data, predicting the outcome height with both predictors, leg_left and leg_right. Before fitting the model and looking at the posterior means, however, consider what we expect. On average, an individual's legs are 45% of his or her height (in these simulated data). So we should expect the beta coefficient that measures the association of a leg with height to end up around the average height (10) divided by 45% of the average height (4.5). This is $10/4.5 \approx 2.2$. Now let's see what happens instead:

R code
5.30

```
m5.8 <- map(
    alist(
        height ~ dnorm( mu , sigma ) ,
        mu <- a + bl*leg_left + br*leg_right ,
        a ~ dnorm( 10 , 100 ) ,
        bl ~ dnorm( 2 , 10 ) ,
        br ~ dnorm( 2 , 10 ) ,
        sigma ~ dunif( 0 , 10 )
    ) ,
    data=d )
precis(m5.8)
```

```
       Mean StdDev  5.5% 94.5%
a      0.70   0.31  0.20  1.20
```

```
bl     -0.43   2.18 -3.92   3.06
br      2.48   2.19 -1.01   5.98
sigma   0.62   0.04  0.55   0.69
```

Those posterior means and standard deviations look crazy. This is a case in which a graphical view of the precis output is more useful, because it displays the posterior means and 89% intervals in a way that allows us with a glance to see that something has gone wrong here:

R code
5.31

```
plot(precis(m5.8))
```

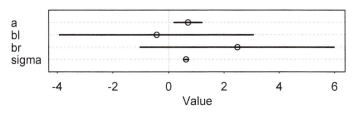

Your numbers and precis plot will not look exactly the same, due to simulation variance. But they will show the same odd result. If both legs have almost identical lengths, and height is so strongly associated with leg length, then why is this posterior distribution so weird? Did the model fitting work correctly?

The model did fit correctly, and the posterior distribution here is the right answer to the question we asked. Recall that a multiple linear regression answers the question: *What is the value of knowing each predictor, after already knowing all of the other predictors?* So in this case, the question becomes: *What is the value of knowing each leg's length, after already knowing the other leg's length?*

The answer to this weird question is equally weird, but perfectly logical. The posterior distribution is the answer to this question, considering every possible combination of the parameters and assigning relative plausibilities to every combination, conditional on this model and these data. It might help to look at the bivariate posterior distribution for bl and br:

R code
5.32

```
post <- extract.samples(m5.8)
plot( bl ~ br , post , col=col.alpha(rangi2,0.1) , pch=16 )
```

The resulting plot is shown on the left of FIGURE 5.8. The posterior distribution for these two parameters is very highly correlated, with all of the plausible values of bl and br lying along a narrow ridge. When bl is large, then br must be small. What has happened here is that since both leg variables contain almost exactly the same information, if you insist on including both in a model, then there will be a practically infinite number of combinations of bl and br that produce the same predictions.

One way to think of this phenomenon is that you have approximated this likelihood:

$$y_i \sim \text{Normal}(\mu_i, \sigma)$$
$$\mu_i = \alpha + \beta_1 x_i + \beta_2 x_i$$

The variable y is the outcome, like height in the example, and x is a single predictor, like the leg lengths in the example. Here x is used twice, which is a perfect example of the problem caused by using the almost-identical leg lengths. From the computer's perspective, this likelihood

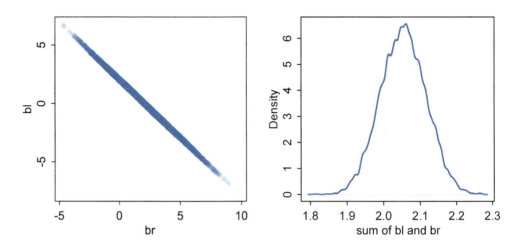

FIGURE 5.8. Left: Posterior distribution of the association of each leg with height, from model m5.8. Since both variables contain almost identical information, the posterior is a narrow ridge of negatively correlated values. Right: The posterior distribution of the sum of the two parameters is centered on the proper association of either leg with height.

is really:

$$y_i \sim \text{Normal}(\mu_i, \sigma)$$
$$\mu_i = \alpha + (\beta_1 + \beta_2)x_i$$

All I've done is factor x_i out of each term. The parameters β_1 and β_2 cannot be pulled apart, because they never separately influence the mean μ. Only their sum, $\beta_1 + \beta_2$, influences μ. So this means the posterior distribution ends up reporting the practically infinite combinations of β_1 and β_2 that make their sum close to the actual association of x with y.

And the posterior distribution in this simulated example has done exactly that: It has produced a good estimate of the sum of bl and br. Here's how you can compute the posterior distribution of their sum, and then plot it:

R code
5.33
```
sum_blbr <- post$bl + post$br
dens( sum_blbr , col=rangi2 , lwd=2 , xlab="sum of bl and br" )
```

And the resulting density plot is shown on the right-hand side of FIGURE 5.8. The posterior mean is in the right neighborhood, a little over 2, and the standard deviation is much smaller than it is for either component of the sum, bl or br. If you fit a regression with only one of the leg length variables, you'll get approximately the same posterior mean:

R code
5.34
```
m5.9 <- map(
    alist(
        height ~ dnorm( mu , sigma ) ,
        mu <- a + bl*leg_left,
        a ~ dnorm( 10 , 100 ) ,
```

```
        bl ~ dnorm( 2 , 10 ) ,
        sigma ~ dunif( 0 , 10 )
    ) ,
    data=d )
precis(m5.9)
```

```
      Mean StdDev 5.5% 94.5%
a     0.74   0.31 0.24  1.23
bl    2.05   0.07 1.94  2.16
sigma 0.63   0.04 0.56  0.70
```

That 2.05 is almost identical to the mean value of sum_blbr.

You'll get slightly different results in your own simulation, due to random variation across simulations. But the basic lesson remains intact across different simulations: *When two predictor variables are very strongly correlated, including both in a model may lead to confusion.* The posterior distribution isn't wrong, in such cases. It's telling you that the question you asked cannot be answered with these data. And that's a great thing for a model to say, that it cannot answer your question. And if you are just interested in prediction, you'll find that this leg model makes fine predictions. It just doesn't make any claims about which leg is more important.

5.3.2. Multicollinear milk. In the leg length example, it's easy to see that including both legs in the model is a little silly. But the problem that arises in real data sets is that we may not anticipate a clash between highly correlated predictors. And therefore we may mistakenly read the posterior distribution to say that neither predictor is important. In this section, we look at an example of this issue with real data.

Let's return to the primate milk data from earlier in the chapter. Let's get back the original data again:

```
library(rethinking)
data(milk)
d <- milk
```

R code
5.35

In these data, we have the variables perc.fat (percent fat) and perc.lactose (percent lactose) that we might use to model the total energy content, kcal.per.g. You're going to use these three variables to explore a natural case of multicollinearity. Note that there are no missing values, NA, in these columns, so there's no need here to extract complete cases.

Start by modeling kcal.per.g as a function of perc.fat and perc.lactose, but in two bivariate regressions:

```
# kcal.per.g regressed on perc.fat
m5.10 <- map(
    alist(
        kcal.per.g ~ dnorm( mu , sigma ) ,
        mu <- a + bf*perc.fat ,
        a ~ dnorm( 0.6 , 10 ) ,
        bf ~ dnorm( 0 , 1 ) ,
        sigma ~ dunif( 0 , 10 )
    ) ,
```

R code
5.36

```
    data=d )

# kcal.per.g regressed on perc.lactose
m5.11 <- map(
    alist(
        kcal.per.g ~ dnorm( mu , sigma ) ,
        mu <- a + bl*perc.lactose ,
        a ~ dnorm( 0.6 , 10 ) ,
        bl ~ dnorm( 0 , 1 ) ,
        sigma ~ dunif( 0 , 10 )
    ) ,
    data=d )

precis( m5.10 , digits=3 )
precis( m5.11 , digits=3 )
```

```
      Mean StdDev  5.5% 94.5%
a     0.301  0.036 0.244 0.358
bf    0.010  0.001 0.008 0.012
sigma 0.073  0.010 0.058 0.089

       Mean StdDev   5.5%  94.5%
a     1.166  0.043  1.098  1.235
bl   -0.011  0.001 -0.012 -0.009
sigma 0.062  0.008  0.049  0.075
```

The posterior mean for bf, the association of percent fat with milk energy, is 0.01, with 89% interval $[0.008, 0.012]$. The posterior mean in the second model for percent lactose is -0.01, with 89% interval $[-0.012, -0.009]$. These posterior means are essentially mirror images of one another, with the posterior mean of bf being as positive as the mean of bl is negative. Both are narrow posterior distributions that lie almost entirely on one side or the other of zero.

Before moving on to see what happens in the multiple regression that includes both predictors, note the parameter values for the slopes bf and bl are small in absolute value. So you might wonder whether these associations really have much influence on the outcome, milk energy. They do. Remember that both predictors are percents, so are potentially large numbers. The absolute magnitude of regression slopes is not always meaningful, because the influence on prediction depends upon the product of the parameter and the data. You have to compute or plot predictions, unless you decide to standardize all of your predictors. And even then you are probably better off always plotting implied predictions than trusting your intuition.

Given the strong association of each predictor with the outcome, we might conclude that both variables are reliable predictors of total energy in milk, across species. The more fat, the more kilocalories in the milk. The more lactose, the fewer kilocalories in milk. But watch what happens when we place both predictor variables in the same regression model:

R code
5.37
```
m5.12 <- map(
    alist(
```

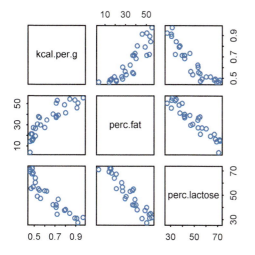

FIGURE 5.9. A pairs plot of the total energy, percent fat, and percent lactose variables from the primate milk data. Percent fat and percent lactose are strongly negatively correlated with one another, providing mostly the same information.

```
    kcal.per.g ~ dnorm( mu , sigma ) ,
    mu <- a + bf*perc.fat + bl*perc.lactose ,
    a ~ dnorm( 0.6 , 10 ) ,
    bf ~ dnorm( 0 , 1 ) ,
    bl ~ dnorm( 0 , 1 ) ,
    sigma ~ dunif( 0 , 10 )
  ) ,
  data=d )
precis( m5.12 , digits=3 )
```

	Mean	StdDev	5.5%	94.5%
a	1.007	0.200	0.688	1.327
bf	0.002	0.002	-0.002	0.006
bl	-0.009	0.002	-0.013	-0.005
sigma	0.061	0.008	0.048	0.074

Now the posterior means of both bf and bl are closer to zero. And the standard deviations for both parameters are twice as large as in the bivariate models (m5.10 and m5.11). In the case of percent fat, the posterior mean is essentially zero.

What has happened here? This is the same phenomenon as in the leg length example. What has happened is that the variables perc.fat and perc.lactose contain much of the same information. They are substitutes for one another. As a result, when you include both in a regression, the posterior distribution ends up describing a long ridge of combinations of bf and bl that are equally plausible.

In the case of the fat and lactose, these two variables form essentially a single axis of variation. The easiest way to see this is to use a pairs plot:

```
pairs( ~ kcal.per.g + perc.fat + perc.lactose ,
    data=d , col=rangi2 )
```

R code
5.38

I display this plot in FIGURE 5.9. Along the diagonal, the variables are labeled. In each scatterplot off the diagonal, the vertical axis variable is the variable labeled on the same row and the horizontal axis variable is the variable labeled in the same column. For example, the two scatterplots in the first row in FIGURE 5.9 are kcal.per.g (vertical) against perc.fat (horizontal) and then kcal.per.g (vertical) against perc.lactose (horizontal). Notice that percent fat is positively correlated with the outcome, while percent lactose is negatively correlated with it. Now look at the right-most scatterplot in the middle row. This plot is the scatter of percent fat (vertical) against percent lactose (horizontal). Notice that the points line up almost entirely along a straight line. These two variables are negatively correlated, and so strongly so that they are nearly redundant. Either helps in predicting kcal.per.g, but neither helps much *once you already know the other*.

You can compute the correlation between the two variables with cor:

R code
5.39
```
cor( d$perc.fat , d$perc.lactose )
```

```
[1] -0.9416373
```

That's a pretty strong correlation. How strong does a correlation have to get, before you should start worrying about multicollinearity? There's no easy answer to that question. Correlations do have to get pretty high before this problem interferes with your analysis. And what matters isn't just the correlation between a pair of variables. Rather, what matters is the correlation that remains after accounting for any other predictors.

But with only two predictors here, we can address the correlation question directly, with a little simulation experiment. Suppose we have only kcal.per.g and perc.fat. Now we construct a random predictor variable, call it x, that is correlated with perc.fat at some predetermined level. Then fit the regression model that tries to predict kcal.per.g using both perc.fat and our random fake variable x. Record the standard error of the estimated effect of perc.fat. Now repeat this procedure many times, at different levels of correlation between perc.fat and x.

If you do this, you'll get a plot like that in FIGURE 5.10. The vertical axis is the average standard deviation across 100 regressions, using a simulated correlated predictor variable x. The horizontal axis shows the intensity of correlation between x and perc.fat. When the two variables are uncorrelated, on the left side of the plot, then the standard deviation of the posterior is small. This means the posterior distribution is piled up a narrow range of values. As the correlation increases—and keep in mind that we aren't adding any information here, just a correlated string of random numbers—the standard deviation inflates. But the effect is far from linear; it accelerates rapidly, as the correlation increases. Above a correlation of 0.9, the standard deviation increases very rapidly, approaching in fact ∞ as the correlation approaches 1. The code for producing this plot contains some techniques not yet explained, so I include it in an optional Overthinking box at the end of this section.

What can be done about multicollinearity? The best thing to do is be aware of it. You can anticipate this problem by checking the predictor variables against one another in a pairs plot. Any pair or cluster of variables with very large correlations, over about 0.9, may be problematic, once included as main effects in the same model. However, it isn't always true that highly correlated variables are completely redundant—other predictors might be correlated with only one of the pair, and so help extract the unique information each predictor provides. So you can't know just from a table of correlations nor from a matrix of scatterplots whether multicollinearity will prevent you from including sets of variables in the same model. Still,

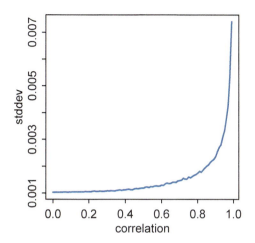

FIGURE 5.10. The effect of correlated predictor variables on the narrowness of the posterior distribution. The vertical axis shows the standard deviation of the posterior distribution of the slope. The horizontal axis is the correlation between the predictor of interest and another predictor added to the model. As the correlation increases, the standard deviation inflates.

you can usually diagnose the problem by looking for a big inflation of standard deviation, when both variables are included in the same model.

Different disciplines have different conventions for dealing with collinear variables. In some fields, it is typical to engage in some kind of data reduction procedure, like PRINCIPLE COMPONENTS or FACTOR ANALYSIS, and then to use the components/factors as predictor variables. In other fields, this is considered voodoo, because principle components and factors are notoriously hard to interpret, and there are usually hidden decisions that go into producing them. It is also difficult to use such constructed variables in prediction, because learning that component 2 is associated with an outcome doesn't immediately tell us how any real measurement is associated with the outcome. With experience, you can cope with these issues, but they remain issues. A popular defensive approach is to show that models using any one member from a cluster of highly correlated variables will produce the same inferences and predictions. For example, sometimes rainfall and soil moisture are highly correlated across sites in an ecological analysis. Using both in a model meant to predict presence/absence of a species of plant may run afoul of multicollinearity, but if you can show that a model with either produces nearly the same predictions, it will be reassuring that both get at the same underlying ecological dimension.

The problem of multicollinearity is really a member of a family of problems with fitting models, a family sometimes known as NON-IDENTIFIABILITY. When a parameter is non-identifiable, it means that the structure of the data and model do not make it possible to estimate the parameter's value. Sometimes this problem arises from mistakes in coding a model, but many important types of models present non-identifiable or weakly identifiable parameters, even when coded completely correctly.

In general, there's no guarantee that the available data contain much information about a parameter of interest. When that's true, your Bayesian machine will return a very wide posterior distribution. That doesn't mean anything is wrong—you got the right answer to the question you asked. But it might lead you to ask a better question. But it's usually helpful to check the covariation among parameters, especially when intervals are extremely wide, so you don't misinterpret what the model is trying to tell you.

Rethinking: Identification guaranteed; comprehension up to you. Technically speaking, *identifia-bility* is not a concern for Bayesian models. The reason is that as long as the posterior distribution is proper—which just means that it integrates to 1—then all of the parameters are identified. But this technical fact doesn't also mean that you can make sense of the posterior distribution. So it's probably better to speak of *weakly identified* parameters in a Bayesian context. There will be several examples as the book progresses.

Overthinking: Simulating collinearity. The code to produce FIGURE 5.10 involves writing a function that generates correlated predictors, fits a model, and returns the standard deviation of the posterior distribution for the slope relating `perc.fat` to `kcal.per.g`. Then the code repeatedly calls this function, with different degrees of correlation as input, and collects the results.

R code
5.40
```
library(rethinking)
data(milk)
d <- milk
sim.coll <- function( r=0.9 ) {
    d$x <- rnorm( nrow(d) , mean=r*d$perc.fat ,
        sd=sqrt( (1-r^2)*var(d$perc.fat) ) )
    m <- lm( kcal.per.g ~ perc.fat + x , data=d )
    sqrt( diag( vcov(m) ) )[2] # stddev of parameter
}
rep.sim.coll <- function( r=0.9 , n=100 ) {
    stddev <- replicate( n , sim.coll(r) )
    mean(stddev)
}
r.seq <- seq(from=0,to=0.99,by=0.01)
stddev <- sapply( r.seq , function(z) rep.sim.coll(r=z,n=100) )
plot( stddev ~ r.seq , type="l" , col=rangi2, lwd=2 , xlab="correlation" )
```

So for each correlation value in `r.seq`, the code generates 100 regressions and returns the average standard deviation from them. This code uses implicit flat priors, which are bad priors. So it does exaggerate the effect of collinear variables. When you use informative priors, the inflation in standard deviation can be much slower.

5.3.3. Post-treatment bias.
It is routine to worry about mistaken inferences that arise from omitting predictor variables. Such mistakes are often called OMITTED VARIABLE BIAS, and the examples from earlier in the chapter illustrate it. It is much less routine to worry about mistaken inferences arising from including variables that are consequences of other variables. We'll call this POST-TREATMENT BIAS.[77]

The language "post-treatment" comes from thinking about experimental designs, but the problem also applies to observational studies. Suppose for example that you are growing some plants in a greenhouse. You want to know the difference in growth under different anti-fungal soil treatments, because fungus on the plants tends to reduce their growth. Plants are initially seeded and sprout. Their heights are measured. Then different soil treatments are applied. Final measures are the height of the plant and the presence of fungus. There are four variables of interest here: initial height, final height, treatment, and presence of fungus. Final height is the outcome of interest. But which of the other variables should be in the model? If your goal is to make a causal inference about the treatment, you shouldn't include the presence of fungus, because it is a *post-treatment* effect.

Let's simulate some data, to make the example more transparent and see what exactly goes wrong when we include a post-treatment variable.

```
# number of plants
N <- 100

# simulate initial heights
h0 <- rnorm(N,10,2)

# assign treatments and simulate fungus and growth
treatment <- rep( 0:1 , each=N/2 )
fungus <- rbinom( N , size=1 , prob=0.5 - treatment*0.4 )
h1 <- h0 + rnorm(N, 5 - 3*fungus)

# compose a clean data frame
d <- data.frame( h0=h0 , h1=h1 , treatment=treatment , fungus=fungus )
```

Now you should have a data frame d with the simulated plant experiment data. Let's fit a model that includes all of the available variables:

```
m5.13 <- map(
    alist(
        h1 ~ dnorm(mu,sigma),
        mu <- a + bh*h0 + bt*treatment + bf*fungus,
        a ~ dnorm(0,100),
        c(bh,bt,bf) ~ dnorm(0,10),
        sigma ~ dunif(0,10)
    ),
    data=d )
precis(m5.13)
```

	Mean	StdDev	5.5%	94.5%
a	5.05	0.49	4.26	5.83
bh	1.00	0.05	0.92	1.08
bt	-0.27	0.22	-0.61	0.08
bf	-3.29	0.25	-3.69	-2.89
sigma	0.95	0.07	0.84	1.05

As usual with these simulation examples, your own estimates will differ due to simulation variance. You should repeat the simulation and analysis to get a sense of this variance. Every sample is special.

Inspecting the summary results, the marginal posterior for bt, the effect of treatment, is actually negative here. But it mainly had very little effect. The other two predictors, h0 and fungus, have important effects. Given that we know the treatment matters, because we built the simulation that way, what happened here?

The problem is that fungus is mostly a consequence of treatment. This is to say that fungus is a post-treatment variable. So when we control for fungus, the model is implicitly answering the question: *Once we already know whether or not a plant developed fungus, does soil treatment matter?* The answer is "no," because soil treatment has its effects on growth through reducing fungus. But we actually want to know, based on the design of the experiment, is the impact of treatment on growth. To measure this properly, we should omit the post-treatment variable `fungus`. Here's what the inference looks like in that case:

R code
5.43

```
m5.14 <- map(
    alist(
        h1 ~ dnorm(mu,sigma),
        mu <- a + bh*h0 + bt*treatment,
        a ~ dnorm(0,100),
        c(bh,bt) ~ dnorm(0,10),
        sigma ~ dunif(0,10)
    ),
    data=d )
precis(m5.14)
```

	Mean	StdDev	5.5%	94.5%
a	4.19	0.80	2.92	5.46
bh	0.94	0.08	0.81	1.07
bt	1.07	0.31	0.57	1.57
sigma	1.56	0.11	1.38	1.73

Now the impact of treatment is strong and positive, as it should be. It makes sense to control for pre-treatment differences, like the initial height h0, that might mask the causal influence of treatment. But including post-treatment variables can actually mask the treatment itself. Of course the fact that including `fungus` greatly reduces the coefficient for `treatment` does suggest that the treatment works for exactly the anticipated reasons. But a correct inference about the treatment still depends upon omitting the post-treatment variable.

The problem of post-treatment variables applies just as well to observational studies as it does to experiments. But in experiments, it can be easy to tell which variables are pre-treatment, like h0, and which are post-treatment, like `fungus`. In observational studies, it is harder to know.

Rethinking: Model comparison doesn't help. In the next chapter, you'll learn about model comparison using information criteria. Like other model comparison and selection schemes, these criteria help in contrasting and choosing model structure. But such approaches are no help in the example presented just above, since the model that includes `fungus` both fits the sample better and would make better out-of-sample predictions. Model m5.13 misleads because it asks the wrong question, not because it would make poor predictions. No statistical procedure can substitute for scientific knowledge and attention to it.

5.4. Categorical variables

A common question for statistical methods is to what extent an outcome changes as a result of presence or absence of a category. For example, consider the different species in the milk energy data again. Some of them are apes, while others are New World monkeys. We might want to ask how predictions should vary when the species is an ape instead of a monkey. Taxonomic group is a CATEGORICAL VARIABLE, because no species can be half-ape and half-monkey. Other common examples of categorical variables include:

- Sex: male, female

- Developmental status: infant, juvenile, adult
- Geographic region: Africa, Europe, Melanesia

Many readers will already know that variables like this, routinely called *factors*, can easily be included in linear models. But what is not widely understood is how these variables are included in a model. This is because the computer does all of the work for us, most of the time. But there are some subtleties here that make exposing the machinery worthwhile. Indeed, interpreting estimates for categorical variables can be harder than for regular continuous variables. Knowing how the machine works removes a lot of this difficulty.

5.4.1. Binary categories. In the simplest case, the variable of interest has only two categories, like *male* and *female*. Let's rewind to the Kalahari data you met in Chapter 4. Back then, we ignored sex when predicting height, but obviously we expect males and females to have different averages. Take a look at the variables available:

R code
5.44

```
data(Howell1)
d <- Howell1
str(d)
```

```
'data.frame': 544 obs. of  4 variables:
 $ height: num  152 140 137 157 145 ...
 $ weight: num  47.8 36.5 31.9 53 41.3 ...
 $ age   : num  63 63 65 41 51 35 32 27 19 54 ...
 $ male  : int  1 0 0 1 0 1 0 1 0 1 ...
```

The `male` variable is our new predictor, an example of a DUMMY VARIABLE. Dummy variables are devices for encoding categories into quantitative models. The purpose of the `male` variable is to indicate when a person in the sample is male. So it takes the value 1 whenever the person is male, but it takes the value 0 when the person is female. It doesn't matter which category—"male" or "female"—is indicated by the 1. The model won't care. But correctly interpreting the model will demand that you remember, so it's a good idea to name the variable after the category assigned the 1 value.

The effect of a dummy variable is to turn a parameter on for those cases in the category. Simultaneously, the variable turns the same parameter off for those cases in another category. This will make more sense, once you see it in the mathematical definition of the model. The model to fit is:

$$h_i \sim \text{Normal}(\mu_i, \sigma)$$
$$\mu_i = \alpha + \beta_m m_i$$
$$\alpha \sim \text{Normal}(178, 100)$$
$$\beta_m \sim \text{Normal}(0, 10)$$
$$\sigma \sim \text{Uniform}(0, 50)$$

where h is height and m is the dummy variable indicating a male individual. The parameter β_m influences prediction only for those cases where $m_i = 1$. When $m_i = 0$, it has no effect on prediction, because it is multiplied by zero inside the linear model, $\alpha + \beta_m m_i$, canceling it out, whatever its value.

To fit this model, use the usual format inside `map`:

R code
5.45
```
m5.15 <- map(
    alist(
        height ~ dnorm( mu , sigma ) ,
        mu <- a + bm*male ,
        a ~ dnorm( 178 , 100 ) ,
        bm ~ dnorm( 0 , 10 ) ,
        sigma ~ dunif( 0 , 50 )
    ) ,
    data=d )
precis(m5.15)
```

```
      Mean StdDev   5.5%   94.5%
a    134.84   1.59 132.29 137.38
bm     7.27   2.28   3.62  10.92
sigma 27.31   0.83  25.99  28.63
```

To interpret these estimates, you have to note that the parameter α (a) is now the average height *among females*. Why? Because when $m_i = 0$, indicating a female, the predicted mean height is just $\mu_i = \alpha + \beta_m(0) = \alpha$. So the estimate says that the expected average female height is 135 cm. The parameter β_m then tells us the average *difference* between males and females, 7.3 cm. So to compute the average male height, you just add these two estimates: $135 + 7.3 = 142.3$.

That's good enough for the posterior mean of average male height, but you'll also need to consider the width of the posterior distribution. And because the parameters α and β_m are correlated with one another, you can't just add together the boundaries in the precis output and get correct boundaries for their sum.[78] But as usual, the most accessible way to derive a percentile interval for average male height is just to sample from the posterior. Then you can just add samples of a and bm together to get the posterior distribution of their sum. Here's all it takes:

R code
5.46
```
post <- extract.samples(m5.15)
mu.male <- post$a + post$bm
PI(mu.male)
```

```
     5%        94%
139.3947 144.7834
```

Working with samples automatically handles the subtle problem that a and bm are correlated with one another. When working with samples, the procedure is the same no matter what the correlation.

Overthinking: Re-parameterizing the model. Before moving on to factors with more than two categories, it may help to fit the same model in a different form, a different *parameterization*. Instead of using a parameter for the difference between males and females, we can make parameters specific to males and females:

$$h_i \sim \text{Normal}(\mu_i, \sigma)$$
$$\mu_i = \alpha_f(1 - m_i) + \alpha_m m_i$$

where α_f is now the average female height and α_m the average male height. Plug in $m_i = 0$ and $m_i = 1$ to verify that either α_f or α_m, but never both, is turned on for any individual case i. Fit the model with:

```
m5.15b <- map(
    alist(
        height ~ dnorm( mu , sigma ) ,
        mu <- af*(1-male) + am*male ,
        af ~ dnorm( 178 , 100 ) ,
        am ~ dnorm( 178 , 100 ) ,
        sigma ~ dunif( 0 , 50 )
    ) ,
    data=d )
```

R code
5.47

It remains an exercise for the reader to confirm that the results are inferentially equivalent to the original m5.13. In fact, you can prove the equivalence with a little algebra. Taking the expression for μ_i above and rearranging:

$$\mu_i = \alpha_f + (\alpha_m - \alpha_f)m_i$$

And there's the definition of the old β_m as the difference between the average male and female, $\alpha_m - \alpha_f$.

5.4.2. Many categories.

Binary categories are easy, because you only need one dummy variable. You just pick one of the categories to be indicated by the dummy. The other category is then measured by the intercept, α.

But when there are more than two categories, you'll need more than one dummy variable. Here's the general rule: To include k categories in a linear model, you require $k - 1$ dummy variables. Each dummy variable indicates, with the value 1, a unique category. The category with no dummy variable assigned to it ends up again as the "intercept" category.

Let's explore an example using the primate milk data again. We're interested now in the clade variable, which encodes the broad taxonomic membership of each species:

```
data(milk)
d <- milk
unique(d$clade)
```

R code
5.48

```
[1] Strepsirrhine    New World Monkey Old World Monkey Ape
Levels: Ape New World Monkey Old World Monkey Strepsirrhine
```

This is a typical case: There are more than two categories and none of them have yet been coded into dummy variables. It's easy to make dummy variables, though. To create a dummy for the New World Monkey category:

```
( d$clade.NWM <- ifelse( d$clade=="New World Monkey" , 1 , 0 ) )
```

R code
5.49

```
 [1] 0 0 0 0 0 1 1 1 1 1 1 1 1 0 0 0 0 0 0 0 0 0 0 0 0 0 0 0 0
```

So only those rows (cases) where clade is New World Monkey get a 1 for this new variable. You can make two more dummy variables with the same strategy:

```
d$clade.OWM <- ifelse( d$clade=="Old World Monkey" , 1 , 0 )
d$clade.S <- ifelse( d$clade=="Strepsirrhine" , 1 , 0 )
```

R code
5.50

There's no need to make another for the category Ape, because it will be the default, "intercept" category. In fact, if you try to include a dummy variable for apes, you'll up with a non-identifiable model (page 149). Can you figure out why?

The model we aim to fit to the data is kcal.per.g regressed on the dummy variables for clade:

$$k_i \sim \text{Normal}(\mu_i, \sigma)$$
$$\mu_i = \alpha + \beta_{\text{NWM}}\text{NWM}_i + \beta_{\text{OWM}}\text{OWM}_i + \beta_\text{S}\text{S}_i$$
$$\alpha \sim \text{Normal}(0.6, 10)$$
$$\beta_{\text{NWM}} \sim \text{Normal}(0, 1)$$
$$\beta_{\text{OWM}} \sim \text{Normal}(0, 1)$$
$$\beta_\text{S} \sim \text{Normal}(0, 1)$$
$$\sigma \sim \text{Uniform}(0, 10)$$

A linear model like this really defines four different linear models, each corresponding to a different category. A table might help you to understand how dummy variables make individual parameters contingent on category:

Category	NWM$_i$	OWM$_i$	S$_i$	μ_i
Ape	0	0	0	$\mu_i = \alpha$
New World monkey	1	0	0	$\mu_i = \alpha + \beta_{\text{NWM}}$
Old World monkey	0	1	0	$\mu_i = \alpha + \beta_{\text{OWM}}$
Strepsirrhine	0	0	1	$\mu_i = \alpha + \beta_\text{S}$

Each category implies a different set of 1s and 0s in the dummy variables, which in turn implies a different equation for μ_i, once simplified.

Fitting the model is straightforward:

R code
5.51

```
m5.16 <- map(
    alist(
        kcal.per.g ~ dnorm( mu , sigma ) ,
        mu <- a + b.NWM*clade.NWM + b.OWM*clade.OWM + b.S*clade.S ,
        a ~ dnorm( 0.6 , 10 ) ,
        b.NWM ~ dnorm( 0 , 1 ) ,
        b.OWM ~ dnorm( 0 , 1 ) ,
        b.S ~ dnorm( 0 , 1 ) ,
        sigma ~ dunif( 0 , 10 )
    ) ,
    data=d )
precis(m5.16)
```

```
        Mean StdDev  5.5% 94.5%
a       0.55   0.04  0.49  0.61
b.NWM   0.17   0.05  0.08  0.25
b.OWM   0.24   0.06  0.15  0.34
b.S    -0.04   0.06 -0.14  0.06
sigma   0.11   0.02  0.09  0.14
```

The estimate a is the average milk energy for apes, and the estimates for the other categories are differences from apes. So to get posterior distributions of the average milk energy in each category, you can again use samples:

```
# sample posterior
post <- extract.samples(m5.16)

# compute averages for each category
mu.ape <- post$a
mu.NWM <- post$a + post$b.NWM
mu.OWM <- post$a + post$b.OWM
mu.S <- post$a + post$b.S

# summarize using precis
precis( data.frame(mu.ape,mu.NWM,mu.OWM,mu.S) )
```

R code
5.52

```
        Mean StdDev |0.89 0.89|
mu.ape 0.55   0.04  0.49  0.61
mu.NWM 0.71   0.04  0.65  0.77
mu.OWM 0.79   0.05  0.71  0.86
mu.S   0.51   0.05  0.43  0.59
```

These tell us that the most plausible (conditional on data and model) average milk energies in each category are 0.55, 0.71, 0.79, and 0.51. The 89% HPDI for each is shown in the two right-most columns.

Once you get accustomed to manipulating estimates in this way, you can effectively re-parameterize your model after you've already fit it to the data. For example, above we fit model m5.14 such that each b parameter is a difference from apes. This makes it hard to reason about the differences among other categories. So suppose you want to know the estimated difference between the two monkey groups. Then just subtract the estimated means to get a difference:

```
diff.NWM.OWM <- mu.NWM - mu.OWM
quantile( diff.NWM.OWM , probs=c(0.025,0.5,0.975) )
```

R code
5.53

```
      2.5%          50%         97.5%
-0.19041506  -0.07374566   0.04737627
```

Those values are the posterior lower 95% boundary, the median, and the upper 95% boundary for the difference between New World and Old World monkeys. Since you're still working with samples from the posterior, you get a posterior distribution for the difference between NWM and OWM. None of the uncertainty in the original posterior distribution is discarded.

Rethinking: Differences and statistical significance. A common error in interpretation of parameter estimates is to suppose that because one parameter is sufficiently far from zero—is "significant"—and another parameter is not—is "not significant"—that the difference between the parameters is also significant. This is not necessarily so.[79] This isn't just an issue for non-Bayesian analysis: If you want to know the distribution of a difference, then you must compute that difference, a *contrast*. It isn't enough to just observe, for example, that a slope among males overlaps a lot with zero while the same slope among females is reliably above zero. You must compute the posterior distribution of the difference in slope between males and females. For example, suppose you have posterior distributions for two parameters, β_f and β_m. β_f's mean and standard deviation is 0.15 ± 0.02, and β_m's is 0.02 ± 0.10. So while β_f is reliably different from zero ("significant") and β_m is not, the difference between the two (assuming they are uncorrelated) is $(0.15 - 0.02) \pm \sqrt{0.02^2 + 0.1^2} \approx 0.13 \pm 0.10$. The distribution of the difference overlaps a lot with zero. In other words, you can be confident that β_f is far from zero, but you cannot be sure that the difference between β_f and β_m is far from zero.

In the context of non-Bayesian significance testing, this phenomenon arises from the fact that statistical significance is inferentially powerful in one dimension: difference from the null. When β_m overlaps with zero, it may also overlap with values very far from zero. It's value is uncertain. So when you then compare β_m to β_f, that comparison is also uncertain, manifesting in the width of the posterior distribution of the difference $\beta_f - \beta_m$. Lurking underneath this example is a more fundamental mistake in interpreting statistical significance: The mistake of accepting the null hypothesis. Whenever an article or book says something like "we found no difference" or "no effect," this usually means that some parameter was not significantly different from zero, and so the authors adopted zero as the estimate. This is both illogical and extremely common.

5.4.3. Adding regular predictor variables.
There is no obstacle now in including other predictor variables, like `perc.fat` or `log(mass)`. Just add them to the equation for the mean, as you normally would. For example, to add the influence of `perc.fat` to the model with dummies for the clades:

$$\mu_i = \alpha + \beta_{\text{NWM}}\text{NWM}_i + \beta_{\text{OWM}}\text{OWM}_i + \beta_S S_i + \beta_F F_i$$

where F_i is the percent fat for the i-th case in the data and β_F is the slope that measures the influence of F on predictions about the mean milk energy.

When you reach the chapter on interactions, we'll return to the relationship between dummy variables and continuous predictors. Since dummy variables turn some fixed adjustment on or off for each row i, they can be used to produce all manner of contingent predictions, depending upon category memberships. Among these contingent effects is the adjustment of slopes themselves, allowing different categories in the data to bear different relationships to a predictor variable.

5.4.4. Another approach: Unique intercepts.
Another way to conceptualize categorical variables is to construct a vector of intercept parameters, one parameter for each category. Then you can create an **INDEX VARIABLE** in your data frame that says which parameter goes with each case. We'll use this approach later on, because it's very common when using multilevel models (Chapter 12).

Here's a quick example. To make an index variable for `clade` in the primate milk data:

R code
5.54

```
( d$clade_id <- coerce_index(d$clade) )
```

```
[1] 4 4 4 4 4 2 2 2 2 2 2 2 2 2 3 3 3 3 3 3 1 1 1 1 1 1 1 1 1 1
```

This variable just gives the number of each unique clade value. There are four different clades, and it doesn't matter which is "1" or "4," because they are unordered. That's why this is an "index." Then you tell map to make a vector of intercepts, one intercept for each unique value in `clade_id`, with:

R code
5.55

```
m5.16_alt <- map(
    alist(
        kcal.per.g ~ dnorm( mu , sigma ) ,
        mu <- a[clade_id] ,
        a[clade_id] ~ dnorm( 0.6 , 10 ) ,
        sigma ~ dunif( 0 , 10 )
    ) ,
    data=d )
precis( m5.16_alt , depth=2 )
```

```
      Mean StdDev 5.5% 94.5%
a[1]  0.55   0.04 0.48  0.61
a[2]  0.71   0.04 0.65  0.78
a[3]  0.79   0.05 0.71  0.86
a[4]  0.51   0.05 0.43  0.59
sigma 0.11   0.02 0.09  0.14
```

Now you get the same averages for each clade that we computed earlier, but this time directly from the fit model. Notice that you'll need to add depth=2 to the precis call in order to print out vector parameters like these. Why? Sometimes there are hundreds or thousands of vector parameters, so they are suppressed by default. You'll get a better sense of how we might end up with hundreds or thousands of parameters, once you begin tackling multilevel models.

5.5. Ordinary least squares and lm

Many readers will know these Gaussian regression models by another label, ORDINARY LEAST SQUARES regression, or OLS for short. OLS is a way of estimating the parameters of a linear regression. Instead of searching for the combination of parameter values that maximizes the posterior probability, OLS instead solves for the parameter values that minimize the sum of the squared residuals. It turns out that this procedure is often functionally equivalent to maximizing the posterior probability or maximizing the likelihood. Indeed, Carl Friedrich Gauss himself invented OLS as a method of computing Bayesian MAP estimates.[80]

In this section, I do not dwell on the mechanics of OLS. There are hundreds of books that derive and discuss OLS estimates. The goal here is just to explain how to use R's basic linear model function, lm, to fit some of the linear regressions you've been using already. Provided you are happy with flat priors, you'll get the same estimates with lm that you got with map.

But the notation that lm uses is very compact, compared to all the detail that map demands. Now that you've mastered linear regression models at the basic probability level required to define one in map, it's safe to move to the convenience of lm, but only when you are okay with flat priors. In the next chapter, you'll see some reasons to not be okay with flat priors. But in many cases, there is so much data that it hardly matters. But even if you never use lm and OLS itself, it's helpful to learn how OLS estimates can be understood from a Bayesian perspective.

5.5.1. Design formulas. The input format for models in lm, and many of the other model fitting functions in R, is a compact form of notation known as a DESIGN FORMULA. Design formulas take the expression for μ_i in the detailed mathematical form of a model and strip out the parameters, leaving only a series of predictor names, separated by + signs. These formulas are sufficient to describe the model's design, provided all priors are flat.

For example, if we have the linear regression:

$$y_i \sim \text{Normal}(\mu_i, \sigma)$$
$$\mu_i = \alpha + \beta x_i$$

then the corresponding design formula is:

$$y ~ 1 + x$$

The leading 1 on the right-hand side indicates the intercept, α. As you add more predictor variables, the design formula expands. For example, this design formula:

$$y ~ 1 + x + z + w$$

corresponds to the model:

$$y_i \sim \text{Normal}(\mu_i, \sigma)$$
$$\mu_i = \alpha + \beta_x x_i + \beta_z z_i + \beta_w w_i$$

Design formulas are intimately related to a common matrix algebra description of linear models. See the Overthinking box on page 124 for a short introduction to this approach.

5.5.2. Using lm. To fit a linear regression using OLS, just provide the design formula and the name of the data frame. So for both examples just above, the calls to lm would be (assuming the data are in a frame named d):

R code
5.56
```
m5.17 <- lm( y ~ 1 + x , data=d )
m5.18 <- lm( y ~ 1 + x + z + w , data=d )
```

The parameter estimates will be named after the predictors, since you didn't provide any explicit parameter names. But otherwise the output will look familiar, and you can sample from the posterior and process estimates just as before. There are some quirks to how lm works, though. Here are some of the important quirks.

5.5.2.1. Intercepts are optional. These two model fits return exactly the same estimates:

R code
5.57
```
m5.17 <- lm( y ~ 1 + x , data=d )
m5.19 <- lm( y ~ x , data=d )
```

When you omit the explicit intercept, lm assumes you wanted one. If you really do not want an intercept—equivalent to fixing $\alpha = 0$—then you can use one of these forms:

R code
5.58
```
m5.20 <- lm( y ~ 0 + x , data=d )
m5.21 <- lm( y ~ x - 1 , data=d )
```

5.5.2.2. Categorical variables. Black-box functions like lm will automatically expand categorical factors into the necessary number of dummy variables. But R is not a mind reader, and sometimes the same variable can be treated as categories or rather as continuous. For example, you might code a variable like season with the values 1, 2, 3, and 4. These could then be categories—implying you don't care about order, but rather about estimating any differences among seasons—or a continuous variable marking passage of time. R is easily confused in these circumstances.

To prevent R from getting confused, and therefore confusing yourself, best to get into the habit of explicitly telling lm when you mean to use a variable as categories. For example:

```
m5.22 <- lm( y ~ 1 + as.factor(season) , data=d )
```
R code
5.59

The function `as.factor` makes sure that lm knows to treat the values in `season` as unordered categories. It'll then make the dummy variables for you, and you'll get a table of estimates reflecting the dummy variables that were produced inside the black box.

You can also just produce the dummy variables yourself and then include them in your design formula. That will allow you precise control over which category corresponds to the intercept.

5.5.2.3. *Transform variables first.* There are many good reasons to transform predictor variables. You've already seen simple transformations like logarithms and squares and cubes, as well as centering and standardizing. Black-box functions like lm sometimes have problems understanding these transformations, when they are included in the design formula. So it's best to make new variables that hold the transformed values. Then include those new variables instead. For example, to fit a cubic regression of y on x:

```
d$x2 <- d$x^2
d$x3 <- d$x^3
m5.23 <- lm( y ~ 1 + x + x2 + x3 , data=d )
```
R code
5.60

You might come across code that uses another mechanism to get the transforms right. This alternative mechanism uses R's "as is" function `I()`. This code will fit the same cubic model as before:

```
m5.24 <- lm( y ~ 1 + x + I(x^2) + I(x^3) , data=d )
```
R code
5.61

Note however that you cannot use `I()` inside map.

5.5.2.4. *No estimate for σ.* Using lm will not provide a posterior distribution for the standard deviation σ—it won't be reported in the table of estimates. If you ask for a summary of the model fit, lm will report the "residual standard error," which is a slightly different estimate of σ, but without any uncertainty information like a standard deviation (or standard error).

Provided you don't care about full prediction intervals, then the lack of an estimate for σ won't bother you, because you won't need an estimate for σ. But if you find yourself needing it, you'll have to fall back on a tool like map, so you can get some sense of the uncertainty of the estimate. As a bonus, you can use informative priors then, as well.

5.5.3. **Building map formulas from lm formulas.** The rethinking package provides a function, glimmer, that translates design formulas into map-style model formulas. For example:

```
data(cars)
glimmer( dist ~ speed , data=cars )
```

```
alist(
    dist ~ dnorm( mu , sigma ),
    mu <- Intercept +
        b_speed*speed,
    Intercept ~ dnorm(0,10),
    b_speed ~ dnorm(0,10),
    sigma ~ dcauchy(0,2)
)
```

The simple linear regression of `dist` on `speed` implies the list of formulas above. The function `glimmer` automatically adds default priors. You can change these to suit your tastes, or remove them all together, if you just want maximum likelihood estimation. See `?glimmer` for additional options.

5.6. Summary

This chapter introduced multiple regression, a way of constructing descriptive models for how the mean of a measurement is associated with more than one predictor variable. The defining question of multiple regression is: *What is the value of knowing each predictor, once we already know the other predictors?* Implicit in this question are: (1) a focus on the value of the predictors for description of the sample, instead of forecasting a future sample; and (2) the assumption that the value of each predictor does not depend upon the values of the other predictors. In the next two chapters, we confront these two issues.

5.7. Practice

Easy.

5E1. Which of the linear models below are multiple linear regressions?

(1) $\mu_i = \alpha + \beta x_i$
(2) $\mu_i = \beta_x x_i + \beta_z z_i$
(3) $\mu_i = \alpha + \beta(x_i - z_i)$
(4) $\mu_i = \alpha + \beta_x x_i + \beta_z z_i$

5E2. Write down a multiple regression to evaluate the claim: *Animal diversity is linearly related to latitude, but only after controlling for plant diversity.* You just need to write down the model definition.

5E3. Write down a multiple regression to evaluate the claim: *Neither amount of funding nor size of laboratory is by itself a good predictor of time to PhD degree; but together these variables are both positively associated with time to degree.* Write down the model definition and indicate which side of zero each slope parameter should be on.

5E4. Suppose you have a single categorical predictor with 4 levels (unique values), labeled A, B, C and D. Let A_i be an indicator variable that is 1 where case i is in category A. Also suppose B_i, C_i, and D_i for the other categories. Now which of the following linear models are inferentially equivalent ways to include the categorical variable in a regression? Models are inferentially equivalent when it's possible to compute one posterior distribution from the posterior distribution of another model.

(1) $\mu_i = \alpha + \beta_A A_i + \beta_B B_i + \beta_D D_i$
(2) $\mu_i = \alpha + \beta_A A_i + \beta_B B_i + \beta_C C_i + \beta_D D_i$

(3) $\mu_i = \alpha + \beta_B B_i + \beta_C C_i + \beta_D D_i$
(4) $\mu_i = \alpha_A A_i + \alpha_B B_i + \alpha_C C_i + \alpha_D D_i$
(5) $\mu_i = \alpha_A(1 - B_i - C_i - D_i) + \alpha_B B_i + \alpha_C C_i + \alpha_D D_i$

Medium.

5M1. Invent your own example of a spurious correlation. An outcome variable should be correlated with both predictor variables. But when both predictors are entered in the same model, the correlation between the outcome and one of the predictors should mostly vanish (or at least be greatly reduced).

5M2. Invent your own example of a masked relationship. An outcome variable should be correlated with both predictor variables, but in opposite directions. And the two predictor variables should be correlated with one another.

5M3. It is sometimes observed that the best predictor of fire risk is the presence of firefighters— States and localities with many firefighters also have more fires. Presumably firefighters do not *cause* fires. Nevertheless, this is not a spurious correlation. Instead fires cause firefighters. Consider the same reversal of causal inference in the context of the divorce and marriage data. How might a high divorce rate cause a higher marriage rate? Can you think of a way to evaluate this relationship, using multiple regression?

5M4. In the divorce data, States with high numbers of Mormons (members of The Church of Jesus Christ of Latter-day Saints, LDS) have much lower divorce rates than the regression models expected. Find a list of LDS population by State and use those numbers as a predictor variable, predicting divorce rate using marriage rate, median age at marriage, and percent LDS population (possibly standardized). You may want to consider transformations of the raw percent LDS variable.

5M5. One way to reason through multiple causation hypotheses is to imagine detailed mechanisms through which predictor variables may influence outcomes. For example, it is sometimes argued that the price of gasoline (predictor variable) is positively associated with lower obesity rates (outcome variable). However, there are at least two important mechanisms by which the price of gas could reduce obesity. First, it could lead to less driving and therefore more exercise. Second, it could lead to less driving, which leads to less eating out, which leads to less consumption of huge restaurant meals. Can you outline one or more multiple regressions that address these two mechanisms? Assume you can have any predictor data you need.

Hard. All three exercises below use the same data, data(foxes) (part of rethinking).[81] The urban fox (*Vulpes vulpes*) is a successful exploiter of human habitat. Since urban foxes move in packs and defend territories, data on habitat quality and population density is also included. The data frame has five columns:

(1) group: Number of the social group the individual fox belongs to
(2) avgfood: The average amount of food available in the territory
(3) groupsize: The number of foxes in the social group
(4) area: Size of the territory
(5) weight: Body weight of the individual fox

5H1. Fit two bivariate Gaussian regressions, using map: (1) body weight as a linear function of territory size (area), and (2) body weight as a linear function of groupsize. Plot the results of these regressions, displaying the MAP regression line and the 95% interval of the mean. Is either variable important for predicting fox body weight?

5H2. Now fit a multiple linear regression with weight as the outcome and both area and groupsize as predictor variables. Plot the predictions of the model for each predictor, holding the other predictor constant at its mean. What does this model say about the importance of each variable? Why do you get different results than you got in the exercise just above?

5H3. Finally, consider the avgfood variable. Fit two more multiple regressions: (1) body weight as an additive function of avgfood and groupsize, and (2) body weight as an additive function of all three variables, avgfood and groupsize and area. Compare the results of these models to the previous models you've fit, in the first two exercises. (a) Is avgfood or area a better predictor of body weight? If you had to choose one or the other to include in a model, which would it be? Support your assessment with any tables or plots you choose. (b) When both avgfood or area are in the same model, their effects are reduced (closer to zero) and their standard errors are larger than when they are included in separate models. Can you explain this result?

6 Overfitting, Regularization, and Information Criteria

Nicolaus Copernicus (1473–1543): Polish astrologer, ecclesiastical lawyer, and blasphemer. Famous for his heliocentric model of the solar system, Copernicus argued for replacing the geocentric model, because the heliocentric model was more "harmonious." This position eventually lead (decades later) to Galileo's famous disharmony with, and trial by, the Church.

This story has become a fable of science's triumph over ideology and superstition. But Copernicus' justification looks poor to us now, ideology aside. There are two problems: The model was neither particularly harmonious nor more accurate than the geocentric model. The Copernican model was very complicated. In fact, it had similar epicycle clutter as the Ptolemaic model (FIGURE 6.1). Copernicus had moved the Sun to the center, but since he still used perfect circles for orbits, he still needed epicycles. And so "harmony" doesn't quite describe the model's appearance. Just like the Ptolemaic model, the Copernican model was effectively a Fourier series, a means of approximating periodic functions. This leads to the second problem: The heliocentric model made exactly the same predictions as the geocentric model. Equivalent approximations can be constructed whether the Earth is stationary or rather moving. So there was no reason to prefer it on the basis of accuracy alone.

Copernicus didn't appeal just to some vague harmony, though. He also argued for the superiority of his model on the basis of needing fewer causes: "We thus follow Nature, who producing nothing in vain or superfluous often prefers to endow one cause with many effects."[82] And it was true that a heliocentric model required fewer circles and epicycles to make the same predictions as a geocentric model. In this sense, it was *simpler* than the geocentric model.

Scholars often prefer simpler theories. This preference is often left vague—a kind of aesthetic preference. Other times we retreat to pragmatism, preferring simpler theories because their simpler models are easier to work with. Frequently, scientists cite a loose principle known as OCKHAM'S RAZOR: *Models with fewer assumptions are to be preferred*. In the case of Copernicus and Ptolemy, the razor makes a clear recommendation. It cannot guarantee that Copernicus was right (he wasn't, after all), but since the heliocentric and geocentric models make the same predictions, at least the razor offers a clear resolution to the dilemma. But the razor can be hard to use more generally, because usually we must choose among models that differ in both their accuracy and their simplicity. How are we to trade these different criteria against one another? The razor offers no guidance.

This chapter describes some of the most commonly used tools for coping with this trade-off. Some notion of simplicity usually features in all of these tools, and so each is commonly compared to Ockham's razor. But each tool is equally about improving predictive accuracy. So they are not like the razor, because they explicitly trade-off accuracy and simplicity.

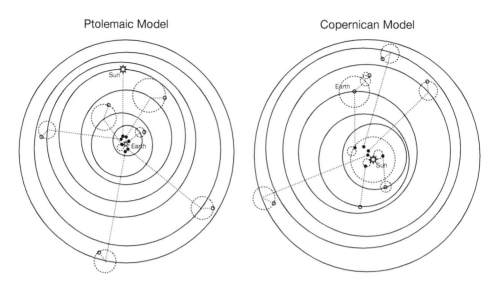

FIGURE 6.1. Ptolemaic (left) and Copernican (right) models of the solar system. Both models use epicycles (circles on circles), and both models produce exactly the same predictions. However, the Copernican model requires fewer circles. (Not all Ptolemaic epicycles are visible in the figure.)

So instead of Ockham's razor, think of Ulysses' compass. Ulysses was the hero of Homer's *Odyssey*. During his voyage, Ulysses had to navigate a narrow straight between the many-headed beast Scylla—who attacked from a cliff face and gobbled up sailors—and the sea monster Charybdis—who pulled boats and men down to a watery grave. Passing too close to either meant disaster. In the context of scientific models, you can think of these monsters as representing two fundamental kinds of statistical error:

(1) The many-headed beast of OVERFITTING, which leads to poor prediction by learning too much from the data
(2) The whirlpool of UNDERFITTING, which leads to poor prediction by learning too little from the data

Our job is to carefully navigate between these monsters. There are two common families of approaches. The first approach is to use a REGULARIZING PRIOR to tell the model not to get too excited by the data. This is the same device that non-Bayesian methods refer to as "penalized likelihood." The second approach is to use some scoring device, like INFORMATION CRITERIA, to model the prediction task and estimate predictive accuracy for some purpose. Both families of approaches are routinely used in the natural and social sciences. Furthermore, they can be—maybe should be—used in combination. So it's worth understanding both, as you're going to need both at some point.

In order to introduce information criteria, this chapter must also introduce INFORMATION THEORY. If this is your first encounter with information theory, it'll probably seem strange. But once you start using these criteria—this chapter describes AIC, DIC, and WAIC—you'll find that implementing them is much easier than understanding them. So most of this chapter aims at conceptual knowledge, with applications to follow.

It's worth noting, before getting started, that this material is hard. If you find yourself confused at any point, you are normal. Any sense of confusion you feel is just your brain correctly calibrating to the subject matter. Over time, confusion is replaced by comprehension for how overfitting, regularization, and information criteria behave in familiar contexts.

The goal remains to construct and criticize statistical golems, none of which are "correct." But regularization and information criteria can help make them more effective and responsible.

> **Rethinking: Stargazing.** The most common form of model selection among practicing scientists is to search for a model in which every coefficient is statistically significant. Statisticians sometimes call this STARGAZING, as it is embodied by scanning for asterisks (**) trailing after estimates. A colleague of mine once called this approach the "Space Odyssey," in honor of A. C. Clarke's novel and film. The model that is full of stars, the thinking goes, is best.
>
> But such a model is not best. Whatever you think about null hypothesis significance testing in general, using it to select among structurally different models is a mistake—p-values are not designed to help you navigate between underfitting and overfitting. As you'll see once you start using AIC and related measures, it is true that predictor variables that do improve prediction are not always statistically significant. It is also possible for variables that are statistically significant to do nothing useful for prediction. Since the conventional 5% threshold is purely conventional, we shouldn't expect it to optimize anything.

> **Rethinking: Is AIC Bayesian?** AIC is not usually thought of as a Bayesian tool. There are both historical and statistical reasons for this. Historically, AIC was originally derived without reference to Bayesian probability. Statistically, AIC uses MAP estimates instead of the entire posterior, and it requires flat priors. So it doesn't look particularly Bayesian. Reinforcing this impression is the existence of another model comparison metric, the BAYESIAN INFORMATION CRITERION (BIC). However, BIC also requires flat priors and MAP estimates, although it's not actually an "information criterion."
>
> Regardless, AIC has a clear and pragmatic interpretation under Bayesian probability, and Akaike and others have long argued for alternative Bayesian justifications of the procedure.[83] And as you'll see later in the book, more obviously Bayesian information criteria like DIC and WAIC provide almost exactly the same results as AIC, when AIC's assumptions are met. In this light, we can fairly regard AIC as a special limit of a Bayesian criterion like WAIC, even if that isn't how AIC was originally derived. All of this is an example of a common feature of statistical procedures: The same procedure can be derived and justified from multiple, sometimes philosophically incompatible, perspectives.

6.1. The problem with parameters

In the previous chapter, we saw how adding variables and parameters to a model can help to reveal hidden effects and improve estimates. You also saw that adding variables can hurt, in particular when predictor variables are highly correlated with one another. But what about when the predictor variables are not highly correlated? Would it be safe to just add them all?

The answer is "no." There are two principle concerns with just adding variables. The first is that adding parameters—making the model more complex—nearly always improves the fit of a model to the data.[84] By "fit" I mean a measure of how well the model can retrodict the data used to fit the model. There are many such measures, each with its own foibles. In the context of linear Gaussian models, R^2 is the most common measure of this kind. Often

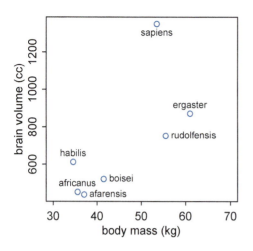

FIGURE 6.2. Average brain volume in cubic centimeters against body mass in kilograms, for six hominin species. What model best describes the relationship between brain size and body size?

described as "variance explained," R^2 is defined as:

$$R^2 = \frac{\text{var}(\text{outcome}) - \text{var}(\text{residuals})}{\text{var}(\text{outcome})} = 1 - \frac{\text{var}(\text{residuals})}{\text{var}(\text{outcome})}$$

Being easy to compute, R^2 is popular. But like other measures of fit to sample, R^2 increases as more predictor variables are added. This is true even when the variables you add to a model are just random numbers, with no relation to the outcome. So it's no good to choose among models using only fit to the data.

Second, while more complex models fit the data better, they often predict new data worse. Models that have many parameters tend to overfit more than do simpler models. This means that a complex model will be very sensitive to the exact sample used to fit it, leading to potentially large mistakes when future data is not exactly like the past data. But simple models, with too few parameters, tend instead to underfit, systematically over-predicting or under-predicting the data, regardless of how well future data resemble past data. So we can't always favor either simple models or complex models.

Let's examine both of these issues in the context of a simple data example.

6.1.1. More parameters always improve fit. OVERFITTING occurs when a model learns too much from the sample. What this means is that there are both *regular* and *irregular* features in every sample. The regular features are the targets of our learning, because they generalize well or answer a question of interest. Regular features are useful, given an objective of our choice. The irregular features are instead aspects of the data that do not generalize and so may mislead us.

Overfitting happens automatically, unfortunately. In the kind of statistical models we've seen so far in this book, adding additional parameters will always improve the fit of a model to the sample. Much later in the book, beginning with Chapter 12, you'll meet models for which adding parameters does not necessarily improve fit to the sample, but may well improve predictive accuracy.

Here's an example of overfitting. The data displayed in FIGURE 6.2 are average brain volumes and body masses for seven hominin species.[85] Let's get these data into R, so you can work with them. I'm going to build these data from direct input, rather than loading a pre-made data frame, just so you see an example of how to build a data frame from scratch.

```
sppnames <- c( "afarensis","africanus","habilis","boisei",
    "rudolfensis","ergaster","sapiens")
brainvolcc <- c( 438 , 452 , 612, 521, 752, 871, 1350 )
masskg <- c( 37.0 , 35.5 , 34.5 , 41.5 , 55.5 , 61.0 , 53.5 )
d <- data.frame( species=sppnames , brain=brainvolcc , mass=masskg )
```

R code
6.1

Now you have a data frame, d, containing the brain size and body size values. It's not unusual for data like this to be highly correlated—brain size is correlated with body size, across
species. A standing question, however, is to what extent particular species have brains that
are larger than we'd expect, after taking body size into account. A very common solution is
to fit a linear regression that models brain size as a linear function of body size. Then the remaining variation in brain size can be modeled as a function of other variables, like ecology
or diet or species age. This is the same "statistical control" strategy explained in the previous
chapter.

But why use a line to relate body size to brain size? It's not clear why nature demands
that the relationship among species be a straight line. Why not consider a curved model,
like a parabola? Indeed, why not a cubic function of body size, or even a quintic model? I
agree that there's no reason yet given to suppose *a priori* that brain size scales only linearly
with body size. Indeed, many readers will prefer to model a linear relationship between log
brain volume and log body mass (an exponential relationship). But that's not the direction
I'm headed with this example. The lesson here will arise, no matter how we transform the
data. Even after a log transform of both variables, there's no reason to insist that linear is the
only imaginable relationship.

Let's fit a series of increasingly complex model families and see which function fits the
data best. We're going to use lm to fit these models, instead of using map. Take a look back at
Section 5.5 (page 159), if you've forgotten or missed how these two methods of fitting linear
models relate to one another. In this case, using lm will get us expediently to the lesson of this
section. We'll also use polynomial regressions, so review Section 4.5 (page 110) if necessary.
Importantly, recall that polynomial regressions are common, but usually a bad idea. In this
example, I will show you that they are a very bad idea.

The simplest model that relates brain size to body size is the linear one. It will be the first
model we consider:

$$v_i \sim \text{Normal}(\mu_i, \sigma)$$
$$\mu_i = \alpha + \beta_1 m_i$$

This is just to say that the average brain volume v_i of species i is a linear function of its body
mass m_i. Priors are necessarily flat here, since we're using lm. Now fit this model to the data
using lm:

```
m6.1 <- lm( brain ~ mass , data=d )
```

R code
6.2

Instead of pausing to plot the fit model, like we did in previous chapters, let's focus on the R^2,
the proportion of variance "explained" by the model. What is really meant here is that the
linear model retrodicts some proportion of the total variation in the outcome data it was fit
to. The remaining variation is just the variation of the residuals (page 127). You can easily
compute R^2 yourself with:

R code
6.3

```
1 - var(resid(m6.1))/var(d$brain)
```

```
[1] 0.490158
```

Take a look at the bottom of the output from summary(m6.1) and you'll find the same number, labeled there "Multiple R-squared."

Let's get some other models to compare to the fit of m6.1. We'll consider five more models, each more complex than the last. Each of these models will just be a polynomial of higher degree. For example, a second-degree polynomial that relates body size to brain size is a parabola. In math form, it is:

$$v_i \sim \text{Normal}(\mu_i, \sigma)$$
$$\mu_i = \alpha + \beta_1 m_i + \beta_2 m_i^2$$

This model family adds one more parameter, β_2, but uses all of the same data as m6.1. Fit this model to the data:

R code
6.4

```
m6.2 <- lm( brain ~ mass + I(mass^2) , data=d )
```

Look back at page 161, if that I(mass^2) thing confuses you.

Now let's fit the rest of the model families. The models m6.3 through m6.6 are just third-degree, fourth-degree, fifth-degree, and sixth-degree polynomials built and fit in the same way. Here is the code to fit all of them to the data:

R code
6.5

```
m6.3 <- lm( brain ~ mass + I(mass^2) + I(mass^3) , data=d )
m6.4 <- lm( brain ~ mass + I(mass^2) + I(mass^3) + I(mass^4) ,
    data=d )
m6.5 <- lm( brain ~ mass + I(mass^2) + I(mass^3) + I(mass^4) +
    I(mass^5) , data=d )
m6.6 <- lm( brain ~ mass + I(mass^2) + I(mass^3) + I(mass^4) +
    I(mass^5) + I(mass^6) , data=d )
```

In FIGURE 6.3, I display the estimates of the average brain size as a function of body size, for all six of the models. Each plot overlays the predicted mean, by extracting the coefficients and plugging them into the formula for μ_i, just like all the previous examples. Each plot also displays R^2 for the fit model. As the degree of the polynomial defining the mean increases, the fit always improves. The fifth-degree polynomial has an R^2 value of 0.99! It almost passes exactly through each point. The sixth-degree polynomial actually does pass through every point, and it has no residual variance. It's a perfect fit, $R^2 = 1$.

However, you can see from looking at the paths of the predicted means that the higher-degree polynomials are increasingly absurd. This absurdity is seen most easily in FIGURE 6.3, panel (f), which shows the most complex model, m6.6. The fit is perfect, but the model is ridiculous. Notice that there is a gap in the body mass data, because there are no fossil hominins with body mass between 55 kg and about 60 kg. In this region, the predicted mean brain size from the high-degree polynomial models has nothing to predict, and so the models pay no price for swinging around wildly in this interval. The swing is so extreme that I had to extend the range of the vertical axis to display the depth at which the predicted mean finally turns back around. At around 58 kg, the model predicts a negative brain size!

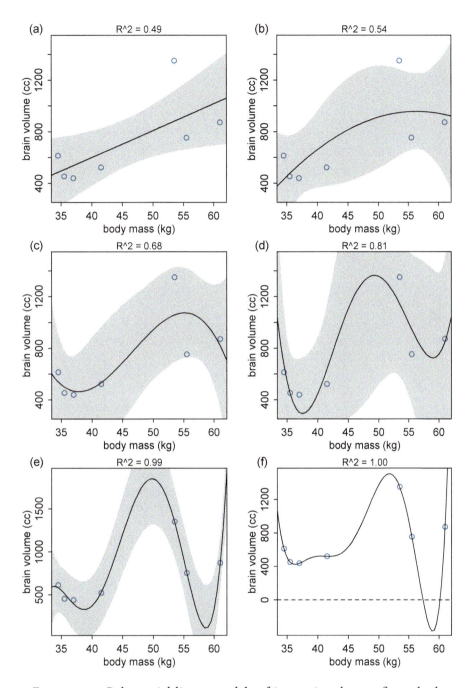

FIGURE 6.3. Polynomial linear models of increasing degree, fit to the hominin data. Each plot shows the predicted mean in black, with 89% interval of the mean shaded. R^2, is displayed above each plot. (a) First-degree polynomial. (b) Second-degree. (c) Third-degree. (d) Fourth-degree. (e) Fifth-degree. (f) Sixth-degree.

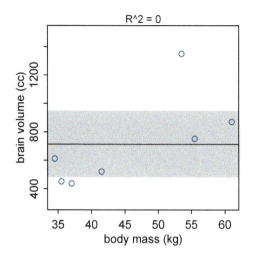

FIGURE 6.4. An underfit model of hominin brain volume. This model ignores any association between body mass and brain volume, producing a horizontal line of predictions. As a result, the model fits badly and (presumably) predicts badly.

The model pays no price (yet) for this absurdity, because there are no cases in the data with body mass near 58 kg.

Why does the sixth-degree polynomial fit perfectly? Because it has enough parameters to assign one to each point of data. The model's equation for the mean has 7 parameters:

$$\mu_i = \alpha + \beta_1 m_i + \beta_2 m_i^2 + \beta_3 m_i^3 + \beta_4 m_i^4 + \beta_5 m_i^5 + \beta_6 m_i^6,$$

and there are 7 species to predict brain sizes for. So effectively, this model assigns a unique parameter to reiterate each observed brain size. This is a general phenomenon: If you adopt a model family with enough parameters, you can fit the data exactly. But such a model will make rather absurd predictions for yet-to-be-observed cases.

Rethinking: Model fitting as compression. Another perspective on the absurd model just above is to consider that model fitting can be considered a form of DATA COMPRESSION. Parameters summarize relationships among the data. These summaries compress the data into a simpler form, although with loss of information ("lossy" compression) about the sample. The parameters can then be used to generate new data, effectively decompressing the data.

When a model has a parameter to correspond to each datum, such as m6.6, then there is actually no compression. The model just encodes the raw data in a different form, using parameters instead. As a result, we learn nothing about the data from such a model. Learning about the data requires using a simpler model that achieves some compression, but not too much. This view of model selection is often known as MINIMUM DESCRIPTION LENGTH (MDL).[86]

6.1.2. Too few parameters hurts, too. The overfit polynomial models manage to fit the data extremely well, but they suffer for this within-sample accuracy by making nonsensical out-of-sample predictions. In contrast, UNDERFITTING produces models that are inaccurate both within and out of sample. They have learned too little, failing to recover regular features of the sample.

For example, consider this model of brain volume:

$$v_i \sim \text{Normal}(\mu, \sigma)$$
$$\mu = \alpha$$

There are no predictor variables here, just the intercept α. Fit this model with:

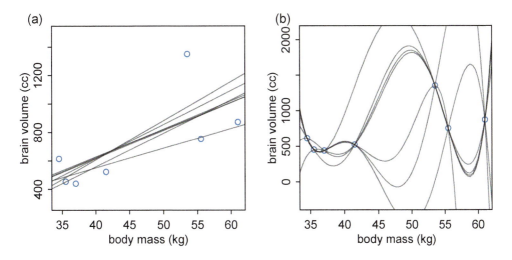

FIGURE 6.5. Underfitting and overfitting as under-sensitivity and over-sensitivity to sample. In both plots, a regression is fit to the seven sets of data made by dropping one row from the original data. (a) An underfit model is insensitive to the sample, changing little as individual points are dropped. (b) An overfit model is sensitive to the sample, changing dramatically as points are dropped.

```
m6.7 <- lm( brain ~ 1 , data=d )
```

R code
6.6

I display this fit in FIGURE 6.4. This model estimates the mean brain volume, ignoring body mass. As a result, the regression line is perfectly horizontal and poorly fits both smaller and larger brain volumes. The confidence region (shaded zone) for the mean is also quite broad, owing to the poor fit of the line to the data. Such a model not only fails to describe the sample. It would also do a poor job for new data.

Another way to conceptualize an underfit model is to notice that it is insensitive to the sample. We could remove any one point from the sample and get pretty much the same regression line. In contrast, the most complex model, m6.6, is very sensitive to the sample. The predicted mean would change course a lot, if we removed any one point from the sample. You can see the truth of this in FIGURE 6.5. In both plots in the figure what I've done is drop each row of the data, one at a time, and refit the model. So each line plotted in (a) is a first-degree polynomial fit to one of the seven possible sets of data constructed from dropping one row. The curves in (b) are instead different fifth-order polynomials fit to the same seven sets of data. Notice that the straight lines hardly vary, while the curves fly about wildly. This is a general contrast between underfit and overfit models: sensitivity to the exact composition of the sample used to fit the model.

Overthinking: Dropping rows. The calculations needed to produce FIGURE 6.5 are made easy by a trick of R's index notation. To drop a row i from a data frame d, just use:

R code
6.7
```
d.new <- d[ -i , ]
```

This means *drop the i-th row and keep all of the columns*. Repeating the regression is then just a matter of looping over the rows, like this:

R code
6.8
```
plot( brain ~ mass , d , col="slateblue" )
for ( i in 1:nrow(d) ) {
    d.new <- d[ -i , ]
    m0 <- lm( brain ~ mass, d.new )
    abline( m0 , col=col.alpha("black",0.5) )
}
```

The same trick works for the overfit model, but you'll have to use another technique to draw the fit. abline only understands bivariate regressions.

Rethinking: Bias and variance. The underfitting/overfitting dichotomy is often described as the BIAS-VARIANCE TRADE-OFF.[87] While not exactly the same distinction, the bias-variance trade-off addresses the same problem. "Bias" is related to underfitting, while "variance" is related to overfitting. These terms are confusing though, because they are used in many different ways in different contexts, even within statistics. The term "bias" also sounds like a bad thing, even though increasing bias often leads to better predictions. For these reasons, this book prefers *underfitting/overfitting*, but you should expect to see the same concepts discussed as *bias/variance*.

6.2. Information theory and model performance

So how do we navigate between the hydra of overfitting and the vortex of underfitting? Whether you end up using regularization or information criteria or both, the first thing you must do is pick a criterion of model performance. What do you want the model to do well at? We'll call this criterion the *target*, and in this section you'll see how information theory provides a common and useful target, the out-of-sample *deviance*.

The path to out-of-sample deviance is twisty, however. Here are the steps ahead. First, we need to establish that joint probability, not average probability, is the right way to judge model accuracy. Second, we need to establish a measurement scale for distance from perfect accuracy. This will require a little *information theory*, as it will provide a natural measurement scale for the distance between two probability distributions. Third, we need to establish *deviance* as an approximation of relative distance from perfect accuracy. Finally, we must establish that it is only deviance out-of-sample that is of interest.

Once you have deviance in hand as a measure model performance, in the sections to follow you'll see how both regularizing priors and information criteria help you improve and estimate the out-of-sample deviance of a model.

6.2.1. Firing the weatherperson.
Accuracy depends upon the definition of the target, and there is no unique best target. In defining a target, there are two major dimensions to worry about:

(1) *Cost-benefit analysis.* How much does it cost when we're wrong? How much do we win when we're right? Most scientists never ask these questions in any formal way, but applied scientists must routinely answer them.

(2) *Accuracy in context.* Some prediction tasks are inherently easier than others. So even if we ignore costs and benefits, we still need a way to judge "accuracy" that accounts for how much a model could possibly improve prediction.

It will help to explore these two dimensions in an example. Suppose in a certain city, a certain weatherperson issues uncertain predictions for rain or shine on each day of the year.[88] The predictions are in the form of probabilities of rain. The currently employed weatherperson predicted these chances of rain over a 10-day sequence, with the actual outcomes shown below each prediction:

Day	1	2	3	4	5	6	7	8	9	10
Prediction	1	1	1	0.6	0.6	0.6	0.6	0.6	0.6	0.6
Observed	☁	☁	☁	☀	☀	☀	☀	☀	☀	☀

A newcomer rolls into town, and this newcomer boasts that he can best the current weatherperson, by always predicting sunshine. Over the same 10 day period, the newcomer's record would be:

Day	1	2	3	4	5	6	7	8	9	10
Prediction	0	0	0	0	0	0	0	0	0	0
Observed	☁	☁	☁	☀	☀	☀	☀	☀	☀	☀

"So by rate of correct prediction alone," the newcomer announces, "I'm the best person for the job."

The newcomer is right. Define *hit rate* as the average chance of a correct prediction. So for the current weatherperson, she gets $3 \times 1 + 7 \times 0.4 = 5.8$ hits in 10 days, for a rate of $5.8/10 = 0.58$ correct predictions per day. In contrast, the newcomer gets $3 \times 0 + 7 \times 1 = 7$, for $7/10 = 0.7$ hits per day. The newcomer wins.

6.2.1.1. *Costs and benefits.* But it's not hard to find another criterion, other than rate of correct prediction, that makes the newcomer look foolish. Any consideration of costs and benefits will suffice. Suppose for example that you hate getting caught in the rain, but you also hate carrying an umbrella. Let's define the cost of getting wet as -5 points of happiness and the cost of carrying an umbrella as -1 point of happiness. Suppose your chance of carrying an umbrella is equal to the forecast probability of rain. Your job is now to maximize your happiness by choosing a weatherperson. Here are your points, following either the current weatherperson or the newcomer:

Day	1	2	3	4	5	6	7	8	9	10
Observed	☁	☁	☁	☀	☀	☀	☀	☀	☀	☀
Points										
Current	−1	−1	−1	−0.6	−0.6	−0.6	−0.6	−0.6	−0.6	−0.6
Newcomer	−5	−5	−5	0	0	0	0	0	0	0

So the current weatherperson nets you $3 \times (-1) + 7 \times (-0.6) = -7.2$ happiness, while the newcomer nets you -15 happiness. So the newcomer doesn't look so clever now. You can play around with the costs and the decision rule, but since the newcomer always gets you caught unprepared in the rain, it's not hard to beat his forecast.

6.2.1.2. *Measuring accuracy.* But even if we ignore costs and benefits of any actual decision based upon the forecasts, there's still ambiguity about which measure of "accuracy" to adopt. There's nothing special about "hit rate." Consider for example computing the probability of predicting the exact sequence of days. This means computing the probability of a correct prediction for each day. Then multiply all of these probabilities together to get the joint probability of correctly predicting the observed sequence. This is the same thing as the joint likelihood, which you've been using up to this point to fit models with Bayes' theorem.

In this light, the newcomer looks even worse. The probability for the current weatherperson is $1^3 \times 0.4^7 \approx 0.005$. For the newcomer, it's $0^3 \times 1^7 = 0$. So the newcomer has zero probability of getting the sequence correct. This is because the newcomer's predictions never expect rain. So even though the newcomer has a high *average* probability of being correct (hit rate), he has a terrible *joint* probability of being correct.

And the joint probability is the measure we want. Why? Because it appears in Bayes' theorem as the likelihood. It's the unique measure that correctly counts up the relative number of ways each event (sequence of rain and shine) could happen. Another way to think of this is to consider what happens when we maximize average probability or joint probability. Maximizing average probability will not also identify the right model. You saw this already with the weatherperson: Assigning zero probability to rain improves hit rate, but it is clearly wrong. In contrast, maximizing joint probability will identify the right model.

But how should we measure distance from the target? A perfect prediction would just report the true probabilities of rain on each day. So when either weatherperson provides a prediction that differs from the target, we can measure the distance of the prediction from the target. But what distance should we adopt?

It's not obvious how to go about answering this question. One reason is that some targets are just easier to hit than other targets. We need a measure of distance that accounts for this. For example, suppose we extend the weather forecast into the winter. Now there are three types of days: rain, sun, and snow. Now there are three ways to be wrong, instead of just two. This has to be reflected in any reasonable measure of distance from the target, because by adding another type of event, the target has gotten harder to hit.

It's like taking a two-dimensional archery bullseye and forcing the archer to hit the target at the right *time*—a third dimension—as well. Now the possible distance between the best archer and the worst archer has grown, because there's another way to miss. And with another way to miss, one might also say that there is another way for an archer to impress. As the potential distance between the target and the shot increases, so too does the potential improvement and ability of a talented archer to impress us.

Rethinking: What is a true model? It's hard to define "true" probabilities, because all models are false. So what does "truth" mean in this context? It means the right probabilities, given our state of ignorance. Our state of ignorance is described by the model. The probability is in the model, not in the world. If we had all of the information relevant to producing a forecast, then rain or sun would be deterministic, and the "true" probabilities would be just 0's and 1's. Absent some relevant information, as in all modeling, outcomes in the small world are uncertain, even though they remain perfectly deterministic in the large world. Because of our ignorance, we can have "true" probabilities between zero and one.

An example might help. Suppose you toss the globe, as in Chapter 2. Before you catch it, the outcome is uncertain. There is a "true" probability of observing *water*, conditional on our assumed model. But if we had enough information about the globe toss—initial conditions, angular momentum vector, and such—then the outcome would be knowable with certainty. No two tosses are ever exactly alike, so the "true" probability of observing *water* must average over the unknown differences to describe the relative plausibility of *water* compared to *land*. There is a right answer to the question this model poses—70% water. It's our ignorance of the physics of the globe toss that leads us to use it as a way to estimate the amount of water on the surface.

6.2.2. Information and uncertainty. One solution to the problem of how to measure distance of a model's accuracy from a target was provided in the late 1940s.[89] Originally applied to problems in communication of messages, such as telegraph, the field of INFORMATION THEORY is now important across the basic and applied sciences, and it has deep connections to Bayesian inference. And like many successful fields, information theory has spawned a large number of bogus applications, as well.[90]

The basic insight is to ask: *How much is our uncertainty reduced by learning an outcome?* Consider the weather forecasts again. Forecasts are issued in advance and the weather is uncertain. When the actual day arrives, the weather is no longer uncertain. The reduction in uncertainty is then a natural measure of how much we have learned, how much "information" we derive from observing the outcome. So if we can develop a precise definition of "uncertainty," we can provide a baseline measure of how hard it is to predict, as well as how much improvement is possible. The measured decrease in uncertainty is the definition of *information* in this context.

> *Information*: The reduction in uncertainty derived from learning an outcome.

To use this definition, what we need is a principled way to quantify the uncertainty inherent in a probability distribution. So suppose again that there are two possible weather events on any particular day: Either it is sunny or it is rainy. Each of these events occurs with some probability, and these probabilities add up to one. What we want is a function that uses the probabilities of shine and rain and produces a measure of uncertainty.

There are many possible ways to measure uncertainty. The most common way begins by naming some properties a measure of uncertainty should possess. These are the three intuitive desiderata:

(1) The measure of uncertainty should be continuous. If it were not, then an arbitrarily small change in any of the probabilities, for example the probability of rain, would result in a massive change in uncertainty.

(2) The measure of uncertainty should increase as the number of possible events increases. For example, suppose there are two cities that need weather forecasts. In the first city, it rains on half of the days in the year and is sunny on the others. In the second, it rains, shines, and hails, each on 1 out of every 3 days in the year. We'd like our measure of uncertainty to be larger in the second city, where there is one more kind of event to predict.

(3) The measure of uncertainty should be additive. What this means is that if we first measure the uncertainty about rain or shine (2 possible events) and then the uncertainty about hot or cold (2 different possible events), the uncertainty over the four combinations of these events—rain/hot, rain/cold, shine/hot, shine/cold—should be the sum of the separate uncertainties.

There is only one function that satisfies these desiderata. This function is usually known as INFORMATION ENTROPY, and has a surprisingly simple definition. If there are n different possible events and each event i has probability p_i, and we call the list of probabilities p, then the unique measure of uncertainty we seek is:

$$H(p) = -\operatorname{E}\log(p_i) = -\sum_{i=1}^{n} p_i \log(p_i) \tag{6.1}$$

In plainer words:

> The uncertainty contained in a probability distribution is the average log-probability of an event.

"Event" here might refer to a type of weather, like rain or shine, or a particular species of bird or even a particular nucleotide in a DNA sequence.

While it's not worth going into the details of the derivation of H, it is worth pointing out that nothing about this function is arbitrary. Every part of it derives from the three requirements above. Still, we accept $H(p)$ as a useful measure of uncertainty not because of the premises that lead to it, but rather because it has turned out to be so useful and productive.

An example will help to demystify the function $H(p)$. To compute the information entropy for the weather, suppose the true probabilities of rain and shine are $p_1 = 0.3$ and $p_2 = 0.7$, respectively. Then:

$$H(p) = -\bigl(p_1 \log(p_1) + p_2 \log(p_2)\bigr) \approx 0.61$$

As an R calculation:

R code
6.9
```
p <- c( 0.3 , 0.7 )
-sum( p*log(p) )
```

```
[1] 0.6108643
```

Suppose instead we live in Abu Dhabi. Then the probabilities of rain and shine might be more like $p_1 = 0.01$ and $p_2 = 0.99$. Now the entropy would be approximately 0.06. Why has the uncertainty decreased? Because in Abu Dhabi it hardly ever rains. Therefore there's much less uncertainty about any given day, compared to a place in which it rains 30% of the time. It's in this way that information entropy measures the uncertainty inherent in a distribution of events. Similarly, if we add another kind of event to the distribution—forecasting into winter, so also predicting snow—entropy tends to increase, due the added dimensionality of the prediction problem. For example, suppose probabilities of sun, rain, and snow are $p_1 = 0.7$, $p_2 = 0.15$, and $p_3 = 0.15$, respectively. Then entropy is about 0.82.

These entropy values by themselves don't mean much to us, though. Instead we can use them to build a measure of accuracy. That comes next.

Overthinking: More on entropy. Above I said that information entropy is the average log-probability. But there's also a -1 in the definition. Multiplying the average log-probability by -1 just makes the entropy H increase from zero, rather than decrease from zero. It's conventional, but not functional.

The logarithms above are natural logs (base e), but changing the base rescales without any effect on inference. Binary logarithms, base 2, are just as common. As long as all of the entropies you compare use the same base, you'll be fine.

The only trick in computing H is to deal with the inevitable question of what to do when $p_i = 0$. The $\log(0) = -\infty$, which won't do. However, L'Hôpital's rule tells us that $\lim_{p_i \to 0} p_i \log(p_i) = 0$. So just assume that $0 \log(0) = 0$, when you compute H. In other words, events that never happen drop out. This is not really a trick. It follows from the definition of a limit. But it isn't obvious. It may make more sense to just remember that when an event never happens, there's no point in keeping it in the model.

Rethinking: The benefits of maximizing uncertainty. Information theory has many applications. A particularly important application is MAXIMUM ENTROPY, also known as MAXENT. Maximum entropy is a family of techniques for finding probability distributions that are most consistent with states of knowledge. In other words, given what we know, what is the *least surprising* distribution? It turns out that one answer to this question maximizes the information entropy, using the prior knowledge as constraint.[91] Maximum entropy features prominently in Chapter 9, where it will help us build generalized linear models (GLMs).

6.2.3. From entropy to accuracy.

It's nice to have a way to quantify uncertainty. H provides this. So we can now say, in a precise way, how hard it is to hit the target. But how can we use information entropy to say how far a model is from the target? The key lies in DIVERGENCE:

> **Divergence:** The additional uncertainty induced by using probabilities from one distribution to describe another distribution.

This is often known as *Kullback-Leibler divergence* or simply K-L divergence, named after the people who introduced it for this purpose.[92]

Suppose for example that the true distribution of events is $p_1 = 0.3, p_2 = 0.7$. If we believe instead that these events happen with probabilities $q_1 = 0.25, q_2 = 0.75$, how much additional uncertainty have we introduced, as a consequence of using $q = \{q_1, q_2\}$ to approximate $p = \{p_1, p_2\}$? The formal answer to this question is based upon H, and has a similarly simple formula:

$$D_{KL}(p, q) = \sum_i p_i \big(\log(p_i) - \log(q_i) \big) = \sum_i p_i \log \left(\frac{p_i}{q_i} \right)$$

In plainer language, the divergence is *the average difference in log probability between the target (p) and model (q)*. This divergence is just the difference between two entropies: The entropy of the target distribution p and the *cross entropy* arising from using q to predict p (see the Overthinking box on the next page for some more detail). When $p = q$, we know the actual probabilities of the events. In that case:

$$D_{KL}(p, q) = D_{KL}(p, p) = \sum_i p_i \big(\log(p_i) - \log(p_i) \big) = 0$$

There is no additional uncertainty induced when we use a probability distribution to represent itself. That's somehow a comforting thought.

But more importantly, as q grows more different from p, the divergence D_{KL} also grows. FIGURE 6.6 displays an example. Suppose the true target distribution is $p = \{0.3, 0.7\}$. Suppose the approximating distribution q can be anything from $q = \{0.01, 0.99\}$ to $q = \{0.99, 0.01\}$. The first of these probabilities, q_1, is displayed on the horizontal axis, and the

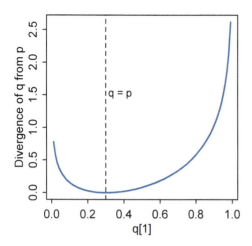

FIGURE 6.6. Information divergence of an approximating distribution q from a true distribution p. Divergence can only equal zero when $q = p$ (dashed line). Otherwise, the divergence is positive and grows as q becomes more dissimilar from p. When we have more than one candidate approximation q, the q with the smallest divergence is the most accurate approximation, in the sense that it induces the least additional uncertainty.

vertical displays the divergence $D_{KL}(p, q)$. Only exactly where $q = p$, at $q_1 = 0.3$, does the divergence achieve a value of zero. Everyplace else, it grows.

What divergence can do for us now is help us contrast different approximations to p. As an approximating function q becomes more accurate, $D_{KL}(p, q)$ will shrink. So if we have a pair of candidate distributions, then the candidate that minimizes the divergence will be closest to the target. Since predictive models specify probabilities of events (observations), we can use divergence to compare the accuracy of models.

Overthinking: Cross entropy and divergence. Deriving divergence is easier than you might think. The insight is in realizing that when we use a probability distribution q to predict events from another distribution p, this defines something known as *cross entropy*: $H(p, q) = -\sum_i p_i \log(q_i)$. The notion is that events arise according the the p's, but they are expected according to the q's, so the entropy is inflated, depending upon how different p and q are. Divergence is defined as the *additional* entropy induced by using q. So it's just the difference between $H(p)$, the actual entropy of events, and $H(p, q)$:

$$D_{KL}(p, q) = H(p, q) - H(p)$$

$$= -\sum_i p_i \log(q_i) - \left(-\sum_i p_i \log(p_i) \right) = -\sum_i p_i \left(\log(q_i) - \log(p_i) \right)$$

So divergence really is measuring how far q is from the target p, in units of entropy. Notice that which is the target matters: $H(p, q)$ does not in general equal $H(q, p)$. For more on that fact, see the Rethinking box that follows.

Rethinking: Divergence depends upon direction. In general, $H(p, q)$ is not equal to $H(q, p)$. The direction matters, when computing divergence. Understanding why this is true is of some value, so here's a contrived teaching example.

Suppose we get in a rocket and head to Mars. But we have no control over our landing spot, once we reach Mars. Let's try to predict whether we land in water or on dry land, using the Earth to provide a probability distribution q to approximate the actual distribution on Mars, p. For the Earth, $q = \{0.7, 0.3\}$, for probability of water and land, respectively. Mars is very dry, but let's say for the sake of the example that there is 1% surface water, so $p = \{0.01, 0.99\}$. If we count the ice caps, that's not too big a lie. Now compute the divergence going from Earth to Mars. It turns out to be

$D_{E \to M} = D_{KL}(p, q) = 1.14$. That's the additional uncertainty induced by using the Earth to predict the Martian landing spot. Now consider going back the other direction. The numbers in p and q stay the same, but we swap their roles, and now $D_{M \to E} = D_{KL}(q, p) = 2.62$. The divergence is more than double in this direction. This result seems to defy comprehension. How can the distance from Earth to Mars be shorter than the distance from Mars to Earth?

Divergence behaves this way as a feature, not a bug. There really is more additional uncertainty induced by using Mars to predict Earth than by using Earth to predict Mars. The reason is that, going from Mars to Earth, Mars has so little water on its surface that we will be very very surprised when we most likely land in water on Earth. In contrast, Earth has good amounts of both water and dry land. So when we use the Earth to predict Mars, we expect both water and land, to some extent, even though we do expect more water than land. So we won't be nearly as surprised when we inevitably arrive on Martian dry land, because 30% of Earth is dry land.

An important practical consequence of this asymmetry, in a model fitting context, is that if we use a distribution with high entropy to approximate an unknown true distribution of events, we will reduce the distance to the truth and therefore the error. This fact will help us build generalized linear models, later on in Chapter 9.

6.2.4. From divergence to deviance.

At this point in the chapter, dear reader, you may be wondering where the chapter is headed. At the start, the goal was to deal with overfitting and underfitting. But now we've spent pages and pages on entropy and other fantasies. It's as if I promised you a day at the beach, but now you find yourself at a dark cabin in the woods, wondering if this is a necessary detour or rather a sinister plot.

It is a necessary detour. The point of all the preceding material about information theory and divergence is to establish both:

(1) How to measure the distance of a model from our target. Information theory gives us the distance measure we need, the K-L divergence.

(2) How to estimate the divergence. Having identified the right measure of distance, we now need a way to estimate it in real statistical modeling tasks.

Item (1) is accomplished. Item (2) remains for last. You're going to see now that the divergence leads to using a measure of model fit known as *deviance*.

To use D_{KL} to compare models, it seems like we would have to know p, the target probability distribution. In all of the examples so far, I've just assumed that p is known. But when we want to find a model q that is the best approximation to p, the "truth," there is usually no way to access p directly. We wouldn't be doing statistical inference, if we already knew p.

But there's an amazing way out of this predicament. It helps that we are only interested in comparing the divergences of different candidates, say q and r. In that case, most of p just subtracts out, because there is a $E \log(p_i)$ term in the divergence of both q and r. This term has no effect on the distance of q and r from one another. So while we don't know where p is, we can estimate how far apart q and r are, and which is closer to the target. It's as if we can't tell how far any particular archer is from hitting the target, but we can tell which archer gets closer and by how much.

All of this also means that all we need to know is a model's average log-probability: $E \log(q_i)$ for q and $E \log(r_i)$ for r. These expressions look a lot like log-probabilities of outcomes you've been using already to simulate implied predictions of a fit model. Indeed, just summing the log-probabilities of each observed case provides an approximation of $E \log(q_i)$.

We don't have to know the p inside the expectation, because nature takes care of presenting the events for us.

So we can compare the average log-probability from each model to get an estimate of the relative distance of each model from the target. This also means that the absolute magnitude of these values will not be interpretable—neither $E \log(q_i)$ nor $E \log(r_i)$ by itself suggests a good or bad model. Only the difference $E \log(q_i) - E \log(r_i)$ informs us about the divergence of each model from the target p.

All of this delivers us to a very common measure of *relative* model fit, one that also turns out to be an approximation of K-L divergence. To approximate the relative value of $E \log(q_i)$, we can use a model's DEVIANCE, which is defined as:

$$D(q) = -2 \sum_i \log(q_i)$$

where i indexes each observation (case), and each q_i is just the likelihood of case i. The -2 in front doesn't do anything important. It's there for historical reasons.[93] Note that deviance is not divided by the number of cases, to make it an average, but rather just summed up across all cases. This doesn't change the underlying relationship to K-L divergence, but it does make it scale with sample size.

You can compute the deviance for any model you've fit already in this book, just by using the MAP estimates to compute a log-probability of the observed data for each row. These probabilities are the q values. Then you add these log-probabilities together and multiply by -2. In many cases, R automates these steps. Most of the standard model fitting functions support logLik, which will do the hard part: compute the sum of log-probabilities, usually known as the log-likelihood of the data. For example:

R code
6.10

```
# fit model with lm
m6.1 <- lm( brain ~ mass , d )

# compute deviance by cheating
(-2) * logLik(m6.1)
```

```
'log Lik.' 94.92499 (df=3)
```

To see how to do the raw calculations yourself, see the Overthinking box below. But note that, because there is uncertainty about the parameters, there is also uncertainty about the deviance of a model. For any specific parameter values, deviance is defined exactly. But since we have a posterior distribution of parameter values, there is also a posterior distribution of the deviance.

Overthinking: Computing deviance. Here's a quick example, using the hominin brain data again.

R code
6.11

```
# standardize the mass before fitting
d$mass.s <- (d$mass-mean(d$mass))/sd(d$mass)
m6.8 <- map(
    alist(
        brain ~ dnorm( mu , sigma ) ,
        mu <- a + b*mass.s
    ) ,
    data=d ,
    start=list(a=mean(d$brain),b=0,sigma=sd(d$brain)) ,
```

```
        method="Nelder-Mead" )

# extract MAP estimates
theta <- coef(m6.8)

# compute deviance
dev <- (-2)*sum( dnorm(
            d$brain ,
            mean=theta[1]+theta[2]*d$mass.s ,
            sd=theta[3] ,
            log=TRUE ) )
dev
```

```
[1] 94.92704
```

That's the same result you'd get with -2*logLik(m6.8). All that's really required is to plug the MAP estimates into the likelihood function. You compute the log-likelihood for each observation. Then you add them all together. Finally, multiply the sum by -2. That yields the deviance. R's logLik function does everything except multiply by -2.

6.2.5. From deviance to out-of-sample. Deviance is a principled way to measure distance from the target. But deviance as computed in the previous section has the same flaw as R^2: It always improves as the model gets more complex, at least for the types of models we have considered so far. Just like R^2, deviance in-sample is a measure of retrodictive accuracy, not predictive accuracy. It is really the deviance on new data that interests us. So before looking at regularization and information criteria as tools for improving and measuring out-of-sample deviance, let's bring the problem into sharper focus by simulating deviance both in and out of sample.

When we usually have data and use it to fit a statistical model, the data comprise a *training sample*. Parameters are estimated from it, and then we can imagine using those estimates to predict outcomes in a new sample, called the *test sample*. R is going to do all of this for you. But here's the full procedure, in outline:

(1) Suppose there's a training sample of size N.
(2) Fit a model to the training sample, and compute the deviance on the training sample. Call this deviance D_{train}.
(3) Suppose another sample of size N from the same process. This is the test sample.
(4) Compute the deviance on the test sample. This means using the MAP estimates from step (2) to compute the deviance for the data in the test sample. Call this deviance D_{test}.

The above is a thought experiment. It allows us to explore the distinction between deviance measured in and out of sample, using a simple prediction scenario.

To visualize the results of the thought experiment, what we'll do now is conduct the above thought experiment 10,000 times, for each of five different linear regression models. The model that generates the data is:

$$y_i \sim \text{Normal}(\mu_i, 1)$$
$$\mu_i = (0.15)x_{1,i} - (0.4)x_{2,i}$$

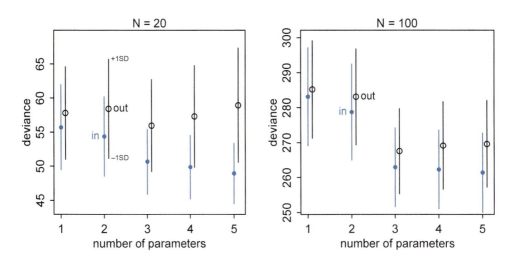

FIGURE 6.7. Deviance in and out of sample. In each plot, models with different numbers of predictor variables are shown on the horizontal axis. Deviance across 10,000 simulations is shown on the vertical. Blue shows deviance in-sample, the training data. Black shows deviance out-of-sample, the test data. Points show means, and the line segments show ±1 standard deviation.

This corresponds to a Gaussian outcome y for which the intercept is $\alpha = 0$ and the slopes for each of two predictors are $\beta_1 = 0.15$ and $\beta_2 = -0.4$. The models for analyzing the data are linear regressions with between 1 and 5 free parameters. The first model, with 1 free parameter to estimate, is just a linear regression with an unknown mean and fixed $\sigma = 1$. Each parameter added to the model adds a predictor variable and its beta-coefficient. Since the "true" model has non-zero coefficients for only the first two predictors, we can say that the true model has 3 parameters. By fitting all five models, with between 1 and 5 parameters, to training samples from the same processes, we can get an impression for how deviance behaves.

FIGURE 6.7 shows the results of 10,000 simulations for each model type, at two different sample sizes. The function that conducts the simulations is sim.train.test in the rethinking package. If you want to conduct more simulations of this sort, see the Overthinking box on the next page for the full code. In the left-hand plot in FIGURE 6.7, both training and test samples contain 20 cases. Blue points and line segments show the mean plus-and-minus one standard deviation of the deviance calculated on the training data. Moving left to right with increasing numbers of parameters, the average deviance declines. A smaller deviance means a better fit. So this decline with increasing model complexity is the same phenomenon you saw earlier in the chapter with R^2.

But now inspect the open points and black line segments. These display the distribution of out-of-sample deviance at each number of parameters. While the training deviance always gets better with an additional parameter, the test deviance is smallest on average for 3 parameters, which is the data-generating model in this case. The deviance out-of-sample gets worse (increases) with the addition of each parameter after the third. These additional parameters fit the noise in the additional predictors. So while deviance keeps improving (declining) in the training sample, it gets worse on average in the test sample. The right-hand plot shows the same relationships for larger samples of $N = 100$ cases.

The size of the standard deviation bars may surprise you. While it is always true on average that deviance out-of-sample is worse than deviance in-sample, any individual pair of train and test samples may reverse the expectation. The reason is that any given training sample may be highly misleading. And any given testing sample may be unrepresentative. Keep this fact in mind as we develop devices for comparing models, because this fact should prevent you from placing too much confidence in analysis of any particular sample. Like all of statistical inference, there are no guarantees here.

On that note, there is also no guarantee that the "true" data-generating model will have the smallest average out-of-sample deviance. You can see a symptom of this fact in the deviance for the 2 parameter model. That model does worse in prediction than the model with only 1 parameter, even though the true model does include the additional predictor. This is because with only $N = 20$ cases, the imprecision of the estimate for the first predictor produces more error than just ignoring it. In the right-hand plot, in contrast, there is enough data to precisely estimate the association between the first predictor and the outcome. Now the deviance for the 2 parameter model is better than that of the 1 parameter model.

Deviance is an assessment of predictive accuracy, not of truth. The true model, in terms of which predictors are included, is not guaranteed to produce the best predictions. Likewise a false model, in terms of which predictors are included, is not guaranteed to produce poor predictions.

The point of this thought experiment is to demonstrate how deviance behaves, in theory. While deviance on training data always improves with additional predictor variables, deviance on future data may or may not, depending upon both the true data-generating process and how much data is available to precisely estimate the parameters. These facts form the basis for understanding both regularizing priors and information criteria.

Overthinking: Simulated training and testing. To reproduce FIGURE 6.7, sim.train.test is run 10,000 (1e4) times for each of the 5 models. This code is sufficient to run all of the simulations:

```
N <- 20
kseq <- 1:5
dev <- sapply( kseq , function(k) {
        print(k);
        r <- replicate( 1e4 , sim.train.test( N=N, k=k ) );
        c( mean(r[1,]) , mean(r[2,]) , sd(r[1,]) , sd(r[2,]) )
    } )
```

R code
6.12

If you are Mac or Linux, you can parallelize the simulations by replacing the replicate line with:

```
    r <- mcreplicate( 1e4 , sim.train.test( N=N, k=k ) , mc.cores=4 )
```

R code
6.13

Set mc.cores to the number of processor cores you want to use for the simulations. Once the simulations complete, dev will be a 4-by-5 matrix of means and standard deviations. To reproduce the plot:

```
plot( 1:5 , dev[1,] , ylim=c( min(dev[1:2,])-5 , max(dev[1:2,])+10 ) ,
    xlim=c(1,5.1) , xlab="number of parameters" , ylab="deviance" ,
```

R code
6.14

```
    pch=16 , col=rangi2 )
mtext( concat( "N = ",N ) )
points( (1:5)+0.1 , dev[2,] )
for ( i in kseq ) {
    pts_in <- dev[1,i] + c(-1,+1)*dev[3,i]
    pts_out <- dev[2,i] + c(-1,+1)*dev[4,i]
    lines( c(i,i) , pts_in , col=rangi2 )
    lines( c(i,i)+0.1 , pts_out )
}
```

By altering this code, you can simulate many different train-test scenarios. See `?sim.train.test` for additional options.

6.3. Regularization

The root of overfitting is a model's tendency to get overexcited by the training sample. When the priors are flat or nearly flat, the machine interprets this to mean that every parameter value is equally plausible. As a result, the model returns a posterior that encodes as much of the training sample—as represented by the likelihood function—as possible.

One way to prevent a model from getting too excited by the training sample is to give it a skeptical prior. By "skeptical," I mean a prior that slows the rate of learning from the sample. The most common skeptical prior is a REGULARIZING PRIOR, which is applied to a beta-coefficient, a "slope" in a linear model. Such a prior, when tuned properly, reduces overfitting while still allowing the model to learn the regular features of a sample. If the prior is too skeptical, however, then regular features will be missed, resulting in underfitting. So the problem is really one of tuning. But as you'll see, even mild skepticism can help a model do better, and doing better is all we can really hope for in the large world, where no model nor prior is optimal.

For example, consider this Gaussian model:

$$y_i \sim \text{Normal}(\mu_i, \sigma)$$
$$\mu_i = \alpha + \beta x_i$$
$$\alpha \sim \text{Normal}(0, 100)$$
$$\beta \sim \text{Normal}(0, 1)$$
$$\sigma \sim \text{Uniform}(0, 10)$$

Assume, as is good practice, that the predictor x is standardized so that its standard deviation is 1 and its mean is zero. Then the prior on α is a nearly flat prior that has no practical effect on inference, as you've seen in earlier chapters.

But the prior on β is narrower and is meant to regularize. The prior $\beta \sim \text{Normal}(0, 1)$ says that, before seeing the data, the machine should be very skeptical of values above 2 and below -2, as a Gaussian prior with a standard deviation of 1 assigns only 5% plausibility to values above and below 2 standard deviations. Because the predictor variable x is standardized, you can interpret this as meaning that a change of 1 standard deviation in x is very unlikely to produce 2 units of change in the outcome.

You can visualize this prior in FIGURE 6.8 as the dashed curve. Since more probability is massed up around zero, estimates are shrunk towards zero—they are conservative. The other curves are narrower priors that are even more skeptical of parameter values far from

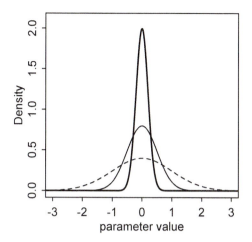

FIGURE 6.8. Regularizing priors, weak and strong. Three Gaussian priors of varying standard deviation. These priors reduce overfitting, but with different strength. Dashed: Normal(0, 1). Thin solid: Normal(0, 0.5). Thick solid: Normal(0, 0.2).

zero. The thin solid curve is a stronger Gaussian prior with a standard deviation of 0.5. The thick solid curve is even stronger, with a standard deviation of only 0.2.

How strong or weak these skeptical priors will be in practice depends upon the data and model. So let's explore a train-test example, similar to what you saw in the previous section (FIGURE 6.7). This time we'll use the regularizing priors pictured in FIGURE 6.8, instead of flat priors. For each of five different models, we simulate 10,000 times for each of the three regularizing priors above. FIGURE 6.9 shows the results. The points are the same flat-prior deviances as in the previous section: blue for training deviance and black for test deviance. The lines show the train and test deviances for the different priors. The blue lines are training deviance and the black lines test deviance. The style of the lines correspond to those in FIGURE 6.8.

Focus on the left-hand plot, where the sample size is $N = 20$, for the moment. The training deviance always increases—gets worse—with tighter priors. The thick blue trend is substantially larger than the others, and this is because the skeptical prior prevents the model from adapting completely to the sample. But the test deviances, out-of-sample, improve (get smaller) with the tighter priors. The model with three parameters is still the best model out-of-sample, and the regularizing priors have little impact on its deviance.

But also notice that as the prior gets more skeptical, the harm done by an overly complex model is greatly reduced. For the Normal(0, 0.2) prior (thick line), the models with 4 and 5 parameters are barely worse than the correct model with 3 parameters. If you can tune the regularizing prior right, then overfitting can be greatly reduced.

Now focus on the right-hand plot, where sample size is $N = 100$. The priors have much less of an effect here, because there is so much more evidence. The priors do help. But overfitting was less of a concern to begin with, and there is enough information in the data to overwhelm even the Normal(0, 0.2) prior (thick line).

Regularizing priors are great, because they reduce overfitting. But if they are too skeptical, they prevent the model from learning from the data. So to use them most effectively, you need some way to tune them. Tuning them isn't always easy. If you have enough data, you can split into "train" and "test" samples and then try different priors and select the one that provides the smallest deviance on the test sample. That is the essence of CROSS-VALIDATION, a common technique for reducing overfitting.

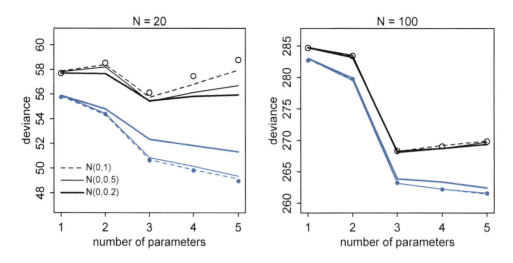

FIGURE 6.9. Regularizing priors and out-of-sample deviance. The points in both plots are the same as in FIGURE 6.7. The lines show training (blue) and testing (black) deviance for the three regularizing priors in FIGURE 6.8. Dashed: Each beta-coefficient is given a Normal$(0, 1)$ prior. Thin solid: Normal$(0, 0.5)$. Thick solid: Normal$(0, 0.2)$.

But if you need to use all of the data to train the model, tuning the prior may not be so easy. It would be nice to have a way to predict a model's out-of-sample deviance, to forecast its predictive accuracy, using only the sample at hand. That's for the next section.

Rethinking: Multilevel models as adaptive regularization. When you encounter multilevel models in Chapter 12, you'll see that their central device is to learn the strength of the prior from the data itself. So you can think of multilevel models as adaptive regularization, where the model itself tries to learn how skeptical it should be.

Rethinking: Ridge regression. Linear models in which the slope parameters use Gaussian priors, centered at zero, are sometimes known as **RIDGE REGRESSION**. Ridge regression typically takes as input a precision λ that essentially describes the narrowness of the prior. $\lambda > 0$ results in less overfitting. However, just as with the Bayesian version, if λ is too large, we risk underfitting.

While not originally developed as Bayesian, ridge regression is another example of how a statistical procedure can be understood from both Bayesian and non-Bayesian perspectives. Ridge regression does not compute a posterior distribution. Instead it uses a modification of OLS that stitches λ into the usual matrix algebra formula for the estimates. The function `lm.ridge`, built into R's MASS library, will fit linear models this way.

6.4. Information criteria

In previous sections, we've used simulations in which a model is fit to one sample, the training sample, and then used to predict a second sample of the same size, the testing sample. Let's look at one of those thought experiments again, but now annotated with the difference between the training deviance (in-sample) and the testing deviance (out-of-sample).

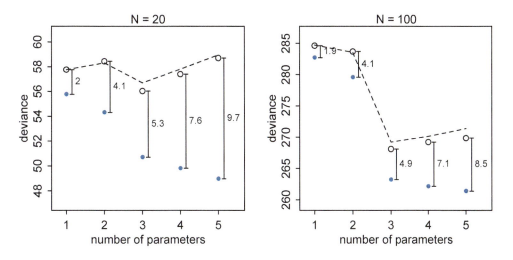

FIGURE 6.10. Deviance in (blue) and out (black) of sample, using flat priors. The vertical segments measure the distance between each pair of deviances. For both $N = 20$ and $N = 100$, this distance is approximately twice the number of parameters. The dashed lines show exactly the deviance insample (training) plus twice the number of parameters on the horizontal axis. These lines therefore show AIC for each model, an approximation of the out-of-sample deviance.

FIGURE 6.10 displays the same blue and black points as earlier in the chapter. But now the vertical line segments measure the average distance between training deviance (blue) and test deviance (open black). Notice that these distances are nearly the same, for each model, at both $N = 20$ (left) and $N = 100$ (right). Each distance is nearly twice the number of parameters, as labeled on the horizontal axis. The dashed lines show exactly the blue points plus twice the number of parameters, tracing closely along the average out-of-sample deviance for each model.

This is the phenomenon behind **INFORMATION CRITERIA**. The most known information criterion is the **AKAIKE INFORMATION CRITERION**, abbreviated **AIC**.[94] AIC provides a surprisingly simple estimate of the average out-of-sample deviance:

$$\text{AIC} = D_{\text{train}} + 2p$$

where p is the number of free parameters to be estimated in the model. This definition reflects the relationship between training and testing deviance in FIGURE 6.10. The dashed lines in FIGURE 6.10 trace out the AIC values of the models.

AIC provides an approximation of predictive accuracy, as measured by out-of-sample deviance. All information criteria aim at this same target, but are derived under more and less general assumptions. AIC is just the oldest and most restrictive. AIC is an approximation that is reliable only when:

(1) The priors are flat or overwhelmed by the likelihood.
(2) The posterior distribution is approximately multivariate Gaussian.
(3) The sample size N is much greater[95] than the number of parameters k.

Since flat priors are hardly ever the best priors, we'll want something more general.

So instead of spending any more time on AIC, this section instead focuses on two common and more-general criteria, DIC and WAIC. The DEVIANCE INFORMATION CRITERION (DIC) accommodates informative priors, but still assumes that the posterior is multivariate Gaussian and that $N \gg k$.[96] The WIDELY APPLICABLE INFORMATION CRITERION (WAIC) is more general yet, making no assumption about the shape of the posterior.[97] We'll take a quick look at both, before working a full data analysis example in the next section. In later chapters, you'll continue to use them.

> **Rethinking: Information criteria and consistency.** As mentioned previously, information criteria like AIC, DIC, and WAIC do not always assign the best expected D_{test} to the "true" model. In statistical jargon, information criteria are not CONSISTENT for model identification. These criteria aim to nominate the model that will produce the best predictions, as judged by out-of-sample deviance, so it shouldn't surprise us that they do not also do something that they aren't designed to do. Other metrics for model comparison are however consistent. So are information criteria broken?
>
> They are not broken, if you care about prediction. Issues like consistency are nearly always evaluated *asymptotically*. This means that we imagine the sample size N approaching infinity. Then we ask how a procedure behaves in this large-data limit. With practically infinite data, AIC/DIC/WAIC will always select the most complex model, so AIC/DIC/WAIC are sometimes accused of "overfitting." But at the large-data limit, the most complex model will make predictions identical to the true model. The reason is that with so much data every parameter can be very precisely estimated. And so using an overly complex model will not hurt prediction. For example, as sample size $N \to \infty$ the model with 5 parameters in FIGURE 6.10 will tell you that the coefficients for predictors after the second are almost exactly zero. Therefore failing to identify the "correct" model does not hurt us, at least not in this sense. Furthermore, in the natural and social sciences the models under consideration are almost never the data-generating models. So it makes little sense to attempt to identify a "true" model.

6.4.1. DIC. The Deviance Information Criterion (DIC) is a widely used and easy to compute Bayesian information criterion. Many software packages provide it, and it's becoming as recognizable as AIC. DIC is essentially a version of AIC that is aware of informative priors. Like AIC, it assumes a multivariate Gaussian posterior distribution. This means if any parameter in the posterior is substantially skewed, and also has a substantial effect on prediction, then DIC like AIC can go horribly wrong.

DIC is calculated from the posterior distribution of the training deviance. What does it mean for the deviance to have a posterior distribution? Well, since the parameters have a posterior distribution, and since the deviance is computed from the parameters, it follows that deviance must also have a posterior distribution. Classical "deviance" is defined at the MAP values, and that's what we've been working with so far. But in principle the posterior distribution provides information about predictive uncertainty. And that uncertainty turns out to help us estimate out-of-sample deviance.

So define D now as the posterior *distribution* of deviance. This means we compute deviance (on the training sample) for each set of sampled parameter values in the posterior distribution. So if we draw 10,000 samples from the posterior, we compute 10,000 deviance values. Let $\bar{D}$ indicate the average of D. Also define $\hat{D}$ as the deviance calculated at the posterior mean. This means we compute the average of each parameter in the posterior distribution. Then we plug those averages into the deviance formula to get $\hat{D}$ out.

Once you have $\bar{D}$ and $\hat{D}$, DIC is calculated as:

$$\text{DIC} = \bar{D} + (\bar{D} - \hat{D}) = \bar{D} + p_D$$

The difference $\bar{D} - \hat{D} = p_D$ is analogous to the number of parameters used in computing AIC. It is an "effective" number of parameters that measures how flexible the model is in fitting the training sample. More flexible models entail greater risk of overfitting. So this p_D term is sometimes called a *penalty term*. It is just the expected distance between the deviance in-sample and the deviance out-of-sample. In the case of flat priors, DIC reduces directly to AIC, because the expected distance is just the number of parameters. But more generally, p_D will be some fraction of the number of parameters, because regularizing priors constrain a model's flexibility.

In most cases, R is going to do all of this for you. The function DIC in the rethinking package will compute DIC for a model fit with map or map2stan (which is introduced in Chapter 8).

6.4.2. WAIC. Even better than DIC is the Widely Applicable Information Criterion (WAIC). WAIC has a more complicated definition, but it is also calculated by taking averages of log-likelihood over the posterior distribution. And it is also just an estimate of out-of-sample deviance. But it does not require a multivariate Gaussian posterior, and it is often more accurate than DIC. There are types of models for which it is hard to define at all, however. We'll discuss that issue more, after defining WAIC.

The distinguishing feature of WAIC is that it is *pointwise*. This means that uncertainty in prediction is considered case-by-case, or point-by-point, in the data. This is useful, because some observations are much harder to predict than others and may also have different uncertainty. In the Gaussian models we've considered so far in this book, it's not easy to appreciate that point. But when we arrive at generalized linear models, it'll be more obvious why this matters. For the moment you can just think of WAIC as handling uncertainty where it actually matters: for each independent observation. It assesses flexibility of a model with respect to fitting each observation, and then sums up across all observations.

Define $\Pr(y_i)$ as the average likelihood of observation i in the training sample. This means we compute the likelihood of y_i for each set of parameters sampled from the posterior distribution. Then we average the likelihoods for each observation i and finally sum over all observations. This produces the first part of WAIC, the log-pointwise-predictive-density, lppd:

$$\text{lppd} = \sum_{i=1}^{N} \log \Pr(y_i)$$

You might say this out loud as:

> *The log-pointwise-predictive-density is the total across observations of the logarithm of the average likelihood of each observation.*

The lppd is just a pointwise analog of deviance, averaged over the posterior distribution. If you multiplied it by -2, it'd be similar to the deviance, in fact.

The second piece of WAIC is the effective number of parameters p_{WAIC}. Define $V(y_i)$ as the variance in log-likelihood for observation i in the training sample. This means we compute the log-likelihood of y_i for each sample from the posterior distribution. Then we take the variance of those values. This is $V(y_i)$. Now p_{WAIC} is defined as:

$$p_{\text{WAIC}} = \sum_{i=1}^{N} V(y_i)$$

Now WAIC is defined as:

$$\text{WAIC} = -2(\text{lppd} - p_{\text{WAIC}})$$

And this value is yet another estimate of out-of-sample deviance.

The function `WAIC` in the `rethinking` package will compute WAIC for a model fit with `map` or `map2stan`. If you want to see a didactic implementation of computing lppd and p_{WAIC}, see the Overthinking box at the end of this section.

Because WAIC requires splitting up the data into independent observations $i = 1...N$, it is sometimes hard to define. Consider for example a model in which each prediction depends upon a previous observation. This happens, for example, in a *time series*. In a time series, a previous observation becomes a predictor variable for the next observation. So it's not easy to think of each observation as independent of, or *exchangeable* with, the others. In such a case, you can of course compute WAIC as if each observation were independent of the others, but it's not clear what the resulting value means.

This caution raises a more general issue with all of these predictive information criteria: Their validity depends upon the predictive task you have in mind. And not all prediction can reasonably take the form that we've been assuming for the train-test simulations in this chapter. When we consider multilevel models, this issue will arise again.

Rethinking: What about BIC? The BAYESIAN INFORMATION CRITERION, abbreviated BIC and also known as the Schwarz criterion,[98] is more commonly juxtaposed with AIC. The choice between BIC or AIC (or neither!) is not about being Bayesian or not. There are both Bayesian and non-Bayesian ways to motivate both, and depending upon how strict one wishes to be, neither may be considered Bayesian.

BIC is related to the logarithm of the *average likelihood* of a linear model. The average likelihood is the denominator in Bayes' theorem, the likelihood averaged over the prior. There is a venerable tradition in Bayesian inference of comparing average likelihoods as a means to comparing models. A ratio of average likelihoods is called a BAYES FACTOR. On the log scale, these ratios are differences, and so comparing differences in average likelihoods resembles comparing differences in information criteria. Since average likelihood is averaged over the prior, more parameters induce a natural penalty on complexity. This helps guard against overfitting, even though the exact penalty is not in general the same as with information criteria.

But some Bayesian statisticians dislike the Bayes factor approach,[99] and all admit that there are technical obstacles to its use. Notably, even when priors are weak and have no influence on estimates within models, priors can have a huge impact on comparisons between models. So while the approach is tremendously valuable, and learning an alternative to information criteria necessarily helps one to understand information criteria even better, a robust treatment of Bayes factors is just beyond the scope of this book. It's important to realize, though, that the choice of Bayesian or not does not also decide between information criteria or Bayes factors. Moreover, there's no need to choose, really. We can always use both and learn from the ways they agree and disagree.

Overthinking: WAIC calculations. To see how the WAIC calculations actually work, consider a simple regression fit with `map`:

R code
6.15
```
data(cars)
m <- map(
    alist(
        dist ~ dnorm(mu,sigma),
```

```
        mu <- a + b*speed,
        a ~ dnorm(0,100),
        b ~ dnorm(0,10),
        sigma ~ dunif(0,30)
    ) , data=cars )
post <- extract.samples(m,n=1000)
```

We'll need the log-likelihood of each observation *i* at each sample *s* from the posterior:

R code
6.16

```
n_samples <- 1000
ll <- sapply( 1:n_samples ,
    function(s) {
        mu <- post$a[s] + post$b[s]*cars$speed
        dnorm( cars$dist , mu , post$sigma[s] , log=TRUE )
    } )
```

You end up with a 50-by-1000 matrix of log-likelihoods, with observations in rows and samples in columns. Now to compute lppd, the Bayesian deviance, we average the samples in each row, take the log, and add all of the logs together. However, to do this with precision, we need to do all of the averaging on the log scale. This is made easy with a function `log_sum_exp`, which computes the log of a sum of exponentiated terms. Then we can just subtract the log of the number of samples. This computes the log of the average.

R code
6.17

```
n_cases <- nrow(cars)
lppd <- sapply( 1:n_cases , function(i) log_sum_exp(ll[i,]) - log(n_samples) )
```

Typing `sum(lppd)` will give you lppd, as defined in the main text. Now for the effective number of parameters, p_{WAIC}. This is more straightforward, as we just compute the variance across samples for each observation, then add these together:

R code
6.18

```
pWAIC <- sapply( 1:n_cases , function(i) var(ll[i,]) )
```

And `sum(pWAIC)` returns p_{WAIC}, as defined in the main text. To compute WAIC:

R code
6.19

```
-2*( sum(lppd) - sum(pWAIC) )
```

```
[1] 421.0367
```

Compare to the output of the WAIC function. There will be simulation variance, because of how the samples are drawn from the map fit. But that variance remains much smaller than the standard error of WAIC itself. You can compute the standard error by computing the square root of number of cases multiplied by the variance over the individual observation terms in WAIC:

R code
6.20

```
waic_vec <- -2*( lppd - pWAIC )
sqrt( n_cases*var(waic_vec) )
```

```
[1] 14.42941
```

As models get more complicated, all that usually changes is how the log-likelihoods, ll, are computed.

Note that each individual observation has its own penalty term in the pWAIC vector we calculated above. This provides an interesting opportunity to study how different observations contribute to overfitting. You can get the same vectorized pointwise output from the WAIC function by using the pointwise=TRUE argument.

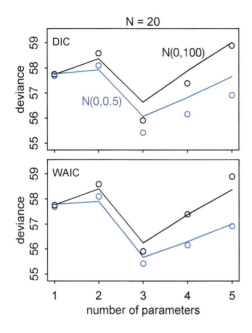

FIGURE 6.11. Out-of-sample deviance as estimated by DIC and WAIC. Points are average out-of-sample deviance over 10,000 simulations. The lines are average DIC (top) and WAIC (bottom) computed from the same simulations. The black points and lines come from simulations with a nearly flat Normal(0, 100) prior. The blue points and lines used a regularizing Normal(0, 0.5) prior.

6.4.3. DIC and WAIC as estimates of deviance.

With definitions of DIC and WAIC in hand, let's review one more simulation exercise. This will let us visualize the estimates of out-of-sample deviance that DIC and WAIC provide, in the same familiar context as earlier sections.

FIGURE 6.11 shows the results of 10,000 simulations each for the five familiar models with between 1 and 5 parameters. These plots display only out-of-sample deviance now, for simplicity. The black points are average out-of-sample deviance resulting from simulations with nearly flat priors. The blue points result from simulations using regularizing Normal(0, 0.5) priors. The black and blue lines show the estimated out-of-sample deviance from DIC (top) and WAIC (bottom), with colors corresponding to groups of points. Both are accurate on average, being within 1 point of deviance of the actual average in most cases. Both DIC and WAIC are useful estimates of the deviance, but WAIC is more accurate in this context.

Also notice that the regularizing prior (blue) still helps, and that DIC and WAIC do track this help. This suggests that using both regularization and information criteria will always beat using only one or the other alone. Regularization, as long as it's not too strong, reduces overfitting for any particular model. Information criteria instead help us measure overfitting across models fit to the same data. These are complementary functions. And since both are very easy to use and widely available, there's no reason to shy away from their use.

Rethinking: Diverse prediction frameworks. The train-test gambit we've been using in this chapter entails predicting a test sample of the same size and nature as the training sample. This most certainly does not mean that information criteria can only be used when we plan to predict a sample of the same size as training. The same size just scales the out-of-sample deviance similarly. In addition, AIC orders models in a way that approximates some forms of cross-validation,[100] and WAIC is explicitly derived as an approximate Bayesian cross-validation. So these criteria do have some claim to generality, because it is the distance between the models that is useful, not the absolute value of the deviance.

But the train-test prediction task is not representative of everything we might wish to do with models. For example, some statisticians prefer to evaluate predictions using a PREQUENTIAL framework, in which models are judged on their accumulated learning error over the training sample.[101] And once you start using multilevel models, "prediction" is no longer uniquely defined, because the test sample can differ from the training sample in ways that forbid use of some the parameter estimates. We'll worry about that issue in Chapter 12.

Perhaps a larger concern is that our train-test thought experiment pulls the test sample from exactly the same process as the training sample. This is a kind of *uniformitarian* assumption, in which future data are expected to come from the same process as past data and have the same rough range of values. This can cause problems. For example, suppose we fit a regression that predicts height using body weight. The training sample comes from a poor town, in which most people are pretty thin. The relationship between height and weight turns out to be positive and strong. Now also suppose our prediction goal is to guess the heights in another, much wealthier, town. Plugging the weights from the wealthy individuals into the model fit to the poor individuals will predict outrageously tall people. The reason is that, once weight becomes large enough, it has essentially no relationship with height. WAIC will not automatically recognize nor solve this problem. Nor will any other isolated procedure. But over repeated rounds of model fitting, attempts at prediction, and model criticism, it is possible to overcome this kind of limitation. As always, statistics is no substitute for science.

6.5. Using information criteria

Let's review the original problem and the road so far. When there are several plausible models for the same set of observations, how should we compare the accuracy of these models? Following the fit to the training sample is no good, because fit will always favor more complex models. From there, many roads diverge. One such road, the path to information criteria, leads to choosing information divergence as a measure of model accuracy. A meta-model of forecasting produces AIC, DIC, and WAIC as estimates of both the average deviance on a new sample and, by proxy, the accuracy of the model. AIC is however a special formula for simple models with flat priors. More general information criteria exist for other circumstances. DIC and WAIC are useful criteria that are easy to compute using samples from the posterior distribution of a model. In addition to information criteria, which estimate accuracy, it is helpful to employ regularizing priors that encode skepticism towards a sample. This reduces overfitting and complements information criteria.

That's the road so far.

But once we have DIC or WAIC calculated for each plausible model, how do we use these values? Since information criteria values provide advice about relative model performance, they can be used in many different ways. Frequently, people discuss MODEL SELECTION, which usually means choosing the model with the lowest AIC/DIC/WAIC value and then discarding the others. But this kind of selection procedure discards the information about relative model accuracy contained in the differences among the AIC/DIC/WAIC values. Why is this information useful? Because sometimes the differences are large and sometimes they are small. Just as relative posterior probability provides advice about how confident we might be about parameters (conditional on the model), relative model accuracy provides advice about how confident we might be about models (conditional on the set of models compared).

So instead of model *selection*, this section provides a brief example of model *comparison* and model *averaging*.

- **MODEL COMPARISON** means using DIC/WAIC in combination with the estimates and posterior predictive checks from each model. It is just as important to understand why a model outperforms another as it is to measure the performance difference. DIC/WAIC alone says very little about such details. But in combination with other information, DIC/WAIC is a big help.
- **MODEL AVERAGING** means using DIC/WAIC to construct a posterior predictive distribution that exploits what we know about relative accuracy of the models. This helps guard against overconfidence in model structure, in the same way that using the entire posterior distribution helps guard against overconfidence in parameter values. What model averaging does not mean is averaging parameter estimates, because parameters in different models have different meanings and should not be averaged, unless you are sure you are in a special case in which it is safe to do so. So it is better to think of model averaging as prediction averaging, because that's what is actually being done.

The section demonstrates how to conduct comparison and averaging, using a simple example with a few predictor variables. Later chapters continue using these tools, and the details of examples do vary. So be wary not to overgeneralize the example that follows.

6.5.1. Model comparison. Recall the primate milk data from the previous chapter. Let's load it into R, remove the NAs, and rescale one of the explanatory variables:

R code
6.21

```
data(milk)
d <- milk[ complete.cases(milk) , ]
d$neocortex <- d$neocortex.perc / 100
dim(d)
```

```
[1] 17  9
```

So your data frame should also have 17 rows (cases) and 9 columns (variables).

By removing the cases with missing values at the start, we accomplish the most important step in model comparison: *Compared models must be fit to exactly the same observations.* If you don't remove the incomplete cases before you begin fitting models, you risk comparing models fit to different numbers of observations. This is a serious risk, because R's more automated model fitting routines, like lm, will automatically and silently drop incomplete cases for you. If one model uses a predictor that has missing values, while another does not, then each model will be fit to a different number of cases. The model fit to fewer observations will almost always have a better deviance and AIC/DIC/WAIC value, because it has been asked to predict less.

We'll repeat the analysis of the previous chapter, predicting kilocalories per gram of milk (kcal.per.g) with the two predictor variables neocortex and the logarithm of mass. But now we'll fit four different models, corresponding to the four simple combinations of linear models with these two predictor variables: (1) a model with both neocortex and log mass, (2) a model with only neocortex, (3) a model with only log mass, and (4) a model with neither predictor (just an intercept). Fitting the four models, using map, is straightforward. The only thing that differs among these models is the equation for μ_i (mu).

But I'm also going to seize this opportunity to introduce a way to constrain the standard deviation of the outcome, σ, to be positive. The trick is to estimate the logarithm of σ, which can be any real number. Then inside the likelihood function we just exponentiate it. Since

$\exp(x) > 0$ for any real number x, this effectively bounds σ to positive reals. Here's what this looks like inside the code:

R code
6.22

```
a.start <- mean(d$kcal.per.g)
sigma.start <- log(sd(d$kcal.per.g))
m6.11 <- map(
    alist(
        kcal.per.g ~ dnorm( a , exp(log.sigma) )
    ) ,
    data=d , start=list(a=a.start,log.sigma=sigma.start) )
m6.12 <- map(
    alist(
        kcal.per.g ~ dnorm( mu , exp(log.sigma) ) ,
        mu <- a + bn*neocortex
    ) ,
    data=d , start=list(a=a.start,bn=0,log.sigma=sigma.start) )
m6.13 <- map(
    alist(
        kcal.per.g ~ dnorm( mu , exp(log.sigma) ) ,
        mu <- a + bm*log(mass)
    ) ,
    data=d , start=list(a=a.start,bm=0,log.sigma=sigma.start) )
m6.14 <- map(
    alist(
        kcal.per.g ~ dnorm( mu , exp(log.sigma) ) ,
        mu <- a + bn*neocortex + bm*log(mass)
    ) ,
    data=d , start=list(a=a.start,bn=0,bm=0,log.sigma=sigma.start) )
```

The priors are all flat above, which is clearly not the best idea. But this will let you get a sense of what the sample alone says, in the absence of regularization, and how WAIC measures overfitting. Then in the problems at the end of the chapter, you'll explore regularization.

Fitting these four models provides answers to questions about how and how much prediction changes when we include either or both predictor variables. With the four sets of estimates, and four deviances, in hand, we'll look quickly at both (1) comparing the models on the basis of WAIC values and (2) on the basis of parameter estimates. These are complementary approaches, aimed at understanding the best model by understanding how predictions and estimates change as predictors are added and subtracted from a model.

6.5.1.1. *Comparing WAIC values.* To compare models using an information criterion, you first compute the criterion. Then you can rank the models from lowest (best) to highest (worst) and also calculate *weights*, which provide a more interpretable measure of the relative distances among the models.

For any particular model, you can compute WAIC or DIC directly. Here is WAIC computed for model m6.14:

R code
6.23

```
WAIC( m6.14 )
```

```
[1] -14.96375
attr(,"lppd")
```

```
[1] 12.34189
attr(,"pWAIC")
[1] 4.860017
attr(,"se")
[1] 7.582438
```

The first value reported is the WAIC value. Note that it is negative in this case. That's fine. There's nothing preventing deviance from being negative. Smaller values are still better. The second value reported is the lppd. The third value is p_{WAIC}. If you subtract pWAIC from lppd and then multiply that difference by -2, you'll get the WAIC value. The final value, se, is the standard error of the WAIC value.[102] This standard error provides rough guidance to the uncertainty in WAIC that arises from sampling. It can be very rough guidance, when the sample size is small. Still, always remember that WAIC is an estimate.

Once you have WAIC (or any other information criterion) calculated for each model, you can begin by ordering the models by their WAIC values. The rethinking package also provides a handy function for ranking models by WAIC (and optionally by DIC, see ?compare):

R code
6.24
```
( milk.models <- compare( m6.11 , m6.12 , m6.13 , m6.14 ) )
```

```
         WAIC pWAIC dWAIC weight   SE  dSE
m6.14 -15.0   4.8   0.0   0.93 7.54   NA
m6.11  -8.3   1.8   6.7   0.03 4.52 7.26
m6.13  -7.9   3.0   7.1   0.03 5.67 5.33
m6.12  -6.2   2.9   8.9   0.01 4.34 7.57
```

The function compare takes fit models as input. It returns a table in which models are ranked from best to worst, with six columns of information.

(1) WAIC is obviously WAIC for each model. Smaller WAIC indicates better estimated out-of-sample deviance, so model m6.14 is ranked first.
(2) pWAIC is the estimated effective number of parameters. This provides a clue as to how flexible each model is in fitting the sample.
(3) dWAIC is the difference between each WAIC and the lowest WAIC. Since only relative deviance matters, this column shows the differences in relative fashion.
(4) weight is the AKAIKE WEIGHT for each model. These values are transformed information criterion values. I'll explain them below.
(5) SE is the standard error of the WAIC estimate. WAIC is an estimate, and provided the sample size N is large enough, its uncertainty will be well approximated by its standard error. So this SE value isn't necessarily very precise, but it does provide a check against overconfidence in differences between WAIC values.
(6) dSE is the standard error of the difference in WAIC between each model and the top-ranked model. So it is missing for the top model. If you want the full set of pairwise model differences, you can extract milk.models@dSE.

And you can plot these values, to provide a possibly more-intuitive presentation:

R code
6.25
```
plot( milk.models , SE=TRUE , dSE=TRUE )
```

This is the result:

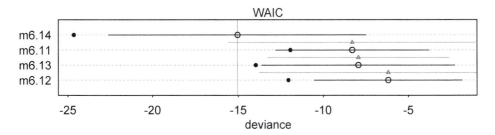

Each row is a model, ordered by WAIC. The filled points are the in-sample deviance of each model, which for WAIC is calculated as $-2 \times \text{lppd}$, which is $2p_{\text{WAIC}}$ from the corresponding WAIC value. The open points are WAIC. The standard error of each WAIC is shown by the dark line segment that passes through each open point. Finally, the standard error of the difference between each WAIC and the top-ranked WAIC is shown by the gray triangles and line segments between the rows. The triangle just above each WAIC point is the difference between that model and the top model. The gray line segment is the standard deviation of that difference.

You need these triangles and gray lines, because the overlap in standard deviations (the dark lines) between models is not also a measure of the standard error in the difference between models. Just like with parameters in the posterior distribution, uncertainty about WAIC is correlated between models. So if you want to know the distribution of the difference in WAIC between models, you have to compute it as a difference, not look at overlap of the uncertainty in individual WAIC values.

Akaike weights help by rescaling. A total weight of 1 is partitioned among the considered models, making it easier to compare their relative predictive accuracy. The weight for a model i in a set of m models is given by:

$$w_i = \frac{\exp(-\frac{1}{2}\text{dWAIC}_i)}{\sum_{j=1}^{m} \exp(-\frac{1}{2}\text{dWAIC}_j)}$$

where dWAIC is the same as `dWAIC` the `compare` table output. This example uses WAIC, but the formula is the same for any other information criterion, since they are all on the deviance scale. The Akaike weight formula might look rather odd, but really all it is doing is putting WAIC on a probability scale, so it just undoes the multiplication by -2 and then exponentiates to reverse the log transformation. Then it standardizes by dividing by the total. So each weight will be a number from 0 to 1, and the weights together always sum to 1. Now larger values are better.

But what do these weights mean? There actually isn't a consensus about that. But here's Akaike's interpretation, which is common.[103]

> *A model's weight is an estimate of the probability that the model will make the best predictions on new data, conditional on the set of models considered.*

Here's the heuristic explanation. First, regard WAIC as the expected deviance of a model on future data. That is to say that WAIC gives us an estimate of $E(D_{\text{test}})$. Akaike weights convert these deviance values, which are log-likelihoods, to plain likelihoods and then standardize them all. This is just like Bayes' theorem uses a sum in the denominator to standardize the product of the likelihood and prior. Therefore the Akaike weights are analogous to posterior probabilities of models, conditional on expected future data. Remember, a probability is a standardized count of all the ways some particular thing might happen, according to our assumptions, among all the things that could happen, also according to our assumptions.

So you can heuristically read each weight as an estimated probability that each model will perform best on future data. In simulation at least, interpreting weights in this way turns out to be appropriate.[104] However, given all the strong assumptions about repeat sampling that go into calculating WAIC, you can't take this heuristic too seriously. The future is unlikely to be exactly like the past, after all. Or in this case, the next batch of mammals we consider may be quite different from these primates.

In this analysis, the best model has more than 90% of the model weight. That's pretty good. But with only 12 cases, the error on the WAIC estimates is substantial, and of course that uncertainty should propagate to the Akaike weights. So don't get too excited. If we take the standard error of the difference from the compare table literally, you can think of the difference as Gaussian distribution centered (for the difference between models m6.14 and m6.11) on 6.7 with a standard deviation of 7.26. If you feel hesitant that you know how to calculate from that the probability that the difference is negative, and so reversed, you can just simulate it:

R code
6.26

```
diff <- rnorm( 1e5 , 6.7 , 7.26 )
sum(diff<0)/1e5
```

```
[1] 0.1773
```

This is only of heuristic value, especially with only 12 cases. On the other hand, for having only 12 cases, this is more than we might have reasonably hoped for.

How you interpret differences in information criteria always depend upon context: sample size, past research, nature of measurements. A principle way that this contextual meaning arises is that Akaike weights are conditional upon the set of models considered. If you add another model, or take one away, all of the weights will change. This is still a *small world* analysis, incapable of considering models that we have yet to imagine or analyze.

So what should we conclude in this context? It's clear that if we restrict ourselves to these four simple linear regressions, we benefit from using both of the predictor variables, at least as far as describing this sample goes. With only 12 cases, the associations between the predictors and the outcome are strong enough and precisely estimated enough to assign 90% of the model weight to m6.14. It's easy to imagine that additional data could reduce the strength of these associations. But the evidence is certainly consistent with the view that neocortex is positively associated with milk energy, controlling for body mass.

Notice as well that either predictor alone is actually expected to do worse than the model without either predictor, m6.11. The expected difference is small, but since neither predictor alone improves the deviance very much, the penalty term knocks down the two models with only one predictor. This is great evidence of a masking effect. In this way, the WAIC comparison echoes the need to measure both brain structure and body mass, if one hopes to understand either.

Rethinking: How big a difference in WAIC is "significant"? Newcomers to information criteria often ask whether a difference between AIC/DIC/WAIC values is "significant." For example, models m6.14 and m6.11 differ by about 6 units of deviance. Is it possible to say whether this difference is big enough to conclude that m6.14 is *significantly* better? In general, it is not possible to provide a principled threshold of difference that makes one model "significantly" better than another, whatever that means. The same is actually true of ordinary significance testing—the 5% convention is just a convention. We could invent some convention for WAIC, but it too would just be a convention. Moreover, we know the models will not make the same predictions—they are different models. So "significance" in this context must have a very different definition than usual.

The attitude this book encourages is to retain and present all models, no matter how big or small the differences in WAIC (or another criterion). The more information in your summary, the more information for peer review, and the more potential for the scholarly community to accumulate information. And keep in mind that averaging model predictions often produces better results than selecting any single model, obviating the "significance" question.

Rethinking: WAIC metaphors. Here are two metaphors to help explain the concepts behind using WAIC (or another information criterion) to compare models.

Think of models as race horses. In any particular race, the best horse may not win. But it's more likely to win than is the worst horse. And when the winning horse finishes in half the time of the second-place horse, you can be pretty sure the winning horse is also the best. But if instead it's a photo-finish, with a near tie between first and second place, then it is much harder to be confident about which is the best horse. WAIC values are analogous to these race times—smaller values are better, and the distances between the horses/models are informative. Akaike weights transform differences in finishing time into probabilities of being the best model/horse on future data/races. But if the track conditions or jockey changes, these probabilities may mislead. Forecasting future racing/prediction based upon a single race/fit carries no guarantees.

Think of models as stones thrown to skip on a pond. No stone will ever reach the other side (perfect prediction), but some sorts of stones make it farther than others, on average (make better test predictions). But on any individual throw, lots of unique conditions avail—the wind might pick up or change direction, a duck could surface to intercept the stone, or the thrower's grip might slip. So which stone will go farthest is not certain. Still, the relative distances reached by each stone therefore provide information about which stone will do best on average. But we can't be too confident about any individual stone, unless the distances between stones is very large.

Of course neither metaphor is perfect. Metaphors never are. But many people find these to be helpful in interpreting information criteria.

6.5.1.2. *Comparing estimates.* In addition to comparing models on the basis of expected test deviance, it is nearly always useful to compare parameter estimates among models. Comparing estimates helps in at least two major ways. First, it is useful to understand *why* a particular model or models have lower WAIC values. Changes in posterior distributions, across models, provide useful hints. Second, regardless of WAIC values, we often want to know whether some parameter's posterior distribution is stable across models. For example, scholars often ask whether a predictor remains important as other predictors are added and subtracted from the model. To address that kind of question, one typically looks for a parameter's posterior distribution to remain stable across models, as well as for all models that contain that parameter to have lower WAIC than those models without it.

In the primate milk example, comparing estimates confirms what you already learned in the previous chapter: The model with both predictors does much better, because each predictor masks the other. In order to demonstrate that in the previous chapter, we actually did fit three of the models currently at hand. Looking at a consolidated table of the MAP estimates makes the comparison a lot easier. The `coeftab` function takes a series of fit models as input and builds such a table:

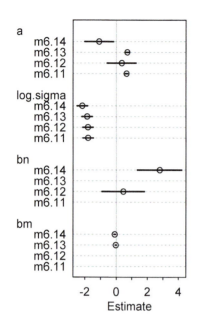

FIGURE 6.12. Comparing the posterior densities of parameters for the four models fit to the primate milk data. Each point is a MAP estimate, and each black line segment is a 89% percentile interval. Estimates are grouped by parameter identity, and each row in a group is a model.

R code
6.27

```
coeftab(m6.11,m6.12,m6.13,m6.14)
```

	m6.11	m6.12	m6.13	m6.14
a	0.66	0.35	0.71	-1.09
log.sigma	-1.79	-1.80	-1.85	-2.16
bn	NA	0.45	NA	2.79
bm	NA	NA	-0.03	-0.10
nobs	17	17	17	17

The nobs at the bottom are the number of observations, just there to help you make sure you fit each model to the same observations. From scanning the table, you can see that the estimates for both bn and bm get farther from zero when they are both present in the model. But standard errors aren't represented here, and seeing how the uncertainty changes is just as important as seeing how the location changes. You can get coeftab to add standard errors to the table (see ?coeftab), but that still doesn't make it easy to appreciate changes in the width of posterior densities. Better to plot these estimates:

R code
6.28

```
plot( coeftab(m6.11,m6.12,m6.13,m6.14) )
```

The result is shown in FIGURE 6.12. Each point is a MAP estimate, and each black line segment is an 89% percentile interval. Each group of estimates corresponds to the same named parameter, across models. Each row in each group is a model, labeled on the left. Now you can quickly scan each group of estimates to see how estimates change or not across models. You can adjust these plots to group by model instead of by parameter and to display only some parameters. See ?coeftab_plot for details.

6.5.2. Model averaging. Way back in Chapter 3, we saw how to preserve the uncertainty about parameters when simulating predictions from a model. Now we have the analogous problem of preserving the uncertainty about models. Treating model weights as heuristic plausibilities that each model will perform best in testing, it makes sense to try to preserve these relative plausibilities when generating predictions. And doing so is mechanically very similar to the procedure with a single model.

To review, let's simulate and plot counterfactual predictions for the minimum-WAIC model, m6.14. Here's the familiar code for simulating the posterior predictive distribution, focusing on counterfactual predictions across the range of neocortex.

R code
6.29

```
# compute counterfactual predictions
# neocortex from 0.5 to 0.8
nc.seq <- seq(from=0.5,to=0.8,length.out=30)
d.predict <- list(
    kcal.per.g = rep(0,30), # empty outcome
    neocortex = nc.seq,      # sequence of neocortex
    mass = rep(4.5,30)       # average mass
)
pred.m6.14 <- link( m6.14 , data=d.predict )
mu <- apply( pred.m6.14 , 2 , mean )
mu.PI <- apply( pred.m6.14 , 2 , PI )

# plot it all
plot( kcal.per.g ~ neocortex , d , col=rangi2 )
lines( nc.seq , mu , lty=2 )
lines( nc.seq , mu.PI[1,] , lty=2 )
lines( nc.seq , mu.PI[2,] , lty=2 )
```

The resulting plot is displayed in FIGURE 6.13. For the moment, ignore the shaded region and focus on the dashed regression line and the dashed 89% percentile interval of the mean. Those are the lines the code above produces. You've seen them before (page 138).

Now let's compute and add model averaged posterior predictions. What we're going to compute is an ENSEMBLE of posterior predictions. Here's the conceptual procedure, and then I'll show you the code that automates it, much like link automates computing μ for each sample in the posterior.

(1) Compute WAIC (or another information criterion) for each model.
(2) Compute the weight for each model.
(3) Compute linear model and simulated outcomes for each model.
(4) Combine these values into an ensemble of predictions, using the model weights as proportions.

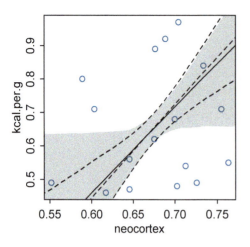

FIGURE 6.13. Model averaged posterior predictive distribution for the primate milk analysis. The dashed regression line and dashed 89% percentile interval correspond to the minimum-WAIC model, m6.14. The solid line and shaded 89% percentile region correspond to the model averaged predictions.

And this is what the function ensemble can do. The ensemble function works a lot like link and sim. In fact, it just calls those functions, for each model you give it, and then combines the results according to Akaike weights. So to build an ensemble according to WAIC weight (the default behavior):

R code
6.30

```
milk.ensemble <- ensemble( m6.11 , m6.12 , m6.13 , m6.14 , data=d.predict )
mu <- apply( milk.ensemble$link , 2 , mean )
mu.PI <- apply( milk.ensemble$link , 2 , PI )
lines( nc.seq , mu )
shade( mu.PI , nc.seq )
```

The solid regression line and shaded region in FIGURE 6.13 display these calculations. The regression line, which shows the average μ at each value horizontal axis, has hardly moved at all. Model m6.14 has about 90% of the weight, recall. So this makes sense.

Even so, model averaging the predictions has a noticeable impact on the intervals of μ. The shaded interval includes a slope of zero. This is because the lower-ranked WAIC models—totaling only about 10% of the Akaike weight—all suggest a near-zero (or exactly zero) slope for neocortex. Retaining the model uncertainty, as estimated by Akaike weight here, helps guard against overconfidence.

There will be many more examples of this procedure in coming chapters. Sometimes model averaging has no practical impact on predictions. Sometimes it has a massive impact. But it is always a conservative procedure that helps to communicate model uncertainty. Model averaging will never make a predictor variable appear more influential than it already appears in any single model.

Rethinking: The Curse of Tippecanoe. One concern with model comparison is, if we try enough combinations and transformations of predictor variables, we might eventually find a model that fits any particular sample very well. But this fit will be a peculiar case of overfitting, unlikely to generalize to new data. And WAIC and similar metrics will be fooled. So it's common to advise against trying every possible model.

Consider by analogy the *Curse of Tippecanoe*.[105] From the year 1840 until 1960, every United States president who was elected in a year ending in the digit 0 (which happens every 20 years, given 4 year terms) has died in office. William Henry Harrison was the first, being elected in 1840 and dying of pneumonia the next year. John F. Kennedy was the last, elected in 1960 and assassinated in 1963. Seven American presidents died in sequence in this pattern. Ronald Reagan was elected in 1980, but despite at least one attempt on his life, he managed to live long after his term was up, breaking the curse. Given enough time and data, a pattern like this can be found for almost any body of data. But without any compelling reason to believe this pattern is meaningful, it is hardly compelling that such patterns exist. Most large sets of data will contain patterns of correlation that are strong and surprising. If we search hard enough, we are bound to find a Curse of Tippecanoe. There are many other patterns in presidential names and dates, and no doubt new ones are being found and circulated all the time.

Fiddling with and constructing many predictor variables is a great way to find coincidences, but not necessarily a great way to evaluate hypotheses. However, fitting many possible models isn't always a dangerous idea, provided some judgment is exercised in weeding down the list of variables at the start. There are two scenarios in which this strategy appears defensible. First, sometimes all one wants to do is explore a set of data, because there are no clear hypotheses to evaluate. This is rightly labeled pejoratively as DATA DREDGING, when one does not admit to it. But when used together with model averaging, and freely admitted, it can be a way to stimulate future investigation. Second, sometimes we need to convince an audience that we have tried all of the combinations of predictors, because none of the variables seem to help much in prediction.

6.6. Summary

This chapter has been a marathon. It began with the problem of overfitting, a universal phenomenon by which models with more parameters fit a sample better, even when the additional parameters are meaningless. Two common tools were introduced to address overfitting: regularizing priors and information criteria. Regularizing priors reduce overfitting during estimation, and information criteria help estimate the degree of overfitting. Practical functions `compare` and `ensemble` in the `rethinking` package were introduced to help analyze collections of models fit to the same data. In the chapters to follow, these tools will be applied to both new and old data examples. In all cases, keep in mind that these tools are heuristic. They provide no guarantees. No statistical procedure will ever substitute for iterative scientific investigation.

6.7. Practice

Easy.

6E1. State the three motivating criteria that define information entropy. Try to express each in your own words.

6E2. Suppose a coin is weighted such that, when it is tossed and lands on a table, it comes up heads 70% of the time. What is the entropy of this coin?

6E3. Suppose a four-sided die is loaded such that, when tossed onto a table, it shows "1" 20%, "2" 25%, "3" 25%, and "4" 30% of the time. What is the entropy of this die?

6E4. Suppose another four-sided die is loaded such that it never shows "4". The other three sides show equally often. What is the entropy of this die?

Medium.

6M1. Write down and compare the definitions of AIC, DIC, and WAIC. Which of these criteria is most general? Which assumptions are required to transform a more general criterion into a less general one?

6M2. Explain the difference between model *selection* and model *averaging*. What information is lost under model selection? What information is lost under model averaging?

6M3. When comparing models with an information criterion, why must all models be fit to exactly the same observations? What would happen to the information criterion values, if the models were fit to different numbers of observations? Perform some experiments, if you are not sure.

6M4. What happens to the effective number of parameters, as measured by DIC or WAIC, as a prior becomes more concentrated? Why? Perform some experiments, if you are not sure.

6M5. Provide an informal explanation of why informative priors reduce overfitting.

6M6. Provide an information explanation of why overly informative priors result in underfitting.

Hard. All practice problems to follow use the same data. Pull out the old Howell !Kung demography data and split it into two equally sized data frames. Here's the code to do it:

R code
6.31
```
library(rethinking)
data(Howell1)
d <- Howell1
d$age <- (d$age - mean(d$age))/sd(d$age)
set.seed( 1000 )
i <- sample(1:nrow(d),size=nrow(d)/2)
d1 <- d[ i , ]
d2 <- d[ -i , ]
```

You now have two randomly formed data frames, each with 272 rows. The notion here is to use the cases in d1 to fit models and the cases in d2 to evaluate them. The set.seed command just ensures that everyone works with the same randomly shuffled data.

Now let h_i and x_i be the height and centered age values, respectively, on row i. Fit the following models to the data in d1:

$$\mathcal{M}_1 : h_i \sim \text{Normal}(\mu_i, \sigma)$$
$$\mu_i = \alpha + \beta_1 x_i$$

$$\mathcal{M}_2 : h_i \sim \text{Normal}(\mu_i, \sigma)$$
$$\mu_i = \alpha + \beta_1 x_i + \beta_2 x_i^2$$

$$\mathcal{M}_3 : h_i \sim \text{Normal}(\mu_i, \sigma)$$
$$\mu_i = \alpha + \beta_1 x_i + \beta_2 x_i^2 + \beta_3 x_i^3$$

$$\mathcal{M}_4 : h_i \sim \text{Normal}(\mu_i, \sigma)$$
$$\mu_i = \alpha + \beta_1 x_i + \beta_2 x_i^2 + \beta_3 x_i^3 + \beta_4 x_i^4$$

$$\mathcal{M}_5 : h_i \sim \text{Normal}(\mu_i, \sigma)$$
$$\mu_i = \alpha + \beta_1 x_i + \beta_2 x_i^2 + \beta_3 x_i^3 + \beta_4 x_i^4 + \beta_5 x_i^5$$

$$\mathcal{M}_6 : h_i \sim \text{Normal}(\mu_i, \sigma)$$
$$\mu_i = \alpha + \beta_1 x_i + \beta_2 x_i^2 + \beta_3 x_i^3 + \beta_4 x_i^4 + \beta_5 x_i^5 + \beta_6 x_i^6$$

Use map to fit these. Use weakly regularizing priors for all parameters.

Note that fitting all of these polynomials to the height-by-age relationship is not a good way to derive insight. It would be better to have a simpler approach that would allow for more insight, like perhaps a piecewise linear model. But the set of polynomial families above will serve to help you practice and understand model comparison and averaging.

6H1. Compare the models above, using WAIC. Compare the model rankings, as well as the WAIC weights.

6H2. For each model, produce a plot with model averaged mean and 97% confidence interval of the mean, superimposed on the raw data. How do predictions differ across models?

6H3. Now also plot the model averaged predictions, across all models. In what ways do the averaged predictions differ from the predictions of the model with the lowest WAIC value?

6H4. Compute the test-sample deviance for each model. This means calculating deviance, but using the data in d2 now. You can compute the log-likelihood of the height data with:

```
sum( dnorm( d2$height , mu , sigma , log=TRUE ) )
```

R code
6.32

where mu is a vector of predicted means (based upon age values and MAP parameters) and sigma is the MAP standard deviation.

6H5. Compare the deviances from **6H4** to the WAIC values. It might be easier to compare if you subtract the smallest value in each list from the others. For example, subtract the minimum WAIC from all of the WAIC values so that the best WAIC is normalized to zero. Which model makes the best out-of-sample predictions in this case? Does WAIC do a good job of estimating the test deviance?

6H6. Consider the following model:

$$h_i \sim \text{Normal}(\mu_i, \sigma)$$
$$\mu_i = \alpha + \beta_1 x_i + \beta_2 x_i^2 + \beta_3 x_i^3 + \beta_4 x_i^4 + \beta_5 x_i^5 + \beta_6 x_i^6$$
$$\beta_1 \sim \text{Normal}(0, 5)$$
$$\beta_2 \sim \text{Normal}(0, 5)$$
$$\beta_3 \sim \text{Normal}(0, 5)$$
$$\beta_4 \sim \text{Normal}(0, 5)$$
$$\beta_5 \sim \text{Normal}(0, 5)$$
$$\beta_6 \sim \text{Normal}(0, 5)$$

and assume flat (or nearly flat) priors on α and σ. This model contains more strongly regularizing priors on the coefficients.

First, fit this model to the data in d1. Report the MAP estimates and plot the implied predictions. Then compute the out-of-sample deviance using the data in d2, using MAP estimates from the model fit to d1 only. How does this model, using regularizing priors, compare to the best WAIC model from earlier? How do you interpret this result?

7 Interactions

The manatee (*Trichechus manatus*) is a slow-moving, aquatic mammal that lives in warm, shallow waters. Manatees have no natural predators, but they do share their waters with motor boats. And motor boats have propellers. While manatees are related to elephants, and so they have very thick skins, propeller blades can and do kill them. A majority of adult manatees bear some kind of scar earned in a collision with a boat (FIGURE 7.1, top).[106]

The Armstrong Whitworth A.W.38 Whitley was a frontline Royal Air Force bomber. During the second World War, the A.W.38 carried bombs and pamphlets into German territory. Unlike the manatee, the A.W.38 has fierce natural enemies: artillery and interceptor fire. Many planes never returned from their missions. And those that survived had the scars to prove it (FIGURE 7.1, bottom).

How is a manatee like an A.W.38 bomber? In both cases—manatee propeller scars and bomber bullet holes—we'd like to do something to improve the odds, to help manatees and bombers survive. Most observers intuit that helping manatees or bombers means reducing the kind of damage we see on them. For manatees, this might mean requiring propeller guards (on the boats, not the manatees). For bombers, it'd mean up-armoring the parts of the plane that show the most damage.

But in both cases, the evidence misleads us. Propellers do not cause most of the injury and death caused to manatees. Rather autopsies confirm that collisions with blunt parts of the boat, like the keel, do far more damage. Similarly, up-armoring the damaged portions of returning bombers did little good. Instead, improving the A.W.38 bomber meant armoring the *undamaged* sections.[107]

The evidence from surviving manatees and bombers is misleading, because it is *conditional* on survival. Manatees and bombers that perished look different. A manatee struck by a keel is less likely to live than another grazed by a propeller. So among the survivors, propeller scars are common. Similarly, bombers that returned home conspicuously lacked damage to the cockpit and engines. They got lucky. Bombers that never returned home were less so. To get the right answer, in either context, we have to realize that the kind of damage seen is conditional on survival.

CONDITIONING is one of the most important principles of statistical inference. Data, like the manatee scars and bomber damage, are conditional on how they get into our sample. Posterior distributions are conditional on the data. All model-based inference is conditional on the model. Every inference is conditional on something, whether we notice it or not.

And a large part of the power of statistical modeling comes from creating devices that allow probability to be conditional of aspects of each case. The linear models you've grown to love are just crude devices that allow each outcome y_i to be conditional on a set of predictors

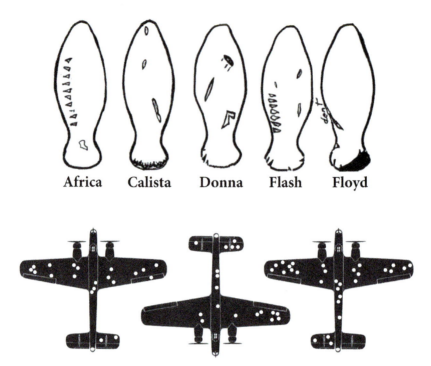

FIGURE 7.1. TOP: Dorsal scars for 5 adult Florida manatees. Rows of short scars, for example on the individuals Africa and Flash, are indicative of propeller laceration. BOTTOM: Three exemplars of damage on A.W.38 bombers returning from missions.

for each case *i*. Like the epicycles of the Ptolemaic and Copernican models (Chapters 4 and 6), linear models give us a way to describe conditionality.

Simple linear models frequently fail to provide enough conditioning, however. Every model so far in this book has assumed that each predictor has an independent association with the mean of the outcome. What if we want to allow the association to be conditional? For example, in the primate milk data from the previous chapters, suppose the relationship between milk energy and brain size varies by taxonomic group (ape, monkey, prosimian). This is the same as suggesting that the influence of brain size on milk energy is conditional on taxonomic group. The linear models of previous chapters cannot address this question.

To model deeper conditionality—where the importance of one predictor depends upon another predictor—we need INTERACTION. Interaction is a kind of conditioning, a way of allowing parameters (really their posterior distributions) to be conditional on further aspects of the data. The simplest kind of interaction, a linear interaction, is built by extending the linear modeling strategy to parameters within the linear model. So it is akin to placing epicycles on epicycles in the Ptolemaic and Copernican models. It is descriptive, but very powerful.

More generally, interactions are central to most statistical models beyond the cozy world of Gaussian outcomes and linear models of the mean. In generalized linear models (GLMs, Chapter 9 and onwards), even when one does not explicitly define variables as interacting, they will always interact to some degree. Moreover, every variable essentially interacts with itself, as the impact of change in its value will depend upon its current value. Say goodbye to simple constant slopes of linear regression. Say hello to impacts that may depend upon the covariation of dozens of predictor variables.

Multilevel models induce similar effects. Common sorts of multilevel models are essentially massive interaction models, in which estimates (intercepts and slopes) are conditional on clusters (person, genus, village, city, galaxy) in the data. Multilevel interaction effects are complex. They're not just allowing the impact of a predictor variable to change depending upon some other variable, but they are also estimating aspects of the *distribution* of those changes. This may sound like genius, or madness, or both. Regardless, you can't have the power of multilevel modeling without it.

Models that allow for complex interactions are easy to fit to data. But they can be considerably harder to understand. And so I spend this chapter reviewing simple interaction effects: how to specify them, how to interpret them, and how to plot them. The chapter starts with a case of an interaction between a single categorical (dummy) variable and a single continuous variable. In this context, it is easy to appreciate the sort of hypothesis that an interaction allows for. Then the chapter moves on to show how manipulating the model of the mean itself can help us understand the behavior of an interaction, before moving on to more complex interactions between multiple continuous predictor variables as well as higher-order interactions among more than two variables. In every section of this chapter, the model predictions are visualized, averaging over uncertainty in parameters.

My hope is that this chapter lays a solid foundation for interpreting generalized linear models and multilevel models in the later chapters.

> **Rethinking: Statistics all-star, Abraham Wald.** The World War II bombers story is the work of Abraham Wald (1902–1950). Wald was born in what is now Romania, but immigrated to the United States after the Nazi invasion of Austria. Wald made many contributions over his short life. Perhaps most germane to the current material, Wald proved that for many types of rules for making statistical decisions, there will exist a Bayesian rule that is at least as good as any non-Bayesian one. Wald proved this, remarkably, beginning with non-Bayesian premises, and so anti-Bayesians could not ignore it. This work was summarized in Wald's 1950 book, published just before his death.[108] Wald died much too young, from a plane crash while touring India.

7.1. Building an interaction

Africa is special. The second largest continent, it is the most culturally and genetically diverse. Africa has about 3 billion fewer people than Asia, but it has just as many living languages. Africa is so genetically diverse that most of the genetic variation outside of Africa is just a subset of the variation within Africa. Africa is also geographically special, in a puzzling way: Bad geography tends to be related to bad economies outside of Africa, but African economies seem immune to bad geography.

To appreciate the puzzle, look at regressions of terrain ruggedness—a particular kind of bad geography—against economic performance (log GDP[109] per capita in the year 2000), both inside and outside of Africa (FIGURE 7.2). Load the data used in this figure, and split it into Africa and non-Africa, with this code:

R code
7.1

```
library(rethinking)
data(rugged)
d <- rugged
```

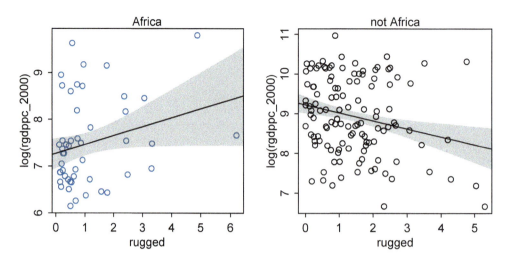

FIGURE 7.2. Separate linear regressions inside and outside of Africa, for log-GDP against terrain ruggedness. The slope is positive inside Africa, but negative outside. How can we recover this reversal of the slope, using the combined data?

```
# make log version of outcome
d$log_gdp <- log( d$rgdppc_2000 )

# extract countries with GDP data
dd <- d[ complete.cases(d$rgdppc_2000) , ]

# split countries into Africa and not-Africa
d.A1 <- dd[ dd$cont_africa==1 , ] # Africa
d.A0 <- dd[ dd$cont_africa==0 , ] # not Africa
```

Each row in these data is a country, and the various columns are economic, geographic, and historical features.[110] The variable rugged is a Terrain Ruggedness Index[111] that quantifies the topographic heterogeneity of a landscape. The outcome variable here is the logarithm of real gross domestic product per capita, from the year 2000, rgdppc_2000. We'll use the logarithm of it, for reasons similar to those we discussed for body mass in Chapter 5. Specifically, the logarithm is the *magnitude* of GDP. Since wealth generates wealth, it tends to be exponentially related to anything that increases it. This is like saying that the absolute distances in wealth grow increasingly large, as nations become wealthier. So when we work with logarithms instead, we can work on a more evenly spaced scale of magnitudes. If this still doesn't quite make sense, see the Rethinking box at the end of this section.

Fit the regression models displayed in FIGURE 7.2 with this code:

R code
7.2
```
# African nations
m7.1 <- map(
    alist(
```

```
        log_gdp ~ dnorm( mu , sigma ) ,
        mu <- a + bR*rugged ,
        a ~ dnorm( 8 , 100 ) ,
        bR ~ dnorm( 0 , 1 ) ,
        sigma ~ dunif( 0 , 10 )
    ) ,
    data=d.A1 )

# non-African nations
m7.2 <- map(
    alist(
        log_gdp ~ dnorm( mu , sigma ) ,
        mu <- a + bR*rugged ,
        a ~ dnorm( 8 , 100 ) ,
        bR ~ dnorm( 0 , 1 ) ,
        sigma ~ dunif( 0 , 10 )
    ) ,
    data=d.A0 )
```

By now, I trust the reader to be able to plot the posterior predictions, as shown in the figure. If you are still unsure how to do this, look back at the examples in Chapters 4 and 5.

So what is going on in FIGURE 7.2? It makes sense that ruggedness is associated with poorer countries, in most of the world. In theory, rugged terrain means transport is difficult, which means market access is hampered, which means reduced gross domestic product. In theory. But the reversed relationship within Africa seems puzzling. Why should difficult terrain be associated with higher GDP per capita?

If this relationship is at all causal, it may be because rugged regions of Africa were protected against the large-scale Atlantic and Indian Ocean slave trades. Slavers preferred to raid easily accessed settlements, with easy routes to the sea. Those regions that suffered under the slave trade understandably continue to suffer economically, long after the decline of slave-trading markets. However, an outcome like GDP has many influences, and is furthermore a strange measure of economic activity. So it is hard to be sure what's going on here.

Regardless of the causal explanation for the reversal, there's a valuable statistical lesson here: How do we discover and describe such a reversal in the context of a regression model? The plots in FIGURE 7.2 cheat, because they split the data into two data frames. But it's not a good idea to split the data in this way. Here are four, of many, reasons.

First, there are usually some parameters, such as σ, that the model says do not depend in any way upon an African identity for each nation. By splitting the data table, you are hurting the accuracy of the estimates for these parameters, because you are essentially making two less-accurate estimates instead of pooling all of the evidence into one estimate. In effect, you have accidentally assumed that variance differs between African and non-African nations. Now, there's nothing wrong with that sort of assumption. But you want to avoid accidental assumptions.

Second, in order to acquire probability statements about the variable you used to split the data, cont_africa in this case, you need to include it in the model. Otherwise, you have only the weakest sort of statistical argument. Isn't there uncertainty about the predictive value of distinguishing between African and non-African nations? Of course there is. Unless you analyze all of the data in a single model, you can't easily quantify that uncertainty. If you just let the posterior distribution do the work for you, you'll have a useful measure of that uncertainty.

Third, we may want to use information criteria or another method to compare models. In order to compare a model that treats all continents the same way to a model that allows different slopes in different continents, we need models that use all of the same data (as explained in Chapter 6). This means we can't split the data, but have to make the model split the data.

Fourth, once you begin using multilevel models (Chapter 12), you'll see that there are advantages to borrowing information across categories like "Africa" and "not Africa." This is especially true when sample sizes vary across categories, such that overfitting risk is higher within some categories. In other words, what we learn about ruggedness outside of Africa should have some effect on our estimate within Africa, and visa versa. Multilevel models borrow information in this way, in order to improve estimates in all categories. When we split the data, this borrowing is impossible.

So let's see how to recover the reversal of slope, within a single model.

7.1.1. Adding a dummy variable doesn't work.
The first thing to realize is that just including the categorical variable (dummy variable) cont_africa won't reveal the reversed slope. It's worth fitting this model to prove it to yourself, though. I'm going to walk through this as a simple model comparison exercise, just so you begin to get some applied examples of concepts you've accumulated from earlier chapters.

The question is to what extent singling out African nations changes predictions. There are two models to fit, to start. The first is just the simple linear regression of log-GDP on ruggedness, but now for the entire data set:

R code
7.3

```
m7.3 <- map(
    alist(
        log_gdp ~ dnorm( mu , sigma ) ,
        mu <- a + bR*rugged ,
        a ~ dnorm( 8 , 100 ) ,
        bR ~ dnorm( 0 , 1 ) ,
        sigma ~ dunif( 0 , 10 )
    ) ,
    data=dd )
```

The second is the model that includes a dummy variable for African nations:

R code
7.4

```
m7.4 <- map(
    alist(
        log_gdp ~ dnorm( mu , sigma ) ,
        mu <- a + bR*rugged + bA*cont_africa ,
        a ~ dnorm( 8 , 100 ) ,
        bR ~ dnorm( 0 , 1 ) ,
        bA ~ dnorm( 0 , 1 ) ,
        sigma ~ dunif( 0 , 10 )
    ) ,
    data=dd )
```

Now to compare these models, using WAIC:

```
compare( m7.3 , m7.4 )
```

```
      WAIC pWAIC dWAIC weight    SE   dSE
m7.4 476.5   4.5   0.0      1 15.29    NA
m7.3 539.6   2.7  63.1      0 13.27 15.05
```

This is a case in which it will be safe to ignore the lower ranked model, as m7.4 gets all the model weight. And while the standard error of the difference in WAIC is 15, the difference itself is 63, implying a 95% interval of 63 ± 30. So the continent predictor variable seems to be picking up something important to the sample, even accounting for expected overfitting.

Now let's plot the posterior predictions for m7.4, so you can see how, despite it's plausible superiority to m7.3, it still doesn't manage different slopes inside and outside of Africa. To sample from the posterior and compute the predicted means and intervals for both African and non-African nations:

```
rugged.seq <- seq(from=-1,to=8,by=0.25)

# compute mu over samples, fixing cont_africa=0
mu.NotAfrica <- link( m7.4 , data=data.frame(cont_africa=0,rugged=rugged.seq) )

# compute mu over samples, fixing cont_africa=1
mu.Africa <- link( m7.4 , data=data.frame(cont_africa=1,rugged=rugged.seq) )

# summarize to means and intervals
mu.NotAfrica.mean <- apply( mu.NotAfrica , 2 , mean )
mu.NotAfrica.PI <- apply( mu.NotAfrica , 2 , PI , prob=0.97 )
mu.Africa.mean <- apply( mu.Africa , 2 , mean )
mu.Africa.PI <- apply( mu.Africa , 2 , PI , prob=0.97 )
```

I show these predictions in FIGURE 7.3. African nations are shown in blue, while nations outside Africa are shown in gray. What you've ended up with here is a rather weak negative relationship between economic development and ruggedness. The African nations do have lower overall economic development, and so the blue regression line is below, but parallel to, the black line. All including a dummy variable for African nations has done is allow the model to predict a lower mean for African nations. It can't do anything to the slope of the line. The fact that WAIC tells you that the model with the dummy variable is hugely better only indicates that African nations on average do have lower GDP.

Rethinking: Why 97%? In the code block just above, and therefore also in FIGURE 7.3, I used 97% intervals of the expected mean. This is a rather non-standard percentile interval. So why use 97%? In this book, I use non-standard percents to constantly remind the reader that conventions like 95% and 5% are arbitrary. Furthermore, boundaries are meaningless. There is a continuous change in probability as we move away from the expected value. So one side of the boundary is almost equally probable as the other side. Also, 97 is a prime number. That doesn't mean it is a better choice than any other number here, but it's no less silly than using a multiple of 5, just because we have five digits on each hand. Resist the tyranny of the Tetrapoda.

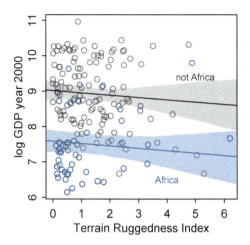

FIGURE 7.3. Including a dummy variable for African nations has no effect on the slope. African nations are shown in blue. Non-African nations are shown in gray. Regression means for each subset of nations are shown in corresponding colors, along with 97% intervals shown by shading.

7.1.2. Adding a linear interaction does work.

How can you recover the change in slope you saw at the start of this section? You need a proper interaction effect. The likelihood for the model you just plotted, in math form, is:

$$Y_i \sim \text{Normal}(\mu_i, \sigma)$$
$$\mu_i = \alpha + \beta_R R_i + \beta_A A_i$$

where Y is log(rgdppc_2000), A is cont_africa, and R is rugged. As you've done since Chapter 4, the linear model is built by replacing the parameter μ in the top line, the likelihood, with a linear equation that is a function of data and new parameters such as α and β.

We'll build interactions by extending this strategy. Now you want to allow the relationship between Y and R to vary as a function of A. Within the model, this relationship is measured by the slope β_R. Following the same strategy of replacing parameters with linear models, the most straightforward way to make β_R depend upon A is just to define the slope β_R as a linear model itself, one that includes A. This approach results in this likelihood (you'll also need priors, which we'll add in a bit):

$$Y_i \sim \text{Normal}(\mu_i, \sigma) \qquad \text{[likelihood]}$$
$$\mu_i = \alpha + \gamma_i R_i + \beta_A A_i \qquad \text{[linear model of } \mu \text{]}$$
$$\gamma_i = \beta_R + \beta_{AR} A_i \qquad \text{[linear model of slope]}$$

This is the first model with two linear models, but its structure is the same as every Gaussian model you've already fit in this book. So you don't need to learn any new tricks for fitting this model to data. The tricks lie entirely in interpreting it.

The first line above is the same Gaussian likelihood you've been using since Chapter 4. The second line is the same kind of additive definition of μ_i that you've seen many times. The third line is the new bit. The new symbol γ_i is just a placeholder for the linear function that defines the slope between GDP and ruggedness. We use "gamma" (γ) here, because it follows "beta" (β) in the Greek alphabet. The equation for γ_i defines the interaction between ruggedness and African nations. It is a *linear interaction effect*, because the equation γ_i is a linear model.

By defining the relationship between GDP and ruggedness in this way, you are explicitly modeling the hypothesis that the slope between GDP and ruggedness depends—is *conditional*—upon whether or not a nation is in Africa. The parameter β_{AR} defines the strength of this dependency. If you set $\beta_{AR} = 0$, then you get the previous model back. If instead $\beta_{AR} > 0$, then African nations have a more positive slope between GDP and ruggedness. If $\beta_{AR} < 0$, African nations have a more negative slope. For any nation not in Africa, $A_i = 0$ and so the interaction parameter β_{AR} has no effect on prediction for that nation. Of course you are going to compute the posterior distribution for β_{AR} from the data. But once you have the posterior distribution, it is only through understanding where the parameter fits into your model that will allow you to interpret it.

To fit this new model, you can just use map as before. Here's the code to fit the model that includes an interaction between ruggedness and being in Africa:

R code
7.7

```
m7.5 <- map(
    alist(
        log_gdp ~ dnorm( mu , sigma ) ,
        mu <- a + gamma*rugged + bA*cont_africa ,
        gamma <- bR + bAR*cont_africa ,
        a ~ dnorm( 8 , 100 ) ,
        bA ~ dnorm( 0 , 1 ) ,
        bR ~ dnorm( 0 , 1 ) ,
        bAR ~ dnorm( 0 , 1 ) ,
        sigma ~ dunif( 0 , 10 )
    ) ,
    data=dd )
```

It looks just as you might expect, with two linear models now. The gamma definition gets evaluated, and then those values are used to evaluate the definition of mu, and finally mu gets used to compute the likelihood. It all just cascades up. For the priors, notice that I'm using weakly regularizing priors for the coefficients, and a very flat prior for the intercept. As in earlier chapters, we usually don't know where the intercept will end up. But if we regularize on the coefficients, then the intercept will be effectively regularized by them.

Before moving on to interpret the estimates and plotting the predictions, let's use WAIC to compare this new model to the previous two.

R code
7.8

```
compare( m7.3 , m7.4 , m7.5 )
```

	WAIC	pWAIC	dWAIC	weight	SE	dSE
m7.5	469.6	5.3	0.0	0.97	15.13	NA
m7.4	476.4	4.4	6.8	0.03	15.35	6.22
m7.3	539.7	2.8	70.1	0.00	13.31	15.22

Model family m7.5 has about 97% of the WAIC-estimated model weight. That's very strong support for including the interaction effect. That probably isn't surprising, given the obvious difference in slope we began this story with. But the modicum of weight given to m7.4 suggests that the posterior means for the slopes in m7.5 are a little overfit. And the standard error of the difference in WAIC between the top two models is almost the same as the difference itself. There are only so many African countries, after all, so the data are sparse as far as estimating the interaction goes.

More importantly, it's not clear what information criteria mean in this data context. We aren't seriously imagining that we will resample from some process that created African nations and non-African nations. So do information criteria make sense at all here? I think they do, but reasonable people can disagree on this point. Here's an argument for how their advice about overfitting is relevant, even when the train/test paradigm behind information criteria cannot be understood literally.

First, no matter what you want to do with the estimates produced by a model, the estimates must be overfit. They must be overfit, because not every feature of the sample is caused by the process of interest. So even if our interest is explanation, rather than prediction, and there will be no new African nations to make a prediction for, overfitting still matters. Whether we use regularizing priors or information criteria or something else, it's always worth worrying about overfitting and measuring it, if possible.

Second, information criteria estimate overfitting from the shape of the posterior distribution calculated on the sample at hand. For example, WAIC's estimate of the overfitting penalty is, as you saw in Chapter 6 (page 191), just the sum variance in log-likelihood of each data point. It's a general measure of a model's flexibility in fitting a sample. As such, it is useful heuristically even when the research context doesn't fit the strict train-test scenario used to derive it. Since regularizing priors help by reducing a model's flexibility in fitting a sample, the ability of information criteria to measure that same flexibility is probably always of some value. This is true if for no other reason than it nudges us towards conservative inference in a scholarly environment rife with false positives.[112]

Overthinking: Conventional form of interaction. Instead of a model definition with more than one linear model, as you saw above, it's conventional to multiply out any interaction effects, so that there is only one linear model. For example, the GDP on ruggedness likelihood could be defined as:

$$Y_i \sim \text{Normal}(\mu_i, \sigma) \qquad \text{[likelihood]}$$

$$\mu_i = \alpha + \beta_R R_i + \beta_{AR} A_i R_i + \beta_A A_i \qquad \text{[linear model of } \mu\text{]}$$

This is equivalent to the form in the main text. The equation for γ has just been substituted into the second line and expanded. This expanded form also works for estimation:

R code
7.9

```
m7.5b <- map(
    alist(
        log_gdp ~ dnorm( mu , sigma ) ,
        mu <- a + bR*rugged + bAR*rugged*cont_africa + bA*cont_africa,
        a ~ dnorm( 8 , 100 ) ,
        bA ~ dnorm( 0 , 1 ) ,
        bR ~ dnorm( 0 , 1 ) ,
        bAR ~ dnorm( 0 , 1 ) ,
        sigma ~ dunif( 0 , 10 )
    ) ,
    data=dd )
```

And the above is actually very common and forms the pattern upon which automated tools like lm expect interactions to be specified. We'll talk about that at the end of this chapter.

7.1.3. Plotting the interaction.
Plotting this model doesn't really require any new tricks. The goal is to make two plots. In the first, we'll display nations in Africa and overlay the posterior mean (MAP) regression line and the 97% interval of that line. In the second, we'll

display nations outside of Africa instead. First, we calculate the necessary posterior mean line and interval, for both plots:

R code
7.10

```
rugged.seq <- seq(from=-1,to=8,by=0.25)

mu.Africa <- link( m7.5 , data=data.frame(cont_africa=1,rugged=rugged.seq) )
mu.Africa.mean <- apply( mu.Africa , 2 , mean )
mu.Africa.PI <- apply( mu.Africa , 2 , PI , prob=0.97 )

mu.NotAfrica <- link( m7.5 , data=data.frame(cont_africa=0,rugged=rugged.seq) )
mu.NotAfrica.mean <- apply( mu.NotAfrica , 2 , mean )
mu.NotAfrica.PI <- apply( mu.NotAfrica , 2 , PI , prob=0.97 )
```

Plotting these calculations is just like previous examples. Here's the code:

R code
7.11

```
# plot African nations with regression
d.A1 <- dd[dd$cont_africa==1,]
plot( log(rgdppc_2000) ~ rugged , data=d.A1 ,
    col=rangi2 , ylab="log GDP year 2000" ,
    xlab="Terrain Ruggedness Index" )
mtext( "African nations" , 3 )
lines( rugged.seq , mu.Africa.mean , col=rangi2 )
shade( mu.Africa.PI , rugged.seq , col=col.alpha(rangi2,0.3) )

# plot non-African nations with regression
d.A0 <- dd[dd$cont_africa==0,]
plot( log(rgdppc_2000) ~ rugged , data=d.A0 ,
    col="black" , ylab="log GDP year 2000" ,
    xlab="Terrain Ruggedness Index" )
mtext( "Non-African nations" , 3 )
lines( rugged.seq , mu.NotAfrica.mean )
shade( mu.NotAfrica.PI , rugged.seq )
```

And the result is shown in FIGURE 7.4. Finally, the slope reverses direction inside and outside of Africa.

7.1.4. Interpreting an interaction estimate. Interpreting interaction estimates is tricky. It's trickier than interpreting ordinary estimates. And for this reason, I usually advise against trying to understand an interaction from tables of numbers alone. Plotting implied predictions does far more for both our own understanding and for our audience's. But often only tables of numbers appear in scientific papers, and so it's worth spending a little time on why the numbers can be so confusing.

There are two basic reasons to be wary of interpreting tables of posterior means and standard deviations as a way to understanding interactions.

(1) When you add an interaction to a model, this changes the meanings of the parameters. A "main effect" coefficient in an interaction model does not mean the same thing as a coefficient of the same name in a model without an interaction. Their distributions cannot usually be directly compared.

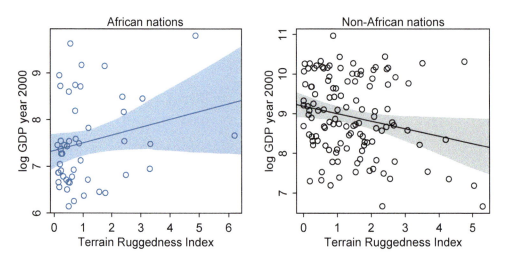

FIGURE 7.4. Posterior predictions for the terrain ruggedness model, including the interaction between Africa and ruggedness. Shaded regions are 97% posterior intervals of the mean.

(2) Tables of numbers don't make it easy to fully incorporate uncertainty in our thinking, since covariance among parameters isn't usually shown. And this gets much harder once the influence of a predictor depends upon multiple parameters.

Let's briefly consider each of these difficulties.

7.1.4.1. *Parameters change meaning.* In a simple linear regression, with no interactions, each coefficient says how much the average outcome, μ, changes when the predictor changes by one unit. And since all of the parameters have independent influences on the outcome, there's no trouble in interpreting each parameter separately. Each slope parameter gives us a direct measure of each predictor variable's influence.

Interaction models ruin this paradise, however. Look at the interaction likelihood again:

$$Y_i \sim \text{Normal}(\mu_i, \sigma) \qquad \text{[likelihood]}$$
$$\mu_i = \alpha + \gamma_i R_i + \beta_A A_i \qquad \text{[linear model of } \mu]$$
$$\gamma_i = \beta_R + \beta_{AR} A_i \qquad \text{[linear model of slope]}$$

Now the change in μ_i that results from a unit change in R_i is given by γ_i. And since γ_i is a function of three things—β_R, β_{AR}, and A_i—we have to know all three in order to know the influence of R_i on the outcome. The only time the slope β_R has its old meaning is when $A_i = 0$, which makes $\gamma_i = \beta_R$. Otherwise, to compute the influence of R_i on the outcome, we have to simultaneously consider two parameters and another predictor variable.

The practical implication of this fact is that you can no longer read the influence of either predictor from the table of estimates. Here are the parameter estimates:

R code
7.12

```
precis(m7.5)
```

```
        Mean StdDev  5.5% 94.5%
a       9.18   0.14  8.97  9.40
bA     -1.85   0.22 -2.20 -1.50
```

```
bR     -0.18   0.08 -0.31 -0.06
bAR     0.35   0.13  0.14  0.55
sigma   0.93   0.05  0.85  1.01
```

Since γ (gamma) doesn't appear in this table—it wasn't estimated—we have to compute it ourselves. It is easy enough to do that at the MAP values (posterior means). For example, the MAP slope relating ruggedness to log-GDP within Africa is:

$$\gamma = \beta_R + \beta_{AR}(1) = -0.18 + 0.35 = 0.17$$

And outside of Africa:

$$\gamma = \beta_R + \beta_{AR}(0) = -0.18$$

So the relationship between ruggedness and log-GDP is essentially reversed inside and outside of Africa.

7.1.4.2. *Incorporating uncertainty.* But that's only at the MAP values. To get some idea of the uncertainty around those γ values, we'll need to use the whole posterior. Since γ depends upon parameters, and those parameters have a posterior distribution, γ must also have a posterior distribution. Read the previous sentence again a few times. It's one of the most important concepts in processing Bayesian model fits. Anything calculated using parameters has a distribution.

To compute the posterior distribution of γ, you could do some integral calculus, or you could just process the samples from the posterior:

```
post <- extract.samples( m7.5 )
gamma.Africa <- post$bR + post$bAR*1
gamma.notAfrica <- post$bR + post$bAR*0
```

R code
7.13

The symbols gamma.Africa and gamma.notAfrica now hold samples from the posterior distributions of γ within Africa and γ outside Africa, respectively. The means of these distributions are just like the calculations we did at the end of the previous section:

```
mean( gamma.Africa)
mean( gamma.notAfrica )
```

R code
7.14

```
[1] 0.1631653
[1] -0.1824547
```

Nearly identical to the MAP values, of course.

But now we also have full distributions of the slopes within and outside of Africa. Let's plot them on the same axis, so we can see their overlap clearly:

```
dens( gamma.Africa , xlim=c(-0.5,0.6) , ylim=c(0,5.5) ,
    xlab="gamma" , col=rangi2 )
dens( gamma.notAfrica , add=TRUE )
```

R code
7.15

And this plot appears as FIGURE 7.5. The posterior distribution of the slope within Africa is shown in blue. The posterior distribution outside of Africa is shown in black.

From here, you could use the samples to ask a bunch of questions, depending upon your interest. For example, what's the probability (according to this model and these data) that the slope within Africa is less than the slope outside of Africa? All we need to do is compute

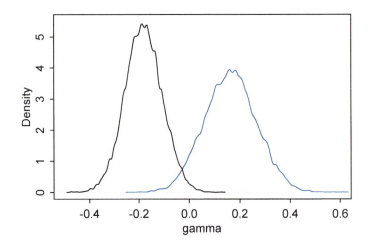

FIGURE 7.5. Posterior distributions of the slope relating terrain ruggedness to log-GDP. Blue: African nations. Black: non-African nations.

the difference between the slopes, for each sample from the posterior, and then ask what proportion of these differences is below zero. This code will do it:

R code
7.16

```
diff <- gamma.Africa - gamma.notAfrica
sum( diff < 0 ) / length( diff )
```

```
[1] 0.0036
```

As always, your answer will be slightly different, due to simulation variance. But this is always going to be a very small number, regardless of the exact samples you get. So conditional on this model and these data, it's highly implausible that the slope association ruggedness with log-GDP is lower inside Africa than outside it.

Also note that this probability, 0.0036, is very tiny compared to the visual overlap of the two distributions in FIGURE 7.5. That is not a mistake. The distributions in the figure are *marginal*, like silhouettes of each distribution, ignoring all of the other dimensions in the posterior. The calculation above is the distribution of the *difference* between the two. The distribution of their difference is not the same as the visual overlap of their marginal distributions. This is also the reason we can't use overlap in confidence intervals of different parameters as an informal test of "significance" of the difference. If you care about the difference, you must compute the distribution of the difference directly.

Rethinking: More on the meaning of posterior probability. Keep in mind that this number, 0.0036, is not the probability of any observable event. For example, in repeat resampling we do not expect 0.36% of countries within Africa to suffer terrain ruggedness worse than a country outside of Africa. Instead, 0.0036 is the relative plausibility that your golem assigns to the question you asked of it, and only for the data you presented to it. So the golem is telling that it is very skeptical of the notion that γ within Africa is lower than γ outside of Africa. How skeptical? Of all the possible states of the world it knows about, only a 0.36% of them are consistent with both the data and the claim that γ in Africa is less than γ outside Africa. Your golem is skeptical, but it's usually a good idea for you to remain skeptical of your golem.

7.2. Symmetry of the linear interaction

Buridan's ass is a toy philosophical problem in which an ass who always moves towards the closest pile of food will starve to death when he finds himself equidistant between two identical piles. The basic problem is one of symmetry: How can the ass decide between two identical options? Like many toy problems, you can't take this one too seriously. Of course the ass will not starve. But thinking about how the symmetry is broken can be productive.

Interactions are like Buridan's ass. Like the two piles of identical food, the linear interaction contains two symmetrical interpretations. Absent some other information, outside the model, there's no logical basis for preferring one over the other. Consider for example the GDP and terrain ruggedness problem. The interaction there has two equally valid phrasings.

(1) How much does the influence of ruggedness (on GDP) depend upon whether the nation is in Africa?

(2) How much does the influence of being in Africa (on GDP) depend upon ruggedness?

While these two possibilities sound different to most humans, your golem thinks they are identical.

In this section, we'll examine this fact analytically. Then we'll plot the ruggedness and GDP example again, but with the reverse phrasing—the influence of Africa depends upon ruggedness.

7.2.1. Buridan's interaction. Consider yet again the mathematical form of the likelihood:

$$Y_i \sim \text{Normal}(\mu_i, \sigma) \qquad \text{[likelihood]}$$
$$\mu_i = \alpha + \gamma_i R_i + \beta_A A_i \qquad \text{[linear model of } \mu\text{]}$$
$$\gamma_i = \beta_R + \beta_{AR} A_i \qquad \text{[linear model of slope]}$$

Let's expand γ_i into the expression for μ_i:

$$\mu_i = \alpha + (\beta_R + \beta_{AR} A_i) R_i + \beta_A A_i$$
$$= \alpha + \beta_R R_i + \beta_{AR} A_i R_i + \beta_A A_i$$

Now factor together the terms with A_i in them:

$$\mu_i = \alpha + \beta_R R_i + \underbrace{(\beta_A + \beta_{AR} R_i)}_{G} A_i$$

The term labeled G looks a lot like γ_i in the original form. This is the same model of μ_i, but re-expressed so that the linear interaction applies to A_i.

The point of the algebra above is to prove that linear interactions are symmetric, just like the choice facing Buridan's ass. Within the model, there's no basis to prefer one interpretation over the other, because in fact they are the same interpretation. But when we reason causally about models, our minds tend to prefer one interpretation over the other, because it's usually easier to imagine manipulating one of the predictor variables instead of the other. In this case, it's hard to imagine manipulating which continent a nation is on. But it's easy to imagine manipulating terrain ruggedness, by flattening hills or blasting tunnels through mountains.[113]

7.2.2. Africa depends upon ruggedness.
It'll be helpful to plot the reverse interpretation: *The influence of being in Africa depends upon terrain ruggedness.* The computation steps are very similar to what we did in the previous section. But now we will display cont_africa on the horizontal axis, while using different lines for different values of rugged.

R code
7.17

```
# get minimum and maximum rugged values
q.rugged <- range(dd$rugged)

# compute lines and confidence intervals
mu.ruggedlo <- link( m7.5 ,
    data=data.frame(rugged=q.rugged[1],cont_africa=0:1) )
mu.ruggedlo.mean <- apply( mu.ruggedlo , 2 , mean )
mu.ruggedlo.PI <- apply( mu.ruggedlo , 2 , PI )

mu.ruggedhi <- link( m7.5 ,
    data=data.frame(rugged=q.rugged[2],cont_africa=0:1) )
mu.ruggedhi.mean <- apply( mu.ruggedhi , 2 , mean )
mu.ruggedhi.PI <- apply( mu.ruggedhi , 2 , PI )

# plot it all, splitting points at median
med.r <- median(dd$rugged)
ox <- ifelse( dd$rugged > med.r , 0.05 , -0.05 )
plot( dd$cont_africa + ox , log(dd$rgdppc_2000) ,
    col=ifelse(dd$rugged>med.r,rangi2,"black") ,
    xlim=c(-0.25,1.25) , xaxt="n" , ylab="log GDP year 2000" ,
    xlab="Continent" )
axis( 1 , at=c(0,1) , labels=c("other","Africa") )
lines( 0:1 , mu.ruggedlo.mean , lty=2 )
shade( mu.ruggedlo.PI , 0:1 )
lines( 0:1 , mu.ruggedhi.mean , col=rangi2 )
shade( mu.ruggedhi.PI , 0:1 , col=col.alpha(rangi2,0.25) )
```

And the entire mess is shown in FIGURE 7.6. The points in the figure are nations. Black points are nations with terrain ruggedness below the median. Blue points are above the median. The horizontal axis is continent, with continents other than Africa on the left and African nations on the right. Notice first that being in Africa is, on average, bad for GDP. Remember colonialism and neo-colonialism? There it is. Now look at the regression lines. The black dashed line and shaded confidence region show the expected reduction in log-GDP when we take a nation with minimum terrain ruggedness (0.003) and change its continent. For low ruggedness, the expected reduction in log-GDP is about 2 full points. The blue regression line and shaded region are instead the expected change for an imaginary nation with maximum observed terrain ruggedness (6.2). Now changing continent has almost no expected effect—the line does slope upwards a tiny amount, but the wide shaded interval should prevent us from getting excited about that fact. For a nation with very high ruggedness, there is almost no negative effect on GDP of being in Africa.

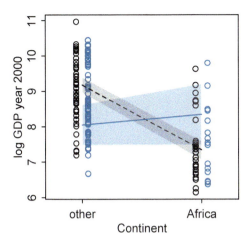

FIGURE 7.6. The other side of the interaction between ruggedness and continent. Blue points are nations with above-median ruggedness. Black points are below the median. Dashed black line: relationship between continent and log-GDP, for an imaginary nation with minimum observed ruggedness (0.003). Blue line: an imaginary nation with maximum observed ruggedness (6.2).

This perspective on the GDP and terrain ruggedness data is completely consistent with the previous perspective. It's simultaneously true in these data (and with this model) that (1) the influence of ruggedness depends upon continent and (2) the influence of continent depends upon ruggedness. Indeed, something is gained by looking at the data in this symmetrical perspective. Just inspecting the first view of the interaction, back on page 220, it's not obvious that African nations are on average nearly always worse off. It's just at very high values of rugged that nations inside and outside of Africa have the same expected log-GDP. This second way of plotting the interaction makes this clearer.

7.3. Continuous interactions

The main point I want to convince the reader of is that interaction effects are difficult to interpret. They are nearly impossible to interpret, using only posterior means and standard deviations. We've already discussed two reasons for this (page 219). A third reason to be wary of using only tables of numbers to interpret interactions is that interactions among continuous variables are especially opaque. It's one thing to make a slope conditional upon a *category*, as in the previous example of ruggedness and being in Africa. In such a context, the model reduces to estimating a different slope for each category. But it's quite a lot harder to understand that a slope varies in a continuous fashion with a continuous variable. Interpretation is much harder in this case, even though the mathematics of the model are essentially the same as in the categorical case.

In pursuit of clarifying the construction and interpretation of CONTINUOUS INTERACTIONS among two or more continuous predictor variables, in this section I develop a simple regression example and show you a way to plot the two-way interaction between two continuous variables. The method I present for plotting this interaction is a *triptych* plot, a panel of three complementary figures that comprise a whole picture of the regression results. There's nothing magic about having three figures—in other cases you might want more or less. Instead, the utility lies in making multiple figures that allow one to see how the interaction alters a slope, across changes in a chosen variable.

This example will also allow me to illustrate two benefits of CENTERING prediction variables. You met centering already (page 99). When a prediction variable is "centered," it is just rescaled so that its mean is zero. Why would you ever do that to the data? There are two common benefits. First, centering the prediction variables can make it much easier to lean on the coefficients alone in understanding the model, especially when you want to compare the estimates from models with and without an interaction. Second, sometimes model fitting has a hard time with uncentered variables. Centering (and possibly also standardizing) the data before fitting the model can help you achieve a faster and more reliable set of estimates.

Still, even with everything centered and double-checked, it can be hard or impossible to really understand continuous interactions from numbers alone. So the message here is to always plot posterior predictions, counterfactual or not, in order to avoid misunderstanding the model fit.

7.3.1. The data.
The data in this example are sizes of blooms from beds of tulips grown in greenhouses, under different soil and light conditions.[114] Load the data with:

R code
7.18
```
library(rethinking)
data(tulips)
d <- tulips
str(d)
```

```
'data.frame': 27 obs. of  4 variables:
 $ bed   : Factor w/ 3 levels "a","b","c": 1 1 1 1 1 1 1 1 1 2 ...
 $ water : int  1 1 1 2 2 2 3 3 3 1 ...
 $ shade : int  1 2 3 1 2 3 1 2 3 1 ...
 $ blooms: num  0 0 111 183.5 59.2 ...
```

The blooms column will be our outcome—what we wish to predict. The water and shade columns will be our predictor variables. water indicates one of three ordered levels of soil moisture, from low (1) to high (3). shade indicates one of three ordered levels of light exposure, from high (1) to low (3). The last column, bed, indicates a cluster of plants from the same section of the greenhouse.

Since both light and water help plants grow and produce blooms, it stands to reason that the independent effect of each will be to produce bigger blooms. But we'll also be interested in the interaction between these two variables. In the absence of light, for example, it's hard to see how water will help a plant—photosynthesis depends upon both light and water. Likewise, in the absence of water, sunlight does a plant little good. One way to model such an interdependency is to use an interaction effect. In the absence of a good mechanistic model of the interaction, one that uses a theory about the plant's physiology to hypothesize the functional relationship between light and water, then a simple linear two-way interaction is a good start. But ultimately it's not close to the best that we could do.

7.3.2. The un-centered models.
While a complete model comparison analysis is possible here, I'm going to simplify the story by focusing on just two models: (1) the model with both water and shade but no interaction and (2) the model with both main effects and the interaction of water with shade. I do so just for the sake of brevity. You can fit the missing models, like those with only one of the two predictor variables, and demonstrate for yourself that the conclusions don't change.

The main effect likelihood is (we'll add priors later):

$$B_i \sim \text{Normal}(\mu_i, \sigma)$$
$$\mu_i = \alpha + \beta_W W_i + \beta_S S_i$$

And the full interaction likelihood is:

$$B_i \sim \text{Normal}(\mu_i, \sigma)$$
$$\mu_i = \alpha + \beta_W W_i + \beta_S S_i + \beta_{WS} W_i S_i$$

where B_i is the value of bloom on row i, W_i is the value of water, and S_i is the value of shade. I'm leaving the categorical variable bed out of this analysis, but I actually think a sincere analysis requires it, and in the practice problems at the end of the chapter, we'll come back and add bed to the analysis. The points I wish to make right now don't depend upon it, however.

Fitting the models with map is just as you might expect. I'll use very flat priors here, so we get results nearly identical to typical maximum likelihood inference. This isn't to imply that this is the best thing to do. It is not the best thing to do. Instead, it's to present another example of how to do it. And in this case, the range of the outcome variable, blooms, is huge. The minimum is zero, and the maximum is 362. So this means that priors that look very flat may not actually be, because "flat" is always relative to the likelihood.

R code
7.19

```
m7.6 <- map(
    alist(
        blooms ~ dnorm( mu , sigma ) ,
        mu <- a + bW*water + bS*shade ,
        a ~ dnorm( 0 , 100 ) ,
        bW ~ dnorm( 0 , 100 ) ,
        bS ~ dnorm( 0 , 100 ) ,
        sigma ~ dunif( 0 , 100 )
    ) ,
    data=d )
m7.7 <- map(
    alist(
        blooms ~ dnorm( mu , sigma ) ,
        mu <- a + bW*water + bS*shade + bWS*water*shade ,
        a ~ dnorm( 0 , 100 ) ,
        bW ~ dnorm( 0 , 100 ) ,
        bS ~ dnorm( 0 , 100 ) ,
        bWS ~ dnorm( 0 , 100 ) ,
        sigma ~ dunif( 0 , 100 )
    ) ,
    data=d )
```

When trying to fit these models, you are likely to get one or both of these errors.

```
Error in map(alist(blooms ~ dnorm(mu, sigma), mu <- a + bW * water + bS *   :
  non-finite finite-difference value [5]
```

This usually indicates poor start values for the parameters. And you might also see:

```
Caution, model may not have converged.
Code 1: Maximum iterations reached.
```

What has happened here is that the model fitting engine, R's optim, searched for a very long time. It searched so long that it reached the time limit, the "maximum iterations." As a result, the estimates are unreliable and should not be used, unless you are sure they are okay. In this case, they are not okay.

It's easy enough to fix the problem, though. Let's fix it, then I'll explain why it happened. There are three basic solutions. We'll adopt the first two.

(1) We can use another method of optimization. There are many different ways to climb the posterior distribution. R's optim knows several. The default for map is called BFGS. This default is very capable, but it sometimes has trouble getting started. Two others to try, when you have trouble, are called Nelder-Mead and (when all else fails) SANN (simulated annealing). Tell map to use another method by adding the parameter method to the call. There will be an example below.

(2) We can tell optim to search longer, so it doesn't reach its maximum iterations. You can tell map to allow a larger number of iterations by passing it a control list. There will be an example below.

(3) We can rescale the data, so that search is not so difficult. The fundamental issue in this example is the scale of the outcome variable, blooms. If we center or standardize it, search will be easier for the computer.

I'll go ahead and use the first two solutions, just so you can see how they are implemented. Typically I'd recommend centering or standardizing the outcome instead (or in addition). But we'll do that in the next subsection.

R code
7.20
```
m7.6 <- map(
    alist(
        blooms ~ dnorm( mu , sigma ) ,
        mu <- a + bW*water + bS*shade ,
        a ~ dnorm( 0 , 100 ) ,
        bW ~ dnorm( 0 , 100 ) ,
        bS ~ dnorm( 0 , 100 ) ,
        sigma ~ dunif( 0 , 100 )
    ) ,
    data=d ,
    method="Nelder-Mead" ,
    control=list(maxit=1e4) )
m7.7 <- map(
    alist(
        blooms ~ dnorm( mu , sigma ) ,
        mu <- a + bW*water + bS*shade + bWS*water*shade ,
        a ~ dnorm( 0 , 100 ) ,
        bW ~ dnorm( 0 , 100 ) ,
        bS ~ dnorm( 0 , 100 ) ,
        bWS ~ dnorm( 0 , 100 ) ,
        sigma ~ dunif( 0 , 100 )
    ) ,
    data=d ,
    method="Nelder-Mead" ,
    control=list(maxit=1e4) )
```

No angry warnings anymore. So let's look at the estimates. I'll use the coeftab function that was introduced in the previous chapter:

R code
7.21

```
coeftab(m7.6,m7.7)
```

```
          m7.6     m7.7
a        53.49   -84.26
bW       76.36   150.96
bS      -38.95    35.00
sigma    57.38    46.27
bWS         NA    -39.5
nobs        27       27
```

Now consider these estimates and try to figure out what the models are telling us about the influence of water and shade on the blooms. First, consider the intercepts a, α. The estimate of the intercept changes a lot from one model to the next, from 53 to -84. What do these values mean? Remember, the intercept is the expected value of the outcome when *all* of the predictor variables take the value zero. In this case, neither of the predictor variables ever takes the value zero within the data. As a result, these intercept estimates are very hard to interpret. This is a very common issue, and most of the regression examples so far in this book have suffered from it.

Now, consider the slope parameters. In the main-effect-only model, m7.6, the MAP value for the main effect of water is positive and the main effect for shade is negative. Take a look at the standard deviations and intervals in precis(m7.6) to verify that both posterior distributions are reliably on one side of zero. You might infer that these posterior distributions suggest that water increases blooms while shade reduces them. For every additional level of soil moisture, blooms increase by 76, on average. For every addition unit of shade, blooms decrease by 42, on average. Those sound reasonable.

But the analogous posterior distributions from the interaction model, m7.7, are quite different. First, assure yourself that the interaction model is indeed a much better model:

R code
7.22

```
compare( m7.6 , m7.7 )
```

```
       WAIC pWAIC dWAIC weight     SE   dSE
m7.7 296.4   6.3   0.0   0.99  10.00    NA
m7.6 305.8   5.2   9.3   0.01   8.94  6.03
```

This comparison assigns nearly all of the weight of evidence to m7.7. So let's consider the posterior distribution from m7.7. Now both main effects are positive, but the new interaction posterior mean is negative. Are you to conclude now that the main effect of shade is to help the tulips? And the negative interaction itself implies that as shade increases, water has a reduced impact on blooms. But reduced by how much?

Sampling from the posterior now and plotting the model's predictions would help immensely with interpretation. And that's the course I want to encourage and provide code for. In general, it's not safe to interpret interactions without plotting them. But for the moment, let's instead look at the value of *centering* the predictor variables and re-estimating the models.

Rethinking: Fighting with your robot. The trouble-shooting in the preceding section is annoying, but it's realistic. These kinds of issues routinely arise in model fitting. With linear models like these, there are ways to compute the posterior distribution that avoid many of these complications. But with non-linear models to come, there is really no way to dodge the issue. In general, how you fit the model is part of the model. So you have to get used to trouble-shooting. And even when the model will fit fine, there may be changes you can make that will help it fit faster and more reliably. There will be more examples as the book progresses.

7.3.3. Center and re-estimate.

To *center* a variable means to create a new variable that contains the same information as the original, but has a new mean of zero. For example, to make centered versions of shade and water, just subtract the mean of the original from each value:

R code
7.23

```
d$shade.c <- d$shade - mean(d$shade)
d$water.c <- d$water - mean(d$water)
```

The new centered variable has the same variance as the original, but now has a mean of zero. In this case, centering recodes the levels of water and shade so that instead of their ranging from 1 to 3, they now range from -1 to 1.

Centering in this analysis will do two things for us. First, it'll fix our previous problem with maximum iterations. Second, it'll make the estimates easier to interpret. Let's re-estimate the two regression models, but now using the new centered variables shade.c and water.c. I'm also going to remove the optim fixes from before, because now we don't need them. But I will add explicit start lists to each model, because the very flat priors we're using here provide terrible random starting locations.

R code
7.24

```
m7.8 <- map(
    alist(
        blooms ~ dnorm( mu , sigma ) ,
        mu <- a + bW*water.c + bS*shade.c ,
        a ~ dnorm( 130 , 100 ) ,
        bW ~ dnorm( 0 , 100 ) ,
        bS ~ dnorm( 0 , 100 ) ,
        sigma ~ dunif( 0 , 100 )
    ) ,
    data=d ,
    start=list(a=mean(d$blooms),bW=0,bS=0,sigma=sd(d$blooms)) )
m7.9 <- map(
    alist(
        blooms ~ dnorm( mu , sigma ) ,
        mu <- a + bW*water.c + bS*shade.c + bWS*water.c*shade.c ,
        a ~ dnorm( 130 , 100 ) ,
        bW ~ dnorm( 0 , 100 ) ,
        bS ~ dnorm( 0 , 100 ) ,
        bWS ~ dnorm( 0 , 100 ) ,
        sigma ~ dunif( 0 , 100 )
    ) ,
    data=d ,
    start=list(a=mean(d$blooms),bW=0,bS=0,bWS=0,sigma=sd(d$blooms)) )
coeftab(m7.8,m7.9)
```

```
        m7.8    m7.9
a       129.00  129.01
bW       74.22   74.96
bS      -40.74  -41.14
sigma    57.35   45.22
bWS         NA  -51.87
nobs        27      27
```

Now when we compare the posterior means across the two models, the main effects are the same. Unlike before, the direction of the association for shade has not changed. More water appears to directly increase blooms, while more shade directly decreases them. Meanwhile, the interaction posterior mean has remained the same as it was in the non-centered model.

Let's explain why centering had these effects.

7.3.3.1. *Estimation worked better.* Estimation failed with the un-centered predictor variables, because there was a long way to go from the start values you provided to map and the MAP values it was seeking. After you centered the variables, there was less distance to go. The primary reason is that, when the predictors are centered, the MAP value for α is just the empirical mean for the outcome, mean(d$blooms). That is also the start value given to map. With un-centered predictors, the MAP value for α lies a long way from the empirical mean. Hence, the long search that failed.

7.3.3.2. *Estimates changed less across models.* Why did centering the predictor variables result in the main effect posterior means remaining the same across the models with and without the interaction? In the un-centered models, the interaction effect is applied to every case, and so none of the parameters in μ makes sense alone. This is because neither of the predictors in those models, shade and water, are ever zero. As a result the interaction parameter always factors into generating a prediction. Consider for example a tulip at the average moisture and shade levels, 2 in each case. The expected blooms for such a tulip is:

$$\mu_i|_{S_i=2,W_i=2} = \alpha + \beta_W(2) + \beta_S(2) + \beta_{WS}(2 \times 2)$$

So to figure out the effect of increasing water by 1 unit, you have to use all of the β parameters. Plugging in the MAP values for the un-centered interaction model, m7.7, we get:

$$\mu_i|_{S_i=2,W_i=2} = -150.8 + 181.5(2) + 64.1(2) - 52.9 \times 2 \times 2$$

You can compute the prediction in R:

R code
7.25

```
k <- coef(m7.7)
k[1] + k[2]*2 + k[3]*2 + k[4]*2*2
```

```
       a
129.6895
```

And by no coincidence at all, this is the same prediction we get from the centered interaction model, m7.9, also using the mean values of both predictors:

R code
7.26

```
k <- coef(m7.9)
k[1] + k[2]*0 + k[3]*0 + k[4]*0*0
```

```
      a
129.008
```

In this case, however, you can arrive at this fact straight away, because the mean value of each predictor is zero, and so all that remains is the intercept:

$$\mu_i|_{S_i=0, W_i=0}2 = \alpha + \beta_W(0) + \beta_S(0) + \beta_{WS}(0 \times 0) = \alpha$$

The intercept actually means something, when you center the predictors. It becomes the grand mean of the outcome variable, mean(d$blooms). This ease of interpretation alone is a good reason to center predictor variables.

So how can we read these improved, centered estimates? Here's the table of estimates:

R code
7.27 `precis(m7.9)`

```
        Mean StdDev    5.5%   94.5%
a     129.01   8.67  115.15 142.87
bW     74.96  10.60   58.02  91.90
bS    -41.14  10.60  -58.08 -24.20
bWS   -51.87  12.95  -72.57 -31.18
sigma  45.22   6.15   35.39  55.06
```

And here are justifiable readings of each:

- The estimate a, α, is the expected value of blooms when both water and shade are at their average values. Their average values are both zero (0), because they were centered before fitting the model.
- The estimate bW, β_W, is the expected change in blooms when water increases by one unit *and* shade is at its average value (of zero). This parameter does not tell you the expected rate of change for any other value of shade. This estimate suggests that when shade is at its average value, increasing water is highly beneficial to blooms.
- The estimate bS, β_S, is the expected change in blooms when shade increases by one unit *and* water is at its average value (of zero). This parameter does not tell you the expected rate of change for any other value of water. This estimate suggests that when water is at its average value, increasing shade is highly detrimental to blooms.
- The estimate bWS, β_{WS}, is the interaction effect. Like all linear interactions, it can be explained in more than one way. First, the estimate tells us the expected change in the influence of water on blooms when increasing shade by one unit. Second, it tells us the expected change in the influence of shade on blooms when increasing water by one unit.

So why is the interaction estimate, bWS, negative? The short answer is that water and shade have opposite effects on blooms, but that each also makes the other more important to the outcome. If you don't see how to read that from the number -52, you are in good company. And that's why the best thing to do is to plot implied predictions.

7.3.4. Plotting implied predictions. Golems (models) have awesome powers of reason, but terrible people skills. The golem provides a posterior distribution of plausibility for combinations of parameter values. But for us humans to understand its implications, we need to decode the posterior into something else. Centered predictors or not, plotting posterior predictions always tells you what the golem is thinking, on the scale of the outcome. That's why we've emphasized plotting so much. But in previous chapters, there were no interactions. As a result, when plotting model predictions as a function of any one predictor, you could hold the other predictors constant at any value you liked. So the choice of which values to set the un-viewed predictor variables to hardly mattered.

Now that'll be different. Once there are interactions in a model, the effect of changing a predictor depends upon the values of the other predictors. Maybe the simplest way to go about plotting such interdependency is to make a frame of multiple bivariate plots. In each plot, you choose different values for the un-viewed variables. Then by comparing the plots to one another, you can see how big of a difference the changes make.

Here's how you might accomplish this visualization, for the tulip data. I'm going to make three plots in a single panel. Such a panel of three plots that are meant to be viewed together is a **TRIPTYCH**, and triptych plots are very handy for understanding the impact of interactions. Here's the strategy. We want each plot to show the bivariate relationship between shade and blooms, as predicted by the model. Each plot will plot predictions for a different value of water. For this example, it is easy to pick which values of water to use, because there are only three values: -1, 0, and 1 (this variable was centered, recall). So the first plot will show the predicted relationship between blooms and shade, holding water constant at -1. The second plot will show the same relationship, but now holding water constant at 0. The final plot will hold water constant at 1. In addition, in each plot, I'll show only the raw data that has the water value appropriate in each case.

You already know how to produce each plot individually, using the same kind of code as previous chapters. I'm going to wrap that kind of code in a loop now, and iterate over the three values of water.c. Here's the code:

R code
7.28

```
# make a plot window with three panels in a single row
par(mfrow=c(1,3)) # 1 row, 3 columns

# loop over values of water.c and plot predictions
shade.seq <- -1:1
for ( w in -1:1 ) {
    dt <- d[d$water.c==w,]
    plot( blooms ~ shade.c , data=dt , col=rangi2 ,
        main=paste("water.c =",w) , xaxp=c(-1,1,2) , ylim=c(0,362) ,
        xlab="shade (centered)" )
    mu <- link( m7.9 , data=data.frame(water.c=w,shade.c=shade.seq) )
    mu.mean <- apply( mu , 2 , mean )
    mu.PI <- apply( mu , 2 , PI , prob=0.97 )
    lines( shade.seq , mu.mean )
    lines( shade.seq , mu.PI[1,] , lty=2 )
    lines( shade.seq , mu.PI[2,] , lty=2 )
}
```

The first line uses par, which manipulates graphical *par*ameter settings. This command tells R's plot window to divide itself up into one row and three columns. This prepares the panel to be a triptych. Then the for statement defines a loop that will assign values -1, 0, and 1 to w on each pass. The rest of the code merely extracts the raw data with the right water.c value for each pass through the loop and then computes and plots the predicted mean and 97% posterior interval of the mean.

FIGURE 7.7 shows the results, for both the model without the interaction (top row) and the model with the interaction (bottom row). In each plot, the horizontal axis is shade level,

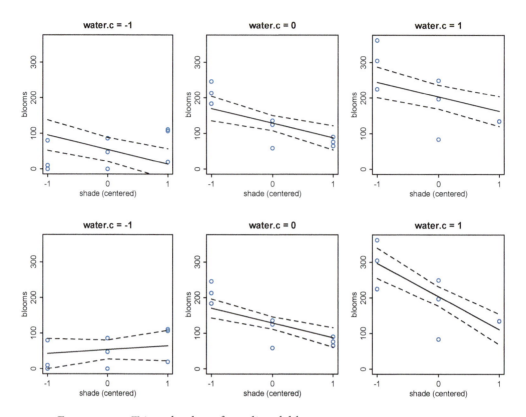

FIGURE 7.7. Triptych plot of predicted blooms across water treatments, without (top row) and with (bottom row) an interaction effect. Blue points in each plot are data. The solid line is the posterior mean and the dashed lines give 97% interval of the mean. Top row: Without the interaction, model m7.8. Each of the three plots of blooms against shade level is for a different water level. The slope of the regression line in each case is exactly the same, because there is no interaction in this model. Bottom row: With the interaction, model m7.9. Now the slope of blooms against shade has a different value in each plot.

from low to high. The vertical axis is blooms. The blue points in each plot are those individual cases that match the water value displayed at the top of each plot. The solid lines are the predicted mean in each case, and the dashed lines are the 97% confidence intervals of those means. Note that the non-interaction model in the top row displays exactly the same slope in each plot within the triptych. The height of that line does change quite a lot across water values. But it always slopes downward at the same rate, in each plot. The interaction model on the bottom row, in contrast, shows a different predicted slope for each value of water. When water is at its lowest, on the left, shade has little effect at all, being nearly flat across all values of shade. There is a very weak positive trend, but substantial uncertainty about it. At the mean water value in the middle plot, increasing shade clearly reduces the size of blooms. Going from the least (−1) to the most shade, blooms are approximately halved in size. At the highest water value, on the right, the slope becomes even more negative, and shade is an even stronger predictor of smaller blooms, as shade increases. Going from the least to the most shade, bloom size declines by about two-thirds.

What is going on here? The likely explanation for these results is that tulips need both water and light to produce blooms. At low water levels, shade can't have much of an effect, because the tulips don't have enough water to produce blooms anyway. At higher water levels, shade can matter more, because the tulips have enough water to produce some blooms. At very high water levels, water is no longer limiting the blooms very much, and so shade can have a much more dramatic impact on the outcome. The same explanation works symmetrically for shade. If there isn't enough light, then more water hardly helps. You could remake FIGURE 7.7 with water on the horizontal axes and shade level varied from left to right, if you'd like to visualize the model predictions that way.

7.4. Interactions in design formulas

To specify interaction models using design formulas, like those used by lm and other automated model fitting functions, you just strip out the parameters from the linear model. Keep in mind that lm forces flat priors, which are hardly ever the best priors.

For example, here's a model with an interaction between x and z:

$$y_i \sim \text{Normal}(\mu_i, \sigma)$$
$$\mu_i = \alpha + \beta_x x_i + \beta_z z_i + \beta_{xz} x_i z_i$$

As a design formula, this is:

```
m7.x <- lm( y ~ x + z + x*z , data=d )
```
R code
7.29

You can also omit the pure x and z terms, the *main effects*, and still get the same model. The following fits the same model as above:

```
m7.x <- lm( y ~ x*z , data=d )
```
R code
7.30

If you do not want the main effects in the model, you must explicitly subtract them from the formula. For example, here's a model with an interaction, but without one of the main effects:

$$y_i \sim \text{Normal}(\mu_i, \sigma)$$
$$\mu_i = \alpha + \beta_x x_i + \beta_{xz} x_i z_i$$

Why would you ever fit this model? When you know a priori that there is no direct effect of z on the outcome, when $\beta_z = 0$ by assumption. As a design formula, this model is:

```
m7.x <- lm( y ~ x + x*z - z , data=d )
```
R code
7.31

There will be an example of such a model later on, in the chapter on count models.

For higher-order interactions, just multiply more predictors together. The design formula always implies all of the lower-order interactions, even though you don't explicitly state them. For example, this model has a three-way interaction between x, z, and w, as well as three two-way interactions and all three main effects:

$$y_i \sim \text{Normal}(\mu_i, \sigma)$$
$$\mu_i = \alpha + \beta_x x_i + \beta_z z_i + \beta_w w_i + \beta_{xz} x_i z_i + \beta_{xw} x_i w_i + \beta_{zw} z_i w_i + \beta_{xzw} x_i z_i w_i$$

As a design formula, this is all that's needed:

R code
7.32
```
m7.x <- lm( y ~ x*z*w , data=d )
```

And you can explicitly subtract any lower-order interactions or main effects to construct a reduced model.

If you are ever in doubt as to how these expansions work, you can get direct access to the function that `lm` uses internally to expand the design formula. Here's an example, expanding a three-way interaction into a full set of terms:

R code
7.33
```
x <- z <- w <- 1
colnames( model.matrix(~x*z*w) )
```

```
[1] "(Intercept)" "x"          "z"          "w"          "x:z"
[6] "x:w"         "z:w"        "x:z:w"
```

The 1 assigned to the symbols x, z, and w is arbitrary. Any number would do. It's just to make some simple variables to feed into the function `model.matrix`. That function takes the design formula, absent the outcome variable, and expands it into a full set of variables for a linear model. This expanded set is the *model matrix*. There's an overthinking box about it on page 124. Finally, those : symbols stand for multiplication.

7.5. Summary

This chapter introduced *interactions*, which allow for the association between a predictor and an outcome to depend upon the value of another predictor. Interactions can be difficult to interpret, and so the chapter also introduced *triptych* plots that help in visualizing the effect of an interaction. No new coding skills were introduced, but the statistical models considered were among the most complicated so far in the book. To go any further, we're going to need a more capable conditioning engine to fit our models to data. That's the topic of the next chapter.

7.6. Practice

Easy.

7E1. For each of the causal relationships below, name a hypothetical third variable that would lead to an interaction effect.

    (1) Bread dough rises because of yeast.
    (2) Education leads to higher income.
    (3) Gasoline makes a car go.

7E2. Which of the following explanations invokes an interaction?

(1) Caramelizing onions requires cooking over low heat and making sure the onions do not dry out.

(2) A car will go faster when it has more cylinders or when it has a better fuel injector.

(3) Most people acquire their political beliefs from their parents, unless they get them instead from their friends.

(4) Intelligent animal species tend to be either highly social or have manipulative appendages (hands, tentacles, etc.).

7E3. For each of the explanations in 7E2, write a linear model that expresses the stated relationship.

Medium.

7M1. Recall the tulips example from the chapter. Suppose another set of treatments adjusted the temperature in the greenhouse over two levels: cold and hot. The data in the chapter were collected at the cold temperature. You find none of the plants grown under the hot temperature developed any blooms at all, regardless of the water and shade levels. Can you explain this result in terms of interactions between water, shade, and temperature?

7M2. Can you invent a regression equation that would make the bloom size zero, whenever the temperature is hot?

7M3. In parts of North America, ravens depend upon wolves for their food. This is because ravens are carnivorous but cannot usually kill or open carcasses of prey. Wolves however can and do kill and tear open animals, and they tolerate ravens co-feeding at their kills. This species relationship is generally described as a "species interaction." Can you invent a hypothetical set of data on raven population size in which this relationship would manifest as a statistical interaction? Do you think the biological interaction could be linear? Why or why not?

Hard.

7H1. Return to the `data(tulips)` example in the chapter. Now include the `bed` variable as a predictor in the interaction model. Don't interact `bed` with the other predictors; just include it as a main effect. Note that `bed` is categorical. So to use it properly, you will need to either construct dummy variables or rather an index variable, as explained in Chapter 6.

7H2. Use WAIC to compare the model from **7H1** to a model that omits `bed`. What do you infer from this comparison? Can you reconcile the WAIC results with the posterior distribution of the `bed` coefficients?

7H3. Consider again the `data(rugged)` data on economic development and terrain ruggedness, examined in this chapter. One of the African countries in that example, Seychelles, is far outside the cloud of other nations, being a rare country with both relatively high GDP and high ruggedness. Seychelles is also unusual, in that it is a group of islands far from the coast of mainland Africa, and its main economic activity is tourism.

One might suspect that this one nation is exerting a strong influence on the conclusions. In this problem, I want you to drop Seychelles from the data and re-evaluate the hypothesis that the relationship of African economies with ruggedness is different from that on other continents.

(a) Begin by using `map` to fit just the interaction model:

$$y_i \sim \text{Normal}(\mu_i, \sigma)$$
$$\mu_i = \alpha + \beta_A A_i + \beta_R R_i + \beta_{AR} A_i R_i$$

where y is log GDP per capita in the year 2000 (log of `rgdppc_2000`); A is `cont_africa`, the dummy variable for being an African nation; and R is the variable `rugged`. Choose your own priors. Compare the inference from this model fit to the data without Seychelles to the same model fit to the full data. Does it still seem like the effect of ruggedness depends upon continent? How much has the expected relationship changed?

(b) Now plot the predictions of the interaction model, with and without Seychelles. Does it still seem like the effect of ruggedness depends upon continent? How much has the expected relationship changed?

(c) Finally, conduct a model comparison analysis, using WAIC. Fit three models to the data without Seychelles:

$$\text{Model 1}: y_i \sim \text{Normal}(\mu_i, \sigma)$$
$$\mu_i = \alpha + \beta_R R_i$$
$$\text{Model 2}: y_i \sim \text{Normal}(\mu_i, \sigma)$$
$$\mu_i = \alpha + \beta_A A_i + \beta_R R_i$$
$$\text{Model 3}: y_i \sim \text{Normal}(\mu_i, \sigma)$$
$$\mu_i = \alpha + \beta_A A_i + \beta_R R_i + \beta_{AR} A_i R_i$$

Use whatever priors you think are sensible. Plot the model-averaged predictions of this model set. Do your inferences differ from those in (b)? Why or why not?

7H4. The values in data(nettle) are data on language diversity in 74 nations.[115] The meaning of each column is given below.

(1) country: Name of the country
(2) num.lang: Number of recognized languages spoken
(3) area: Area in square kilometers
(4) k.pop: Population, in thousands
(5) num.stations: Number of weather stations that provided data for the next two columns
(6) mean.growing.season: Average length of growing season, in months
(7) sd.growing.season: Standard deviation of length of growing season, in months

Use these data to evaluate the hypothesis that language diversity is partly a product of food security. The notion is that, in productive ecologies, people don't need large social networks to buffer them against risk of food shortfalls. This means ethnic groups can be smaller and more self-sufficient, leading to more languages per capita. In contrast, in a poor ecology, there is more subsistence risk, and so human societies have adapted by building larger networks of mutual obligation to provide food insurance. This in turn creates social forces that help prevent languages from diversifying.

Specifically, you will try to model the number of languages per capita as the outcome variable:

R code
7.34
```
d$lang.per.cap <- d$num.lang / d$k.pop
```

Use the logarithm of this new variable as your regression outcome. (A count model would be better here, but you'll learn those later, in Chapter 10.)

This problem is open ended, allowing you to decide how you address the hypotheses and the uncertain advice the modeling provides. If you think you need to use WAIC anyplace, please do. If you think you need certain priors, argue for them. If you think you need to plot predictions in a certain way, please do. Just try to honestly evaluate the main effects of both mean.growing.season and sd.growing.season, as well as their two-way interaction, as outlined in parts (a), (b), and (c) below. If you are not sure which approach to use, try several.

(a) Evaluate the hypothesis that language diversity, as measured by log(lang.per.cap), is positively associated with the average length of the growing season, mean.growing.season. Consider log(area) in your regression(s) as a covariate (not an interaction). Interpret your results.

(b) Now evaluate the hypothesis that language diversity is negatively associated with the standard deviation of length of growing season, sd.growing.season. This hypothesis follows from uncertainty in harvest favoring social insurance through larger social networks and therefore fewer languages. Again, consider log(area) as a covariate (not an interaction). Interpret your results.

(c) Finally, evaluate the hypothesis that mean.growing.season and sd.growing.season interact to synergistically reduce language diversity. The idea is that, in nations with longer average growing seasons, high variance makes storage and redistribution even more important than it would be otherwise. That way, people can cooperate to preserve and protect windfalls to be used during the droughts. These forces in turn may lead to greater social integration and fewer languages.

8 Markov Chain Monte Carlo

For most of Western history, chance has been a villain. In classic Roman civilization, chance was personified by Fortuna, goddess of cruel fate, with her spinning wheel of luck. Opposed to her sat Minerva, goddess of wisdom and understanding. Only the desperate would pray to Fortuna, while everyone implored Minerva for aid. Certainly science was the domain of Minerva, a realm with no useful role for Fortuna to play.

But by the beginning of the 20th century, the opposition between Fortuna and Minerva had changed to a collaboration. Scientists, servants of Minerva, began publishing books of random numbers, instruments of chance to be used for learning about the world. Now, chance and wisdom share a cooperative relationship, and few of us are any longer bewildered by the notion that an understanding of chance could help us acquire wisdom. Everything from weather forecasting to finance to evolutionary biology is dominated by the study of stochastic processes.[116]

This chapter introduces one of the more marvelous examples of how Fortuna and Minerva cooperate: the estimation of posterior probability distributions using a stochastic process known as MARKOV CHAIN MONTE CARLO (MCMC) estimation. Unlike in every earlier chapter in this book, here we'll produce samples from the joint posterior of a model without maximizing anything. Instead of having to lean on quadratic and other approximations of the shape of the posterior, now we'll be able to sample directly from the posterior without assuming a Gaussian, or any other, shape for it.

The cost of this power is that it may take much longer for our estimation to complete, and usually more work is required to specify the model as well. But the benefit is escaping the awkwardness of assuming multivariate normality. Equally important is the ability to directly estimate models, such as the generalized linear and multilevel models of later chapters. Such models routinely produce non-Gaussian posterior distributions, and sometimes they cannot be estimated at all with the techniques of earlier chapters.

The good news is that tools for building and inspecting MCMC estimates are getting better all the time. In this chapter you'll meet a convenient way to convert the map formulas you've used so far into Markov chains. The engine that makes this possible is STAN (free and online at: mc-stan.org). Stan's creators describe it as "a probabilistic programming language implementing statistical inference." You won't be working directly in Stan to begin with—the rethinking package provides tools that hide it from you for now. But as you move on to more advanced techniques, you'll be able to generate Stan versions of the models you already understand. Then you can tinker with them and witness the power of a fully armed and operational Stan.

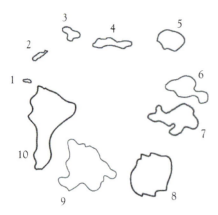

FIGURE 8.1. Good King Markov's island kingdom. Each of the 10 islands has a population proportional to its number, 1 through 10. The King's goal is to visit each island, in the long run, in proportion to its population size. This can be accomplished by the *Metropolis algorithm*.

Rethinking: Stan was a man. The Stan programming language is not an abbreviation or acronym. Rather, it is named after Stanislaw Ulam (1909–1984). Ulam is credited as one of the inventors of Markov chain Monte Carlo. Together with Ed Teller, Ulam applied it to designing fusion bombs. But he and others soon applied the general Monte Carlo method to diverse problems of less monstrous nature. Ulam made important contributions in pure mathematics, chaos theory, and molecular and theoretical biology, as well.

8.1. Good King Markov and His island kingdom

For the moment, forget about posterior densities and MCMC. Consider instead the tale of Good King Markov.[117] King Markov was a benevolent autocrat of an island kingdom, a circular archipelago, with 10 islands. Each island was neighbored by two others, and the entire archipelago formed a ring. The islands were of different sizes, and so had different sized populations living on them. The second island was about twice as populous as the first, the third about three times as populous as the first, and so on, up to the largest island, which was 10 times as populous as the smallest. The good king's island kingdom is displayed in FIGURE 8.1, with the islands numbered by their relative population sizes.

The Good King was an autocrat, but he did have a number of obligations to His people. Among these obligations, King Markov agreed to visit each island in His kingdom from time to time. Since the people love their king, each island would prefer that he visit them more often. And so everyone agreed that the king should visit each island in proportion to its population size, visiting the largest island 10 times as often as the smallest, for example.

The Good King Markov, however, wasn't one for schedules or bookkeeping, and so he wanted a way to fulfill his obligation without planning his travels months ahead of time. Also, since the archipelago was a ring, the King insisted that he only move among adjacent islands, to minimize time spent on the water—like many citizens of his kingdom, the king believes there are sea monsters in the middle of the archipelago.

The king's advisor, a Mr Metropolis, engineered a clever solution to these demands. We'll call this solution the *Metropolis algorithm*. Here's how it works.

(1) Wherever the King is, each week he decides between staying put for another week or moving to one of the two adjacent islands. To decide his next move, he flips a coin.

(2) If the coin turns up heads, the King considers moving to the adjacent island clockwise around the archipelago. If the coin turns up tails, he considers instead moving counterclockwise. Call the island the coin nominates the *proposal* island.

(3) Now, to see whether or not he moves to the proposal island, King Markov counts out a number of seashells equal to the relative population size of the proposal island. So for example, if the proposal island is number 9, then he counts out 9 seashells. Then he also counts out a number of stones equal to the relative population of the current island. So for example, if the current island is number 10, then King Markov ends up holding 10 stones, in addition to the 9 seashells.

(4) When there are more seashells than stones, King Markov always moves to the proposal island. But if there are fewer shells than stones, he discards a number of stones equal to the number of shells. So for example, if there are 4 shells and 6 stones, he ends up with 4 shells and $6 - 4 = 2$ stones. Then he places the shells and the remaining stones in a bag. He reaches in and randomly pulls out one object. If it is a shell, he moves to the proposal island. Otherwise, he stays put another week. As a result, the probability that he moves is equal to the number of shells divided by the original number of stones.

This procedure may seem baroque and, honestly, a bit crazy. But it does work. The king will appear to move around the islands randomly, sometimes staying on one island for weeks, other times bouncing around without apparent pattern. But in the long run, this procedure guarantees that the king will be found on each island in proportion to its population size.

You can prove this to yourself, by simulating King Markov's journey. Here's a short piece of code to do this, storing the history of the king's island positions in the vector `positions`:

R code
8.1

```
num_weeks <- 1e5
positions <- rep(0,num_weeks)
current <- 10
for ( i in 1:num_weeks ) {
    # record current position
    positions[i] <- current

    # flip coin to generate proposal
    proposal <- current + sample( c(-1,1) , size=1 )
    # now make sure he loops around the archipelago
    if ( proposal < 1 ) proposal <- 10
    if ( proposal > 10 ) proposal <- 1

    # move?
    prob_move <- proposal/current
    current <- ifelse( runif(1) < prob_move , proposal , current )
}
```

I've added comments to this code, to help you decipher it. The first three lines just define the number of weeks to simulate, an empty history vector, and a starting island position (the biggest island, number 10). Then the `for` loop steps through the weeks. Each week, it records the king's current position. Then it simulates a coin flip to nominate a proposal island. The only trick here lies in making sure that a proposal of "11" loops around to island 1 and a

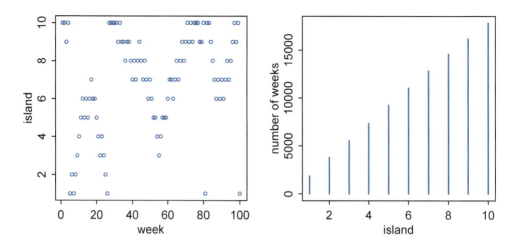

FIGURE 8.2. Results of the king following the Metropolis algorithm. The left-hand plot shows the king's position (vertical axis) across weeks (horizontal axis). In any particular week, it's nearly impossible to say where the king will be. The right-hand plot shows the long-run behavior of the algorithm, as the time spent on each island turns out to be proportional to its population size.

proposal of "0" loops around to island 10. Finally, a random number between zero and one is generated (runif(1)), and the king moves, if this random number is less than the ratio of the proposal island's population to the current island's population (proposal/current).

You can see the results of this simulation in FIGURE 8.2. The left-hand plot shows the king's location across the first 100 weeks of his simulated travels. As you move from the left to the right in this plot, the points show the king's location through time. The king travels among islands, or sometimes stays in place for a few weeks. This plot demonstrates the seemingly pointless path the Metropolis algorithm sends the king on. The right-hand plot shows that the path is far from pointless, however. The horizontal axis is now islands (and their relative populations), while the vertical is the number of weeks the king is found on each. After the entire 100,000 weeks (almost 2000 years) of the simulation, you can see that the proportion of time spent on each island converges to be almost exactly proportional to the relative populations of the islands.

The algorithm will still work in this way, even if we allow the king to be equally likely to propose a move to any island from any island, not just among neighbors. As long as King Markov still uses the ratio of the proposal island's population to the current island's population as his probability of moving, in the long run, he will spend the right amount of time on each island. The algorithm would also work for any size archipelago, even if the king didn't know how many islands were in it. All he needs to know at any point in time is the population of the current island and the population of the proposal island. Then, without any forward planning or backwards record keeping, King Markov can satisfy his royal obligation to visit his people proportionally.

8.2. Markov chain Monte Carlo

The precise algorithm King Markov used is a special case of the general Metropolis algorithm from the real world.[118] And this algorithm is an example of Markov chain Monte Carlo. In real applications, the goal is of course not to help an autocrat schedule his journeys, but instead to draw samples from an unknown and usually complex target distribution, like a posterior probability distribution.

- The "islands" in our objective are parameter values, and they need not be discrete, but can instead take on a continuous range of values as usual.
- The "population sizes" in our objective are the posterior probabilities at each parameter value.
- The "weeks" in our objective are samples taken from the joint posterior of the parameters in the model.

Provided the way we choose our proposed parameter values at each step is symmetric—so that there is an equal chance of proposing from A to B and from B to A—then the Metropolis algorithm will eventually give us a collection of samples from the joint posterior. We can then use these samples just like all the samples you've already used in this book.

The Metropolis algorithm is the grandparent of several different strategies for getting samples from unknown posterior distributions. In the remainder of this section, I briefly explain the concepts behind two of the most important in contemporary Bayesian inference: Gibbs sampling and Hamiltonian (aka Hybrid, aka HMC) Monte Carlo. Both are pretty common in applied Bayesian statistics. This book will use HMC, but a casual understanding of both algorithms is helpful for appreciating the advantages of each. After becoming acquainted with them, we'll turn in the second half of the chapter to using Hamiltonian Monte Carlo to do some new things with linear regression models.

8.2.1. Gibbs sampling. The Metropolis algorithm works whenever the probability of proposing a jump to B from A is equal to the probability of proposing A from B, when the proposal distribution is symmetric. There is a more general method, known as Metropolis-Hastings,[119] that allows asymmetric proposals. This would mean, in the context of King Markov's fable, that the King's coin were biased to lead him clockwise on average.

Why would we want an algorithm that allows asymmetric proposals? One reason is that it makes it easier to handle parameters, like standard deviations, that have boundaries at zero. A better reason, however, is that it allows us to generate savvy proposals that explore the posterior distribution more efficiently. By "more efficiently," I mean that we can acquire an equally good image of the posterior distribution in fewer steps.

The most common way to generate savvy proposals is a technique known as Gibbs sampling.[120] Gibbs sampling is a variant of the Metropolis-Hastings algorithm that uses clever proposals and is therefore more efficient. By "efficient," I mean that you can get a good estimate of the posterior from Gibbs sampling with many fewer samples than a comparable Metropolis approach. The improvement arises from *adaptive proposals* in which the distribution of proposed parameter values adjusts itself intelligently, depending upon the parameter values at the moment.

How Gibbs sampling computes these adaptive proposals depends upon using particular combinations of prior distributions and likelihoods known as *conjugate pairs*. Conjugate pairs have analytical solutions for the posterior distribution of an individual parameter. And these solutions are what allow Gibbs sampling to make smart jumps around the joint posterior distribution of all parameters.

In practice, Gibbs sampling can be very efficient, and it's the basis of popular Bayesian model fitting software like BUGS (Bayesian inference Using Gibbs Sampling) and JAGS (Just Another Gibbs Sampler). In these programs, you compose your statistical model using definitions very similar to what you've been doing so far in this book. The software automates the rest, to the best of its ability.

But there are some severe limitations to Gibbs sampling. First, maybe you don't want to use conjugate priors. Some conjugate priors seem silly, and choosing a prior so that the model fits efficiently isn't really a strong argument from a scientific perspective. Second, as models become more complex and contain hundreds or thousands or tens of thousands of parameters, Gibbs sampling becomes shockingly inefficient. In those cases, there are other algorithms.

8.2.2. Hamiltonian Monte Carlo.

> It appears to be a quite general principle that, whenever there is a randomized way of doing something, then there is a nonrandomized way that delivers better performance but requires more thought. —E. T. Jaynes[121]

The Metropolis algorithm and Gibbs sampling are both highly random procedures. They try out new parameter values and see how good they are, compared to the current values. But Gibbs sampling gains efficiency by reducing this randomness and exploiting knowledge of the target distribution. This seems to fit Jaynes' suggestion, quoted above, that when there is a random way of accomplishing some calculation, there is probably a less random way that is better. This less random way may require a lot more thought, however.

HAMILTONIAN MONTE CARLO (or Hybrid Monte Carlo, HMC) pushes Jaynes' principle further. HMC is much more computationally costly than are Metropolis or Gibbs sampling. But its proposals are typically much more efficient. As a result, it doesn't need as many samples to describe the posterior distribution. And as models become more complex—thousands or tens of thousands of parameters—HMC can really outshine other algorithms.

We're going to be using HMC on and off for the remainder of this book. You won't have to implement it yourself. But understanding some of the concept behind it will help you grasp how it outperforms Metropolis and Gibbs sampling and also why it is not a universal solution to all MCMC problems.

Suppose King Markov's cousin Monty is King on the mainland. Monty's kingdom is not a discrete set of islands. Instead, it is a continuous territory stretched out along a narrow valley. But the King has a similar obligation: to visit his citizens in proportion to their local density. Like Markov, Monty doesn't wish to bother with schedules and calculations. So likewise he's not going to take a full census and solve for some optimal travel schedule.

Also like Markov, Monty has a highly educated and mathematically gifted advisor. His name is Hamilton. Hamilton realized that a much more efficient way to visit the citizens in the continuous Kingdom is to travel back and forth along its length. In order to spend more time in densely settled areas, they should slow the royal vehicle down when houses grow more dense. Likewise, they should speed up when houses grow more sparse. This strategy requires knowing how quickly population density is changing, at their current location. But it doesn't require remembering where they've been or knowing the population distribution anyplace else. And a major benefit of this strategy compared to that of Metropolis is that the King makes a full sweep of the kingdom before revisiting anyone.

This story is analogous to how Hamiltonian Monte Carlo works. In statistical applications, the royal vehicle is the current vector of parameter values. Let's consider the single parameter case, just to keep things simple. In that case, the log-posterior is like a bowl, with the MAP at its nadir. Then the job is to sweep across the surface of the bowl, adjusting speed in proportion to how high up we are.

HMC really does run a physics simulation, pretending the vector of parameters gives the position of a little frictionless particle. The log-posterior provides a surface for this particle to glide across. When the log-posterior is very flat, because there isn't much information in the likelihood and the priors are rather flat, then the particle can glide for a long time before the slope (gradient) makes it turn around. When instead the log-posterior is very steep, because either the likelihood or the priors are very concentrated, then the particle doesn't get far before turning around.

All of this sounds, and is, very complex. But what is gained from all of this complexity is very efficient sampling of complex models. In cases where ordinary Metropolis or Gibbs sampling wander slowly through parameter space, Hamiltonian Monte Carlo remains efficient. This is especially true when working with multilevel models with hundreds or thousands of parameters. So HMC is becoming a popular conditioning engine.

As always, there are some limitations. HMC requires continuous parameters. It can't glide through a discrete parameter. In practice, this means that certain advanced techniques, like the imputation of discrete missing data, are not possible with HMC alone. And there are types of models that remain difficult for any MCMC strategy. HMC isn't a magic formula.

In practice, a big limitation of HMC is that it needs to be tuned to a particular model and its data—the frictionless particle does need mass, so it can acquire momentum, and the choice of mass can have big effects on efficiency. There are also a number of other parameters that define the HMC algorithm, but no the statistical model, that can change how efficiently the Markov chain samples. Tuning all of those parameters by hand is a pain. That's where an engine like Stan (mc-stan.org) comes in. Stan automates much of that tuning.[122] In the next section, you'll see how to use Stan to fit the models from earlier chapters. As the book continues, you'll encounter models that cannot be fit without some MCMC approach, so HMC and Stan will grow increasingly important.

> **Rethinking: The MCMC horizon.** While the ideas behind Markov chain Monte Carlo are not new, widespread use dates only to the last decade of the 20th century.[123] New variants of and improvements to MCMC algorithms arise all the time. We might anticipate that interesting advances are coming, and that the current crop of tools—Gibbs sampling and HMC for example—will look rather pedestrian in another 20 years. At least we can hope.

8.3. Easy HMC: map2stan

The rethinking package provides a convenient interface, map2stan, to compile lists of formulas, like the lists you've been using so far to construct map estimates, into Stan HMC code. A little more housekeeping is needed to use map2stan: You need to preprocess any variable transformations, and you need to construct a clean data frame with only the variables you will use. But otherwise installing Stan on your computer is the hardest part. And once you get comfortable with interpreting samples produced in this way, you go peek inside and see exactly how the model formulas you already understand correspond to the code that drives the Markov chain.

To see how it's done, let's revisit the terrain ruggedness example from Chapter 7. This code will load the data and reduce it down to cases (nations) that have the outcome variable of interest:

```
library(rethinking)
data(rugged)
d <- rugged
d$log_gdp <- log(d$rgdppc_2000)
dd <- d[ complete.cases(d$rgdppc_2000) , ]
```

So you remember the old way, we're going to repeat the procedure for fitting the interaction model. This model aims to predict log-GDP with terrain ruggedness, continent, and the interaction of the two. Here's the way to do it with map, just like before.

```
m8.1 <- map(
    alist(
        log_gdp ~ dnorm( mu , sigma ) ,
        mu <- a + bR*rugged + bA*cont_africa + bAR*rugged*cont_africa ,
        a ~ dnorm(0,100),
        bR ~ dnorm(0,10),
        bA ~ dnorm(0,10),
        bAR ~ dnorm(0,10),
        sigma ~ dunif(0,10)
    ) ,
    data=dd )
precis(m8.1)
```

```
       Mean StdDev  5.5% 94.5%
a      9.22   0.14  9.00  9.44
bR    -0.20   0.08 -0.32 -0.08
bA    -1.95   0.22 -2.31 -1.59
bAR    0.39   0.13  0.19  0.60
sigma  0.93   0.05  0.85  1.01
```

Just as you saw in the previous chapter.

8.3.1. Preparation. But now we'll also fit this model using Hamiltonian Monte Carlo. This means there will be no more quadratic approximation—if the posterior distribution is non-Gaussian, then we'll get whatever non-Gaussian shape it has. You can use exactly the same formula list as before, but you need to do two additional things.

   (1) Preprocess all variable transformations. If the outcome is transformed somehow, like by taking the logarithm, then do this before fitting the model by constructing a new variable in the data frame. Likewise, if any predictor variables are transformed, including squaring and cubing and such to build polynomial models, then compute these transformed values before fitting the model.
   (2) Once you've got all the variables ready, make a new trimmed down data frame that contains only the variables you will actually use to fit the model. Technically, you don't have to do this. But doing so avoids common problems. For example, if any of the unused variables have missing values, NA, then Stan will refuse to work.

Here's how to do both for the terrain ruggedness model.

```
dd.trim <- dd[ , c("log_gdp","rugged","cont_africa") ]
str(dd.trim)
```
R code
8.4

```
'data.frame': 170 obs. of  3 variables:
 $ log_gdp    : num  7.49 8.22 9.93 9.41 7.79 ...
 $ rugged     : num  0.858 3.427 0.769 0.775 2.688 ...
 $ cont_africa: int  1 0 0 0 0 0 0 0 0 1 ...
```

The data frame dd.trim contains only the three variables we're using.

8.3.2. Estimation. Now provided you have the rstan package installed (mc-stan.org), you can get samples from the posterior distribution with this code:

```
m8.1stan <- map2stan(
    alist(
        log_gdp ~ dnorm( mu , sigma ) ,
        mu <- a + bR*rugged + bA*cont_africa + bAR*rugged*cont_africa ,
        a ~ dnorm(0,100),
        bR ~ dnorm(0,10),
        bA ~ dnorm(0,10),
        bAR ~ dnorm(0,10),
        sigma ~ dcauchy(0,2)
    ) ,
    data=dd.trim )
```
R code
8.5

There is one change to note here, but to explain later. The uniform prior on sigma has been changed to a half-CAUCHY prior. The Cauchy distribution is a useful thick-tailed probability distribution related to the Student t distribution. There's an Overthinking box about it later in the chapter (page 260). You can think of it as a weakly regularizing prior for standard deviations.[124] We'll use it again later in the chapter. And you'll have many chances to get used to it, as the book continues.

But note that it is not necessary to use a half-Cauchy. The uniform prior will still work, and a simple exponential prior is also appropriate. In this example, as in many, there is so much data that the prior hardly matters. There is a practice problem at the end of this chapter to guide you in comparing these priors.

After messages about translating, compiling, and sampling (see the Overthinking box later in this section for some explanations of these messages), map2stan returns an object that contains a bunch of summary information, as well as samples from the posterior distribution of all parameters. You can compare estimates:

```
precis(m8.1stan)
```
R code
8.6

	Mean	StdDev	lower 0.89	upper 0.89	n_eff	Rhat
a	9.24	0.14	9.03	9.47	291	1
bR	-0.21	0.08	-0.32	-0.07	306	1
bA	-1.97	0.23	-2.31	-1.58	351	1
bAR	0.40	0.13	0.20	0.63	350	1
sigma	0.95	0.05	0.86	1.03	566	1

These estimates are very similar to the quadratic approximation. But note a few new things. First, the interval boundaries in the table just above are highest posterior density intervals (HPDI, page 56), not ordinary percentile intervals (PI). Second, there are two new columns, n_eff and Rhat. These columns provide MCMC diagnostic criteria, to help you tell how well estimation worked. We'll discuss them in detail later in the chapter. For now, it's enough to know that n_eff is a crude estimate of the number of independent samples you managed to get. Rhat is a complicated estimate of the convergence of the Markov chains to the target distribution. It should approach 1.00 from above, when all is well.

8.3.3. Sampling again, in parallel.

The example so far is a very easy problem for MCMC. So even the default 1000 samples is enough for accurate inference. But often the default won't be enough. There will be specific advice in Section 8.4 (page 255).

For now, it's worth noting that once you have compiled your Stan model with map2stan, you can draw more samples from it anytime, running as many independent Markov chains as you like. And you can easily parallelize those chains, as well. To run four independent Markov chains for the model above, and to distribute them across separate processors in your computer, just pass the previous fit back to map2stan:

R code
8.7
```
m8.1stan_4chains <- map2stan( m8.1stan , chains=4 , cores=4 )
precis(m8.1stan_4chains)
```

	Mean	StdDev	lower 0.89	upper 0.89	n_eff	Rhat
a	9.23	0.14	9.01	9.45	1029	1
bR	-0.20	0.08	-0.33	-0.08	1057	1
bA	-1.94	0.22	-2.29	-1.58	1154	1
bAR	0.39	0.13	0.17	0.59	1144	1
sigma	0.95	0.05	0.87	1.03	1960	1

The resample function will also recompute DIC and WAIC for you, using only the new samples. You can also add the cores argument to any original map2stan call. It will automatically parallelize the chains.

8.3.4. Visualization.

By plotting the samples, you can get a direct appreciation for how Gaussian (quadratic) the actual posterior density has turned out to be. To pull out the samples, use:

R code
8.8
```
post <- extract.samples( m8.1stan )
str(post)
```

```
List of 5
 $ a    : num [1:1000(1d)] 9.3 9.34 9.21 9.43 9.28 ...
 $ br   : num [1:1000(1d)] -0.133 -0.214 -0.215 -0.229 -0.209 ...
 $ bA   : num [1:1000(1d)] -1.91 -2.25 -2.26 -1.98 -1.99 ...
 $ brA  : num [1:1000(1d)] 0.133 0.367 0.533 0.254 0.468 ...
 $ sigma: num [1:1000(1d)] 0.988 0.949 0.904 0.976 0.934 ...
```

Note that post here is a list, not a data.frame. This fact will be very useful to use later on, when we encounter multilevel models. For now, if it doesn't behave like you expect it to, you can coerce it to a data frame with post<-as.data.frame(post). There are only 1000 samples for each parameter, because that's the default. In this case, it's enough. There's a lot

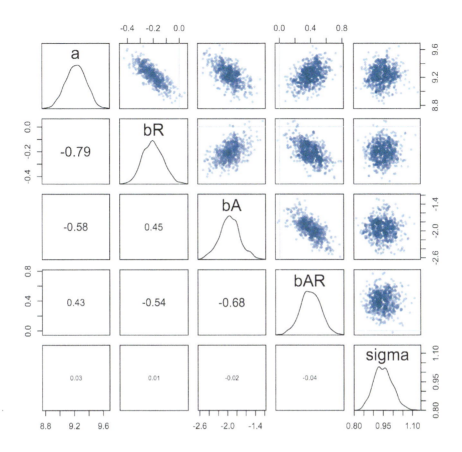

FIGURE 8.3. Pairs plot of the samples produced by Stan. The diagonal shows a density estimate for each parameter. Below the diagonal, correlations between parameters are shown.

more said about number of samples and such in the next major section of this chapter. Put those worries aside for the moment.

To plot all these samples at once, provided there aren't too many parameters, you can use the standard plotting function pairs:

```
pairs(post)
```
R code
8.9

Or use pairs directly on the fit model, so that R knows to display parameter names and parameter correlations:

```
pairs(m8.1stan)
```
R code
8.10

FIGURE 8.3 shows the resulting plot. This is a pairs plot, so it's still a matrix of bivariate scatter plots. But now along the diagonal the smoothed histogram of each parameter is shown, along with its name. And in the lower triangle of the matrix, below the diagonal, the correlation between each pair of parameters is shown, with stronger correlations indicated by relative size.

For this model and these data, the resulting posterior distribution is quite nearly multi-variate Gaussian. The density for `sigma` is certainly skewed in the expected direction. But otherwise the quadratic approximation does almost as well as Hamiltonian Monte Carlo. This is a very simple kind of model structure of course, with Gaussian priors, so an approximately quadratic posterior should be no surprise. Later, we'll see some more exotic posterior distributions.

Overthinking: Stan messages. When you fit a model using `map2stan`, R will first translate your model formula into a Stan language model. Then it sends that model to Stan. The messages you see in your R console are status updates from Stan. Stan first again translates the model, this time into C++ code. That code is then sent to a C++ compiler, to build an executable file is a specialized sampling engine for your model. Then Stan feeds the data and starting values to that executable file, and if all goes well, sampling begins. You will see Stan count through the iterations. During sampling, you might occasionally see a scary looking warning something like this:

```
Informational Message: The current Metropolis proposal is about to be rejected
because of the following issue: Error in function stan::prob::multi_normal_log
(N4stan5agrad3varE):Covariance matrix is not positive definite. Covariance
matrix(0,0) is 0:0. If this warning occurs sporadically, such as for highly
constrained variable types like covariance matrices, then the sampler is
fine, but if this warning occurs often then your model may be either severely
ill-conditioned or misspecified.
```

Severely ill-conditioned or misspecified? That certainly sounds bad. But rarely does this message indicate a serious problem. As long as it happens only a handful of times, and especially if it only happens during warmup, then odds are very good the chain is fine. You should still always check the chain for problems, of course. Just don't panic when you see this message. Keep calm and sample on.

8.3.5. Using the samples. Once you have samples in an object like `post`, you work with them just as you've already learned to do. If you have the samples from the posterior and you know the model, you can do anything: simulate predictions, compute differences between parameters, and calculate DIC and WAIC.

By default `map2stan` computes DIC and WAIC for you. You can extract them with `DIC(m8.1stan)` and `WAIC(m8.1stan)`. DIC and WAIC are also reported in the default show output for a `map2stan` model fit.

R code
8.11

```
show(m8.1stan)
```

```
map2stan model fit
1000 samples from 1 chain

Formula:
log_gdp ~ dnorm(mu, sigma)
mu <- a + bR * rugged + bA * cont_africa + bAR * rugged * cont_africa
a ~ dnorm(0, 100)
bR ~ dnorm(0, 10)
bA ~ dnorm(0, 10)
bAR ~ dnorm(0, 10)
sigma ~ dcauchy(0, 2)
```

```
Log-likelihood at expected values: -229.43
Deviance: 458.85
DIC: 468.73
Effective number of parameters (pD): 4.94

WAIC (SE): 469.3 (14.8)
pWAIC: 5.14
```

This report just reiterates the formulas used to define the Markov chain and then reports information criteria.

For computing predictions, the functions `postcheck`, `link`, and `sim` work on map2stan models just as they do on `map` models. For model comparison, `compare` and `ensemble` also work the same way. Regardless of how you fit and get samples from a model, once you have those samples, the logic is always the same: Process the samples to address questions about the relative plausibility of different parameter values and implied predictions.

8.3.6. Checking the chain. Provided the Markov chain is defined correctly—and it is here—then it is guaranteed to converge in the long run to the answer we want, the posterior distribution. But the machine does sometimes malfunction. In the next major section, we'll dwell on causes of and solutions to malfunction.

For now, let's meet the most broadly useful tool for diagnosing malfunction, a TRACE PLOT. A trace plot merely plots the samples in sequential order, joined by a line. It's King Markov's path through the islands, in the metaphor at the start of the chapter. Looking at the trace plot of each parameter can help to diagnose many common problems. And once you come to recognize a healthy, functioning Markov chain, quick checks of trace plots provide a lot of peace of mind. A trace plot isn't the last thing analysts do to inspect MCMC output. But it's nearly always the first.

In the terrain ruggedness example, the trace plot shows a very healthy chain. View it with:

```
plot(m8.1stan)
```

R code
8.12

The result is shown in FIGURE 8.4. Each plot in this figure is similar to what you'd get if you just used, for example, `plot(post$a,type="l")`, but with some extra information and labeling to help out. You can think of the zig-zagging trace of each parameter as the path the chain took through each dimension of parameter space.

The gray region in each plot, the first 1000 samples, marks the *adaptation* samples. During adaptation, the Markov chain is learning to more efficiently sample from the posterior distribution. So these samples are not necessarily reliable to use for inference. They are automatically discarded by `extract.samples`, which returns only the samples shown in the white regions of FIGURE 8.4.

Now, how is this chain a healthy one? Typically we look for two things in these trace plots: stationarity and good mixing. Stationarity refers to the path staying within the posterior distribution. Notice that these traces, for example, all stick around a very stable central tendency, the center of gravity of each dimension of the posterior. Another way to think of this is that the mean value of the chain is quite stable from beginning to end.

A well-mixing chain means that each successive sample within each parameter is not highly correlated with the sample before it. Visually, you can see this by the rapid zig-zag

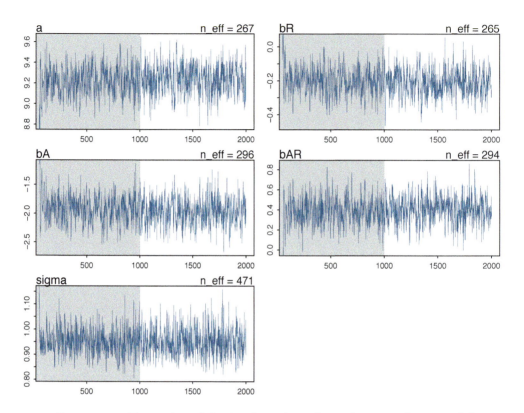

FIGURE 8.4. Trace plot of the Markov chain from the ruggedness model, m8.1stan. This is a clean, healthy Markov chain, both stationary and well-mixing. The gray region is warmup, during which the Markov chain was adapting to improve sampling efficiency. The white region contains the samples used for inference.

motion of each path, as the trace traverses the posterior distribution without getting mired anyplace.

To really understand these points, though, you'll have to see some trace plots for unhealthy chains. That's the project of the next section.

Overthinking: Raw Stan model code. All map2stan does is translate a list of formulas into Stan's modeling language. Then Stan does the rest. Learning how to write Stan code is not necessary for most of the models in this book. But other models do require some direct interaction with Stan, because it is capable of much more than map2stan allows you to express. And even for simple models, you'll gain additional comprehension and control, if you peek into the machine. You can always access the raw Stan code that map2stan produces by using the function stancode. For example, stancode(m8.1stan) prints out the Stan code for the ruggedness model. Before you're familiar with Stan's language, it'll look long and weird. So let's focus on just the most important part, the "model block":

```
model{
    vector[N] mu;
    sigma ~ cauchy( 0 , 2 );
    bAR ~ normal( 0 , 10 );
    bA ~ normal( 0 , 10 );
    bR ~ normal( 0 , 10 );
```

```
    a ~ normal( 0 , 100 );
    for ( i in 1:N ) {
        mu[i] <- a + bR * rugged[i] + bA * cont_africa[i] +
                bAR * rugged[i] * cont_africa[i];
    }
    log_gdp ~ normal( mu , sigma );
}
```

This is Stan code, not R code. It is essentially the formula list you provided to map2stan, but in reverse order. The first line, vector[N] mu; names a symbol to hold the linear model, the kind of explicit housekeeping many computer languages require but that R does not. The rest of the code, however, just reiterates your formulas, beginning with priors and then computing the value of mu for each observation (nation). Finally, the last line comprises the likelihood. Given this kind of model definition, Stan will define and sample from the HMC chain or chains you ask for.

8.4. Care and feeding of your Markov chain

Markov chain Monte Carlo is a highly technical and usually automated procedure. Most people who use it don't really understand what it is doing. That's okay, up to a point. Science requires division of labor, and if every one of us had to write our own Markov chains from scratch, a lot less research would get done in the aggregate.

But as with many technical and powerful procedures, it's natural to feel uneasy about MCMC and maybe even a little superstitious. Something magical is happening inside the computer, and unless we make the right sacrifices and say the right words, the magic might become a curse. The good news is that HMC, unlike Gibbs sampling and ordinary Metropolis, makes it easy to tell when the magic goes wrong.

8.4.1. How many samples do you need? You can control the number of samples from the chain by using the iter and warmup parameters. The defaults are 2000 for iter and warmup is set to iter/2, which gives you 1000 warmup samples and 1000 real samples to use for inference. But these defaults are just meant to get you started, to make sure the chain gets started okay. Then you can decide on other values for iter and warmup.

So how many samples do we need for accurate inference about the posterior distribution? It depends. First, what really matters is the *effective* number of samples, not the raw number. The effective number of samples is an estimate of the number of independent samples from the posterior distribution. Markov chains are typically *autocorrelated*, so that sequential samples are not entirely independent. Stan chains tend to be less autocorrelated than those produced by other engines, but there is always some autocorrelation. Stan provides an estimate of effective number of samples as n_eff, and you'll see examples a bit later in the chapter.

Second, what do you want to know? If all you want are posterior means, it doesn't take many samples at all to get very good estimates. Even a couple hundred samples will do. But if you care about the exact shape in the extreme tails of the posterior, the 99th percentile or so, then you'll need many many more. So there is no universally useful number of samples to aim for. In most typical regression applications, you can get a very good estimate of the posterior mean with as few as 200 effective samples. And if the posterior is approximately Gaussian, then all you need in addition is a good estimate of the variance, which can be had with one order of magnitude more, in most cases. For highly skewed posteriors, you'll have to think more about which region of the distribution interests you.

The warmup setting is more subtle. On the one hand, you want to have the shortest warmup period necessary, so you can get on with real sampling. But on the other hand, more warmup can mean more efficient sampling. With Stan models, typically you can devote as much as half of your total samples, the `iter` value, to warmup and come out very well. But for simple models like those you've fit so far, much less warmup is really needed. Models can vary a lot in the shape of their posterior distributions, so again there is no universally best answer. But if you are having trouble, you might try increasing the warmup. If not, you might try reducing it. There's a practice problem at the end of the chapter that guides you in experimenting with the amount of warmup.

> **Rethinking: Warmup is not burn-in.** Other MCMC algorithms and software often discuss BURN-IN. With a sampling strategy like ordinary Metropolis, it is conventional and useful to trim off the front of the chain, the "burn-in" phase. This is done because it is unlikely that the chain has reached stationarity within the first few samples. Trimming off the front of the chain hopefully removes any influence of which starting value you chose for a parameter.[125]
>
> But Stan's sampling algorithms use a different approach. What Stan does during warmup is quite different from what it does after warmup. The warmup samples are used to adapt sampling, and so are not actually part of the target posterior distribution at all, no matter how long warmup continues. They are not burning in, but rather more like cycling the motor to heat things up and get ready for sampling. When real sampling begins, the samples will be immediately from the target distribution, assuming adaptation was successful. Still, you can usually tell if adaptation was successful because the warmup samples will come to look very much like the real samples. But that isn't always the case. For bad chains, the warmup will often look pretty good, but then actual sampling will demonstrate severe problems. You'll see examples a bit later in the chapter.

8.4.2. How many chains do you need?

It is very common to run more than one Markov chain, when estimating a single model. To do this with `map2stan` or `stan` itself, the `chains` argument specifies the number of independent Markov chains to sample from. And the optional `cores` argument lets you distribute the chains across different processors, so they can run simultaneously, rather than sequentially. All of the non-warmup samples from each chain will be automatically combined in the resulting inferences.

So the question naturally arises: How many chains do we need? There are three answers to this question. First, when debugging a model, use a single chain. Then when deciding whether the chains are valid, you need more than one chain. Third, when you begin the final run that you'll make inferences from, you only really need one chain. But using more than one chain is fine, as well. It just doesn't matter, once you're sure it's working. I'll briefly explain these answers.

The first time you try to sample from a chain, you might not be sure whether the chain is working right. So of course you will check the trace plot. Having more than one chain during these checks helps to make sure that the Markov chains are all converging to the same distribution. Sometimes, individual chains look like they've settled down to a stable distribution, but if you run the chain again, it might settle down to a different distribution. When you run multiple Markov chains, and see that all of them end up in the same region of parameter space, it provides a check that the machine is working correctly. Using 3 or 4 chains is conventional, and quite often more than enough to reassure us that the sampling is working properly.

But once you've verified that the sampling is working well, and you have a good idea of how many warmup samples you need, it's perfectly safe to just run one long chain. For example, suppose we learn that we need 1000 warmup samples and about 9000 real samples in total. Should we run one chain, with `warmup=1000` and `iter=10000`, or rather 3 chains, with `warmup=1000` and `iter=4000`? It doesn't really matter, in terms of inference.

But it might matter in efficiency, because the 3 chains cost you an extra 2000 samples of warmup that just get thrown away. And since warmup is typically the slowest part of the chain, these extra 2000 samples cost a disproportionate amount of your computer's time. On the other hand, if you run the chains on different computers or processor cores within a single computer, then you might prefer 3 chains, because you can spread the load and finish the whole job faster.

There are exotic situations in which all of the advice above must be modified. But for typical regression models, you can live by the motto *four short chains to check, one long chain for inference*. Things may still go wrong—you'll see some examples in the next sections, so you know what to look for. And once you know what to look for, you can fix any problems before running a long final Markov chain.

One of the perks of using HMC and Stan is that when sampling isn't working right, it's usually very obvious. As you'll see in the sections to follow, bad chains tend to have conspicuous behavior. Other methods of MCMC sampling, like Gibbs sampling and ordinary Metropolis, aren't so easy to diagnose.

> **Rethinking: Convergence diagnostics.** The default diagnostic output from Stan includes two metrics, `n_eff` and `Rhat`. The first is a measure of the effective number of samples. The second is the Gelman-Rubin convergence diagnostic, $\hat{R}$.[126] When `n_eff` is much lower than the actual number of iterations (minus warmup) of your chains, it means the chains are inefficient, but possibly still okay. When `Rhat` is above 1.00, it usually indicates that the chain has not yet converged, and probably you shouldn't trust the samples. If you draw more iterations, it could be fine, or it could never converge. See the Stan user manual for more details. It's important however not to rely too much on these diagnostics. Like all heuristics, there are cases in which they provide poor advice. For example, `Rhat` can reach 1.00 even for an invalid chain. So view it perhaps as a signal of danger, but never of safety. For conventional models, these metrics typically work well.

8.4.3. Taming a wild chain.

One common problem with some models is that there are broad, flat regions of the posterior density. This happens most often, as you might guess, when one uses flat priors. The problem this can generate is a wild, wandering Markov chain that erratically samples extremely positive and extremely negative parameter values.

Let's look at a simple example. The code below tries to estimate the mean and standard deviation of the two Gaussian observations -1 and 1. But it uses totally flat priors.

R code
8.13

```
y <- c(-1,1)
m8.2 <- map2stan(
    alist(
        y ~ dnorm( mu , sigma ) ,
        mu <- alpha
    ) ,
    data=list(y=y) , start=list(alpha=0,sigma=1) ,
    chains=2 , iter=4000 , warmup=1000 )
```

Now let's look at the `precis` output:

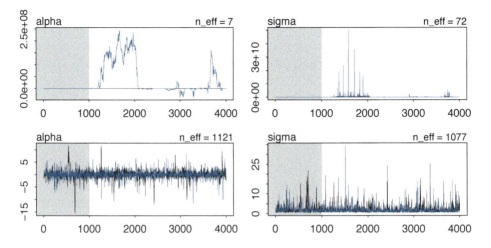

FIGURE 8.5. Diagnosing and healing a sick Markov chain. Top row: Trace plot from two independent chains defined by model m8.2. These chains are not stationary and should not be used for inference. Bottom row: Adding weakly informative priors (see m8.3) clears up the condition right away. These chains are fine to use for inference.

R code
8.14
```
precis(m8.2)
```

```
               Mean         StdDev    lower 0.89 upper 0.89 n_eff Rhat
alpha    21583691     54448550 -19611287.92  129922812     7 1.36
sigma  139399593   1147514738         29.06  185868167    72 1.02
```

Whoa! Those estimates can't be right. The mean of −1 and 1 is zero, so we're hoping to get a mean value for alpha around zero. Instead we get crazy values and implausibly wide confidence intervals. Inference for sigma is no better. You can also see that the diagnostic criteria indicate unreliable estimates. The number of effective samples, n_eff, is very small. And Rhat should approach 1.00 in a healthy set of chains. Even a value of 1.01 is suspicious. An Rhat of 1.10 indicates a catastrophe.

Take a look at the trace plot for this fit, plot(m8.2). It's shown in the top row of FIG-URE 8.5. The reason for the weird estimates is that the Markov chains seem to drift around and spike occasionally to extreme values. This is not a stationary pair of chains, and they do not provide useful samples.

It's easy to tame this particular chain by using weakly informative priors. The reason the model above drifts wildly in both dimensions is that there is very little data, just two observations, and flat priors. The flat priors say that every possible value of the parameter is equally plausible, a priori. For parameters that can take a potentially infinite number of values, like alpha, this means the Markov chain needs to occasionally sample some pretty extreme and implausible values, like negative 30 million. These extreme drifts overwhelm the chain. If the likelihood were stronger, then the chain would be fine, because it would stick closer to zero.

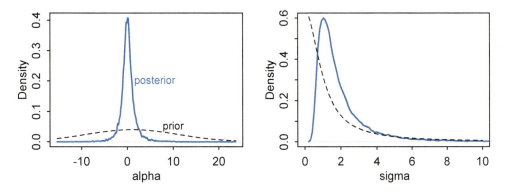

FIGURE 8.6. Prior (dashed) and posterior (blue) for the model with weakly informative priors, m8.3. Even with only two observations, the likelihood easily overcomes these priors. Yet the model cannot be successfully estimated without them.

But it doesn't take much information in the prior to stop this foolishness, even without a stronger likelihood. Let's use this model:

$$y_i \sim \text{Normal}(\mu, \sigma)$$
$$\mu = \alpha$$
$$\alpha \sim \text{Normal}(1, 10)$$
$$\sigma \sim \text{HalfCauchy}(0, 1)$$

I've just added weakly informative priors for α and σ. We'll plot these priors in a moment, so you will be able to see just how weak they are. But let's re-estimate first:

R code
8.15

```
m8.3 <- map2stan(
    alist(
        y ~ dnorm( mu , sigma ) ,
        mu <- alpha ,
        alpha ~ dnorm( 1 , 10 ) ,
        sigma ~ dcauchy( 0 , 1 )
    ) ,
    data=list(y=y) , start=list(alpha=0,sigma=1) ,
    chains=2 , iter=4000 , warmup=1000 )
precis(m8.3)
```

	Mean	StdDev	lower 0.89	upper 0.89	n_eff	Rhat
alpha	-0.01	1.60	-1.98	2.37	1121	1
sigma	1.98	1.91	0.47	3.45	1077	1

That's much better. Take a look at the bottom row in FIGURE 8.5. This trace plot looks healthy. Both chains are stationary around the same values, and mixing is good. No more wild detours off to negative 20 million.

To appreciate what has happened, take a look at the priors (dashed) and posteriors (blue) in FIGURE 8.6. Both the Gaussian prior for α and the Cauchy prior for σ contain very gradual downhill slopes. They are so gradual, that even with only two observations, as in this example, the likelihood almost completely overcomes them. The mean of the prior for α is 1, but the mean of the posterior is zero, just as the likelihood says it should be. The prior for σ is maximized at zero. But the posterior has its median around 1.4. The standard deviation of the data is 1.4.

These weakly informative priors have helped by providing a very gentle nudge towards reasonable values of the parameters. Now values like 30 million are no longer equally plausible as small values like 1 or 2. Lots of problematic chains want subtle priors like these, designed to tune estimation by assuming a tiny bit of prior information about each parameter. And even though the priors end up getting washed out right away—two observations were enough here—they still have a big effect on inference, by allowing us to get an answer. That answer is also a good answer.

Overthinking: Cauchy distribution. The models in this chapter, and in many chapters to follow, use half-Cauchy priors for standard deviations. The CAUCHY (ko-shee) distribution gives the distribution of the ratio of two random Gaussian draws. Its parameters are a *location* x_0 and a *scale* γ. The location says where the center is, and the scale defines how stretched out the distribution is. Its probability density is:

$$p(x|x_0, \gamma) = \left(\pi\gamma \left[1 + \left(\frac{x-x_0}{\gamma} \right)^2 \right] \right)^{-1}$$

Note however that the Cauchy has no defined mean nor variance, so the location and scale are not its mean and, say, standard deviation. The reason the Cauchy has no mean and variance is that it is a very thick-tailed distribution. At any moment in a Cauchy sampling process, it is possible to draw an extreme value that overwhelms all of the previous draws. The consequence of this fact is that the sequence never converges to a stable mean and variance. It just keeps moving. You can prove this to yourself with a little simulation. The code below samples 10,000 values from a Cauchy distribution. Then it computes and plots the running mean at each sample. Run this simulation a few times to see how the trace of the mean is highly unpredictable.

R code
8.16
```
y <- rcauchy(1e4,0,5)
mu <- sapply( 1:length(y) , function(i) sum(y[1:i])/i )
plot(mu,type="l")
```

The Cauchy distributions in the model definitions are implicitly half-Cauchy, a Cauchy defined over the positive reals only. This is because they are applied to a parameter, usually σ, that is strictly positive. Stan figures out that you meant for it to be half-Cauchy. If you are curious how it knows this, check the raw Stan code with `stancode` and look for the `<lower=0>` constraint in the definition of the parameter `sigma`.

8.4.4. Non-identifiable parameters.
Back in Chapter 5, you met the problem of highly correlated predictors and the non-identifiable parameters they can create. Here you'll see what such parameters look like inside of a Markov chain. You'll also see how you can identify them, in principle, by using a little prior information. Most importantly, the badly behaving chains produced in this example will exhibit characteristic bad behavior, so when you see the same pattern in your own models, you'll have a hunch about the cause.

To construct a non-identifiable model, we first simulate 100 observations from a Gaussian distribution with mean zero and standard deviation 1.

R code
8.17

```
y <- rnorm( 100 , mean=0 , sd=1 )
```

By simulating the data, we know the right answer. Then we fit this model:

$$y_i \sim \text{Normal}(\mu, \sigma)$$
$$\mu = \alpha_1 + \alpha_2$$
$$\sigma \sim \text{HalfCauchy}(0, 1)$$

The linear model contains two parameters, α_1 and α_2, which cannot be identified. Only their sum can be identified, and it should be about zero, after estimation.

Let's run the Markov chain and see what happens. This chain is going to take much longer than the previous ones. But it should still finish after a few minutes.

R code
8.18

```
m8.4 <- map2stan(
    alist(
        y ~ dnorm( mu , sigma ) ,
        mu <- a1 + a2 ,
        sigma ~ dcauchy( 0 , 1 )
    ) ,
    data=list(y=y) , start=list(a1=0,a2=0,sigma=1) ,
    chains=2 , iter=4000 , warmup=1000 )
precis(m8.4)
```

	Mean	StdDev	lower 0.89	upper 0.89	n_eff	Rhat
a1	-1194.76	1344.19	-2928.62	1053.52	1	2.83
a2	1194.81	1344.19	-1054.86	2927.39	1	2.83
sigma	0.92	0.07	0.81	1.02	17	1.13

Those estimates look suspicious, and the n_eff and Rhat values are terrible. The means for a1 and a2 are almost exactly the same distance from zero, but on opposite sides. And the standard deviations of the chains are massive. This is of course a result of the fact that we cannot simultaneously estimate a1 and a2, but only their sum.

Looking at the trace plot reveals more. The left column in FIGURE 8.7 shows two Markov chains from the model above. These chains do not look like they are stationary, nor do they seem to be mixing very well. Indeed, when you see a pattern like this, it is reason to worry. Don't use these samples.

Again, weak priors can rescue us. Now the model fitting code is:

R code
8.19

```
m8.5 <- map2stan(
    alist(
        y ~ dnorm( mu , sigma ) ,
        mu <- a1 + a2 ,
        a1 ~ dnorm( 0 , 10 ) ,
        a2 ~ dnorm( 0 , 10 ) ,
        sigma ~ dcauchy( 0 , 1 )
    ) ,
    data=list(y=y) , start=list(a1=0,a2=0,sigma=1) ,
    chains=2 , iter=4000 , warmup=1000 )
precis(m8.5)
```

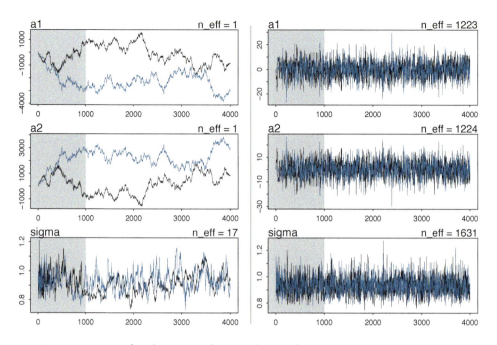

FIGURE 8.7. Left column: A chain with wandering parameters, a1 and a2, generated by m8.4. Right column: Same model but now with weakly informative priors, m8.5.

	Mean	StdDev	lower 0.89	upper 0.89	n_eff	Rhat
a1	-0.23	6.97	-11.25	10.95	1223	1
a2	0.28	6.97	-10.85	11.36	1224	1
sigma	0.93	0.07	0.82	1.04	1631	1

The estimates for a1 and a2 are better identified now. And take a look at the right column traces in FIGURE 8.7. Notice also that the model sampled a lot faster. With flat priors, m8.4, sampling may take 8 times as long as it does for m8.5. Often, a model that is very slow to sample is under-identified. This is an aspect of something Bayesian statistician Andrew Gelman calls the **FOLK THEOREM OF STATISTICAL COMPUTING**: When you are having trouble fitting a model, it often indicates a bad model.

In the end, adding some weakly informative priors saves this model. You might think you'd never accidentally try to fit an unidentified model. But you'd be wrong. Even if you don't make obvious mistakes, complex models can easily become unidentified or nearly so. With many predictors, and especially with interactions, correlations among parameters can be large. Just a little prior information telling the model "none of these parameters can be 30 million" often helps, and it has no effect on estimates. A flat prior really is flat, all the way to infinity. Unless you believe infinity is a reasonable estimate, don't use a flat prior.

Additionally, adding weak priors can speed up sampling, because the Markov chain won't feel that it has to run out to extreme values that you, but not your model, already know are highly implausible.

8.5. Summary

This chapter has been an informal introduction to Markov chain Monte Carlo (MCMC) estimation. The goal has been to introduce the purpose and approach MCMC algorithms. The major algorithms introduced were the Metropolis, Gibbs sampling, and Hamiltonian Monte Carlo algorithms. Each has its advantages and disadvantages. A function in the rethinking package, map2stan, was introduced that uses the Stan (mc-stan.org) Hamiltonian Monte Carlo engine to fit models as they are defined in this book. General advice about diagnosing poor MCMC fits was introduced by the use of a couple of pathological examples.

8.6. Practice

Easy.

8E1. Which of the following is a requirement of the simple Metropolis algorithm?

    (1) The parameters must be discrete.
    (2) The likelihood function must be Gaussian.
    (3) The proposal distribution must be symmetric.

8E2. Gibbs sampling is more efficient than the Metropolis algorithm. How does it achieve this extra efficiency? Are there any limitations to the Gibbs sampling strategy?

8E3. Which sort of parameters can Hamiltonian Monte Carlo not handle? Can you explain why?

8E4. Explain the difference between the effective number of samples, n_eff as calculated by Stan, and the actual number of samples.

8E5. Which value should Rhat approach, when a chain is sampling the posterior distribution correctly?

8E6. Sketch a good trace plot for a Markov chain, one that is effectively sampling from the posterior distribution. What is good about its shape? Then sketch a trace plot for a malfunctioning Markov chain. What about its shape indicates malfunction?

Medium.

8M1. Re-estimate the terrain ruggedness model from the chapter, but now using a uniform prior and an exponential prior for the standard deviation, sigma. The uniform prior should be dunif(0,10) and the exponential should be dexp(1). Do the different priors have any detectible influence on the posterior distribution?

8M2. The Cauchy and exponential priors from the terrain ruggedness model are very weak. They can be made more informative by reducing their scale. Compare the dcauchy and dexp priors for progressively smaller values of the scaling parameter. As these priors become stronger, how does each influence the posterior distribution?

8M3. Re-estimate one of the Stan models from the chapter, but at different numbers of warmup iterations. Be sure to use the same number of sampling iterations in each case. Compare the n_eff values. How much warmup is enough?

Hard.

8H1. Run the model below and then inspect the posterior distribution and explain what it is accomplishing.

R code
8.20
```
mp <- map2stan(
    alist(
        a ~ dnorm(0,1),
        b ~ dcauchy(0,1)
    ),
    data=list(y=1),
    start=list(a=0,b=0),
    iter=1e4, warmup=100 , WAIC=FALSE )
```

Compare the samples for the parameters a and b. Can you explain the different trace plots, using what you know about the Cauchy distribution?

8H2. Recall the divorce rate example from Chapter 5. Repeat that analysis, using map2stan this time, fitting models m5.1, m5.2, and m5.3. Use compare to compare the models on the basis of WAIC. Explain the results.

8H3. Sometimes changing a prior for one parameter has unanticipated effects on other parameters. This is because when a parameter is highly correlated with another parameter in the posterior, the prior influences both parameters. Here's an example to work and think through.

Go back to the leg length example in Chapter 5. Here is the code again, which simulates height and leg lengths for 100 imagined individuals:

R code
8.21
```
N <- 100                        # number of individuals
height <- rnorm(N,10,2)         # sim total height of each
leg_prop <- runif(N,0.4,0.5)    # leg as proportion of height
leg_left <- leg_prop*height +   # sim left leg as proportion + error
    rnorm( N , 0 , 0.02 )
leg_right <- leg_prop*height +  # sim right leg as proportion + error
    rnorm( N , 0 , 0.02 )
                                # combine into data frame
d <- data.frame(height,leg_left,leg_right)
```

And below is the model you fit before, resulting in a highly correlated posterior for the two beta parameters. This time, fit the model using map2stan:

R code
8.22
```
m5.8s <- map2stan(
    alist(
        height ~ dnorm( mu , sigma ) ,
        mu <- a + bl*leg_left + br*leg_right ,
        a ~ dnorm( 10 , 100 ) ,
        bl ~ dnorm( 2 , 10 ) ,
        br ~ dnorm( 2 , 10 ) ,
        sigma ~ dcauchy( 0 , 1 )
    ) ,
    data=d, chains=4,
    start=list(a=10,bl=0,br=0,sigma=1) )
```

Compare the posterior distribution produced by the code above to the posterior distribution produced when you change the prior for br so that it is strictly positive:

```
m5.8s2 <- map2stan(                                                          R code
    alist(                                                                   8.23
        height ~ dnorm( mu , sigma ) ,
        mu <- a + bl*leg_left + br*leg_right ,
        a ~ dnorm( 10 , 100 ) ,
        bl ~ dnorm( 2 , 10 ) ,
        br ~ dnorm( 2 , 10 ) & T[0,] ,
        sigma ~ dcauchy( 0 , 1 )
    ) ,
    data=d, chains=4,
    start=list(a=10,bl=0,br=0,sigma=1) )
```

Note that `T[0,]` on the right-hand side of the prior for `br`. What the `T[0,]` does is *truncate* the normal distribution so that it has positive probability only above zero. In other words, that prior ensures that the posterior distribution for `br` will have no probability mass below zero.

Compare the two posterior distributions for `m5.8s` and `m5.8s2`. What has changed in the posterior distribution of both beta parameters? Can you explain the change induced by the change in prior?

8H4. For the two models fit in the previous problem, use DIC or WAIC to compare the effective numbers of parameters for each model. Which model has more effective parameters? Why?

8H5. Modify the Metropolis algorithm code from the chapter to handle the case that the island populations have a different distribution than the island labels. This means the island's number will not be the same as its population.

8H6. Modify the Metropolis algorithm code from the chapter to write your own simple MCMC estimator for globe tossing data and model from Chapter 2.

9 Big Entropy and the Generalized Linear Model

Most readers of this book will share the experience of fighting with tangled electrical cords. Whether behind a desk or stuffed in a box, cords and cables tend toward tying themselves in knots. Why is this? There is of course real physics at work. But at a descriptive level, the reason is entropy: There are vastly more ways for cords to end up in a knot than for them to remain untied.[127] So if I were to carefully lay a dozen cords in a box and then seal the box and shake it, we should bet that at least some of the cords will be tangled together when I again open the box. We don't need to know anything about the physics of cords or knots. We just have to bet on entropy. Events that can happen vastly more ways are more likely.

Exploiting entropy is not going to untie your cords. But it will help you solve some problems in choosing distributions. Statistical models force many choices upon us. Some of these choices are distributions that represent uncertainty. We must choose, for each parameter, a prior distribution. And we must choose a likelihood function, which serves as a distribution of data. There are conventional choices, such as wide Gaussian priors and the Gaussian likelihood of linear regression. These conventional choices work unreasonably well in many circumstances. But very often the conventional choices are not the best choices. Inference can be more powerful when we use all of the information, and doing so usually requires going beyond convention.

To go beyond convention, it helps to have some principles to guide choice. When an engineer wants to make an unconventional bridge, engineering principles help guide choice. When a researcher wants to build an unconventional model, entropy provides one useful principle to guide choice of probability distributions: Bet on the distribution with the biggest entropy. Why? There are three sorts of justifications.

First, the distribution with the biggest entropy is the widest and least informative distribution. Choosing the distribution with the largest entropy means spreading probability as evenly as possible, while still remaining consistent with anything we think we know about a process. In the context of choosing a prior, it means choosing the least informative distribution consistent with any partial scientific knowledge we have about a parameter. In the context of choosing a likelihood, it means selecting the distribution we'd get by counting up all the ways outcomes could arise, consistent with the constraints on the outcome variable. In both cases, the resulting distribution embodies the least information while remaining true to the information we've provided.

Second, nature tends to produce empirical distributions that have high entropy. Back in Chapter 4, I introduced the Gaussian distribution by demonstrating how any process that repeatedly adds together fluctuations will tend towards an empirical distribution with the distinctive Gaussian shape. That shape is the one that contains no information about the underlying process except its location and variance. As a result, it has maximum entropy.

Natural processes other than addition also tend to produce maximum entropy distributions. But they are not Gaussian, because they retain different information about the underlying process.

Third, regardless of why it works, it tends to work. Mathematical procedures are effective even when we don't understand them. There are no guarantees that any logic in the small world (Chapter 2) will be useful in the large world. We use logic in science because it has a strong record of effectiveness in addressing real world problems. This is the historical justification: The approach has solved difficult problems in the past. This is no guarantee that it will work on your problem. But no approach can guarantee that.

This chapter serves as a conceptual introduction to GENERALIZED LINEAR MODELS and the principle of MAXIMUM ENTROPY. A generalized linear model (GLM) is much like the linear regressions of previous chapters. It is a model that replaces a parameter of a likelihood function with a linear model. But GLMs need not use Gaussian likelihoods. Any likelihood function can be used, and linear models can be attached to any or all of the parameters that describe its shape. The principle of maximum entropy helps us choose likelihood functions, by providing a way to use stated assumptions about constraints on the outcome variable to choose the likelihood function that is the most conservative distribution compatible with the known constraints. Using this principle recovers all the most common likelihood functions of many statistical approaches, Bayesian or not, while simultaneously providing a clear rationale for choice among them.

The chapters to follow this one build computational skills for working with different flavors of GLM. Chapter 10 addresses models for count variables. Chapter 11 explores more complicated models, such as ordinal outcomes and mixtures. Portions of these chapters are specialized by model type. So you can skip sections that don't interest you at the moment. The multilevel chapters, beginning with Chapter 12, make use of binomial count models, however. So some familiarity with the material in Chapter 10 will be helpful.

> **Rethinking: Bayesian updating and maximum entropy.** Another kind of probability distribution, the posterior distribution deduced by Bayesian updating, is also a case of maximizing entropy. The posterior distribution has the greatest entropy relative to the prior (the smallest cross entropy) among all distributions consistent with the assumed constraints and the observed data.[128] This fact won't change how you calculate. But it should provide a deeper appreciation of the fundamental connections between Bayesian inference and information theory. Notably, Bayesian updating is just like maximum entropy in that it produces the least informative distribution that is still consistent with our assumptions. Or you might say that the posterior distribution has the smallest divergence from the prior that is possible while remaining consistent with the constraints and data.

9.1. Maximum entropy

In Chapter 6, you met the basics of information theory. In brief, we seek a measure of uncertainty that satisfies three criteria: (1) the measure should be continuous; (2) it should increase as the number of possible events increases; and (3) it should be additive. The resulting unique measure of the uncertainty of a probability distribution p with probabilities p_i for each possible event i turns out to be just the average log-probability:

$$H(p) = -\sum_i p_i \log p_i$$

This function is known as *information entropy*.

The principle of maximum entropy applies this measure of uncertainty to the problem of choosing among probability distributions. Perhaps the simplest way to state the maximum entropy principle is:

> The distribution that can happen the most ways is also the distribution with the biggest information entropy. The distribution with the biggest entropy is the most conservative distribution that obeys its constraints.

There's nothing intuitive about this idea, so if it seems weird, you are normal.

To begin to understand maximum entropy, forget about information and probability theory for the moment. Imagine instead 5 buckets and a pile of 10 individually numbered pebbles. You stand and toss all 10 pebbles such that each pebble is equally likely to land in any of the 5 buckets. This means that every particular arrangement of the 10 individual pebbles is equally likely—it's just as likely to get all 10 in bucket 3 as it is to get pebble 1 in bucket 2, pebbles 2–9 in bucket 3, and pebble 10 in bucket 4.

But some kinds of arrangements are much more likely. Some arrangements look the same, because they show the same number of pebbles in the same individual buckets. These are distributions of pebbles. FIGURE 9.1 illustrates 5 such distributions. So for example there is only 1 way to arrange the individual pebbles so that all of them are in bucket 3 (plot A). But there are 90 ways to arrange the individual pebbles so that 2 of them are in bucket 2, 8 in bucket 3, and 2 in bucket 4 (plot B). Plots C, D, and E show that the number of unique arrangements corresponding to a distribution grows very rapidly as the distribution places a more equal number of pebbles in each bucket. By the time there are 2 pebbles in each bucket (plot E), there are 113400 ways to realize this distribution. There is no other distribution of the pebbles that can be realized a greater number of ways.

Let's put each distribution of pebbles in a list:

```
p <- list()
p$A <- c(0,0,10,0,0)
p$B <- c(0,1,8,1,0)
p$C <- c(0,2,6,2,0)
p$D <- c(1,2,4,2,1)
p$E <- c(2,2,2,2,2)
```

R code
9.1

And let's normalize each such that it is a probability distribution. This means we just divide each count of pebbles by the total number of pebbles:

```
p_norm <- lapply( p , function(q) q/sum(q))
```

R code
9.2

Since these are now probability distributions, we can compute the information entropy of each. The only trick here is to remember L'Hôpital's rule (see page 179):

```
( H <- sapply( p_norm , function(q) -sum(ifelse(q==0,0,q*log(q))) ) )
```

R code
9.3

```
        A         B         C         D         E
0.0000000 0.6390319 0.9502705 1.4708085 1.6094379
```

So distribution E, which can realized by far the greatest number of ways, also has the biggest entropy. This is no coincidence. To see why, let's compute the logarithm of number of ways

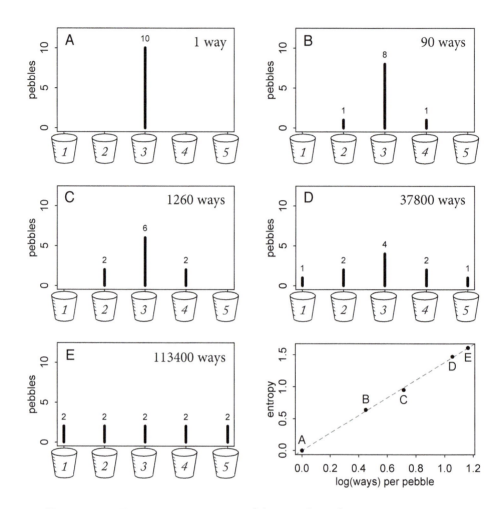

FIGURE 9.1. Entropy as a measure of the number of unique arrangements of a system that produce the same distribution. Plots A through E show the numbers of unique ways to arrange 10 pebbles into each of 5 different distributions. Bottom-right: The entropy of each distribution plotted against the log number of ways per pebble to produce it.

each distribution can be realized, then divide that logarithm by 10, the number of pebbles. This gives us the log ways per pebble for each distribution:

R code
9.4

```
ways <- c(1,90,1260,37800,113400)
logwayspp <- log(ways)/10
```

The bottom-right plot in FIGURE 9.1 displays these logwayspp values against the information entropies H. These two sets of values contain the same information, as information entropy is an approximation of the log ways per pebble (see the Overthinking box at the end for details). As the number of pebbles grows larger, the approximation gets better. It's already extremely good, for just 10 pebbles. Information entropy is a way of counting how many unique arrangements correspond to a distribution.

This is useful, because the distribution that can happen the greatest number of ways is the most plausible distribution. Call this distribution the MAXIMUM ENTROPY DISTRIBUTION. As you might guess from the pebble example, the number of ways corresponding to the maximum entropy distribution eclipses that of any other distribution. And the numbers of ways for each distribution most similar to the maximum entropy distribution eclipse those of less similar distributions. And so on, such that the vast majority of unique arrangements of pebbles produce either the maximum entropy distribution or rather a distribution very similar to it. And that is why it's often effective to bet on maximum entropy: It's the center of gravity for the highly plausible distributions.

Its high plausibility is conditional on our assumptions, of course. To grasp the role of assumptions—constraints and data—in maximum entropy, we'll explore two examples. First, we'll derive the Gaussian distribution as the solution to an entropy maximization problem. Second, we'll derive the binomial distribution, which we used way back in Chapter 2 to draw marbles and toss globes, as the solution to a different entropy maximization problem. These derivations will not be mathematically rigorous. Rather, they will be graphical and aim to deliver a conceptual appreciation for what this thing called *entropy* is doing. The Overthinking boxes in this section provide connections to the mathematics, for those who are interested.

But the most important thing is to be patient with yourself. Understanding of and intuition for probability theory comes with experience. You can usefully apply the principle of maximum entropy before you fully understand it. Indeed, it may be that no one fully understands it. Over time, and within the contexts that you find it useful, the principle will become more intuitive.

Rethinking: What good is intuition? Like many aspects of information theory, maximum entropy is not very intuitive. But note that intuition is just a guide to developing methods. When a method works, it hardly matters whether our intuition agrees. This point is important, because some people still debate statistical approaches on the basis of philosophical principles and intuitive appeal. Philosophy does matter, because it influences development and application. But it is a poor way to judge whether or not an approach is useful. Results are what matter. For example, the three criteria used to derive information entropy, back in Chapter 6, are not also the justification for using information entropy. The justification is rather that it has worked so well on so many problems where other methods have failed.

Overthinking: The Wallis derivation. Intuitively, we can justify maximum entropy just based upon the definition of information entropy. But there's another derivation, attributed to Graham Wallis,[129] that doesn't invoke "information" at all. Here's a short version of the argument. Suppose there are M observable events, and we wish to assign a plausibility to each. We know some constraints about the process that produces these events, such as its expected value or variance. Now imagine setting up M buckets and tossing a large number N of individual stones into them at random, in such a way that each stone is equally likely to land in any of the M buckets. After all the stones have landed, we count up the number of stones in each bucket i and use these counts n_i to construct a candidate probability distribution defined by $p_i = n_i/N$. If this candidate distribution is consistent with our constraints, we add it to a list. If not, we empty the buckets and try again. After many rounds of this, the distribution that has occurred the most times is the fairest—in the sense that no bias was involved in tossing the stones into buckets—that still obeys the constraints that we imposed.

If we could employ the population of a large country in tossing stones every day for years on end, we could do this empirically. Luckily, the procedure can be studied mathematically. The probability of any particular candidate distribution is just its multinomial probability, the probability of the observed stone counts under uniform chances of landing in each bucket:

$$\Pr(n_1, n_2, ..., n_m) = \frac{N!}{n_1! n_2! ... n_m!} \prod_{i=1}^{M} \left(\frac{1}{M}\right)^{n_i} = \frac{N!}{n_1! n_2! ... n_m!} \left(\frac{1}{M}\right)^N = W \left(\frac{1}{M}\right)^N$$

The distribution that is realized most often will have the largest value of that ugly fraction W with the factorials in it. Call W the *multiplicity*, because it states the number of different ways a particular set of counts could be realized. For example, landing all stones in the first bucket can happen only one way, by getting all the stones into that bucket and none in any of the other buckets. But there are many more ways to evenly distribute the stones in the buckets, because order does not matter. We care about this multiplicity, because we are seeking the distribution that would happen most often. So by selecting the distribution that maximizes this multiplicity, we can accomplish that goal.

We're almost at entropy. It's easier to work with $\frac{1}{N} \log(W)$, which will be maximized by the same distribution as W. Also note that $n_i = N p_i$. These changes give us:

$$\frac{1}{N} \log W = \frac{1}{N} \left(\log N! - \sum_i \log[(N p_i)!] \right)$$

Now since N is very large, we can approximate $\log N!$ with Stirling's approximation, $N \log N - N$:

$$\frac{1}{N} \log W \approx \frac{1}{N} \left(N \log N - N - \sum_i (N p_i \log(N p_i) - N p_i) \right) = -\sum_i p_i \log p_i$$

And that's the exact same formula as Shannon's information entropy. Among distributions that satisfy our constraints, the distribution that maximizes the expression above is the distribution that spreads out probability as evenly as possible, while still obeying the constraints.

This result generalizes easily to the case in which there is not an equal chance of each stone landing in each bucket.[130] If we have prior information specified as a probability q_i that a stone lands in bucket i, then the quantity to maximize is instead:

$$\frac{1}{N} \log \Pr(n_1, n_2, ..., n_m) \approx -\sum_i p_i \log(p_i/q_i)$$

You may recognize this as KL divergence from Chapter 6, just with a negative in front. This reveals that the distribution that maximizes entropy is also the distribution that minimizes the information distance from the prior, among distributions consistent with the constraints. When the prior is flat, maximum entropy gives the flattest distribution possible. When the prior is not flat, maximum entropy updates the prior and returns the distribution that is most like the prior but still consistent with the constraints. This procedure is often called *minimum cross-entropy*. Furthermore, Bayesian updating itself can be expressed as the solution to a maximum entropy problem in which the data represent constraints.[131] Therefore Bayesian inference can be seen as producing a posterior distribution that is most similar to the prior distribution as possible, while remaining logically consistent with the stated information.

9.1.1. Gaussian.

When I introduced the Gaussian distribution in Chapter 4 (page 72), it emerged from a generative process in which 1000 people repeatedly flipped coins and took steps left (heads) or right (tails) with each flip. The addition of steps led inevitably to a distribution of positions resembling the Gaussian bell curve. This process represents the most basic generative dynamic that leads to Gaussian distributions in nature. When many small factors add up, the ensemble of sums tends towards Gaussian.

But obviously many other distributions are possible. The coin-flipping dynamic could place all 1000 people on the same side of the soccer field, for example. So why don't we see

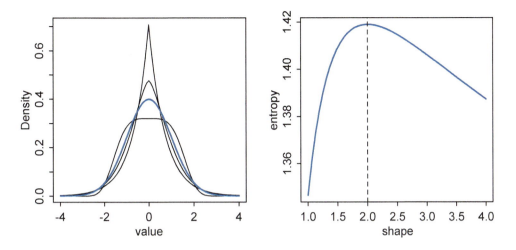

FIGURE 9.2. Maximum entropy and the Gaussian distribution. Left: Comparison of Gaussian (blue) and several other continuous distributions with the same variance. Right: Entropy is maximized when curvature of a generalized normal distribution matches the Gaussian, where shape is equal to 2.

those other distributions in nature? Because for every sequence of coin flips that can produce such an imbalanced outcome, there are vastly many more that can produce an approximately balanced outcome. The bell curve emerges, empirically, because there are so many different detailed states of the physical system that can produce it. Whatever does happen, it's bound to produce an ensemble that is approximately Gaussian. So if all you know about a collection of continuous values is its variance (or that it has a finite variance, even if you don't know it yet), the safest bet is that the collection ends up in one of these vastly many bell-shaped configurations.[132]

And maximum entropy just seeks the distribution that can arise the largest number of ways, so it does a good job of finding limiting distributions like this. But since entropy is maximized when probability is spread out as evenly as possible, maximum entropy also seeks the distribution that is most even, while still obeying its constraints. In order to visualize how the Gaussian is the most even distribution for any given variance, let's consider a family of generalized distributions with equal variance. A *generalized normal distribution* is defined by the probability density:

$$\mathrm{Pr}(y|\mu, \alpha, \beta) = \frac{\beta}{2\alpha\Gamma(1/\beta)} e^{-\left(\frac{|y-\mu|}{\alpha}\right)^{\beta}}$$

We want to compare a regular Gaussian distribution with variance σ^2 to several generalized normals with the same variance.[133]

The left-hand plot in FIGURE 9.2 presents one Gaussian distribution, in blue, together with three generalized normal distributions with the same variance. All four distributions have variance $\sigma^2 = 1$. Two of the generalized distributions are more peaked, and have thicker tails, than the Gaussian. Probability has been redistributed from the middle to the tails, keeping the variance constant. The third generalized distribution is instead thicker in the middle and thinner in the tails. It again keeps the variance constant, this time by redistributing probability from the tails to the center. The blue Gaussian distribution sits between these extremes.

In the right-hand plot of FIGURE 9.2, β is called "shape" and varies from 1 to 4, and entropy is plotted on the vertical axis. The generalized normal is perfectly Gaussian where $\beta = 2$, and that's exactly where entropy is maximized. All of these distributions are symmetrical, but that doesn't affect the result. There are other generalized families of distributions that can be skewed as well, and even then the bell curve has maximum entropy. See the Overthinking box at the bottom of this page, if you want a more satisfying proof.

To appreciate why the Gaussian shape has the biggest entropy for any continuous distribution with this variance, consider that entropy increases as we make a distribution flatter. So we could easily make up a probability distribution with larger entropy than the blue distribution in FIGURE 9.2: Just take probability from the center and put it in the tails. The more uniform the distribution looks, the higher its entropy will be. But there are limits on how much of this we can do and maintain the same variance, $\sigma^2 = 1$. A perfectly uniform distribution would have infinite variance, in fact. So the variance constraint is actually a severe constraint, forcing the high-probability portion of the distribution to a small area around the mean. Then the Gaussian distribution gets its shape by being as spread out as possible for a distribution with fixed variance.

The take-home lesson from all of this is that, if all we are willing to assume about a collection of measurements is that they have a finite variance, then the Gaussian distribution represents the most conservative probability distribution to assign to those measurements. But very often we are comfortable assuming something more. And in those cases, provided our assumptions are good ones, the principle of maximum entropy leads to distributions other than the Gaussian.

Overthinking: Proof of Gaussian maximum entropy. Proving that the Gaussian has the largest entropy of any distribution with a given variance is easier than you might think. Here's the shortest proof I know.[134] Let $p(x) = (2\pi\sigma^2)^{-1/2} \exp(-(x - \mu)^2/(2\sigma^2))$ stand for the Gaussian probability density function. Let $q(x)$ be some other probability density function with the same variance σ^2. The mean μ doesn't matter here, because entropy doesn't depend upon location, just shape.

The entropy of the Gaussian is $H(p) = -\int p(x)\log p(x)dx = \frac{1}{2}\log(2\pi e \sigma^2)$. We seek to prove that no distribution $q(x)$ can have higher entropy than this, provided they have the same variance and are both defined on the entire real number line, from $-\infty$ to $+\infty$. We can accomplish this by using our old friend, from Chapter 6, KL divergence:

$$D_{\mathrm{KL}}(q, p) = \int_{-\infty}^{\infty} q(x)\log\left(\frac{q(x)}{p(x)}\right) dx = -H(q, p) - H(q)$$

$H(q) = -\int q(x)\log q(x)dx$ is the entropy of $q(x)$ and $H(q, p) = \int q(x)\log p(x)dx$ is the cross-entropy of the two. Why use D_{KL} here? Because it is always positive (or zero), which guarantees that $-H(q, p) \geq H(q)$. So while we can't compute $H(q)$, it turns out that we can compute $H(q, p)$. And as you'll see, that solves the whole problem. So let's compute $H(q, p)$. It's defined as:

$$H(q, p) = \int_{-\infty}^{\infty} q(x)\log p(x)dx = \int_{-\infty}^{\infty} q(x)\log\left[(2\pi\sigma^2)^{-1/2}\exp\left(-\frac{(x - \mu)^2}{2\sigma^2}\right)\right] dx$$

This will be conceptually easier if we remember that the integral above just takes the average over x. So we can rewrite the above as:

$$H(q, p) = \mathrm{E}\log\left[(2\pi\sigma^2)^{-1/2}\exp\left(-\frac{(x - \mu)^2}{2\sigma^2}\right)\right] = -\frac{1}{2}\log(2\pi\sigma^2) - \frac{1}{2\sigma^2}\mathrm{E}\left((x - \mu)^2\right)$$

Now the term on the far right is just the average squared deviation from the mean, which is the very definition of variance. Since the variance of the unknown function $q(x)$ is constrained to be σ^2:

$$H(q,p) = -\frac{1}{2}\log(2\pi\sigma^2) - \frac{1}{2\sigma^2}\sigma^2 = -\frac{1}{2}\left(\log(2\pi\sigma^2) + 1\right) = -\frac{1}{2}\log(2\pi e\sigma^2)$$

And that is exactly $-H(p)$. So since $-H(q,p) \geq H(q)$ by definition, and since $H(p) = -H(q,p)$, it follows that $H(p) \geq H(q)$. The Gaussian has the highest entropy possible for any continuous distribution with variance σ^2.

9.1.2. Binomial. Way back in Chapter 2, I introduced Bayesian updating by drawing blue and white marbles from a bag. I showed that the likelihood—the relative plausibility of an observation—arises from counting the numbers of ways that a given observation could arise, according to our assumptions. The resulting distribution is known as the BINOMIAL DISTRIBUTION. If only two things can happen (blue or white marble, for example), and there's a constant chance p of each across n trials, then the probability of observing y events of type 1 and $n - y$ events of type 2 is:

$$\Pr(y|n,p) = \frac{n!}{y!(n-y)!}p^y(1-p)^{n-y}$$

It may help to note that the fraction with the factorials is just saying how many different ordered sequences of n outcomes have a count y. So a more elementary view is that the probability of any unique sequence of binary events y_1 through y_n is just:

$$\Pr(y_1, y_2, ..., y_n|n,p) = p^y(1-p)^{n-y}$$

For the moment, we'll work with this elementary form, because it will make it easier to appreciate the basis for treating all sequences with the same count y as the same outcome.

Now we want to demonstrate that this same distribution has the largest entropy of any distribution that satisfies these constraints: (1) only two unordered events, and (2) constant expected value. To develop some intuition for the result, let's explore two examples in which we fix the expected value. In both examples, we have to assign probability to each possible outcome, while keeping the expected value of the distribution constant. And in both examples, the unique distribution that maximizes entropy is the binomial distribution with the same expected value.

Here's the first example. Suppose again, like in Chapter 2, that we have a bag with an unknown number of blue and white marbles within it. We draw two marbles from the bag, with replacement. There are therefore four possible sequences: (1) two white marbles, (2) one blue and then one white, (3) one white and then one blue, and (4) two blue marbles. Our task is to assign probabilities to each of these possible outcomes. Suppose we know that the expected number of blue marbles over two draws is exactly 1. This is the expected value constraint on the distributions we'll consider.

We seek the distribution with the biggest entropy. Let's consider four candidate distributions, shown in FIGURE 9.3. Here are the probabilities that define each distribution:

Distribution	ww	bw	wb	bb
A	1/4	1/4	1/4	1/4
B	2/6	1/6	1/6	2/6
C	1/6	2/6	2/6	1/6
D	1/8	4/8	2/8	1/8

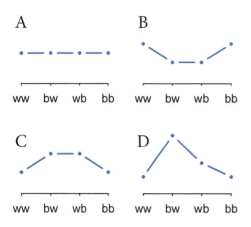

FIGURE 9.3. Four different distributions with the same expected value, 1 blue marble in 2 draws. The outcomes on the horizontal axes correspond to 2 white marbles (ww), 1 blue and then 1 white (bw), 1 white and then 1 blue (wb), and 2 blue marbles (bb).

Distribution A is the binomial distribution with $n = 2$ and $p = 0.5$. The outcomes bw and wb are usually collapsed into the same outcome type. But in principle they are different outcomes, whether we care about the order of outcomes or not. So the corresponding binomial probabilities are $\Pr(ww) = (1 - p)^2$, $\Pr(bw) = p(1 - p)$, $\Pr(wb) = (1 - p)p$, and $\Pr(bb) = p^2$. Since $p = 0.5$ in this example, all four probabilities evaluate to 1/4.

The other distributions—B, C, and D—have the same expected value, but none of them is binomial. We can expediently verify this by placing them inside a list and passing each to an expected value formula:

R code
9.5

```
# build list of the candidate distributions
p <- list()
p[[1]] <- c(1/4,1/4,1/4,1/4)
p[[2]] <- c(2/6,1/6,1/6,2/6)
p[[3]] <- c(1/6,2/6,2/6,1/6)
p[[4]] <- c(1/8,4/8,2/8,1/8)

# compute expected value of each
sapply( p , function(p) sum(p*c(0,1,1,2)) )
```

```
[1] 1 1 1 1
```

And likewise we can quickly compute the entropy of each distribution:

R code
9.6

```
# compute entropy of each distribution
sapply( p , function(p) -sum( p*log(p) ) )
```

```
[1] 1.386294 1.329661 1.329661 1.213008
```

Distribution A, the binomial distribution, has the largest entropy among the four. To appreciate why, consider that information entropy increases as a probability distribution becomes more even. Distribution A is a flat line, as you can see in FIGURE 9.3. It can't be made any more even, and each of the other distributions is clearly less even. That's why they have smaller entropies. And since distribution A is consistent with the constraint that the expected value be 1, it follows that distribution A, which is binomial, has the maximum entropy of any distribution with these constraints.

This example is too special to demonstrate the general case, however. It's special because when the expected value is 1, the distribution over outcomes can be flat and remain consistent with the constraint. But what about when the expected value constraint is not 1? Suppose for our second example that the expected value must be instead 1.4 blue marbles in two draws. This corresponds to $p = 0.7$. So you can think of this as 7 blue marbles and 3 white marbles hidden inside the bag. The binomial distribution with this expected value is:

```
p <- 0.7
( A <- c( (1-p)^2 , p*(1-p) , (1-p)*p , p^2 ) )
```
R code
9.7

```
[1] 0.09 0.21 0.21 0.49
```

This distribution is definitely not flat. So to appreciate how this distribution has maximum entropy—is the flattest distribution with expected value 1.4—we'll simulate a bunch of distributions with the same expected value and then compare entropies. The entropy of the distribution above is just:

```
-sum( A*log(A) )
```
R code
9.8

```
[1] 1.221729
```

So if we randomly generate thousands of distributions with expected value 1.4, we expect that none will have a larger entropy than this.

We can use a short R function to simulate random probability distributions that have any specified expected value. The code below will do the job. Don't worry about how it works (unless you want to[135]).

```
sim.p <- function(G=1.4) {
    x123 <- runif(3)
    x4 <- ( (G)*sum(x123)-x123[2]-x123[3] )/(2-G)
    z <- sum( c(x123,x4) )
    p <- c( x123 , x4 )/z
    list( H=-sum( p*log(p) ) , p=p )
}
```
R code
9.9

This function generates a random distribution with expected value G and then returns its entropy along with the distribution. We want to invoke this function a large number of times. Here is how to call it 100000 times and then plot the distribution of resulting entropies:

```
H <- replicate( 1e5 , sim.p(1.4) )
dens( as.numeric(H[1,]) , adj=0.1 )
```
R code
9.10

The list H now holds 100000 probability distributions and their calculated entropies. The distribution of entropies is shown in the left-hand plot in FIGURE 9.4. The letters A, B, C, and D mark different example entropies. The distributions corresponding to each are shown in the right-hand part of the figure. The distribution A with the largest observed entropy is nearly identical to the binomial we calculated earlier. And its entropy is nearly identical as well.

You don't have to take my word for it. Let's split out the entropies and distributions, so that it's easier to work with them:

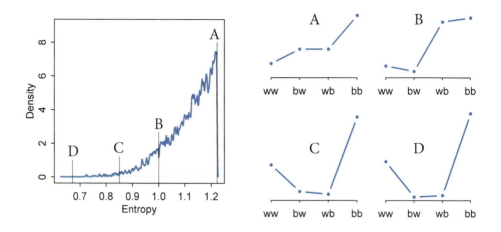

FIGURE 9.4. Left: Distribution of entropies from randomly simulated distributions with expected value 1.4. The letters A, B, C, and D mark the entropies of individual distributions shown on the right. Right: Individual probability distributions. As entropy decreases, going from A to D, the distribution becomes more uneven. The distribution marked A is the binomial distribution with $np = 1.4$.

R code
9.11
```
entropies <- as.numeric(H[1,])
distributions <- H[2,]
```

Now we can ask what the largest observed entropy was:

R code
9.12
```
max(entropies)
```

```
[1] 1.221728
```

That value is nearly identical to the entropy of the binomial distribution we calculated before. And the distribution with that entropy is:

R code
9.13
```
distributions[ which.max(entropies) ]
```

```
[[1]]
[1] 0.08981599 0.21043116 0.20993686 0.48981599
```

And that's almost exactly $\{0.09, 0.21, 0.21, 0.49\}$, the distribution we calculated earlier.

The other distributions in FIGURE 9.4—B, C, and D—are all less even than A. They demonstrate how as entropy declines the probability distributions become progressively less even. All four of these distributions really do have expected value 1.4. But among the infinite distributions that satisfy this constraint, it is only the most even distribution, the exact one nominated by the binomial distribution, that has greatest entropy.

So what? There are a few conceptual lessons to take away from this example. First, hopefully it reinforces the maximum entropy nature of the binomial distribution. When only two un-ordered outcomes are possible—such as blue and white marbles—and the expected numbers of each type of event are assumed to be constant, then the distribution that is most consistent with these constraints is the binomial distribution. This distribution spreads probability out as evenly and conservatively as possible.

Second, of course usually we do not know the expected value, but wish to estimate it. But this is actually the same problem, because assuming the distribution has a constant expected value leads to the binomial distribution as well, but with unknown expected value np, which must be estimated from the data. (You'll learn how to do this in Chapter 10.) If only two un-ordered outcomes are possible and you think the process generating them is invariant in time—so that the expected value remains constant at each combination of predictor values—then the distribution that is most conservative is the binomial. This is analogous to how the Gaussian distribution is the most conservative distribution for a continuous outcome variable with finite variance. Variables with different constraints get different maximum entropy distributions, but the underlying principle remains the same.

Third, back in Chapter 2, we derived the binomial distribution just by counting how many paths through the garden of forking data were consistent with our assumptions. For each possible composition of the bag of marbles—which corresponds here to each possible expected value—there is a unique number of ways to realize any possible sequence of data. The likelihoods derived in that way turn out to be exactly the same as the likelihoods we get by maximizing entropy. This is not a coincidence. Entropy counts up the number of different ways a process can produce a particular result, according to our assumptions. The garden of forking data did only the same thing—count up the numbers of ways a sequence could arise, given assumptions.

Entropy maximization, like so much in probability theory, is really just counting. But it's abbreviated counting that allows us to generalize lessons learned in one context to new problems in new contexts. Instead of having to tediously draw out a garden of forking data, we can instead map constraints on an outcome to a probability distribution. There is no guarantee that this is the best probability distribution for the real problem you are analyzing. But there is a guarantee that no other distribution more conservatively reflects your assumptions.

That's not everything, but nor is it nothing. Any other distribution implies hidden constraints that are unknown to us, reflecting phantom assumptions. A full and honest accounting of assumptions is helpful, because it aids in understanding how a model misbehaves. And since all models misbehave sometimes, it's good to be able to anticipate those times before they happen, as well as to learn from those times when they inevitably do.

Rethinking: Conditional independence. All this talk of constant expected value brings up an important question: Do these distributions necessarily assume that each observation is uncorrelated with every other observation? Not really. What is usually meant by "independence" in a probability distribution is just that each observation is uncorrelated with the others, once we know the corresponding predictor values. This is usually known as CONDITIONAL INDEPENDENCE, the claim that observations are independent after accounting for differences in predictors, through the model. It's a modeling assumption. What this assumption doesn't cover is a situation in which an observed event directly causes the next observed event. For example, if you buy the next Nick Cave album because I buy the next Nick Cave album, then your behavior is not independent of mine, even after conditioning on the fact that we both like that sort of music.

Overthinking: Binomial maximum entropy. The usual way to derive a maximum entropy distribution is to state the constraints and then use a mathematical device called the *Lagrangian* to solve for the probability assignments that maximize entropy. But instead we'll extend the strategy used in the overthinking box on page 274. As a bonus, this strategy will allow us to derive the constraints that are necessary for a distribution, in this case the binomial, to be a maximum entropy distribution.

Let p be the binomial distribution, and let p_i be the probability of a sequence of observations i with number of successes x_i and number of failures $n - x_i$. Let q be some other discrete distribution defined over the same set of observable sequences. As before, KL divergence tells us that:

$$-H(q, p) \geq H(q) \implies -\sum_i q_i \log p_i \geq -\sum_i q_i \log q_i$$

What we're going to do now is work with $H(q, p)$ and simplify it until we can isolate the constraint that defines the class of distributions for which p has maximum entropy. Let $\lambda = \sum_i p_i x_i$ be the expected value of p. Then from the definition of $H(q, p)$:

$$-H(q, p) = -\sum_i q_i \log \left[\left(\frac{\lambda}{n} \right)^{x_i} \left(1 - \frac{\lambda}{n} \right)^{n - x_i} \right] = -\sum_i q_i \left(x_i \log \left[\frac{\lambda}{n} \right] + (n - x_i) \log \left[1 - \frac{\lambda}{n} \right] \right)$$

After some algebra:

$$-H(q, p) = -\sum_i q_i \left(x_i \log \left[\frac{\lambda}{n - \lambda} \right] + n \log \left[\frac{n - \lambda}{n} \right] \right) = -n \log \left[\frac{n - \lambda}{n} \right] - \log \left[\frac{\lambda}{n - \lambda} \right] \underbrace{\sum_i q_i x_i}_{\bar{q}}$$

The term on the far right labeled $\bar{q}$ is the expected value of the distribution q. If we knew it, we could complete the calculation, because no other term depends upon q_i. This means that expected value is the constraint that defines the class of distributions for which the binomial p has maximum entropy. If we now set the expected value of q equal to λ, then $H(q) = H(p)$. For any other expected value of q, $H(p) > H(q)$.

Finally, notice the term $\log[\lambda/(n - \lambda)]$. This term is the log of the ratio of the expected number of successes to the expected number of failures. That ratio is the "odds" of a success, and its logarithm is called "log odds." This quantity will feature prominently in models we construct from the binomial distribution, in Chapter 11.

9.2. Generalized linear models

The Gaussian models of previous chapters worked by first assuming a Gaussian distribution over outcomes. Then, we replaced the parameter that defines the mean of that distribution, μ, with a linear model. This resulted in likelihood definitions of the sort:

$$y_i \sim \text{Normal}(\mu_i, \sigma)$$
$$\mu_i = \alpha + \beta x_i$$

For an outcome variable that is continuous and far from any theoretical maximum or minimum, this sort of Gaussian model has maximum entropy.

But when the outcome variable is either discrete or bounded, a Gaussian likelihood is not the most powerful choice. Consider for example a count outcome, such as the number of blue marbles pulled from a bag. Such a variable is constrained to be zero or a positive integer. Using a Gaussian model with such a variable won't result in a terrifying explosion. But it can't be trusted to do much more than estimate the average count. It certainly can't be trusted to produce sensible predictions, because while you and I know that counts can't be negative, a linear regression model does not. So it would happily predict negative values, whenever the mean count is close to zero.

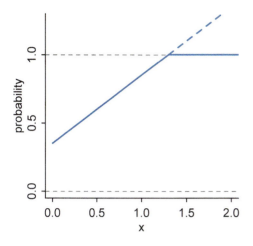

FIGURE 9.5. Why we need link functions. The solid blue line is a linear model of a probability mass. It increases linearly with a predictor, x, on the horizontal axis. But when it reaches the maximum probability mass of 1, at the dashed boundary, it will happily continue upwards, as shown by the dashed blue line. In reality, further increases in x could not further increase probability, as indicated by the horizontal continuation of the solid trend.

Luckily, it's easy to do better. By using all of our prior knowledge about the outcome variable, usually in the form of constraints on the possible values it can take, we can appeal to maximum entropy for the choice of distribution. Then all we have to do is generalize the linear regression strategy—replace a parameter describing the shape of the likelihood with a linear model—to probability distributions other than the Gaussian.

This is the essence of a **GENERALIZED LINEAR MODEL**.[136] And it results in models that look like this:

$$y_i \sim \text{Binomial}(n, p_i)$$
$$f(p_i) = \alpha + \beta x_i$$

There are only two changes here from the familiar Gaussian model. The first is principled—the principle of maximum entropy. The second is an epicycle—a modeling trick that works descriptively but not causally—but a quite successful one. I'll briefly explain each, before moving on in the remainder of the section to describe all of the most common distributions used to construct generalized linear models. Later chapters show you how to implement them.

First, the likelihood is binomial instead of Gaussian. For a count outcome y for which each observation arises from n trials and with constant expected value np, the binomial distribution has maximum entropy. So it's the least informative distribution that satisfies our prior knowledge of the outcomes y. If the outcome variable had different constraints, it could be a different maximum entropy distribution.

Second, there is now a funny little f at the start of the second line of the model. This represents a **LINK FUNCTION**, to be determined separately from the choice of distribution. Generalized linear models need a link function, because rarely is there a "μ", a parameter describing the average outcome, and rarely are parameters unbounded in both directions, like μ is. For example, the shape of the binomial distribution is determined, like the Gaussian, by two parameters. But unlike the Gaussian, neither of these parameters is the mean. Instead, the mean outcome is np, which is a function of both parameters. Since n is usually known (but not always), it is most common to attach a linear model to the unknown part, p. But p is a probability mass, so p_i must lie between zero and one. But there's nothing to stop the linear model $\alpha + \beta x_i$ from falling below zero or exceeding one. FIGURE 9.5 plots an example.

The link function f provides a solution to this common problem. This chapter will introduce the two most common link functions. Then you'll see how to use them in the chapters that follow.

> **Rethinking: The scourge of Histomancy.** One strategy for choosing an outcome distribution is to plot the histogram of the outcome variable and, by gazing into its soul, decide what sort of distribution function to use. Call this strategy HISTOMANCY, the ancient art of divining likelihood functions from empirical histograms. This sorcery is used, for example, when testing for normality before deciding whether or not to use a non-parametric procedure. Histomancy is a false god, because even perfectly good Gaussian variables may not look Gaussian when displayed as a histogram. Why? Because at most what a Gaussian likelihood assumes is not that the aggregated data look Gaussian, but rather that the *residuals*, after fitting the model, look Gaussian. So for example the combined histogram of male and female body weights is certainly not Gaussian. But it is (approximately) a mixture of Gaussian distributions, so after conditioning on sex, the residuals may be quite normal. Other times, people decide not to use a Poisson model, because the variance of the aggregate outcome exceeds its mean (see Chapter 10). But again, at most what a Poisson likelihood assumes is that the variance equals the mean after conditioning on predictors. It may very well be that a Gaussian or Poisson likelihood is a poor assumption in any particular context. But this can't easily be decided via Histomancy. This is why we need principles, whether maximum entropy or otherwise.

9.2.1. Meet the family.

The most common distributions used in statistical modeling are members of a family known as the EXPONENTIAL FAMILY. Every member of this family is a maximum entropy distribution, for some set of constraints. And conveniently, just about every other statistical modeling tradition employs the exact same distributions, even though they arrive at them via justifications other than maximum entropy.

FIGURE 9.6 illustrates the representative shapes of the most common exponential family distributions used in GLMs. The horizontal axis in each plot represents values of a variable, and the vertical axis represents probability density (for the continuous distributions) or probability mass (for the discrete distributions). For each distribution, the figure also provides the notation (above each density plot) and the name of R's corresponding built-in distribution function (below each density plot). The gray arrows in FIGURE 9.6 indicate some of the ways that these distributions are dynamically related to one another. These relationships arise from generative processes that can convert one distribution to another. You do not need to know these relationships in order to successfully use these distributions in your modeling. But the generative relationships do help to demystify these distributions, by tying them to causation and measurement.

Two of these distributions, the Gaussian and binomial, are already familiar to you. Together, they comprise the most commonly used outcome distributions in applied statistics, through the procedures of linear regression (Chapter 4) and logistic regression (Chapter 10). There are also three new distributions that deserve some commentary.

The EXPONENTIAL DISTRIBUTION (center) is constrained to be zero or positive. It is a fundamental distribution of distance and duration, kinds of measurements that represent displacement from some point of reference, either in time or space. If the probability of an event is constant in time or across space, then the distribution of events tends towards exponential. The exponential distribution has maximum entropy among all non-negative continuous distributions with the same average displacement. Its shape is described by a single parameter, the rate of events λ, or the average displacement λ^{-1}. This distribution is the core of survival and event history analysis, which is not covered in this book.

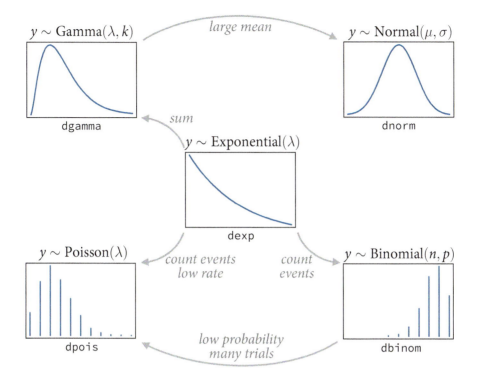

FIGURE 9.6. Some of the exponential family distributions, their notation, and some of their relationships. Center: exponential distribution. Clockwise, from top-left: gamma, normal (Gaussian), binomial and Poisson distributions.

The GAMMA DISTRIBUTION (top-left) is also constrained to be zero or positive. It too is a fundamental distribution of distance and duration. But unlike the exponential distribution, the gamma distribution can have a peak above zero. If an event can only happen after two or more exponentially distributed events happen, the resulting waiting times will be gamma distributed. For example, age of cancer onset is approximately gamma distributed, since multiple events are necessary for onset.[137] The gamma distribution has maximum entropy among all distributions with the same mean and same average logarithm. Its shape is described by two parameters, but there are at least three different common descriptions of these parameters, so some care is required when working with it. The gamma distribution is common in survival and event history analysis, as well as some contexts in which a continuous measurement is constrained to be positive.

The POISSON DISTRIBUTION (bottom-left) is a count distribution like the binomial. It is actually a special case of the binomial, mathematically. If the number of trials n is very large (and usually unknown) and the probability of a success p is very small, then a binomial distribution converges to a Poisson distribution with an expected rate of events per unit time of $\lambda = np$. Practically, the Poisson distribution is used for counts that never get close to any theoretical maximum. As a special case of the binomial, it has maximum entropy under exactly the same constraints. Its shape is described by a single parameter, the rate of events λ. Poisson GLMs are detailed in the next chapter.

There are many other exponential family distributions, and many of them are useful. But don't worry that you need to memorize them all. You can pick up new distributions, and the sorts of generative processes they correspond to, as needed. It's also not important that an outcome distribution be a member of the exponential family—if you think you have good reasons to use some other distribution, then use it. But you should also check its performance, just like you would any modeling assumption.

> **Rethinking: A likelihood is a prior.** In traditional statistics, likelihood functions are "objective" and prior distributions "subjective." However, likelihoods are themselves prior probability distributions: They are priors for the data, conditional on the parameters. And just like with other priors, there is no correct likelihood. But there are better and worse likelihoods, depending upon the context. Useful inference does not require that the data (or residuals) be actually distributed according to the likelihood anymore than it requires the posterior distribution to be like the prior.

9.2.2. Linking linear models to distributions.
To build a regression model from any of the exponential family distributions is just a matter of attaching one or more linear models to one or more of the parameters that describe the distribution's shape. But as hinted at earlier, usually we require a **LINK FUNCTION** to prevent mathematical accidents like negative distances or probability masses that exceed 1. So for any outcome distribution, say for example the exotic "Zaphod" distribution,[138] we write:

$$y_i \sim \text{Zaphod}(\theta_i, \phi)$$
$$f(\theta_i) = \alpha + \beta x_i$$

where f is a link function.

But what function should f be? A link function's job is to map the linear space of a model like $\alpha + \beta x_i$ onto the non-linear space of a parameter like θ. So f is chosen with that goal in mind. Most of the time, for most GLMs, you can use one of two exceedingly common links, a *logit link* or a *log link*. Let's introduce each, and you'll work with both in later chapters.

The **LOGIT LINK** maps a parameter that is defined as a probability mass, and therefore constrained to lie between zero and one, onto a linear model that can take on any real value. This link is extremely common when working with binomial GLMs. In the context of a model definition, it looks like this:

$$y_i \sim \text{Binomial}(n, p_i)$$
$$\text{logit}(p_i) = \alpha + \beta x_i$$

And the logit function itself is defined as the *log-odds*:

$$\text{logit}(p_i) = \log \frac{p_i}{1 - p_i}$$

The "odds" of an event are just the probability it happens divided by the probability it does not happen. So really all that is being stated here is:

$$\log \frac{p_i}{1 - p_i} = \alpha + \beta x_i$$

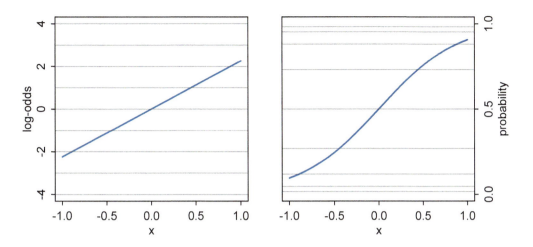

FIGURE 9.7. The logit link transforms a linear model (left) into a probability (right). This transformation compresses the geometry far from zero, such that a unit change on the linear scale (left) means less and less change on the probability scale (right).

So to figure out the definition of p_i implied here, just do a little algebra and solve the above equation for p_i:

$$p_i = \frac{\exp(\alpha + \beta x_i)}{1 + \exp(\alpha + \beta x_i)}$$

The above function is usually called the LOGISTIC. In this context, it is also commonly called the INVERSE-LOGIT, because it inverts the logit transform.

What all of this means is that when you use a logit link for a parameter, you are defining the parameter's value to be the logistic transform of the linear model. FIGURE 9.7 illustrates the transformation that takes place when using a logit link. On the left, the geometry of the linear model is shown, with horizontal lines indicating unit changes in the value of the linear model as the value of a predictor x changes. This is the log-odds space, which extends continuously in both positive and negative directions. On the right, the linear space is transformed and is now constrained entirely between zero and one. The horizontal lines have been compressed near the boundaries, in order to make the linear space fit within the probability space. This compression produces the characteristic logistic shape of the transformed linear model shown in the right-hand plot.

This compression does affect interpretation of parameter estimates, because no longer does a unit change in a predictor variable produce a constant change in the mean of the outcome variable. Instead, a unit change in x_i may produce a larger or smaller change in the probability p_i, depending upon how far from zero the log-odds are. For example, in FIGURE 9.7, when $x = 0$ the linear model has a value of zero on the log-odds scale. A half-unit increase in x results in about a 0.25 increase in probability. But each addition half-unit will produce less and less of an increase in probability, until any increase is vanishingly small. And if you think about it, a good model of probability needs to behave this way. When an

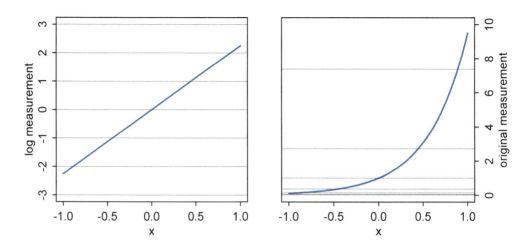

FIGURE 9.8. The log link transforms a linear model (left) into a strictly positive measurement (right). This transform results in an exponential scaling of the linear model, with a unit change on the linear scale mapping onto increasingly larger changes on the outcome scale.

event is almost guaranteed to happen, its probability cannot increase very much, no matter how important the predictor may be.

You'll find examples of this compression phenomenon in later chapters. The key lesson for now is just that no regression coefficient, such as β, from a GLM ever produces a constant change on the outcome scale. Recall that we defined interaction (Chapter 7) as a situation in which the effect of a predictor depends upon the value of another predictor. Well now every predictor essentially interacts with itself, because the impact of a change in a predictor depends upon the value of the predictor before the change. More generally, every predictor variable effectively interacts with every other predictor variable, whether you explicitly model them as interactions or not. This fact makes the visualization of counter-factual predictions even more important for understanding what the model is telling you.

The second very common link function is the **LOG LINK**. This link function maps a parameter that is defined over only positive real values onto a linear model. For example, suppose we want to model the standard deviation σ of a Gaussian distribution so it is a function of a predictor variable x. The parameter σ must be positive, because a standard deviation cannot be negative nor can it be zero. The model might look like:

$$y_i \sim \text{Normal}(\mu, \sigma_i)$$
$$\log(\sigma_i) = \alpha + \beta x_i$$

In this model, the mean μ is constant, but the standard deviation scales with the value x_i. A log link is both conventional and useful in this situation. It prevents σ from taking on a negative value.

What the log link effectively assumes is that the parameter's value is the exponentiation of the linear model. Solving $\log(\sigma_i) = \alpha + \beta x_i$ for σ_i yields the inverse link:

$$\sigma_i = \exp(\alpha + \beta x_i)$$

The impact of this assumption can be seen in FIGURE 9.8. Using a log link for a linear model (left) implies an exponential scaling of the outcome with the predictor variable (right). Another way to think of this relationship is to remember that logarithms are *magnitudes*. An increase of one unit on the log scale means an increase of an order of magnitude on the untransformed scale. And this fact is reflected in the widening intervals between the horizontal lines in the right-hand plot of FIGURE 9.8.

While using a log link does solve the problem of constraining the parameter to be positive, it may also create a problem when the model is asked to predict well outside the range of data used to fit it. Exponential relationships grow, well, exponentially. Just like a linear model cannot be linear forever, an exponential model cannot be exponential forever. Human height cannot be linearly related to weight forever, because very heavy people stop getting taller and start getting wider. Likewise, the property damage caused by a hurricane may be approximately exponentially related to wind speed for smaller storms. But for very big storms, damage may be capped by the fact that everything gets destroyed.

Rethinking: When in doubt, play with assumptions. Link functions do amount to assumptions. And like all assumptions, they are useful in different contexts. The conventional logit and log links are widely useful, but they can sometimes distort inference. If you ever have doubts, and want to reassure yourself that your conclusions are not sensitive to choice of link function, then do what you'd do for any other modeling assumption: SENSITIVITY ANALYSIS. A sensitivity analysis explores how changes in assumptions influence inference. If none of the alternative assumptions you consider have much impact on inference, that's worth reporting. Likewise, if the alternatives you consider do have an important impact on inference, that's also worth reporting. The same sort of advice follows for other modeling assumptions: likelihoods, linear models, priors, and even how the model is fit to data. As with many machines, exploring how a model behaves under extreme conditions helps us understand how it behaves under ordinary conditions.

Some people are nervous about sensitivity analysis, because it feels like fishing for results, or "p-hacking."[139] The goal of sensitivity analysis is really the opposite of p-hacking. In p-hacking, many justifiable analyses are tried, and the one that attains statistical significance is reported. In sensitivity analysis, many justifiable analyses are tried, and all of them are described.

Overthinking: Parameters interacting with themselves. We can find some further clarity on the claim that GLMs force every predictor variable to interact with itself by mathematically computing the rate of change in the outcome for a given change in the value of the predictor. First, recall that in a classic Gaussian model the mean is modeled like:

$$\mu = \alpha + \beta x$$

So the rate of change in μ with respect to x is just $\partial \mu / \partial x = \beta$. And that's constant. It doesn't matter what value x has. But now consider the rate of change in a binomial probability p with respect to a predictor x:

$$p = \frac{\exp(\alpha + \beta x)}{1 + \exp(\alpha + \beta x)}$$

And now taking the derivative with respect to x yields:

$$\frac{\partial p}{\partial x} = \frac{\beta}{2\big(1 + \cosh(\alpha + \beta x)\big)}$$

Since x appears in this answer, the impact of a change in x depends upon x. That's an interaction with itself.

9.2.3. Absolute and relative differences. There is an important practical consequence of the way that a link function compresses and expands different portions of the linear model's range: Parameter estimates do not by themselves tell you the importance of a predictor on the outcome. The reason is that each parameter represents a *relative* difference on the scale of the linear model, ignoring other parameters, while we are really interested in *absolute* differences in outcomes that must incorporate all parameters.

This point will come up again in the context of data examples in later chapters, when it will be easier to illustrate its importance. For now, just keep in mind that a big beta-coefficient may not correspond to a big effect on the outcome.

9.2.4. GLMs and information criteria. What you learned in Chapter 6 about information criteria and regularizing priors applies also to GLMs. But with all these new outcome distributions at your command, it is tempting to use information criteria to compare models with different likelihood functions. Is a Gaussian or binomial better? Can't we just let WAIC sort it out?

Unfortunately, WAIC (or any other information criterion) cannot sort it out. The problem is that deviance is part normalizing constant. The constant affects the absolute magnitude of the deviance, but it doesn't affect fit to data. Since information criteria are all based on deviance, their magnitude also depends upon these constants. That is fine, as long as all of the models you compare use the same outcome distribution type—Gaussian, binomial, exponential, gamma, Poisson, or another. In that case, the constants subtract out when you compare models by their differences. But if two models have different outcome distributions, the constants don't subtract out, and you can be misled by a difference in AIC/DIC/WAIC.

Really all you have to remember is to only compare models that all use the same type of likelihood. Of course it is possible to compare models that use different likelihoods, just not with information criteria. Luckily, the principle of maximum entropy ordinarily motivates an easy choice of likelihood, at least for ordinary regression models. So there is no need to lean on information criteria for this modeling choice.

There are a few nuances with WAIC and individual GLM types. These nuances will arise as examples of each GLM are worked, in later chapters.

9.3. Maximum entropy priors

The principle of maximum entropy helps us to make modeling choices. When pressed to choose an outcome distribution—a likelihood—maximum entropy nominates the least informative distribution consistent with the constraints on the outcome variable. Applying the principle in this way leads to many of the same distributional choices that are commonly regarded as just convenient assumptions or useful conventions.

Another way that the principle of maximum entropy helps with choosing distributions arises when choosing priors. GLMs are easy to use with conventional weakly informative priors of the sort you've been using up to this point in the book. Such priors are nice, because they allow the data to dominate inference while also taming some of the pathologies of unconstrained estimation. There were some striking examples of their "soft power" in Chapter 8.

But sometimes, rarely, some of the parameters in a GLM refer to things we might actually have background information about. When that's true, maximum entropy provides a way to generate a prior that embodies the background information, while assuming as little else as possible. This makes them appealing, conservative choices.

We won't be using maximum entropy to choose priors in this book, but when you come across an analysis that does, you can interpret the principle in the same way as you do with likelihoods and understand the approach as an attempt to include relevant background information about parameters, while introducing no other assumptions by accident.

9.4. Summary

This chapter has been a conceptual, not practical, introduction to maximum entropy and generalized linear models. The principle of maximum entropy provides an empirically successful way to choose likelihood functions. Information entropy is essentially a measure of the number of ways a distribution can arise, according to stated assumptions. By choosing the distribution with the biggest information entropy, we thereby choose a distribution that obeys the constraints on outcome variables, without importing additional assumptions. Generalized linear models arise naturally from this approach, as extensions of the linear models in previous chapters. The necessity of choosing a link function to bind the linear model to the generalized outcome introduces new complexities in model specification, estimation, and interpretation. You'll become comfortable with these complexities through examples in later chapters.

10 Counting and Classification

All over the world, every day, scientists throw away information. Sometimes this is through the removal of "outliers," cases in the data that offend the model and so are exiled. More routinely, counted things are converted to proportions before analysis. Why does analysis of proportions throw away information? Because 10/20 and 1/2 are the same proportion, one-half, but have very different sample sizes. Once converted to proportions, and treated as outcomes in a linear regression, the information about sample size has been destroyed.

It's easy to retain the information about sample size. All that is needed is to model what has actually been observed, the counts instead of the proportions. No one has ever observed a proportion. Instead we observe counts. And this chapter is about forms of regression that allow us to retain the sample size information embodied in count data, while still modeling complex associations between suites of predictors and the outcome.

The fundamental friction of passing between count data and inference about proportion never fades, however. Extra care has to be observed in interpretation, as the parameters will no longer exist on the same scale as the outcome variable, as they do with ordinary linear regression. You'll get used to this friction quickly enough. But it's normal to find it confusing and frustrating at first.

We will work complete examples of the two most common count regressions.

(1) BINOMIAL REGRESSION is the name we'll use for a family of related procedures that all model a binary classification—alive/dead, accept/reject, left/right—for which the total of both categories is known. This is just like the marble and globe tossing examples from Chapter 2. But now you get to incorporate predictor variables.

(2) POISSON REGRESSION is a GLM that models a count outcome without a known maximum—number of elephants in Kenya, number of people who apply to a physics PhD program, number of significance tests in an issue of *Psychological Science*. As described in Chapter 9, the Poisson model can also be conceived of as a binomial model with a very large maximum but a very small probability per trial.

At the end, the chapter describes some other count regressions, but does not fully work examples.

All of the examples in this chapter, and the chapters to come, use all of the tools introduced in previous chapters. Regularizing priors, information criteria, and MCMC estimation are woven into the data analysis examples. So as you work through the examples that introduced each new type of GLM, you'll also get to practice and better understand previous lessons.

10.1. Binomial regression

By this point in the book, the binomial distribution is familiar. In the context of defining a model, the binomial distribution is denoted:

$$y \sim \text{Binomial}(n, p)$$

where y is a count (zero or a positive whole number), p is the probability any particular "trial" is a success, and n is the number of trials. As the basis for a generalized linear model, the binomial distribution has maximum entropy when each trial must result in one of two events and the expected value is constant. This much was described at length in Chapter 9.

Now we'll define and really work examples of binomial regression, using predictor variables and interactions and information criteria. There are two common flavors of GLM that use binomial likelihood functions:

(1) LOGISTIC REGRESSION is the common name when the data are organized into single-trial cases, such that the outcome variable can only take values 0 and 1.
(2) When individual trials with the same covariate values are instead aggregated together, it is common to speak of an AGGREGATED BINOMIAL REGRESSION. In this case, the outcome can take the value zero or any positive integer up to n, the number of trials.

Both flavors use the same logit link function (page 284), so both may sometimes be called "logistic" regression, as the inverse of the logit function is the logistic. Either form of binomial regression can be converted into the other by aggregating (logistic to aggregated) or exploding (aggregated to logistic) the outcome variable. We'll fully work an example of each.

Like other GLMs, binomial regression is never guaranteed to produce a nice multivariate Gaussian posterior distribution. So MAP estimation is not always satisfactory. We'll work each example using map, but we'll also check the inferences against MCMC sampling, using map2stan. The reason to do it both ways is so you can get a sense of both how often MAP estimation works, even when in principle it should not, and why it fails in particular contexts.

10.1.1. Logistic regression: Prosocial chimpanzees. The data for this example come from an experiment[140] aimed at evaluating the prosocial tendencies of chimpanzees (*Pan troglodytes*). The experimental structure mimics many common experiments conducted on human students (*Homo sapiens studiensis*) by economists and psychologists. A focal chimpanzee sits at one end of a long table with two levers, one on the left and one on the right in FIGURE 10.1. On the table are four dishes which may contain desirable food items. The two dishes on the right side of the table are attached by a mechanism to the right-hand lever. The two dishes on the left side are similarly attached to the left-hand lever.

When either the left or right lever is pulled by the focal animal, the two dishes on the same side slide towards opposite ends of the table. This delivers whatever is in those dishes to the opposite ends. In all experimental trials, both dishes on the focal animal's side contain food items. But only one of the dishes on the other side of the table contains a food item. Therefore while both levers deliver food to the focal animal, only one of the levers delivers food to the other side of the table.

There are two experimental conditions. In the *partner* condition, another chimpanzee is seated at the opposite end of the table, as pictured in FIGURE 10.1. In the control condition, the other side of the table is empty. Finally, two counterbalancing treatments alternate which side, left or right, has a food item for the other side of the table. This helps detect any handedness preferences for individual focal animals.

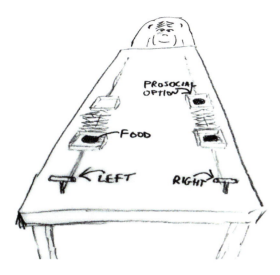

FIGURE 10.1. Chimpanzee prosociality experiment, as seen from the perspective of the focal animal. The left and right levers are indicated in the foreground. Pulling either expands an accordion device in the center, pushing the food trays towards both ends of the table. Both food trays close to the focal animal have food in them. Only one of the food trays on the other side contains food. The partner condition means another animal, as pictured, sits on the other end of the table. Otherwise, the other end was empty.

When human students participate in an experiment like this, they nearly always choose the lever linked to two pieces of food, the *prosocial* option, but only when another student sits on the opposite side of the table. The motivating question is whether a focal chimpanzee behaves similarly, choosing the prosocial option more often when another animal is present. In terms of linear models, we want to estimate the interaction between condition (presence or absence of another animal) and option (which side is prosocial).

Load the data from the `rethinking` package:

R code
10.1

```
library(rethinking)
data(chimpanzees)
d <- chimpanzees
```

Take a look at the built-in help, `?chimpanzees`, for details on all of the available variables. We're going to focus on `pulled_left` as the outcome to predict, with `prosoc_left` and `condition` as predictor variables. The outcome `pulled_left` is a 0 or 1 indicator that the focal animal pulled the left-hand lever. The predictor `prosoc_left` is a 0/1 indicator that the left-hand lever was (1) or was not (0) attached to the prosocial option, the side with two pieces of food. The `condition` predictor is another 0/1 indicator, with value 1 for the partner condition and value 0 for the control condition.

Now we're ready to fit a model. The model implied by the research question is, in mathematical form first:

$$L_i \sim \text{Binomial}(1, p_i)$$
$$\text{logit}(p_i) = \alpha + (\beta_P + \beta_{PC}C_i)P_i$$
$$\alpha \sim \text{Normal}(0, 10)$$
$$\beta_P \sim \text{Normal}(0, 10)$$
$$\beta_{PC} \sim \text{Normal}(0, 10)$$

Here L indicates `pulled_left`, P indicates `prosoc_left`, and C indicates `condition`. The tricky part of the model above is the linear model for $\text{logit}(p_i)$. It is an interaction model, in which the association between P_i and the log-odds that $L_i = 1$ depends upon the value of C_i. But note that there is no main effect of C_i itself, no plain beta-coefficient for condition.

Why? Because there is no reason to hypothesize that the presence of absence of another animal creates a tendency to pull the left-hand lever. This is equivalent to assuming that the main effect of condition is exactly zero. You can check this assumption later, if you like.

The priors above are chosen for lack of informativeness—they are very gently regularizing, but will be overwhelmed by even moderate evidence. So the estimates we'll get from this model will no doubt be overfit to sample somewhat. To get some comparative measure of that overfitting, we'll also fit two other models with fewer predictors. First a model with just the intercept:

$$L_i \sim \text{Binomial}(1, p_i)$$
$$\text{logit}(p_i) = \alpha$$
$$\alpha \sim \text{Normal}(0, 10)$$

And also a model that tries to predict lever pulls using only `prosoc_left`, but ignoring `condition`:

$$L_i \sim \text{Binomial}(1, p_i)$$
$$\text{logit}(p_i) = \alpha + \beta_P P_i$$
$$\alpha \sim \text{Normal}(0, 10)$$
$$\beta_P \sim \text{Normal}(0, 10)$$

Before charging ahead and fitting all three of these models, let's just fit the simplest one, the intercept-only model, and talk through the log-odds estimate it provides. The map code looks a lot like the mathematical specification:

R code
10.2
```
m10.1 <- map(
    alist(
        pulled_left ~ dbinom( 1 , p ) ,
        logit(p) <- a ,
        a ~ dnorm(0,10)
    ) ,
    data=d )
precis(m10.1)
```

```
  Mean StdDev 5.5% 94.5%
a 0.32   0.09 0.18  0.46
```

Now to interpret the estimate for a (α), we have to remember that the parameters in a logistic regression are on the scale of log-odds. To get them back onto the probability scale, we have to use the inverse link function. In this case, the inverse link is *logistic*, provided as `logistic` in the `rethinking` package. So the above summary implies a MAP probability of pulling the left lever of `logistic(0.32)` ≈ 0.58, with a 89% interval of 0.54 to 0.61:

R code
10.3
```
logistic( c(0.18,0.46) )
```

```
[1] 0.5448789 0.6130142
```

So we have to note, before considering any predictors, that the chimpanzees exhibited a preference for the left-hand lever. Later we'll see that individual chimpanzees vary a lot in their handedness, so this overall average is misleading.

Now fit the next two models, with no surprises in the code:

R code
10.4

```
m10.2 <- map(
    alist(
        pulled_left ~ dbinom( 1 , p ) ,
        logit(p) <- a + bp*prosoc_left ,
        a ~ dnorm(0,10) ,
        bp ~ dnorm(0,10)
    ) ,
    data=d )
m10.3 <- map(
    alist(
        pulled_left ~ dbinom( 1 , p ) ,
        logit(p) <- a + (bp + bpC*condition)*prosoc_left ,
        a ~ dnorm(0,10) ,
        bp ~ dnorm(0,10) ,
        bpC ~ dnorm(0,10)
    ) ,
    data=d )
```

And to compare the three models, use the compare function introduced in Chapter 6:

R code
10.5

```
compare( m10.1 , m10.2 , m10.3 )
```

```
      WAIC pWAIC dWAIC weight   SE  dSE
m10.2 680.6     2   0.0   0.70 9.30   NA
m10.3 682.4     3   1.8   0.28 9.34 0.81
m10.1 688.0     1   7.4   0.02 7.13 6.14
```

The model that includes condition doesn't do best, but does get more than 25% of the WAIC weight. You can also plot the compare results, which look like this:

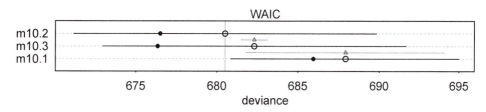

Notice that even though m10.2 isn't hugely better than m10.3, the difference has a small standard error. Even doubling the standard error to get a 95% interval, the order of the two models would not change. So on the basis of information criteria, even though model m10.3 of course fits the sample better than model m10.2 (because it has more parameters), it does not fit sufficiently better to overcome the expected overfitting.

This is no reason to ignore m10.3, however. After all, it's the model that reflects the structure of the experiment. It's not enough to show some other model beats it by WAIC comparison. We also want to understand why m10.3 compares poorly with m10.2. So let's look at the estimates for m10.3:

R code
10.6

```
precis(m10.3)
```

```
       Mean StdDev  5.5% 94.5%
a      0.05   0.13 -0.15  0.25
bp     0.61   0.23  0.25  0.97
bpC   -0.10   0.26 -0.53  0.32
```

The estimated interaction effect `bpC` is negative, with a rather wide posterior on both sides of zero. So regardless of the information theory ranking, the estimates suggest that the chimpanzees did not care much about the other animal's presence. But they do prefer to pull the prosocial option, as indicated by the estimate for `bp`.

To get a better sense of this impact of the estimate 0.61 for `bp`, we have to distinguish between **ABSOLUTE EFFECT** and **RELATIVE EFFECT**. The absolute effect is the change in the probability of the outcome. So it depends upon all of the parameters, and it tells us the practical impact of a change in a predictor. The relative effect is instead just a proportional change induced by a change in the predictor. Let's work through both, mainly in an attempt to convince you that relative effects can be highly misleading, because they ignore all of the other parameters. However, both absolute and relative effects matter, depending upon context.

First, let's consider the relative effect size of `prosoc_left` and its parameter `bp`. The customary measure of relative effect for a logistic model is the **PROPORTIONAL CHANGE IN ODDS**. You can compute the proportional odds by merely exponentiating the parameter estimate. Remember, odds are the ratio of the probability an event happens to the probability it does not happen. So in this case the relevant odds are the odds of pulling the left-hand lever (the outcome variable). If changing the predictor `prosoc_left` from 0 to 1 increases the log-odds of pulling the left-hand lever by 0.61 (the MAP estimate above), then this also implies that the odds are multiplied by:

R code
10.7
```
exp(0.61)
```

```
[1] 1.840431
```

You can read this a proportional increase of 1.84 in the odds of pulling the left-hand lever. What this means is that the odds increase by 84%.

The major difficulty with the proportional odds is that the actual change in probability will also depend upon the intercept, α, as well as any other predictor variables. Remember, GLMs like this logistic regression induce interactions among all variables. In this case, you can think of these interactions as resulting from both ceiling and floor effects: If the intercept is large enough to guarantee a pull, then increasing the odds by 84% isn't going to make it any more guaranteed. For example, suppose α were estimated to have a value of 4. Then the probability of a pull, ignoring everything else, would be:

R code
10.8
```
logistic( 4 )
```

```
[1] 0.9820138
```

Adding in an increase of 0.61 (the estimate for `bp`) changes this to:

R code
10.9
```
logistic( 4 + 0.61 )
```

```
[1] 0.9901462
```

So that's a difference, on the absolute scale, of less than 1%, despite being an 84% increase in proportional odds. Likewise, if the intercept is very negative, then the probability of a pull is almost zero. An increase in odds of 84% may not be enough to get the probability up off the floor. Relative effects, as measured by proportional odds or anything else, can be misleading. It's always wise to simultaneously consider absolute effects, the prediction scale.

Let's consider the model-averaged posterior predictive check now, to get a sense of the absolute effect of each treatment on the probability of pulling the left-hand lever. To do this, we follow the same steps as in previous chapters, using ensemble to build a mixture of predictions across the models we've fit, each model weighted according to its WAIC score.

R code
10.10

```
# dummy data for predictions across treatments
d.pred <- data.frame(
    prosoc_left = c(0,1,0,1),   # right/left/right/left
    condition = c(0,0,1,1)      # control/control/partner/partner
)

# build prediction ensemble
chimp.ensemble <- ensemble( m10.1 , m10.2 , m10.3 , data=d.pred )

# summarize
pred.p <- apply( chimp.ensemble$link , 2 , mean )
pred.p.PI <- apply( chimp.ensemble$link , 2 , PI )
```

For plotting, first we'll make an empty plot, then draw the trend for each individual chimpanzee, then finally superimpose the prediction ensemble you just calculated. A new trick in this code is the use of the function by to compute averages for each chimpanzee. See the Overthinking box at the end of this subsection for an explanation of how it works.

R code
10.11

```
# empty plot frame with good axes
plot( 0 , 0 , type="n" , xlab="prosoc_left/condition" ,
    ylab="proportion pulled left" , ylim=c(0,1) , xaxt="n" ,
    xlim=c(1,4) )
axis( 1 , at=1:4 , labels=c("0/0","1/0","0/1","1/1") )

# plot raw data, one trend for each of 7 individual chimpanzees
# will use by() here; see Overthinking box for explanation
p <- by( d$pulled_left ,
    list(d$prosoc_left,d$condition,d$actor) , mean )
for ( chimp in 1:7 )
    lines( 1:4 , as.vector(p[,,chimp]) , col=rangi2 , lwd=1.5 )

# now superimpose posterior predictions
lines( 1:4 , pred.p )
shade( pred.p.PI , 1:4 )
```

FIGURE 10.2 shows the result. The blue lines display the empirical averages for each of the seven chimpanzees who participated in the experiment. The black line shows the average predicted probability of pulling the left-hand lever, across treatments. The zig-zag pattern arises from more left-hand pulls when the prosocial option is on the left. So the chimpanzees were, at least on average, attracted to the prosocial option. But the partner condition, shown

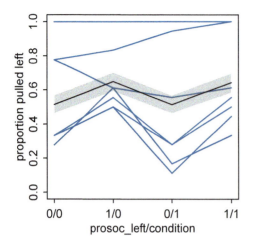

FIGURE 10.2. Model averaged posterior predictive check for the chimpanzee models. Along the horizontal axis, treatments are laid out. The vertical is the proportion of left lever pulls predicted (black) and observed (blue). Each blue trend is an individual chimpanzee. The shaded region is the 89% percentile interval of the mean prediction.

by the last two treatment on the far right of the figure, are no higher than the first two treatments from the control condition. So it made little difference whether or not another animal was present to receive the food on the other side of the table.

But note that those are rather poor predictions. The model did what we asked it to do: Estimate the average across all chimpanzees. But there is a lot of variation among individuals in the blue lines in FIGURE 10.2. In principle, individual variation could mask the association of interest. That isn't much of a risk here—four individuals show the same zig-zag pattern as the prediction, just offset towards the right-hand lever. The other three prefer the left lever overall, but only one of them actually pulls the left lever more when both the prosocial option is on the left and a partner is present. Still, it'll be useful to see how to model the individual variation, even if it isn't going to change the effective inference in this context.

But first, let's check that the quadratic approximation for the posterior distribution is okay in this case. With GLMs recall there is no guarantee of a Gaussian posterior distribution, even if all your priors are Gaussian. So let's quickly compare the estimates above to the same model fit using MCMC, via Stan. This is made easy by map2stan, which was introduced in Chapter 8. It'll harvest the model formula from m10.3 and build the MCMC code from it.

R code
10.12
```
# clean NAs from the data
d2 <- d
d2$recipient <- NULL

# re-use map fit to get the formula
m10.3stan <- map2stan( m10.3 , data=d2 , iter=1e4 , warmup=1000 )
precis(m10.3stan)
```

```
      Mean StdDev lower 0.89 upper 0.89 n_eff Rhat
a     0.05   0.13      -0.15       0.25  3284    1
bp    0.62   0.22       0.28       0.98  3032    1
bpC  -0.11   0.26      -0.53       0.29  3184    1
```

If you glance back at the quadratic approximate posterior estimated from map, you'll see these numbers are almost exactly the same. The pairs plot confirms that the posterior is multivariate Gaussian.

R code
10.13

```
pairs(m10.3stan)
```

I don't show this plot here, but please do go run this for yourself and confirm that the posterior is very well approximated by a multivariate Gaussian. Once we modify the model to estimate individual variation, this will no longer be true.

Now back to modeling individual variation. There is plenty of evidence of handedness in these data. Four of the individuals tend to pull the right-hand lever, across all treatments. Three individuals tend to pull the left across all treatments. One individual, actor number 2, always pulled the left-hand lever, regardless of treatment. That's the horizontal blue line at the top in FIGURE 10.2.

Think of handedness here as a masking variable. If we can model it well, maybe we can get a better picture of what happened across treatments. So what we wish to do is estimate handedness as a distinct intercept for each individual, each actor. You could do this using a dummy variable for each individual. But it'll be more convenient to use a vector of intercepts, one for each actor. This form is equivalent to making dummy variables, but it is more compact and mirrors the structure we'll use later (in Chapter 12) to build multilevel models. It was introduced briefly at the end of Chapter 5 (page 158).

Here is the mathematical form of the model, followed by explanation:

$$L_i \sim \text{Binomial}(1, p_i)$$
$$\text{logit}(p_i) = \alpha_{\text{ACTOR}[i]} + (\beta_P + \beta_{PC}C_i)P_i$$
$$\alpha_{\text{ACTOR}} \sim \text{Normal}(0, 10)$$
$$\beta_P \sim \text{Normal}(0, 10)$$
$$\beta_{PC} \sim \text{Normal}(0, 10)$$

The only change is to add a little ACTOR[i] as a subscript to the intercept α, and then in its prior the ACTOR alone appears again. The notation above is one common convention for defining a vector of parameters, one for each value that the variable ACTOR can take. Implicitly, ACTOR can take the values 1 through 7, because there are 7 individuals in the data, and the variable actor in the data indicates which individual was the focal for each row in the data. The notation $\alpha_{\text{ACTOR}[i]}$ indicates using the value of actor that is found on row i, and this would appear as actor[i] in R code. You can say ACTOR[i] out loud as "the value of ACTOR for case i." And the prior for α_{ACTOR} just states that every element of the vector α gets the same prior.

We'll go straight to fitting this model with MCMC, since the posterior distribution will turn out to have some skew in it. Here's the code:

R code
10.14

```
m10.4 <- map2stan(
    alist(
        pulled_left ~ dbinom( 1 , p ) ,
        logit(p) <- a[actor] + (bp + bpC*condition)*prosoc_left ,
        a[actor] ~ dnorm(0,10),
        bp ~ dnorm(0,10),
        bpC ~ dnorm(0,10)
    ) ,
    data=d2 , chains=2 , iter=2500 , warmup=500 )
```

Notice that the model code above uses [actor] in both the linear model and the prior definition. The code recognizes that there are 7 unique values in the actor variable. You can confirm this for yourself:

```
unique( d$actor )
```

```
[1] 1 2 3 4 5 6 7
```

Then it makes a vector named a of length 7 to hold the parameters. And so, after sampling, you get an estimate for all 7 of them:

```
precis( m10.4 , depth=2 )
```

```
        Mean StdDev lower 0.89 upper 0.89 n_eff Rhat
a[1] -0.74   0.27      -1.19      -0.32   2626    1
a[2] 10.91   5.42       3.45      18.37   1126    1
a[3] -1.05   0.28      -1.52      -0.61   2102    1
a[4] -1.05   0.29      -1.54      -0.62   2781    1
a[5] -0.74   0.27      -1.15      -0.30   2294    1
a[6]  0.21   0.27      -0.21       0.66   2569    1
a[7]  1.83   0.40       1.17       2.43   2470    1
bp    0.84   0.28       0.38       1.25   1546    1
bpC  -0.13   0.30      -0.60       0.36   2239    1
```

The depth=2 in the call to precis is needed to show vector parameters. You'll appreciate this feature a lot once we reach multilevel models, which can have hundreds or thousands of vector parameters, most of which you don't want to inspect directly in a summary table of this kind.

Notice that this posterior is definitely not entirely Gaussian—look at the interval for a[2]. To make this easier to see, let's plot the marginal density for a[2]. First, extract the samples:

```
post <- extract.samples( m10.4 )
str( post )
```

```
List of 3
 $ a  : num [1:2000, 1:7] -0.687 -0.236 -1.131 -0.489 -0.423 ...
 $ bp : num [1:2000(1d)] 0.63 1.119 0.839 0.627 0.628 ...
 $ bpC: num [1:2000(1d)] -0.0979 -0.5319 0.0285 0.2954 -0.0226 ...
```

I've displayed the structure above so you can notice that the samples for the intercepts are held in a matrix named a, a matrix with 2000 rows and 7 columns. Each column is a parameter, a[1] through a[7], and each row is a sample from the posterior distribution. So to plot the density for a[2], we ask for all the rows from column 2:

```
dens( post$a[,2] )
```

The result is shown in FIGURE 10.3. You can see that there is strong skew here. Plausible values of a[2] are always positive, indicating a left-hand bias. But the range of plausible values is truly enormous. What has happened here is that many very large positive values are plausible, because actor number 2 always pulled the left-hand lever (as seen in FIGURE 10.2).

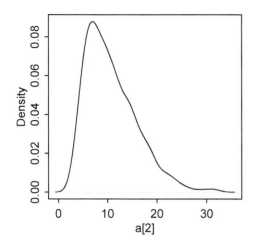

FIGURE 10.3. Marginal posterior density of a[2], the intercept for actor number 2. Very many very large values are plausible. Since actor number 2 always pulled the left-hand lever, the data cannot discriminate among these large values, as all of them guarantee pulling the left-hand lever.

As long as a[2] is large enough to get the probability close to 1, just about any value will lead to the same predictions. The only reason the posterior distribution falls off at all is because we used a weak regularizing prior in the model definition. There is just no information in the data itself to discriminate among all those very high log-odds values, all of which ensure pulling the left-hand lever every time.

You can best appreciate the way these individual intercepts influence fit by plotting posterior predictions again. The code below just modifies the code from earlier to show only a single individual, the one specified by the first line. Just change the value of chimp to plot any of the 7 actors in the data.

R code
10.19

```
chimp <- 1
d.pred <- list(
    pulled_left = rep( 0 , 4 ), # empty outcome
    prosoc_left = c(0,1,0,1),   # right/left/right/left
    condition = c(0,0,1,1),     # control/control/partner/partner
    actor = rep(chimp,4)
)
link.m10.4 <- link( m10.4 , data=d.pred )
pred.p <- apply( link.m10.4 , 2 , mean )
pred.p.PI <- apply( link.m10.4 , 2 , PI )

plot( 0 , 0 , type="n" , xlab="prosoc_left/condition" ,
    ylab="proportion pulled left" , ylim=c(0,1) , xaxt="n" ,
    xlim=c(1,4) , yaxp=c(0,1,2) )
axis( 1 , at=1:4 , labels=c("0/0","1/0","0/1","1/1") )
mtext( paste( "actor" , chimp ) )

p <- by( d$pulled_left ,
    list(d$prosoc_left,d$condition,d$actor) , mean )
lines( 1:4 , as.vector(p[,,chimp]) , col=rangi2 , lwd=2 )

lines( 1:4 , pred.p )
```

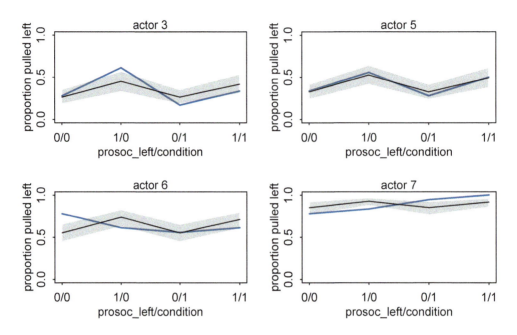

FIGURE 10.4. Posterior prediction check for model m10.4, the chimpanzee model that includes a unique intercept for each individual. Each plot shows the empirical proportion of left pulls in each treatment (blue), for a single individual. The black line and shaded region show the average predicted probability and its 89% interval. Only four of the seven individuals shown.

```
shade( pred.p.PI , 1:4 )
```

Actors 3, 5, 6, and 7 are shown in FIGURE 10.4. Notice that these individual intercepts do help the model fit the overall level for each chimpanzee. But they do not change the basic zig-zag prediction pattern across treatments.

There are a number of loose ends with this analysis. Does model m10.4, with its 6 additional parameters, still look good after estimating overfitting with WAIC? What do the estimates for m10.4 look like under the quadratic approximation that map uses? These ends are tied up in the practice problems at the end of this chapter.

Overthinking: Using the by function. The code above used by to compute proportion pulled_left for each combination of prosoc_left, condition, and actor. The way by works is a little jumbled, but once you understand it, you can use it to compute many different summary statistics, or even to repeat an analysis on subsets of your data.

The first parameter you pass to by is a variable you want to summarize. The second is a list of variables to slice up the first variable by. The third parameter in by is the function to apply to the first parameter. So you can read the call to by in the section above like "calculate the mean of pulled_left for each combination of values in the three variables prosoc_left, condition, and actor." The result is an array with a dimension for each variable used in subsetting. So above we ended up with an array with 3 dimensions: prosoc_left, condition, and actor. So we could pull out the averages for an individual chimpanzee with p[,,chimp]. If you instead wanted all of the proportions for the partner condition, you could type p[,1,].

10.1.2. Aggregated binomial: Chimpanzees again, condensed. In the chimpanzees data context, the models all calculated the likelihood of observing either zero or one pulls of the left-hand lever. The models did so, because the data were organized such that each row describes the outcome of a single pull. But in principle the same data could be organized differently. As long as we don't care about the order of the individual pulls, the same information is contained in a count of how many times each individual pulled the left-hand lever, for each combination of predictor variables.

For example, to calculate the number of times each chimpanzee pulled the left-hand lever, for each combination of predictor values:

R code
10.20

```
data(chimpanzees)
d <- chimpanzees
d.aggregated <- aggregate( d$pulled_left ,
    list(prosoc_left=d$prosoc_left,condition=d$condition,actor=d$actor) ,
    sum )
```

Here are the results for the first two chimpanzees:

```
  prosoc_left condition actor   x
1           0         0     1   6
2           1         0     1   9
3           0         1     1   5
4           1         1     1  10
5           0         0     2  18
6           1         0     2  18
7           0         1     2  18
8           1         1     2  18
```

The x column on the right is the count of times each actor pulled the left-hand lever for trials with the values of the predictors shown on each row. Notice that there are four different combinations of the two predictors, so there are four rows for each actor now. Also recall that actor number 2 always pulled the left-hand lever. As a result, the counts for actor 2 are all 18—there were 18 trials for each animal for each treatment. Now we can get exactly the same inferences as before, just by defining the following model:

R code
10.21

```
m10.5 <- map(
    alist(
        x ~ dbinom( 18 , p ) ,
        logit(p) <- a + (bp + bpC*condition)*prosoc_left ,
        a ~ dnorm(0,10) ,
        bp ~ dnorm(0,10) ,
        bpC ~ dnorm(0,10)
    ) ,
    data=d.aggregated )
```

Take note of the 18 in the spot where a 1 used to be. Now there are 18 trials on each row, and the likelihood defines the probability of each count x out of 18 trials. Inspect the precis output. You'll see that the posterior distribution is the same as the one from model m10.3.

10.1.3. Aggregated binomial: Graduate school admissions. Often the number of trials on each row is not a constant. So then in place of the "18" we insert a variable from the data. Let's work through an example. First, load the data:

R code
10.22
```
library(rethinking)
data(UCBadmit)
d <- UCBadmit
```

This data table only has 12 rows, so let's look at the entire thing:

```
   dept applicant.gender admit reject applications
1     A             male   512    313          825
2     A           female    89     19          108
3     B             male   353    207          560
4     B           female    17      8           25
5     C             male   120    205          325
6     C           female   202    391          593
7     D             male   138    279          417
8     D           female   131    244          375
9     E             male    53    138          191
10    E           female    94    299          393
11    F             male    22    351          373
12    F           female    24    317          341
```

These are graduate school applications to 6 different academic departments at UC Berkeley.[141] The `admit` column indicates the number offered admission. The `reject` column indicates the opposite decision. The `applications` column is just the sum of `admit` and `reject`. Each application has a 0 or 1 outcome for admission, but since these outcomes have been aggregated by department and gender, there are only 12 rows. These 12 rows however represent 4526 applications, the sum of the `applications` column. So there is a lot of data here—counting the rows in the data table is no longer a sensible way to assess sample size. We could split these data apart into 0/1 Bernoulli trials, like in the original `chimpanzees` data. Then there would be 4526 rows in the data.

Our job is to evaluate whether these data contain evidence of gender bias in admissions. We will model the admission decisions, focusing on applicant gender as a predictor variable. So we want to fit at least two models:

(1) A binomial regression that models `admit` as a function of each applicant's gender. This will estimate the association between gender and probability of admission.

(2) A binomial regression that models `admit` as a constant, ignoring gender. This will allow us to get a sense of any overfitting committed by the first model.

This is what the first model looks like, in mathematical form:

$$n_{\text{admit},i} \sim \text{Binomial}(n_i, p_i)$$
$$\text{logit}(p_i) = \alpha + \beta_m m_i$$
$$\alpha \sim \text{Normal}(0, 10)$$
$$\beta_m \sim \text{Normal}(0, 10)$$

The variable n_i indicates `applications[i]`, the number of applications on row i. The predictor m_i is a dummy that indicates "male." We'll construct it just before fitting both models, like this:

```
d$male <- ifelse( d$applicant.gender=="male" , 1 , 0 )
m10.6 <- map(
    alist(
        admit ~ dbinom( applications , p ) ,
        logit(p) <- a + bm*male ,
        a ~ dnorm(0,10) ,
        bm ~ dnorm(0,10)
    ) ,
    data=d )
m10.7 <- map(
    alist(
        admit ~ dbinom( applications , p ) ,
        logit(p) <- a ,
        a ~ dnorm(0,10)
    ) ,
    data=d )
```

R code
10.23

A quick WAIC comparison verifies that the `male` predictor variable improves expected out-of-sample deviance by a very large amount:

```
compare( m10.6 , m10.7 )
```

R code
10.24

```
        WAIC pWAIC dWAIC weight    SE   dSE
m10.6 5954.9     2   0.0      1 34.98    NA
m10.7 6046.3     1  91.5      0 29.93 19.13
```

This comparison suggests that gender matters a lot. To see how it matters, we have to look at estimates for m10.6:

```
precis(m10.6)
```

R code
10.25

```
    Mean StdDev  5.5% 94.5%
a  -0.83   0.05 -0.91 -0.75
bm  0.61   0.06  0.51  0.71
```

Seems like being male is an advantage in this context. You can compute the relative difference in admission odds as $\exp(0.61) \approx 1.84$. This means that a male applicant's odds were 184% of a female applicant's. On the absolute scale, which is what matters, the difference in probability of admission is:

```
post <- extract.samples( m10.6 )
p.admit.male <- logistic( post$a + post$bm )
p.admit.female <- logistic( post$a )
diff.admit <- p.admit.male - p.admit.female
quantile( diff.admit , c(0.025,0.5,0.975) )
```

R code
10.26

```
     2.5%       50%     97.5%
0.1132778 0.1413527 0.1693274
```

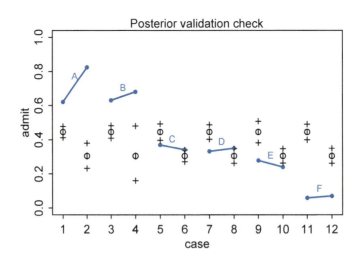

FIGURE 10.5. Posterior validation for model m10.6. Blue points are observed proportions admitted for each row in the data, with points from the same department connected by a blue line. Open points, the tiny vertical black lines within them, and the crosses are expected proportions, 89% intervals of the expectation, and 89% interval of simulated samples, respectively.

This means that the median estimate of the male advantage is about 14%, with a 95% interval from 11% to almost 17%. You may also want to inspect the density plot: dens(diff.admit) (not shown).

Before moving on to speculate on the cause of the male advantage, let's plot posterior predictions for the model. We'll use the default posterior validation check function, postcheck, and then dress it up a little by adding lines to connect data points from the same department.

R code
10.27
```
postcheck( m10.6 , n=1e4 )
# draw lines connecting points from same dept
for ( i in 1:6 ) {
    x <- 1 + 2*(i-1)
    y1 <- d$admit[x]/d$applications[x]
    y2 <- d$admit[x+1]/d$applications[x+1]
    lines( c(x,x+1) , c(y1,y2) , col=rangi2 , lwd=2 )
        text( x+0.5 , (y1+y2)/2 + 0.05 , d$dept[x] , cex=0.8 , col=rangi2 )
}
```

The result is shown as FIGURE 10.5. Those are pretty terrible predictions. There are only two departments in which females had a lower rate of admission than males (C and E), and yet the model says that females should expect to have a 14% lower chance of admission.

Sometimes a fit this bad is the result of a coding mistake. In this case, it is not. The model did correctly answer the question we asked of it: *What are the average probabilities of admission for females and males, across all departments?* The problem in this case is that males and females do not apply to the same departments, and departments vary in their rates of admission. This makes the answer misleading. You can see the steady decline in admission probability for both males and females from department A to department F. Females in these data tended not to apply to departments like A and B, which had high overall admission rates. Instead they applied in large numbers to departments like F, which admitted less than 10% of applicants.

So while it is true overall that females had a lower probability of admission in these data, it is clearly not true within most departments. And note that just inspecting the posterior distribution alone would never have revealed that fact to us. We had to appeal to something outside the fit model. In this case, it was a simple posterior validation check.

Instead of asking *"What are the average probabilities of admission for females and males across all departments?"* we want to ask *"What is the average difference in probability of admission between females and males within departments?"* In order to ask the second question, we estimate a unique female admission rate in each department—an intercept—and then an average male difference. Here's a model that asks this new question:

$$n_{\text{admit},i} \sim \text{Binomial}(n_i, p_i)$$
$$\text{logit}(p_i) = \alpha_{\text{DEPT}[i]} + \beta_m m_i$$
$$\alpha_{\text{DEPT}} \sim \text{Normal}(0, 10)$$
$$\beta_m \sim \text{Normal}(0, 10)$$

where DEPT indexes department. So now each department gets its own log-odds of admission, α_{DEPT}, but the model still estimates a universal adjustment—same in all departments—for a male application, β_m.

Fitting this model, along with a version that omits male, is straightforward. We'll use the indexing notation again to construct an intercept for each department. But first, we also need to construct a numerical index that numbers the departments 1 through 6. The function coerce_index can do this for us, using the dept factor as input. Here's the code to construct the index and fit both models:

R code
10.28

```
# make index
d$dept_id <- coerce_index( d$dept )

# model with unique intercept for each dept
m10.8 <- map(
    alist(
        admit ~ dbinom( applications , p ) ,
        logit(p) <- a[dept_id] ,
        a[dept_id] ~ dnorm(0,10)
    ) , data=d )

# model with male difference as well
m10.9 <- map(
    alist(
        admit ~ dbinom( applications , p ) ,
        logit(p) <- a[dept_id] + bm*male ,
        a[dept_id] ~ dnorm(0,10) ,
        bm ~ dnorm(0,10)
    ) , data=d )
```

Now to compare all four models from this section:

```
compare( m10.6 , m10.7 , m10.8 , m10.9 )
```

```
            WAIC pWAIC dWAIC weight    SE   dSE
m10.8 5200.9     6   0.0   0.56 57.02    NA
m10.9 5201.4     7   0.5   0.44 57.06  2.48
m10.6 5954.8     2 753.9   0.00 34.98 48.53
m10.7 6046.3     1 845.4   0.00 29.95 52.37
```

The new models fit much better, unsurprisingly. But now the model without male is ranked first. Still, the WAIC difference between m10.8 and m10.9 is tiny—both models get about half the Akaike weight. I'd call this a tie. So there's modest support for some effect of gender, even if it is overfit a little. So let's look at the estimates from m10.9 and see how the estimated association of gender with admission has changed:

```
precis( m10.9 , depth=2 )
```

```
       Mean StdDev  5.5% 94.5%
a[1]   0.68   0.10  0.52  0.84
a[2]   0.64   0.12  0.45  0.82
a[3]  -0.58   0.07 -0.70 -0.46
a[4]  -0.61   0.09 -0.75 -0.48
a[5]  -1.06   0.10 -1.22 -0.90
a[6]  -2.62   0.16 -2.88 -2.37
bm    -0.10   0.08 -0.23  0.03
```

The estimate for bm goes in the opposite direction now. On the proportional odds scale, the estimate becomes: $\exp(-0.1) \approx 0.9$. So a male in this sample has about 90% the odds of admission as a female, comparing within departments.

We can also plot posterior predictions again and see how the department intercepts capture the variation in overall admission rates. You can use the same code as earlier, just replacing m10.6 inside the call to postcheck with m10.9. The result is shown as FIGURE 10.6.

Before moving on, let's check the quadratic approximation. Since it turned out in the chimpanzees data that individual intercepts caused problems for quadratic approximation, it's worth checking here as well. To fit model m10.9 with MCMC:

```
m10.9stan <- map2stan( m10.9 , chains=2 , iter=2500 , warmup=500 )
precis(m10.9stan,depth=2)
```

```
       Mean StdDev lower 0.89 upper 0.89 n_eff Rhat
a[1]   0.68   0.10       0.52       0.84  1463    1
a[2]   0.64   0.12       0.45       0.82  1510    1
a[3]  -0.58   0.07      -0.70      -0.47  2267    1
a[4]  -0.61   0.09      -0.75      -0.48  2166    1
a[5]  -1.06   0.10      -1.22      -0.91  2755    1
a[6]  -2.64   0.17      -2.89      -2.36  3012    1
bm    -0.10   0.08      -0.23       0.03  1224    1
```

These estimates are practically identical to those from m10.9. Go ahead and inspect the pairs plot, too. You'll see that the quadratic approximation has done a very good job. This

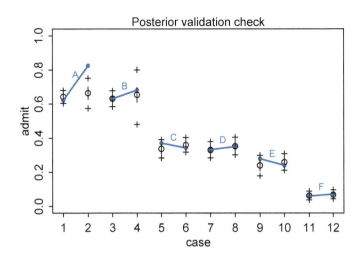

FIGURE 10.6. Posterior validation for m10.9. The unique intercepts for each department, A through F, capture variation in overall admission rates among departments. This allows the model to compare male and female admission rates, controlling for heterogeneity across departments.

is typical for binomial regression, as long as none of the intercepts are very far from zero, such that they push against the ceiling or floor, and none of the predictors are strongly associated with the outcome. Then the quadratic approximation can be very accurate. But since it's easy to check, it's usually worth a check.

Rethinking: Simpson's paradox is not a paradox. This empirical example is a famous one in statistical teaching. It is often used to illustrate a phenomenon known as SIMPSON'S PARADOX.[142] Like most paradoxes, there is no violation of logic, just of intuition. And since different people have different intuition, Simpson's paradox means different things to different people. The poor intuition being violated in this case is that a positive association in the entire population should also hold within each department. Overall, females in these data did have a harder time getting admitted to graduate school. But that arose because females applied to the hardest departments for anyone, male or female, to gain admission to.

Perhaps a little more paradoxical is that this phenomenon can repeat itself indefinitely within a sample. Any association between an outcome and a predictor can be nullified or reversed when another predictor is added to the model. All that we can do about this is to remain skeptical of models and try to imagine ways they might be deceiving us. Thinking causally about these settings sometimes helps.[143] But also see the box about causal inference on page 120.

Overthinking: WAIC and aggregated binomial models. The WAIC function in rethinking detects aggregated binomial models and automatically splits them apart into 0/1 Bernoulli trials, for the purpose of calculating WAIC. It does this, because WAIC is computed point by point (see Chapter 6). So what you define as a "point" affects WAIC's value. In an aggregated binomial each "point" is a bunch of independent trials that happen to share the same predictor values. In order for the disaggregated and aggregated models to agree, it makes sense to use the disaggregated representation.

A consequence of this is that DIC and WAIC for an aggregated binomial will look very different from one another, even though they will produce similar model rankings. Try for example comparing compare(m10.8,m10.9,func=DIC) to the WAIC table for the same two models. The reason the absolute magnitudes of WAIC and DIC end up so different in these cases is because the binomial distribution, in aggregated form, has a leading coefficient that is the *multiplicity*, the number of ways that the sequence of 0/1 outcomes could be reordered. This coefficient doesn't change inference, since it isn't a function of the parameters. But it does change the value of the log-likelihood and therefore the deviance.

10.1.4. Fitting binomial regressions with glm. R's standard glm function allows fitting a variety of generalized linear models, as long as you are okay with and cautious of flat priors. An aggregated binomial uses cbind to build the outcome variable. For example, this code will yield similar results as the map approach in the previous section:

R code
10.32
```
m10.7glm <- glm( cbind(admit,reject) ~ 1 , data=d , family=binomial )
m10.6glm <- glm( cbind(admit,reject) ~ male , data=d , family=binomial )
m10.8glm <- glm( cbind(admit,reject) ~ dept , data=d , family=binomial )
m10.9glm <- glm( cbind(admit,reject) ~ male + dept , data=d ,
    family=binomial )
```

When the outcome is instead coded as 0/1, the input looks like a linear regression formula:

R code
10.33
```
data(chimpanzees)
m10.4glm <- glm(
    pulled_left ~ as.factor(actor) + prosoc_left * condition - condition ,
    data=chimpanzees , family=binomial )
```

Note the necessity of subtracting condition to remove that main effect from the model.

And you can use glimmer to build a map-style model from one of these linear model formulas. For example:

R code
10.34
```
glimmer( pulled_left ~ prosoc_left * condition - condition ,
    data=chimpanzees , family=binomial )
```

```
alist(
    pulled_left ~ dbinom( 1 , p ),
    logit(p) <- Intercept +
        b_prosoc_left*prosoc_left +
        b_prosoc_left_X_condition*prosoc_left_X_condition,
    Intercept ~ dnorm(0,10),
    b_prosoc_left ~ dnorm(0,10),
    b_prosoc_left_X_condition ~ dnorm(0,10)
)
```

The parameter names are inelegant, but you can edit the above to your liking.

Notice that glimmer above inserts weakly regularizing priors by default (see ?glimmer for options). Sometimes, the implicit flat priors of glm lead to nonsense estimates. For example, consider the following simple data and model context:

R code
10.35
```
# outcome and predictor almost perfectly associated
y <- c( rep(0,10) , rep(1,10) )
```

```
x <- c( rep(-1,9) , rep(1,11) )
# fit binomial GLM
m.bad <- glm( y ~ x , data=list(y=y,x=x) , family=binomial )
precis(m.bad)
```

```
            Mean  StdDev    5.5%    94.5%
(Intercept) -9.13 2955.06 -4731.89 4713.63
x           11.43 2955.06 -4711.33 4734.19
```

Those intervals should worry us. What has happened here is that the outcome is so strongly associated with the predictor that the slope on x tries to grow very large. At large log-odds, almost any value is just about as good as any other. So the uncertainty is asymmetric, and the flat prior does nothing to calm inference down on the high end of it. The estimates above, read naively, suggest that there is no association between y and x, even though they are strongly associated.

This is easy to fix up with a very weakly informative prior, but it's worrisome that glm does nothing to even warn us about this behavior. Here's a quick fix, using map:

R code
10.36

```
m.good <- map(
    alist(
        y ~ dbinom( 1 , p ),
        logit(p) <- a + b*x,
        c(a,b) ~ dnorm(0,10)
    ) , data=list(y=y,x=x) )
precis(m.good)
```

```
   Mean StdDev  5.5% 94.5%
a -1.73   2.78 -6.16  2.71
b  4.02   2.78 -0.42  8.45
```

Of course, the uncertainty is not symmetric in this case, so the quadratic assumption is misleading. Better yet would be to take a quick look at the MCMC samples:

R code
10.37

```
m.good.stan <- map2stan( m.good )
pairs(m.good.stan)
```

Inspecting the pairs plot (not shown) demonstrates just how subtle even simple models can be, once we start working with GLMs. I don't say this to scare the reader. But it's true that even simple models can behave in complicated ways. How you fit the model is part of the model, and in principle no GLM is safe for MAP estimation.

10.2. Poisson regression

When a binomial distribution has a very small probability of an event p and a very large number of trials n, then it takes on a special shape. The expected value of a binomial distribution is just np, and its variance is $np(1 - p)$. But when n is very large and p is very small, then these are approximately the same.

For example, suppose you own a monastery that is in the business, like many monasteries before the invention of the printing press, of copying manuscripts. You employ 1000 monks, and on any particular day about 1 of them finishes a manuscript. Since the monks are working independently of one another, and manuscripts vary in length, some days produce 3 or more manuscripts, and many days produce none. Since this is a binomial process, you can calculate the variance across days as $np(1 - p) = 1000(0.001)(1 - 0.001) \approx 1$. You can simulate this, for example over 10,000 (1e5) days:

R code
10.38
```
y <- rbinom(1e5,1000,1/1000)
c( mean(y) , var(y) )
```

```
[1] 0.9968400 0.9928199
```
The mean and the variance are nearly identical. This is a special shape of the binomial. This special shape is known as the Poisson distribution, and it is useful because it allows us to model binomial events for which the number of trials n is unknown or uncountably large. Suppose for example that you come to own, through imperial drama, another monastery. You don't know how many monks toil within it, but your advisors tell you that it produces, on average, 2 manuscripts per day. With this information alone, you can infer the entire distribution of numbers of manuscripts completed each day.

To build models with a Poisson likelihood, the model form is even simpler than it is for a binomial or Gaussian model. This simplicity arises from the Poisson's having only one parameter that describes its shape, resulting in a likelihood definition like this:

$$y \sim \text{Poisson}(\lambda)$$

The parameter λ is the expected value of the outcome y.

To build a GLM with this likelihood, we also need a link function. The conventional link function for a Poisson model is the log link, as introduced in the previous chapter (page 286). So to embed a linear model, we use:

$$y_i \sim \text{Poisson}(\lambda_i)$$
$$\log(\lambda_i) = \alpha + \beta x_i$$

The log link ensures that λ_i is always positive, which is required of the expected value of a count outcome. But as mentioned in the previous chapter, it also implies an exponential relationship between predictors and the expected value. Exponential relationships grow very quickly, and few natural phenomena can remain exponential for long. So one thing to always check with a log link is whether it makes sense at all ranges of the predictor variables.

The parameter λ is the expected value, but it's also commonly thought of as a rate. Both interpretations are correct, and realizing this allows us to make Poisson models for which the *exposure* varies across cases i. Suppose for example that a neighboring monastery performs weekly totals of completed manuscripts while your monastery does daily totals. If you come into possession of both sets of records, how could you analyze both in the same model, given that the counts are aggregated over different amounts of time, different exposures?

Here's how. Implicitly, λ is equal to an expected number of events, μ, per unit time or distance, τ. This implies that $\lambda = \mu/\tau$, which lets us redefine the link:

$$y_i \sim \text{Poisson}(\lambda_i)$$
$$\log \lambda_i = \log \frac{\mu_i}{\tau_i} = \alpha + \beta x_i$$

Since the logarithm of a ratio is the same as a difference of logarithms, we can also write:

$$\log \lambda_i = \log \mu_i - \log \tau_i = \alpha + \beta x_i$$

These τ values are the "exposures." So if different observations i have different exposures, then this implies that the expected value on row i is given by:

$$\log \mu_i = \log \tau_i + \alpha + \beta x_i$$

When $\tau_i = 1$, then $\log \tau_i = 0$ and we're back where we started. But when the exposure varies across cases, then τ_i does the important work of correctly scaling the expected number of events for each case i. So you can model cases with different exposures just by writing a model like:

$$y_i \sim \text{Poisson}(\mu_i)$$
$$\log \mu_i = \log \tau_i + \alpha + \beta x_i$$

where τ is a column in the data. So this is just like adding a predictor, the logarithm of the exposure, without adding a parameter for it. There will be an example later in this section.

10.2.1. Example: Oceanic tool complexity. Here's an example of Poisson GLM analysis. The island societies of Oceania provide a natural experiment in technological evolution. Different historical island populations possessed tool kits of different size. These kits include fish hooks, axes, boats, hand plows, and many other types of tools. A number of theories predict that larger populations will both develop and sustain more complex tool kits. So the natural variation in population size induced by natural variation in island size in Oceania provides a natural experiment to test these ideas. It's also suggested that contact rates among populations effectively increase population size, as it's relevant to technological evolution. So variation in contact rates among Oceanic societies is also relevant.

Here are the data we'll work with:[144]

R code
10.39

```
library(rethinking)
data(Kline)
d <- Kline
d
```

	culture	population	contact	total_tools	mean_TU
1	Malekula	1100	low	13	3.2
2	Tikopia	1500	low	22	4.7
3	Santa Cruz	3600	low	24	4.0
4	Yap	4791	high	43	5.0
5	Lau Fiji	7400	high	33	5.0
6	Trobriand	8000	high	19	4.0
7	Chuuk	9200	high	40	3.8
8	Manus	13000	low	28	6.6
9	Tonga	17500	high	55	5.4
10	Hawaii	275000	low	71	6.6

And that's the entire data set. You can see the location of these societies in the Pacific Ocean in FIGURE 10.7. As before, keep in mind that the number of rows is not clearly the same as the "sample size" in a count model. Still, there isn't a lot of data here, because there just aren't that many historic Oceanic societies for which reliable data can be gathered. So we'll want to use regularization to damp down overfitting. But as you'll see, a lot can still be learned from these data.

The total_tools variable will be the outcome variable. We'll model the idea that:

FIGURE 10.7. Locations of societies in the `Kline` data. The Equator and International Date Line are shown.

(1) The number of tools increases with the log `population` size. Why log? Because that's what the theory says, that it is the order of magnitude of the population that matters, not the absolute size of it. So we'll look for a positive association between `total_tools` and log `population`.
(2) The number of tools increases with the `contact` rate. Islands that are better network acquire or sustain more tool types.
(3) The impact of `population` on tool counts is increased by high `contact`. This is to say that the association between `total_tools` and log `population` depends upon `contact`. So we will look for a positive interaction between log `population` and `contact`.

Let's build now. First, we make some new columns with the log of `population` and a dummy variable for high `contact`:

R code
10.40

```
d$log_pop <- log(d$population)
d$contact_high <- ifelse( d$contact=="high" , 1 , 0 )
```

The model that conforms to the research hypothesis includes an interaction between log-population and contact rate. In math form, it is:

$$T_i \sim \text{Poisson}(\lambda_i)$$
$$\log \lambda_i = \alpha + \beta_P \log P_i + \beta_C C_i + \beta_{PC} C_i \log P_i$$
$$\alpha \sim \text{Normal}(0, 100)$$
$$\beta_P \sim \text{Normal}(0, 1)$$
$$\beta_C \sim \text{Normal}(0, 1)$$
$$\beta_{PC} \sim \text{Normal}(0, 1)$$

where P is `population` and C is `contact_high`. I've used more strongly regularizing priors on the β parameters, because the sample is small, so we should fear overfitting more. Indeed, those Normal$(0, 1)$ priors are probably not conservative enough. And since the predictors are not centered—more on that a little later—there's no telling where α should end up, so I've assigned an essentially flat prior to it.

And now to fit the model to the data, we can use `map` as usual:

```
m10.10 <- map(
    alist(
        total_tools ~ dpois( lambda ),
        log(lambda) <- a + bp*log_pop +
            bc*contact_high + bpc*contact_high*log_pop,
        a ~ dnorm(0,100),
        c(bp,bc,bpc) ~ dnorm(0,1)
    ),
    data=d )
```

R code
10.41

Let's glance at the estimates, just to remind ourselves that when the model includes an interaction, and especially when the predictors are not centered, we can't tell from the table of estimates alone what is going on. I'll show the dotchart for the estimates, as well.

```
precis(m10.10,corr=TRUE)
plot(precis(m10.10))
```

R code
10.42

	Mean	StdDev	5.5%	94.5%	a	bp	bc	bpc
a	0.94	0.36	0.37	1.52	1.00	-0.98	-0.13	0.07
bp	0.26	0.03	0.21	0.32	-0.98	1.00	0.12	-0.08
bc	-0.09	0.84	-1.43	1.25	-0.13	0.12	1.00	-0.99
bpc	0.04	0.09	-0.10	0.19	0.07	-0.08	-0.99	1.00

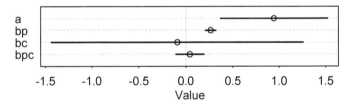

I've used the corr=TRUE option to include the correlations among the parameters. But first notice that the main effect of log-population, bp, is positive and precise, but that both bc and bpc overlap zero substantially. So you might think it's safe to say that log-population is reliably associated with the total tools, but that contact rate has no impact on prediction in this model.

You might think that, but you'd be wrong. As always, it's very easy to be mislead by tables of estimates, especially when an interaction is involved. To prove that contact rate is having an important effect on prediction in this model, let's just compute some counterfactual predictions. Consider two islands, both with log-population of 8, but one with high contact and the other with low. Let's calculate λ, the expected tool count, for each. This is accomplished just by drawing samples from the posterior, plugging those samples into the linear model, and then inverting the link function to get back to the scale of the outcome variable. In this case, inverting the link means exponentiating with exp:

```
post <- extract.samples(m10.10)
lambda_high <- exp( post$a + post$bc + (post$bp + post$bpc)*8 )
lambda_low <- exp( post$a + post$bp*8 )
```

R code
10.43

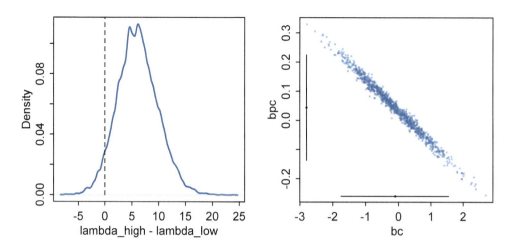

FIGURE 10.8. Left: The distribution of plausible difference in average tool count between identical islands at log-population of 8 but with different contact rates. Even though both the main effect of contact, bc, and the interaction with log-population, bpc, overlap zero, together they imply a 95% plausibility that an island with high contact has more expected tools than an island with low contact. Right: Joint posterior distribution of bc and bpc, showing a strong negative correlation. Black line segments show the marginal posterior distributions of both parameters. Neither parameter is conventionally "significant," but together they imply that contact rate consistently influences prediction.

Since the posterior is a distribution, the contents of lambda_high and lambda_low are also distributions. Now let's compute the difference between these two distributions, to get the distribution of plausible differences in tools between an island with high contact and one with low contact:

R code
10.44
```
diff <- lambda_high - lambda_low
sum(diff > 0)/length(diff)
```

```
[1] 0.9527
```

That's a 95% plausibility that the high-contact island has more tools than the low-contact islands. This distribution is plotted in FIGURE 10.8. How can this be, when both bc and bpc are, conventionally speaking, "non-significant"? One reason is because the uncertainty in the parameters is correlated. Take a look back at the precis table for m10.10. The correlation between bc and bpc is strongly negative, as seen in the right-hand plot in FIGURE 10.8. So when bc is small, bpc is large. As a result, you can't just inspect the *marginal* uncertainty in each parameter, which is what is shown in the table of estimates, and get an accurate understanding of the impact of the *joint* uncertainty on prediction.

A better way to assess whether a predictor, like contact_high, is expected to improve prediction is to use model comparison. Since model comparisons are done on the scale of predicted outcomes, they automatically take account of these correlations. So let's fit some other models, each with fewer terms. This will let us both estimate overfitting and get a sense for the improvement in expected prediction that each term in the model provides.

First, consider a model that omits the interaction:

R code
10.45

```
# no interaction
m10.11 <- map(
    alist(
        total_tools ~ dpois( lambda ),
        log(lambda) <- a + bp*log_pop + bc*contact_high,
        a ~ dnorm(0,100),
        c(bp,bc) ~ dnorm( 0 , 1 )
    ), data=d )
```

Now consider two more models, each with only one of the predictor variables:

R code
10.46

```
# no contact rate
m10.12 <- map(
    alist(
        total_tools ~ dpois( lambda ),
        log(lambda) <- a + bp*log_pop,
        a ~ dnorm(0,100),
        bp ~ dnorm( 0 , 1 )
    ), data=d )

# no log-population
m10.13 <- map(
    alist(
        total_tools ~ dpois( lambda ),
        log(lambda) <- a + bc*contact_high,
        a ~ dnorm(0,100),
        bc ~ dnorm( 0 , 1 )
    ), data=d )
```

And finally consider a kind of "null" model with only the intercept. Note that this is not a null model in the classical sense, as we'll evaluate it on par with the other models.

R code
10.47

```
# intercept only
m10.14 <- map(
    alist(
        total_tools ~ dpois( lambda ),
        log(lambda) <- a,
        a ~ dnorm(0,100)
    ), data=d )

# compare all using WAIC
# adding n=1e4 for more stable WAIC estimates
# will also plot the comparison
( islands.compare <- compare(m10.10,m10.11,m10.12,m10.13,m10.14,n=1e4) )
plot(islands.compare)
```

	WAIC	pWAIC	dWAIC	weight	SE	dSE
m10.11	79.0	4.2	0.0	0.62	11.19	NA
m10.10	80.1	4.9	1.2	0.35	11.42	1.28
m10.12	84.6	3.8	5.6	0.04	8.91	8.47
m10.14	141.5	8.2	62.5	0.00	31.53	34.42
m10.13	149.8	16.7	70.8	0.00	43.96	46.01

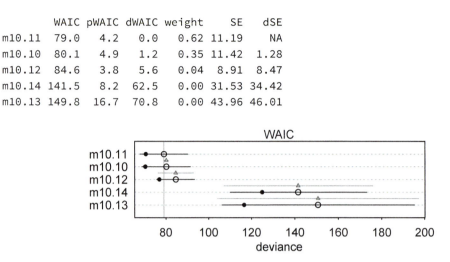

The top two models include both predictors, but the top model, m10.11, excludes the interaction between them. There's a lot of model weight assigned to both, however. This suggests the interaction estimate is probably overfit. But this model set is decent evidence that contact rate matters, although it influences prediction much less than log-population does. Keep in mind that all forms of model comparison, information theoretic or not, do not provide any guarantees. They are models themselves of out-of-sample prediction. They are not forms of divination.

To get a better sense of what these models imply, let's plot some counterfactual predictions, using an ensemble of the top three models, which together have all of the Akaike weight. This is a big chunk of code, but it's heavily commented and there are no new tricks.

R code
10.48

```
# make plot of raw data to begin
# point character (pch) indicates contact rate
pch <- ifelse( d$contact_high==1 , 16 , 1 )
plot( d$log_pop , d$total_tools , col=rangi2 , pch=pch ,
    xlab="log-population" , ylab="total tools" )

# sequence of log-population sizes to compute over
log_pop.seq <- seq( from=6 , to=13 , length.out=30 )

# compute trend for high contact islands
d.pred <- data.frame(
    log_pop = log_pop.seq,
    contact_high = 1
)
lambda.pred.h <- ensemble( m10.10 , m10.11 , m10.12 , data=d.pred )
lambda.med <- apply( lambda.pred.h$link , 2 , median )
lambda.PI <- apply( lambda.pred.h$link , 2 , PI )

# plot predicted trend for high contact islands
lines( log_pop.seq , lambda.med , col=rangi2 )
shade( lambda.PI , log_pop.seq , col=col.alpha(rangi2,0.2) )

# compute trend for low contact islands
```

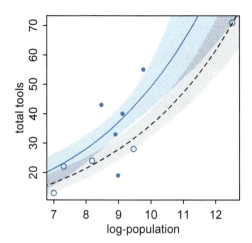

FIGURE 10.9. Ensemble posterior predictions for the islands model set. Filled points are islands with high contact rate. The blue trend and confidence region are counterfactual predictions for islands with high contact, across values of log-population on the horizontal axis. The dashed trend and gray region are predictions for low contact islands.

```
d.pred <- data.frame(
    log_pop = log_pop.seq,
    contact_high = 0
)
lambda.pred.l <- ensemble( m10.10 , m10.11 , m10.12 , data=d.pred )
lambda.med <- apply( lambda.pred.l$link , 2 , median )
lambda.PI <- apply( lambda.pred.l$link , 2 , PI )

# plot again
lines( log_pop.seq , lambda.med , lty=2 )
shade( lambda.PI , log_pop.seq , col=col.alpha("black",0.1) )
```

The result is shown in FIGURE 10.9. The shaded blue region is for islands with high contact. The gray region is for islands with low contact. Notice that both trends curve dramatically upwards as log-population increases. The impact of contact rate can be seen by the distance between the blue and gray predictions. There is plenty of overlap, especially at low and high log-population values, where there are no islands with high contact rate.

10.2.2. MCMC islands. Let's verify that the MAP estimates in the previous section are accurately describing the shape of the posterior distribution. Remember, in principle the posterior distribution for a GLM may not be multivariate Gaussian, even if all your priors are Gaussian. Checking is easily accomplished by passing any of the map models to map2stan:

```
m10.10stan <- map2stan( m10.10 , iter=3000 , warmup=1000 , chains=4 )
precis(m10.10stan)
```

R code
10.49

	Mean	StdDev	lower 0.89	upper 0.89	n_eff	Rhat
a	0.93	0.37	0.32	1.51	2077	1
bp	0.27	0.04	0.21	0.32	2054	1
bc	-0.08	0.86	-1.51	1.24	2169	1
bpc	0.04	0.09	-0.11	0.19	2164	1

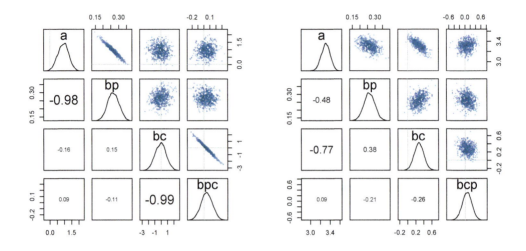

FIGURE 10.10. Posterior distributions for models m10.10stan (left) and m10.10stan.c (right).

These estimates and intervals are the same as before. So yes, the posterior is approximately Gaussian. But take a look at the pairs plot on the left-hand side of FIGURE 10.10. Note the very strong correlations between two pairs of parameters. Hamiltonian Monte Carlo is very good at handling this sort of thing, but it would still be better to avoid such strong correlations, when possible. The reason is that even HMC is going to be less efficient, when there are strong correlations in the posterior distribution.

So this is a good opportunity to show how centering predictors can aid in inference, by reducing correlations among parameters. Let's center log_pop and re-estimate, to see how much more efficient the Markov chain becomes:

R code
10.50

```
# construct centered predictor
d$log_pop_c <- d$log_pop - mean(d$log_pop)

# re-estimate
m10.10stan.c <- map2stan(
    alist(
        total_tools ~ dpois( lambda ) ,
        log(lambda) <- a + bp*log_pop_c + bc*contact_high +
            bcp*log_pop_c*contact_high ,
        a ~ dnorm(0,10) ,
        bp ~ dnorm(0,1) ,
        bc ~ dnorm(0,1) ,
        bcp ~ dnorm(0,1)
    ) ,
    data=d , iter=3000 , warmup=1000 , chains=4 )
precis(m10.10stan.c)
```

	Mean	StdDev	lower 0.89	upper 0.89	n_eff	Rhat
a	3.31	0.09	3.17	3.46	2574	1

```
bp   0.26   0.04        0.20          0.32   3154     1
bc   0.28   0.12        0.09          0.47   2837     1
bcp  0.07   0.18       -0.21          0.34   3838     1
```

The estimates are going to look different, because of the centering, but the predictions remain the same. But we're interested in the shape of the posterior distribution. Look now at the right-hand pairs plot in FIGURE 10.10. Now those strong correlations are gone. And the Markov chain also was more efficient, resulting in a greater number of *effective samples*. When the chain explores the posterior well, n_eff will be larger. That means you'll need fewer samples to get a good picture of the posterior distribution. In these models, they sample so quickly that you might not care—you could take 100,000 samples in under a minute. But with more data and more complex models, you might not want to wait the hours it takes to get a good picture. So recoding the data pays off.

10.2.3. Example: Exposure and the offset. For the last Poisson example, we'll look at a case where the exposure varies across observations. When the length of observation, area of sampling, or intensity of sampling varies, the counts we observe also naturally vary. Since a Poisson distribution assumes that the rate of events is constant in time (or space), it's easy to handle this. All we need to do, as explained on page 312, is to add the logarithm of the exposure to the linear model. The term we add is typically called an *offset*.

We'll simulate for this example, both to provide another example of dummy-data simulation as well as to ensure we get the right answer from the offset approach. Suppose, as we did earlier, that you own a monastery. The data available to you about the rate at which manuscripts are completed is totaled up each day. Suppose the true rate is $\lambda = 1.5$ manuscripts per day. We can simulate a month of daily counts:

R code
10.51
```
num_days <- 30
y <- rpois( num_days , 1.5 )
```

So now y holds 30 days of simulated counts of completed manuscripts.

Also suppose that your monastery is turning a tidy profit, so you are considering purchasing another monastery. Before purchasing, you'd like to know how productive the new monastery might be. Unfortunately, the current owners don't keep daily records, so a head-to-head comparison of the daily totals isn't possible. Instead, the owners keep weekly totals. Suppose the daily rate at the new monastery is actually $\lambda = 0.5$ manuscript per day. To simulate data on a weekly basis, we just multiply this average by 7, the exposure:

R code
10.52
```
num_weeks <- 4
y_new <- rpois( num_weeks , 0.5*7 )
```

And new y_new holds four weeks of counts of completed manuscripts.

To analyze both y, totaled up daily, and y_new, totaled up weekly, we just add the logarithm of the exposure to linear model. First, let's build a data frame to organize the counts and help you see the exposure for each case:

R code
10.53
```
y_all <- c( y , y_new )
exposure <- c( rep(1,30) , rep(7,4) )
monastery <- c( rep(0,30) , rep(1,4) )
d <- data.frame( y=y_all , days=exposure , monastery=monastery )
```

Take a look at d and confirm that there are three columns: The observed counts are in y, the number of days each count was totaled over are in days, and the new monastery is indicated by monastery.

To fit the model, and estimate the rate of manuscript production at each monastery, we just compute the log of each exposure and then include that variable in linear model. This code will do the job:

R code
10.54

```
# compute the offset
d$log_days <- log( d$days )

# fit the model
m10.15 <- map(
    alist(
        y ~ dpois( lambda ),
        log(lambda) <- log_days + a + b*monastery,
        a ~ dnorm(0,100),
        b ~ dnorm(0,1)
    ),
    data=d )
```

To compute the posterior distributions of λ in each monastery, we sample from the posterior and then just use the linear model, but without the offset now. We don't use the offset again, when computing predictions, because the parameters are already on the daily scale, for both monasteries.

R code
10.55

```
post <- extract.samples( m10.15 )
lambda_old <- exp( post$a )
lambda_new <- exp( post$a + post$b )
precis( data.frame( lambda_old , lambda_new ) )
```

```
              Mean StdDev |0.89 0.89|
lambda_old 1.61    0.23   1.26  1.99
lambda_new 0.59    0.14   0.37  0.81
```

Your estimates will be slightly different, because you got different randomly simulated data. But the comparison should be qualitatively the same: The new monastery produces about half as many manuscripts per day. So you aren't going to pay that much for it.

10.3. Other count regressions

In this final section of the chapter, you'll meet four other common count regressions: multinomial, geometric, negative-binomial, and beta-binomial. The first two are also maximum entropy distributions, under their own unique constraints. The third and fourth are actually mixture models. Mixture models are discussed in the next chapter. But you'll get a quick introduction to the negative-binomial and beta-binomial here, just so you see them coming and can relate them more easily to the basic binomial GLM.

We won't work full examples in this section. Instead the goal is to provide enough information to enable you to recognize when these outcome models are appropriate, as well as when others use them inappropriately. I will however present very short model specifications for each, just to demystify them. In more advanced books, you will find fully worked examples.

10.3.1. Multinomial. The binomial distribution is relevant when there are only two things that can happen, and we count those things. Other times, more than two things can happen. For example, recall the bag of marbles from way back in Chapter 2. It contained only blue and white marbles. But suppose we introduce red marbles, as well. Now each draw from the bag can be one of three categories, and the count that accumulates is across all three categories. So we end up with a count of blue, white, and red marbles.

When more than two types of unordered events are possible, and the probability of each type of event is constant across trials, then the maximum entropy distribution is the MULTINOMIAL DISTRIBUTION. You already met the multinomial, implicitly, in Chapter 9 when we tossed pebbles into buckets as an introduction to maximum entropy. The binomial is really a special case of this distribution. And so its distribution formula resembles the binomial, just extrapolated out to three or more types of events. If there are K types of events with probabilities $p_1, ..., p_K$, then the probability of observing $y_1, ..., y_K$ events of each type out of n total trials is:

$$\Pr(y_1, ..., y_K | n, p_1, ..., p_K) = \frac{n!}{\prod_i y_i!} \prod_{i=1}^{K} p_i^{y_i}$$

The fraction with $n!$ on top just expresses the number of different orderings that give the same counts $y_1, ..., y_K$.

A model built on a multinomial distribution may also be called a CATEGORICAL regression, usually when each event is isolated on a single row, like with logistic regression. In machine learning, this model type is sometimes known as the MAXIMUM ENTROPY CLASSIFIER. Building a generalized linear model from a multinomial likelihood is complicated, because as the event types multiply, so too do your modeling choices. And there are two different approaches to constructing the likelihoods, as well. The first is based directly on the multinomial likelihood and uses a generalization of the logit link. I'll show you an example of this approach, which I'll call the *explicit* approach. The second approach transforms the multinomial likelihood into a series of Poisson likelihoods, oddly enough. I'll introduce this approach at the end of the section.

10.3.1.1. Explicit multinomial models. The conventional and natural link in this context is the *multinomial logit*. This link function takes a vector of *scores*, one for each of K event types, and computes the probability of a particular type of event k as:

$$\Pr(k | s_1, s_2, ..., s_K) = \frac{\exp(s_k)}{\sum_{i=1}^{K} \exp(s_i)}$$

The rethinking package provides this link as the softmax function. Combined with this conventional link, this type of GLM is often called *multinomial logistic regression*.

In principle, a multinomial GLM isn't so hard to build. But in practice, such models are much harder for beginners than are other types of GLMs. New questions have to be answered, just to specify a multinomial GLM. And interpretation becomes harder as well. The biggest issue is what to do with the multiple linear models. In a binomial GLM, you can pick either of the two possible events and build a single linear model for its log odds. The other event is handled automatically. But in a multinomial (or categorical) GLM, you need

$K - 1$ linear models for K types of events. In each of these, you can use any predictors and parameters you like—they don't have to be the same, and there are often good reasons for them to be different. In the special case of two types of events, none of these choices arise, because there is only one linear model. And that's why the binomial GLM is so much easier.

There are two basic cases: (1) predictors have different values for different types of events, and (2) parameters are distinct for each type of event. The first case is useful when each type of event has its own quantitative *traits*, and you want to estimate the association between those traits and the probability each type of event appears in the data. The second case is useful when you are interested instead in features of some entity that produces each event, whatever type it turns out to be. Let's consider each case separately and talk through an empirically motivated example of each. You can mix both cases in the same model. But it'll be easier to grasp the distinction in pure examples of each.

For example, suppose you are modeling choice of career for a number of young adults. One of the relevant predictor variables is expected income. In that case, the same parameter β_{INCOME} appears in each linear model, in order to estimate the impact of the income trait on the probability a career is chosen. But a different income value multiplies the parameter in each linear model.

Here's a simulated example in R code. This code simulates career choice from three different careers, each with its own income trait. These traits are used to assign a *score* to each type of event. Then when the model is fit to the data, one of these scores is held constant, and the other two scores are estimated, using the known income traits. It is a little confusing, yes. Step through the implementation, and it'll make more sense. First, we simulate fake career choices:

R code
10.56

```
# simulate career choices among 500 individuals
N <- 500                  # number of individuals
income <- 1:3             # expected income of each career
score <- 0.5*income       # scores for each career, based on income
# next line converts scores to probabilities
p <- softmax(score[1],score[2],score[3])

# now simulate choice
# outcome career holds event type values, not counts
career <- rep(NA,N)       # empty vector of choices for each individual
# sample chosen career for each individual
for ( i in 1:N ) career[i] <- sample( 1:3 , size=1 , prob=p )
```

To fit the model to these fake data, we use the dcategorical likelihood, which is the multinomial logistic regression distribution. It works when each value in the outcome variable, here career, contains the individual event types on each row. To convert all the scores to probabilities, we'll use the multinomial logit link, which is called softmax. We also have to pick one of the event types to be the reference type. We'll use the first one. Instead of getting a linear model, that type is assigned a constant value. Then the other types get linear models that contain parameters relative to the reference type.

R code
10.57

```
# fit the model, using dcategorical and softmax link
m10.16 <- map(
    alist(
```

```
    career ~ dcategorical( softmax(0,s2,s3) ),
    s2 <- b*2,    # linear model for event type 2
    s3 <- b*3,    # linear model for event type 3
    b ~ dnorm(0,5)
  ) ,
  data=list(career=career) )
```

Notice that there are no intercepts in the linear model.

Be aware that the estimates you get from these models are extraordinarily difficult to interpret. You absolutely must convert them to a vector of probabilities, to make much sense of them. The principle reason is that the estimates swing around wildly, depending upon which event type you assign a constant score. In the example above, I chose the first event type, the first career. If you choose another, you'll get different estimates, but the same predictions.

Now consider an example of the second case. Suppose you are still modeling career choice. But now you want to estimate the association between each person's family income and which career he or she chooses. So the predictor variable must have the same value in each linear model, for each row in the data. But now there is a unique parameter multiplying it in each linear model. This provides an estimate of the impact of family income on choice, for each type of career.

R code
10.58

```
N <- 100
# simulate family incomes for each individual
family_income <- runif(N)
# assign a unique coefficient for each type of event
b <- (1:-1)
career <- rep(NA,N)  # empty vector of choices for each individual
for ( i in 1:N ) {
    score <- 0.5*(1:3) + b*family_income[i]
    p <- softmax(score[1],score[2],score[3])
    career[i] <- sample( 1:3 , size=1 , prob=p )
}

m10.17 <- map(
    alist(
        career ~ dcategorical( softmax(0,s2,s3) ),
        s2 <- a2 + b2*family_income,
        s3 <- a3 + b3*family_income,
        c(a2,a3,b2,b3) ~ dnorm(0,5)
    ) ,
    data=list(career=career,family_income=family_income) )
```

Again, computing implied predictions is the safest way to interpret these models. They do a great job of classifying discrete, unordered events. But the parameters are on a scale that is very hard to interpret.

10.3.1.2. *Multinomial in disguise as Poisson.* Another way to fit a multinomial likelihood is to refactor it into a series of Poisson likelihoods.[145] That should sound a bit crazy. But it's actually both principled and commonplace to model multinomial outcomes this way. It's principled, because the mathematics justifies it. And it's commonplace, because it is usually computationally easier to use Poisson rather than multinomial likelihoods. Here I'll give an example of an implementation. For the mathematical details of the transformation, see the Overthinking box at the end.

I appreciate that this kind of thing—modeling the same data different ways but getting the same inferences—is exactly the kind of thing that makes statistics maddening for scientists. So I'll begin by taking a binomial example from earlier in the chapter and doing it over as a Poisson regression. Since the binomial is just a special case of the multinomial, the approach extrapolates to any number of event types. Think again of the UC Berkeley admissions data. Let's load it again:

R code
10.59

```
library(rethinking)
data(UCBadmit)
d <- UCBadmit
```

Now let's use a Poisson regression to model both the rate of admission and the rate of rejection. And we'll compare the inference to the binomial model's probability of admission. Here are both the binomial and Poisson models:

R code
10.60

```
# binomial model of overall admission probability
m_binom <- map(
    alist(
        admit ~ dbinom(applications,p),
        logit(p) <- a,
        a ~ dnorm(0,100)
    ),
    data=d )

# Poisson model of overall admission rate and rejection rate
d$rej <- d$reject # 'reject' is a reserved word
m_pois <- map2stan(
    alist(
        admit ~ dpois(lambda1),
        rej ~ dpois(lambda2),
        log(lambda1) <- a1,
        log(lambda2) <- a2,
        c(a1,a2) ~ dnorm(0,100)
    ),
    data=d , chains=3 , cores=3 )
```

Let's consider just the posterior means, for the sake of simplicity. But keep in mind that the entire posterior is what matters. First, the inferred binomial *probability* of admission, across the entire data set, is:

R code
10.61

```
logistic(coef(m_binom))
```

```
        a
0.3877596
```

And in the Poisson model, the implied probability of admission is given by:

$$p_{\text{ADMIT}} = \frac{\lambda_1}{\lambda_1 + \lambda_2} = \frac{\exp(a_1)}{\exp(a_1) + \exp(a_2)}$$

In code form:

R code
10.62

```
k <- as.numeric(coef(m_pois))
exp(k[1])/(exp(k[1])+exp(k[2]))
```

```
[1] 0.3879816
```

And that's the same inference as in the binomial model.

Overthinking: Multinomial-Poisson transformation. The Poisson distribution was introduced earlier in this chapter. The Poisson probability of y_1 events of type 1, assuming a rate λ_1, is given by:

$$\Pr(y_1|\lambda_1) = \frac{e^{-\lambda_1}\lambda_1^{y_1}}{y_1!}$$

I'll show you a magic trick for extracting this expression from the multinomial probability expression. The multinomial probability is just an extrapolation of the binomial to more than two types of events. So we'll work here with the binomial distribution, but in multinomial form, just to make the derivation a little easier. The probability of counts y_1 and y_2 for event types 1 and 2 with probabilities p_1 and p_2, respectively, out of n trials, is:

$$\Pr(y_1, y_2|n, p_1, p_2) = \frac{n!}{y_1! y_2!}p_1^{y_1} p_2^{y_2}$$

We need some definitions now. Let $\Lambda = \lambda_1 + \lambda_2$, $p_1 = \lambda_1/\Lambda$, and $p_2 = \lambda_2/\Lambda$. Substituting these into the binomial probability:

$$\Pr(y_1, y_2|n, \lambda_1, \lambda_2) = \frac{n!}{y_1! y_2!} \left(\frac{\lambda_1}{\Lambda}\right)^{y_1} \left(\frac{\lambda_2}{\Lambda}\right)^{y_2} = \frac{n!}{\Lambda^{y_1} \Lambda^{y_2}} \frac{\lambda_1^{y_1} \lambda_2^{y_2}}{y_1! \, y_2!} = \frac{n!}{\Lambda^n} \frac{\lambda_1^{y_1} \lambda_2^{y_2}}{y_1! \, y_2!}$$

Now we simultaneously multiply and divide by both $e^{-\lambda_1}$ and $e^{-\lambda_2}$, then perform some strategic rearrangement:

$$\Pr(y_1, y_2|n, \lambda_1, \lambda_2) = \frac{n!}{\Lambda^n} \frac{e^{-\lambda_1}}{e^{-\lambda_1}} \frac{\lambda_1^{y_1}}{y_1!} \frac{e^{-\lambda_2}}{e^{-\lambda_2}} \frac{\lambda_2^{y_2}}{y_2!} = \frac{n!}{\Lambda^n e^{-\lambda_1}e^{-\lambda_2}} \frac{e^{-\lambda_1}\lambda_1^{y_1}}{y_1!} \frac{e^{-\lambda_2}\lambda_2^{y_2}}{y_2!}$$

$$= \frac{n!}{\underbrace{e^{-\Lambda}\Lambda^n}_{\Pr(n)^{-1}}} \underbrace{\frac{e^{-\lambda_1}\lambda_1^{y_1}}{y_1!}}_{\Pr(y_1)} \underbrace{\frac{e^{-\lambda_2}\lambda_2^{y_2}}{y_2!}}_{\Pr(y_2)}$$

The final expression is the product of the Poisson probabilities $\Pr(y_1)$ and $\Pr(y_2)$, divided by the Poisson probability of n, $\Pr(n)$. It makes sense that the product is divided by $\Pr(n)$, because this is a conditional probability for y_1 and y_2. All of this means that if there are k event types, you can model multinomial probabilities $p_1, ..., p_k$ using Poisson rate parameters $\lambda_1, ..., \lambda_k$. And you can recover the multinomial probabilities using the definition $p_i = \lambda_i/\sum_j \lambda_j$.

10.3.2. Geometric. Sometimes a count variable is a number of events up until something happened. Call this "something" the terminating event. Often we want to model the probability of that event, a kind of analysis known as EVENT HISTORY ANALYSIS or SURVIVAL ANALYSIS. When the probability of the terminating event is constant through time (or distance), and the units of time (or distance) are discrete, a common likelihood function is the

GEOMETRIC DISTRIBUTION. This distribution has the form:

$$\Pr(y|p) = p(1 - p)^{y-1}$$

where y is the number of time steps (events) until the terminating event occurred and p is the probability of that event in each time step. This distribution has maximum entropy for unbounded counts with constant expected value. But it's easier to remember its appropriate domain by thinking about predicting the number of events until a particular event of interest.

Here's a quick simulation example.

R code
10.63

```
# simulate
N <- 100
x <- runif(N)
y <- rgeom( N , prob=logistic( -1 + 2*x ) )

# estimate
m10.18 <- map(
    alist(
        y ~ dgeom( p ),
        logit(p) <- a + b*x,
        a ~ dnorm(0,10),
        b ~ dnorm(0,1)
    ),
    data=list(y=y,x=x) )
precis(m10.18)
```

```
   Mean StdDev  5.5% 94.5%
a -1.00   0.22 -1.34 -0.65
b  1.75   0.43  1.06  2.43
```

And the above model will map2stan just fine, allowing you to later build multilevel models (Chapter 12) from it, if you need to.

10.3.3. Negative-binomial and beta-binomial. Let's think back to our bag of marbles from Chapter 2. Inside the bag were blue and white marbles, and each draw from the bag gave us some information about the bag's composition. Suppose instead that there were multiple bags, each with a different composition of blue and white marbles. Each set of draws, a path through the garden, comes from the same bag. But different sets come from different bags. As a result, the counts of blue marbles vary more than they would, if we used only one bag.

Processes of this kind are often called *mixtures*, because they mix together different maximum entropy distributions. We'll spend more time on mixtures in the next chapter. For now, it's worth noting that the two most common generalizations of count GLMs are actually mixtures: the BETA-BINOMIAL and NEGATIVE-BINOMIAL. Both are used when counts are thought to be OVER-DISPERSED. This means that variation in the counts exceeds what we'd expect from a pure binomial or Poisson process.

We'll leave the details until the next chapter.

10.4. Summary

This chapter described some of the most common generalized linear models, those used to model counts. It is important to never convert counts to proportions before analysis, because doing so destroys information about sample size. A fundamental difficulty with these models is that parameters are on a different scale, typically log-odds (for binomial) or log-rate (for Poisson), than the outcome variable they describe. Therefore computing implied predictions is even more important than before.

10.5. Practice

Easy.

10E1. If an event has probability 0.35, what are the log-odds of this event?

10E2. If an event has log-odds 3.2, what is the probability of this event?

10E3. Suppose that a coefficient in a logistic regression has value 1.7. What does this imply about the proportional change in odds of the outcome?

10E4. Why do Poisson regressions sometimes require the use of an *offset*? Provide an example.

Medium.

10M1. As explained in the chapter, binomial data can be organized in aggregated and disaggregated forms, without any impact on inference. But the likelihood of the data does change when the data are converted between the two formats. Can you explain why?

10M2. If a coefficient in a Poisson regression has value 1.7, what does this imply about the change in the outcome?

10M3. Explain why the logit link is appropriate for a binomial generalized linear model.

10M4. Explain why the log link is appropriate for a Poisson generalized linear model.

10M5. What would it imply to use a logit link for the mean of a Poisson generalized linear model? Can you think of a real research problem for which this would make sense?

10M6. State the constraints for which the binomial and Poisson distributions have maximum entropy. Are the constraints different at all for binomial and Poisson? Why or why not?

Hard.

10H1. Use `map` to construct a quadratic approximate posterior distribution for the chimpanzee model that includes a unique intercept for each actor, `m10.4` (page 299). Compare the quadratic approximation to the posterior distribution produced instead from MCMC. Can you explain both the differences and the similarities between the approximate and the MCMC distributions?

10H2. Use WAIC to compare the chimpanzee model that includes a unique intercept for each actor, `m10.4` (page 299), to the simpler models fit in the same section.

10H3. The data contained in `library(MASS);data(eagles)` are records of salmon pirating attempts by Bald Eagles in Washington State. See `?eagles` for details. While one eagle feeds, sometimes another will swoop in and try to steal the salmon from it. Call the feeding eagle the "victim" and the thief the "pirate." Use the available data to build a binomial GLM of successful pirating attempts.

(a) Consider the following model:

$$y_i \sim \text{Binomial}(n_i, p_i)$$

$$\log \frac{p_i}{1 - p_i} = \alpha + \beta_P P_i + \beta_V V_i + \beta_A A_i$$

$$\alpha \sim \text{Normal}(0, 10)$$

$$\beta_P \sim \text{Normal}(0, 5)$$

$$\beta_V \sim \text{Normal}(0, 5)$$

$$\beta_A \sim \text{Normal}(0, 5)$$

where y is the number of successful attempts, n is the total number of attempts, P is a dummy variable indicating whether or not the pirate had large body size, V is a dummy variable indicating whether or not the victim had large body size, and finally A is a dummy variable indicating whether or not the pirate was an adult. Fit the model above to the `eagles` data, using both `map` and `map2stan`. Is the quadratic approximation okay?

(b) Now interpret the estimates. If the quadratic approximation turned out okay, then it's okay to use the `map` estimates. Otherwise stick to `map2stan` estimates. Then plot the posterior predictions. Compute and display both (1) the predicted **probability** of success and its 89% interval for each row (i) in the data, as well as (2) the predicted success **count** and its 89% interval. What different information does each type of posterior prediction provide?

(c) Now try to improve the model. Consider an interaction between the pirate's size and age (immature or adult). Compare this model to the previous one, using WAIC. Interpret.

10H4. The data contained in `data(salamanders)` are counts of salamanders (*Plethodon elongatus*) from 47 different 49-m² plots in northern California.[146] The column `SALAMAN` is the count in each plot, and the columns `PCTCOVER` and `FORESTAGE` are percent of ground cover and age of trees in the plot, respectively. You will model `SALAMAN` as a Poisson variable.

(a) Model the relationship between density and percent cover, using a log-link (same as the example in the book and lecture). Use weakly informative priors of your choosing. Check the quadratic approximation again, by comparing `map` to `map2stan`. Then plot the expected counts and their 89% interval against percent cover. In which ways does the model do a good job? In which ways does it do a bad job?

(b) Can you improve the model by using the other predictor, `FORESTAGE`? Try any models you think useful. Can you explain why `FORESTAGE` helps or does not help with prediction?

11 Monsters and Mixtures

In Hawaiian legend, Nanaue was the son of a shark who fell in love with a human women. He grew into a murderous man with a shark mouth in the middle of his back. In Greek legend, the minotaur was a man with the head of a bull. He was the spawn of a human woman and a great white bull. The gryphon is a legendary monster that is part eagle and part lion. Maori legends speak of *Taniwha*, monsters with features of serpents and birds and even sharks, much like the dragons of Chinese and European mythology.

By piecing together parts of different creatures, it's easy to make a monster. Many monsters are hybrids. Many statistical models are too. This chapter is about constructing likelihood and link functions by piecing together the simpler components of previous chapters. Like legendary monsters, these hybrid likelihoods contain pieces of other model types. Endowed with some properties of each piece, they help us model outcome variables with inconvenient, but common, properties. Being monsters, these models are both powerful and dangerous. They are often harder to estimate and to understand. But with some knowledge and caution, they are important tools.

We'll consider two common and useful examples. The first type is the ORDERED CATEGORICAL model, useful for categorical outcomes with a fixed ordering. This model is built by merging a categorical likelihood function with a special kind of link function, usually a CUMULATIVE LINK. The second type is a family of ZERO-INFLATED and ZERO-AUGMENTED models, each of which mixes a binary event with an ordinary GLM likelihood like a Poisson or binomial.

Both types of models help us transform our modeling to cope with the inconvenient realities of measurement, rather than transforming measurements to cope with the constraints of our models. There are lots of other model types that arise for this purpose and in this way, by mixing bits of simpler models together. We can't possibly cover them all. But when you encounter a new type, at least you'll have a framework in which to understand it. And if you ever need to construct your own unique monster, feel free to do so. Just be sure to validate it by simulating dummy data and then recovering the data-generating process through fitting the model to the dummy data.

11.1. Ordered categorical outcomes

It is very common in the social sciences, and occasional in the natural sciences, to have an outcome variable that is discrete, like a count, but in which the values merely indicate different ordered *levels* along some dimension. For example, if I were to ask you how much you like to eat fish, on a scale from 1 to 7, you might say 5. If I were to ask 100 people the same question, I'd end up with 100 values between 1 and 7. In modeling each outcome value,

I'd have to keep in mind that these values are *ordered*, because 7 is greater than 6, which is greater than 5, and so on. But unlike a count, the differences in value are not necessarily equal. It might be much harder to move someone's preference for fish from 1 to 2 than it is to move it from 5 to 6, for example.

In principle, an ordered categorical variable is just a multinomial prediction problem (page 323). But the constraint that the categories be ordered demands a special treatment. What we'd like is for any associated predictor variable, as it increases, to move predictions progressively through the categories in sequence. So for example if preference for ice cream is positively associated with years of age, then the model should sequentially move predictions upwards as age increases: 3 to 4, 4 to 5, 5 to 6, etc. This presents a challenge: how to ensure that the linear model maps onto the outcomes in the right order.

The conventional solution is to use a CUMULATIVE LINK function.[147] The cumulative probability of a value is the probability of that value *or any smaller value*. In the context of ordered categories, the cumulative probability of 3 is the sum of the probabilities of 3, 2, and 1. Ordered categories by convention begin at 1, so a result less than 1 has no probability at all.

By linking a linear model to cumulative probability, it is possible to guarantee the ordering of the outcomes. I'll explain why in two steps. Step 1 is to explain how to parameterize a distribution of outcomes on the scale of log-cumulative-odds. Step 2 is to introduce a predictor (or more than one predictor) to these log-cumulative-odds values, allowing you to model associations between predictors and the outcome while obeying the ordered nature of prediction.

Both steps will be unfolded in context of a data example, to make the discussion more concrete. So next you meet some data.

11.1.1. Example: Moral intuition.
The data for this example come from a series of experiments conducted by philosophers.[148] Yes, philosophers do sometimes conduct experiments. In this case, the experiments aim to collect empirical evidence relevant to debates about *moral intuition*, the forms of reasoning through which people develop judgments about the moral goodness and badness of actions. These debates are relevant to all of the social sciences, because they touch on broader issues of reasoning, the role of emotions in decision making, and theories of moral development, both in individuals and groups.

These experiments get measurements of moral judgment by using scenarios known as "trolley problems." The classic version invokes a runaway trolley, but what these scenarios share is that they have proved vexing or paradoxical to moral philosophers. Here's a traditional example, using a "boxcar" in place of a "trolley":

> Standing by the railroad tracks, Dennis sees an empty, out-of-control boxcar about to hit five people. Next to Dennis is a lever that can be pulled, sending the boxcar down a side track and away from the five people. But pulling the lever will also lower the railing on a footbridge spanning the side track, causing one person to fall off the footbridge and onto the side track, where he will be hit by the boxcar. If Dennis pulls the lever the boxcar will switch tracks and not hit the five people, and the one person to fall and be hit by the boxcar. If Dennis does not pull the lever the boxcar will continue down the tracks and hit five people, and the one person will remain safe above the side track.

How morally permissible is it for Dennis to pull the lever?

The reason these scenarios can be philosophically vexing is that the analytical content of two scenarios can be identical, and yet people reliably reach different judgments about the moral permissibility of the same action in the different scenarios. Previous research has lead to at least three important principles of unconscious reasoning that may explain variations in judgment. These principles are:

The action principle: Harm caused by action is morally worse than equivalent harm caused by omission.

The intention principle: Harm intended as the means to a goal is morally worse than equivalent harm foreseen as the side effect of a goal.

The contact principle: Using physical contact to cause harm to a victim is morally worse than causing equivalent harm to a victim without using physical contact.

The experimental context within which we'll explore these principles comprises stories that vary the principles, while keeping many of the basic objects and actors the same. For example, the version of the boxcar story quoted just above implies the *action* principle, but not the others. Since the actor (Dennis) had to do something to create the outcome, rather than remain passive, this is an action scenario. However, the harm caused to the one man who will fall is not necessary, or intended, in order to save the five. Thus it is not an example of the intention principle. And there is no direct contact, so it is also not an example of the contact principle.

You can construct a boxcar story with the same outline, but now with both the action principle and the intention principle. That is, in this version, the actor both does something to change the outcome and the action must cause harm to the one person in order to save the other five:

> Standing by the railroad tracks, Evan sees an empty, out-of-control boxcar about to hit five people. Next to Evan is a lever that can be pulled, lowering the railing on a footbridge that spans the main track, and causing one person to fall off the footbridge and onto the main track, where he will be hit by the boxcar. The boxcar will slow down because of the one person, therefore preventing the five from being hit. If Evan pulls the lever the one person will fall and be hit by the boxcar, and therefore the boxcar will slow down and not hit the five people. If Evan does not pull the lever the boxcar will continue down the tracks and hit the five people, and the one person will remain safe above the main track.

Most people judge that, if Even pulls the lever, it is morally worse (less permissible) than when Dennis pulls the lever. You'll see by how much, as we analyze these data. Load the data with:

R code
11.1

```
library(rethinking)
data(Trolley)
d <- Trolley
```

There are 12 columns and 9930 rows, comprising data for 331 unique individuals. The outcome we'll be interested in is `response`, which is an integer from 1 to 7 indicating how morally permissible the participant found the action to be taken (or not) in the story. Since this type of rating is categorical and ordered, it's exactly the right type of problem for our ordered model.

11.1.2. Describing an ordered distribution with intercepts. First, let's see how to describe a distribution of discrete ordered values. Take a look at the overall distribution, the histogram, of the outcome variable.

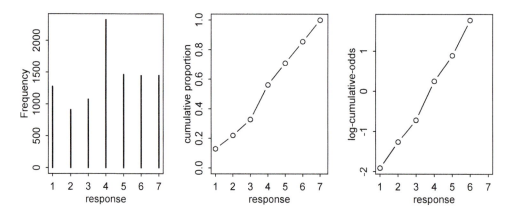

FIGURE 11.1. Re-describing a discrete distribution using log-cumulative-odds. Left: Histogram of discrete response in the sample. Middle: Cumulative proportion of each response. Right: Logarithm of cumulative odds of each response. Note that the log-cumulative-odds of response value 7 is infinity, so it is not shown.

R code
11.2

```
simplehist( d$response , xlim=c(1,7) , xlab="response" )
```

The result is shown in the left-hand plot in FIGURE 11.1.

Our goal is to re-describe this histogram on the log-cumulative-odds scale. This just means constructing the odds of a cumulative probability and then taking a logarithm. Why do this arcane thing? Because this is the cumulative analog of the logit link we used in previous chapters. The logit is log-odds, and cumulative logit is log-cumulative-odds. Both are designed to constrain the probabilities to the 0/1 interval. Then when we decide to add predictor variables, we can safely do so on the cumulative logit scale. The link function takes care of converting the parameter estimates to the proper probability scale.

The first step in the conversion is to compute cumulative probabilities from the histogram:

R code
11.3

```
# discrete proportion of each response value
pr_k <- table( d$response ) / nrow(d)

# cumsum converts to cumulative proportions
cum_pr_k <- cumsum( pr_k )

# plot
plot( 1:7 , cum_pr_k , type="b" , xlab="response" ,
ylab="cumulative proportion" , ylim=c(0,1) )
```

And the result is shown as the middle plot in FIGURE 11.1.

Then to re-describe the histogram as log-cumulative odds, we'll need a series of intercept parameters. Each intercept will be on the log-cumulative-odds scale and stand in for the cumulative probability of each outcome. So this is just the application of the link function. The log-cumulative-odds that a response value y_i is equal-to-or-less-than some possible outcome value k is:

$$\log \frac{\Pr(y_i \le k)}{1 - \Pr(y_i \le k)} = \alpha_k \tag{11.1}$$

where α_k is an "intercept" unique to each possible outcome value k. We can compute these intercept parameters directly:

R code
11.4

```
logit <- function(x) log(x/(1-x)) # convenience function
( lco <- logit( cum_pr_k ) )
```

```
        1          2          3          4          5          6          7
-1.9160912 -1.2666056 -0.7186340  0.2477857  0.8898637  1.7693809        Inf
```

These values are plotted in the right-hand panel of FIGURE 11.1. Notice that the cumulative logit of the largest response, 7, is infinity. This is because $\log(1/(1 - 1)) = \infty$. Since the largest response value always has a cumulative probability of 1, we effectively do not need a parameter for it. We get it for free, from the law of total probability. So for $K = 7$ possible response values, we only need $K - 1 = 6$ intercepts.

All of the above is very nice, but what we really want is the posterior distribution of these intercepts. This will allow us to take into account sample size and prior information, as well as insert predictor variables (in the next section). To use Bayes' theorem to compute the posterior distribution of these intercepts, we'll need to compute the likelihood of each possible response value. So the last step in constructing the basic model fitting engine for ordered categorical outcomes is to use the cumulative probabilities, $\Pr(y_i \le k)$, to compute likelihood, $\Pr(y_i = k)$.

FIGURE 11.2 illustrates how this is done. Each intercept α_k implies a cumulative probability for each k. You just use the inverse link to translate from log-cumulative-odds back to cumulative probability. So when we observe k and need its likelihood, we can get the likelihood by subtraction:

$$p_k = \Pr(y_i = k) = \Pr(y_i \le k) - \Pr(y_i \le k - 1) \tag{11.2}$$

The blue line segments in FIGURE 11.2 are these likelihoods, computed by subtraction. With these in hand, the posterior distribution is computed the usual way.

Let's go ahead and see how it's done, in code form. Conventions for writing mathematical forms of the ordered logit vary a lot. We'll use this convention:

$$R_i \sim \text{Ordered}(\mathbf{p}) \qquad\qquad \text{[likelihood]}$$
$$\text{logit}(p_k) = \alpha_k \qquad\qquad \text{[cumulative link and linear model]}$$
$$\alpha_k \sim \text{Normal}(0, 10) \qquad\qquad \text{[common prior for each intercept]}$$

The Ordered distribution is really just a categorical distribution that takes a vector $\mathbf{p} = \{p_1, p_2, p_3, p_4, p_5, p_6\}$ of probabilities of each response value below the maximum response (7 in this example). Each response value k in this vector is defined by its link to an intercept parameter, α_k. Finally, some weakly regularizing priors are placed on these intercepts. In this example, there is a lot of data, so just about any prior will be overwhelmed. As always, in small sample contexts, you'll have to think much harder about priors. Consider for example that we know $\alpha_1 < \alpha_2$, before we even see the data.

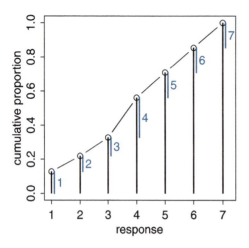

FIGURE 11.2. Cumulative probability and ordered likelihood. The horizontal axis displays possible observable outcomes, from 1 through 7. The vertical axis displays cumulative probability. The gray bars over each outcome show cumulative probability. These keep growing with each successive outcome value. The blue line segments show the discrete probability of each individual outcome. These are the likelihoods that go into Bayes' theorem.

In code form for map and map2stan, the link function will be embedded in the likelihood function already. This makes the calculations more efficient and prevents headaches. So to fit the basic model, incorporating no predictor variables:

R code
11.5
```
m11.1 <- map(
    alist(
        response ~ dordlogit( phi , c(a1,a2,a3,a4,a5,a6) ),
        phi <- 0,
        c(a1,a2,a3,a4,a5,a6) ~ dnorm(0,10)
    ) ,
    data=d ,
    start=list(a1=-2,a2=-1,a3=0,a4=1,a5=2,a6=2.5) )
```

The phi in the model code is a placeholder for the linear model that incorporates predictor variables. You'll use it shortly. It is a constant zero for now, because only the intercept parameters are of interest. The start values for the intercepts are chosen just to start them in the right order. The exact values aren't important, but their ordering, on the log-cumulative-odds scale, is important. Their posterior distribution is also on the log-cumulative-odds scale:

R code
11.6
```
precis(m11.1)
```

```
      Mean StdDev   5.5% 94.5%
a1 -1.92    0.03  -1.96 -1.87
a2 -1.27    0.02  -1.31 -1.23
a3 -0.72    0.02  -0.75 -0.68
a4  0.25    0.02   0.22  0.28
a5  0.89    0.02   0.85  0.93
a6  1.77    0.03   1.72  1.81
```

Since there is a lot of data here, the posterior for each intercept is quite precisely estimated, as you can see from the tiny standard deviations. To get cumulative probabilities back:

R code
11.7

```
logistic(coef(m11.1))
```

```
        a1        a2        a3        a4        a5        a6
0.1283005 0.2198398 0.3276948 0.5616311 0.7088609 0.8543786
```

And of course those are the same as the values in cum_pr_k that we computed earlier. But now we also have a posterior distribution around these values, and we're ready to add predictor variables in the next section.

To fit the same model using Stan's HMC engine, it is better to use an explicit vector of intercept parameters:

R code
11.8

```
# note that data with name 'case' not allowed in Stan
# so will pass pruned data list
m11.1stan <- map2stan(
    alist(
        response ~ dordlogit( phi , cutpoints ),
        phi <- 0,
        cutpoints ~ dnorm(0,10)
    ) ,
    data=list(response=d$response),
    start=list(cutpoints=c(-2,-1,0,1,2,2.5)) ,
    chains=2 , cores=2 )

# need depth=2 to show vector of parameters
precis(m11.1stan,depth=2)
```

```
             Mean StdDev lower 0.89 upper 0.89 n_eff Rhat
cutpoints[1] -1.92   0.03      -1.97      -1.87  1012    1
cutpoints[2] -1.27   0.02      -1.31      -1.23  1461    1
cutpoints[3] -0.72   0.02      -0.75      -0.68  1845    1
cutpoints[4]  0.25   0.02       0.22       0.28  2000    1
cutpoints[5]  0.89   0.02       0.85       0.92  2000    1
cutpoints[6]  1.77   0.03       1.72       1.81  1851    1
```

The individual cutpoints parameters correspond to each α_k from earlier.

11.1.3. Adding predictor variables.

This flurry of computation has gotten us very little so far, aside from a Bayesian estimate of a histogram. But all of it has been necessary in order to prepare the model for the addition of predictor variables that obey the ordered constraint on the outcomes.

To include predictor variables, we define the log-cumulative-odds of each response k as a sum of its intercept α_k and a typical linear model. Suppose for example we want to add a predictor x to the model. We'll do this by defining a linear model $\phi_i = \beta x_i$. Then each cumulative logit becomes:

$$\log \frac{\Pr(y_i \le k)}{1 - \Pr(y_i \le k)} = \alpha_k - \phi_i$$

$$\phi_i = \beta x_i$$

This form automatically ensures the correct ordering of the outcome values, while still morphing the likelihood of each individual value as the predictor x_i changes value. Why is the linear model ϕ subtracted from each intercept? Because if we decrease the log-cumulative-odds of every outcome value k below the maximum, this necessarily shifts probability mass upwards towards higher outcome values.

For example, suppose we take the MAP estimates from m11.1 and subtract 0.5 from each. The function dordlogit makes the calculation of likelihoods straightforward:

R code
11.9

```
( pk <- dordlogit( 1:7 , 0 , coef(m11.1) ) )
```

```
[1] 0.12830051 0.09153931 0.10785502 0.23393627 0.14722982 0.14551766 0.14562142
```

These probabilities imply an average outcome value of:

R code
11.10

```
sum( pk*(1:7) )
```

```
[1] 4.199294
```

And now subtracting 0.5 from each:

R code
11.11

```
( pk <- dordlogit( 1:7 , 0 , coef(m11.1)-0.5 ) )
```

```
[1] 0.08195550 0.06401015 0.08221206 0.20910054 0.15897033 0.18438530 0.21936612
```

Compare these to the likelihoods just above and notice that the values on the left have diminished while the values on the right have increased. The expected value is now:

R code
11.12

```
sum( pk*(1:7) )
```

```
[1] 4.72974
```

And that's why we subtract ϕ, the linear model βx_i, from each intercept, rather than add it. This way, a positive β value indicates that an increase in the predictor variable x results in an increase in the average response.

Now we can turn back to our "trolley" data and include predictor variables to help explain variation in responses. The predictor variables of interest are going to be action, intention, and contact, each a dummy variable corresponding to each principle outlined earlier. The log-cumulative-odds of each response k will now be:

$$\log \frac{\Pr(y_i \leq k)}{1 - \Pr(y_i \leq k)} = \alpha_k - \phi_i$$
$$\phi_i = \beta_A A_i + \beta_I I_i + \beta_C C_i$$

where A_i indicates the value of action on row i, I_i indicates the value of intention on row i, and C_i indicates the value of contact on row i. What we've done here is define the log-odds of each possible response to be an additive model of the features of the story corresponding to each response.

You fit this model just as you'd expect, by adding the slopes and predictor variables to the phi parameter inside dordlogit. Here's a working model:

R code
11.13

```
m11.2 <- map(
    alist(
        response ~ dordlogit( phi , c(a1,a2,a3,a4,a5,a6) ) ,
        phi <- bA*action + bI*intention + bC*contact,
```

```
        c(bA,bI,bC) ~ dnorm(0,10),
        c(a1,a2,a3,a4,a5,a6) ~ dnorm(0,10)
    ) ,
    data=d ,
    start=list(a1=-1.9,a2=-1.2,a3=-0.7,a4=0.2,a5=0.9,a6=1.8) )
```

The parameter phi now contains the additive function with slope parameters and predictor variables. Notice that there is no link function around phi, because the link is really inside dordlogit already (hence "logit" in its name). Notice also that I've adopted the approximate MAP estimates from the previous model, m11.1, as starting values for the intercepts. This helps map find the new MAP estimates more quickly.

Let's fit one more model of interest, before comparing them and turning to plotting model-averaged predictions. It's possible that the variables action and intention interact. This would mean that a story containing both action and intention is non-additively worse (or better) than what you'd expect by just adding together the separate estimates for action and intention. It's just as possible that contact and intention interact. The variables action and contact cannot interact, because *contact* is just a type of *action* in these data. But that leaves us with two interaction effects to model.

Fitting the interaction model follows the pattern you are accustomed to by now.

R code
11.14

```
m11.3 <- map(
    alist(
        response ~ dordlogit( phi , c(a1,a2,a3,a4,a5,a6) ) ,
        phi <- bA*action + bI*intention + bC*contact +
            bAI*action*intention + bCI*contact*intention ,
        c(bA,bI,bC,bAI,bCI) ~ dnorm(0,10),
        c(a1,a2,a3,a4,a5,a6) ~ dnorm(0,10)
    ) ,
    data=d ,
    start=list(a1=-1.9,a2=-1.2,a3=-0.7,a4=0.2,a5=0.9,a6=1.8) )
```

No new tricks here. The above just adds two interaction terms and two interaction parameters, bAI and bCI.

Now let's compare these three models. You can use coeftab to get a quick comparison of estimates:

R code
11.15

```
coeftab(m11.1,m11.2,m11.3)
```

```
        m11.1   m11.2   m11.3
a1      -1.92   -2.84   -2.63
a2      -1.27   -2.16   -1.94
a3      -0.72   -1.57   -1.34
a4       0.25   -0.55   -0.31
a5       0.89    0.12    0.36
a6       1.77    1.02    1.27
bA         NA   -0.71   -0.47
bI         NA   -0.72   -0.28
bC         NA   -0.96   -0.33
```

```
bAI         NA       NA     -0.45
bCI         NA       NA     -1.27
nobs      9930     9930     9930
```

Whatever do these estimates mean? The first six rows, from a1 to a6, are just the α intercepts, one for each value below the maximum of "7". These really can't be interpreted on their own, unless you are very used to reading log-odds values. But they do define the relative frequencies of the outcomes, when all predictor variables are set to zero. So they are "intercepts" as in simpler models.

The next 5 rows, from bA to bCI, are the various slope parameters: three main effects and two interactions. These are interpretable on their own, to a limited extent. It makes sense to ask, first, if they are very far from zero. You can check the standard errors and intervals with precis and verify that all of the slope estimates are quite reliably negative. Second, all of the slopes are negative, which implies that each factor/interaction *reduces* the average response. Including action, intention or contact in a story leads people to judge it as less morally permissible. But by how much? Remember, these parameters are part of a function defining cumulative log-odds, so they can be interpreted as changes in cumulative log-odds. But unless you are very comfortable thinking about log-odds and cumulative probability distributions, that doesn't help you much. It also doesn't help that this change applies to the cumulative log-odds of *every* value of the response variable, aside from the maximum one (which is fixed at cumulative log-odds ∞).

So what to do? First, let's compare these models using WAIC. Computing predictions for these models is relatively slow, compared to earlier examples in the book. So I'll turn on the progress display here, using the refresh argument:

R code
11.16

```
compare( m11.1 , m11.2 , m11.3 , refresh=0.1 )
```

```
              WAIC pWAIC dWAIC weight    SE   dSE
m11.3 36929.4  11.2   0.0      1 81.29    NA
m11.2 37090.3   9.2 160.8      0 76.22 25.80
m11.1 37854.7   6.1 925.2      0 57.71 62.69
```

Now, since model m11.3 absolutely dominates based upon WAIC—note the dWAIC of 161 ± 26—we can safely proceed here, ignoring model uncertainty. But if the WAICs had turned out to be less decisive, the only thing you'd change in the code that follows is passing a list of models to ensemble, as you've done in previous chapters.

Now, to get a good idea of what m11.3 implies, let's plot implied predictions. There is no perfect way to plot the predictions of these log-cumulative-odds models. Why? Because each prediction is really a vector of probabilities, one for each possible outcome value. So as a predictor variable changes value, the entire vector changes. This kind of thing can be visualized in several different ways.

One common and useful way is to use the horizontal axis for a predictor variable and the vertical axis for cumulative probability. Then you can plot a curve for each response value, as it changes across values of the predictor variable. After plotting a curve for each response value, you'll end up mapping the distribution of responses, as it changes across values of the predictor variable.

So let's do that. First, we compute:

```
post <- extract.samples( m11.3 )
```

Then make an empty plot:

```
plot( 1 , 1 , type="n" , xlab="intention" , ylab="probability" ,
    xlim=c(0,1) , ylim=c(0,1) , xaxp=c(0,1,1) , yaxp=c(0,1,2) )
```

Now loop over the first 100 samples in `post` and plot their predictions, across values of `intention`:

```
kA <- 0      # value for action
kC <- 1      # value for contact
kI <- 0:1    # values of intention to calculate over
for ( s in 1:100 ) {
    p <- post[s,]
    ak <- as.numeric(p[1:6])
    phi <- p$bA*kA + p$bI*kI + p$bC*kC +
        p$bAI*kA*kI + p$bCI*kC*kI
    pk <- pordlogit( 1:6 , a=ak , phi=phi )
    for ( i in 1:6 )
        lines( kI , pk[,i] , col=col.alpha(rangi2,0.1) )
}
mtext( concat( "action=",kA,", contact=",kC ) )
```

The first three lines in the code above define the values of `action` (kA), `intention` (kI), and `contact` (kC) to use for the calculations and plotting. The value assigned to kI is a vector, because that is the variable we are varying on the horizontal axis. If it had more values than 0 and 1, then more values would appear in this vector. Then the code loops over the first 100 samples from the naive posterior, computing cumulative probabilities for each response value, for each sample. The function `pordlogit` computes cumulative ordered logit probabilities, just as `dordlogit` computes likelihoods. Finally, each response boundary is plotted, using `lines`.

By modifying the above code to change the values in kA and kC, you can make a triptych (page 225) for model `m11.3`. The results are shown in FIGURE 11.3, with a little extra decoration, to show the outcome values between each predicted boundary. In each plot, the blue lines indicate the boundaries between response values, numbered 1 through 7, bottom to top. The thickness of the blue lines corresponds to the variation in predictions due to variation in samples from the posterior. Since there is so much data in this example, the path of the predicted boundaries is quite certain. The horizontal axis represents values of `intention`, either zero or one. The change in height of each boundary going from left to right in each plot indicates the predicted impact of changing a story from non-intention to intention. Finally, each plot sets the other two predictor variables, `action` and `contact`, to either zero or one. In the upper-left, both are set to zero. This plot shows the predicted effect of taking a story with no-action, no-contact, and no-intention and adding intention to it. In the upper-right, `action` is now set to one. This plot shows the predicted impact of taking a story with action and no-intention (action and contact never go together in this experiment, recall) and adding intention. This upper-right plot demonstrates the interaction between

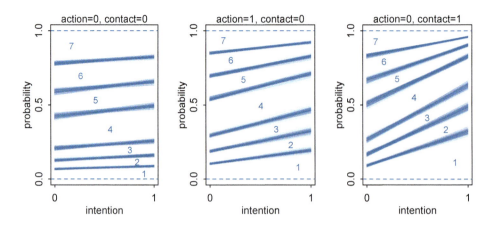

FIGURE 11.3. Posterior predictions of the ordered categorical model with interactions, m11.3. Each plot shows how the distribution of predicted responses varies by intention. Left: Effect of intention when action and contact are both zero. The other two plots each change either action or contact to one.

action and intention. Finally, in the lower-left, contact is set to one. This plot shows the predicted impact of taking a story with contact and no-intention and adding intention to it. This plot shows the large interaction effect between contact and intention, the largest estimated effect in the model.

Rethinking: Staring into the abyss. The plotting code for ordered logistic models is complicated, compared to that of models from previous chapters. But as models become more monstrous, so too does the code needed to compute predictions and display them. With power comes hardship. It's better to see the guts of the machine than to live in awe or fear of it. Software can be and often is written to hide all the monstrosity from us. But this doesn't make it go away. Instead, it just makes the models forever mysterious. For some users, mystery translates into awe. For others, it translates into skepticism. Neither condition is necessary, as long as we're willing to learn the structure of the models we are using.

And if you aren't willing to learn the structure of the models, then don't do your own statistics. Instead, collaborate with or hire a statistician.

11.2. Zero-inflated outcomes

Very often, the things we can measure are not emissions from any pure process. Instead, they are *mixtures* of multiple processes. Whenever there are different causes for the same observation, then a MIXTURE MODEL may be useful. A mixture model uses more than one simple probability distribution to model a mixture of causes. In effect, these models use more than one likelihood for the same outcome variable.

Count variables are especially prone to needing a mixture treatment. The reason is that a count of zero can often arise more than one way. A "zero" means that nothing happened, and nothing can happen either because the rate of events is low or rather because the process that generates events failed to get started. If we are counting scrub jays in the woods, we might record a zero because there were no scrub jays in the woods or rather because we scared them all off before we starting looking. Either way, the data contains a zero.

So in this section you'll see how to construct simple zero-inflated models. You'll be able to use the same components from earlier models, but they'll be assembled in a different way. So even if you never need to use or interpret a zero-inflated model, seeing how they are constructed should expand your modeling imagination.

> **Rethinking: Breaking the law.** In the sciences, there is sometimes a culture of anxiety surrounding statistical inference. It used to be that researchers couldn't easily construct and study their own custom models, because they had to rely upon statisticians to properly study the models first. This led to concerns about unconventional models, concerns about breaking the laws of statistics. But statistical computing is much more capable now. Now you can imagine your own generative process, simulate data from it, write the model, and verify that it recovers the true parameter values. You don't have to wait for a mathematician to legalize the model you need.

11.2.1. Example: Zero-inflated Poisson. Back in Chapter 10, I introduced Poisson GLMs by using the example of a monastery producing manuscripts. Each day, a large number of monks finish copying a small number of manuscripts. The process is essentially binomial, but with a large number of trials and very low probability, so the distribution tends towards Poisson.

Now imagine that the monks take breaks on some days. On those days, no manuscripts are completed. Instead, the wine cellar is opened and more earthly delights are practiced. As the monastery owner, you'd like to know how often the monks drink. The obstacle for inference is that there will be zeros on honest non-drinking days, as well, just by chance. So how can you estimate the number of days spent drinking?

Let's make a mixture to solve this problem.[149] We want to consider that any zero in the data can arise from two processes: (1) the monks spent the day drinking and (2) they worked that day but nevertheless failed to complete any manuscripts. Let p be the probability the monks spend the day drinking. Let λ be the mean number of manuscripts completed, when the monks work.

To get this model going, we need to define a likelihood function that mixes these two processes. To grasp how we can construct such a monster, think of the monks' drinking as resulting from a coin flip (FIGURE 11.4). The "coin" shows a cask of wine on one side and a quill on the other. The probability the wine cask shows is p, which could be any value from 0 to 1. Depending upon the outcome of the coin flip, the monks either begin drinking or rather begin copying. Drinking monks always produce zero completed manuscripts. Working monks produce a Poisson number of completed manuscripts with some average rate λ. So it is possible still to observe a zero, even when the monks work.

With these assumptions, the likelihood of observing a zero is:

$$\Pr(0|p, \lambda) = \Pr(\text{drink}|p) + \Pr(\text{work}|p) \times \Pr(0|\lambda)$$
$$= p + (1 - p)\exp(-\lambda)$$

Since the Poisson likelihood of y is $\Pr(y|\lambda) = \lambda^y \exp(-\lambda)/y!$, the likelihood of $y = 0$ is just $\exp(-\lambda)$. The above is just the mathematics for:

> *The probability of observing a zero is the probability that the monks didn't drink OR (+) the probability that the monks worked AND (×) failed to finish anything.*

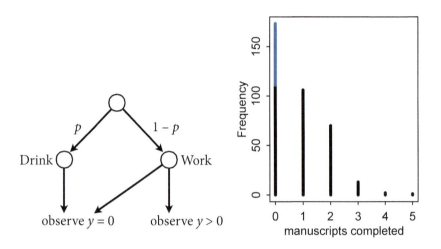

FIGURE 11.4. Left: Structure of the zero-inflated likelihood calculation. Beginning at the top, the monks drink p of the time or instead work $1 - p$ of the time. Drinking monks always produce an observation $y = 0$. Working monks may produce either $y = 0$ or $y > 0$. Right: Frequency distribution of zero-inflated observations. The blue line segment over zero shows the $y = 0$ observations that arose from drinking. In real data, we typically cannot see which zeros come from which process.

And the likelihood of a non-zero value y is:

$$\Pr(y|p, \lambda) = \Pr(\text{drink}|p)(0) + \Pr(\text{work}|p)\Pr(y|\lambda)$$
$$= (1 - p)\frac{\lambda^y \exp(-\lambda)}{y!}$$

Since drinking monks never produce $y > 0$, the expression above is just the chance the monks both work, $1 - p$, and finish y manuscripts.

Define ZIPoisson as the distribution above, with parameters p (probability of a zero) and λ (mean of Poisson) to describe its shape. Then a zero-inflated Poisson regression takes the form:

$$y_i \sim \text{ZIPoisson}(p_i, \lambda_i)$$
$$\text{logit}(p_i) = \alpha_p + \beta_p x_i$$
$$\log(\lambda_i) = \alpha_\lambda + \beta_\lambda x_i$$

Notice that there are two linear models and two link functions, one for each process in the ZIPoisson. The parameters of the linear models differ, because any predictor such as x may be associated differently with each part of the mixture. In fact, you don't even have to use the same predictors in both models—you can construct the two linear models however you wish, depending upon your hypothesis.

We have everything we need now, except for some actual data. So let's simulate the monks' drinking and working. Then you'll see the code used to recover the parameter values used in the simulation.

```
# define parameters                                                             R code
prob_drink <- 0.2 # 20% of days                                                 11.20
rate_work <- 1    # average 1 manuscript per day

# sample one year of production
N <- 365

# simulate days monks drink
drink <- rbinom( N , 1 , prob_drink )

# simulate manuscripts completed
y <- (1-drink)*rpois( N , rate_work )
```

The outcome variable we get to observe is y, which is just a list of counts of completed manuscripts, one count for each day of the year. Take a look at the outcome variable:

```
simplehist( y , xlab="manuscripts completed" , lwd=4 )                          R code
zeros_drink <- sum(drink)                                                        11.21
zeros_work <- sum(y==0 & drink==0)
zeros_total <- sum(y==0)
lines( c(0,0) , c(zeros_work,zeros_total) , lwd=4 , col=rangi2 )
```

This plot is shown on the right-hand side of FIGURE 11.4. The zeros produced by drinking are shown in blue. Those from work are shown in black. The total number of zeros is inflated, relative to a typical Poisson distribution.

And to fit the model, the rethinking package provides the zero-inflated Poisson likelihood as dzipois. For more detail on how it relates to the mathematics above, see the Overthinking box at the end of this section. Using dzipois is straightforward:

```
m11.4 <- map(                                                                   R code
    alist(                                                                      11.22
        y ~ dzipois( p , lambda ),
        logit(p) <- ap,
        log(lambda) <- al,
        ap ~ dnorm(0,1),
        al ~ dnorm(0,10)
    ) ,
    data=list(y=y) )
precis(m11.4)
```

```
    Mean StdDev  5.5% 94.5%
ap -1.23   0.29 -1.69 -0.77
al  0.06   0.08 -0.08  0.19
```

On the natural scale, those MAP estimates are:

```
logistic(-1.39) # probability drink                                             R code
exp(0.05)        # rate finish manuscripts, when not drinking                   11.23
```

```
[1] 0.1994078
[1] 1.051271
```

Notice that we can get an accurate estimate of the proportion of days the monks drink, even though we can't say for any particular day whether or not they drank.

This example is the simplest possible. In real problems, you might have predictor variables that are associated one or both processes inside the zero-inflated Poisson mixture. In that case, you just add those variables and their slope parameters to either or both linear models.

Overthinking: Zero-inflated Poisson distribution function. The function dzipois is implemented in a way that guards against some kinds of numerical error. So its code looks confusing—just type "dzipois" at the R prompt and see. But really all it's doing is implementing the likelihood formula defined in the section above. Here's a more transparent version of the function, in which it's much easier to see the mixture definition:

R code
11.24

```
dzip <- function( x , p , lambda , log=TRUE ) {
    ll <- ifelse(
        x==0 ,
        p + (1-p)*exp(-lambda) ,
        (1-p)*dpois(x,lambda,FALSE)
    )
    if ( log==TRUE ) ll <- log(ll)
    return(ll)
}
```

You can replace dzipois with dzip in the map code above for m12.12 and get identical estimates. Why do we need the complexity inside dzipois then? It's nearly always better to do probability calculations on the log scale. Things work okay in this example, even if we don't. But that isn't always the case.

11.3. Over-dispersed outcomes

All statistical models omit something. The question is only whether that something is necessary for making useful inferences. One symptom that something important has been omitted from a count model is OVER-DISPERSION. The variance of a variable is sometimes called its *dispersion*. For a counting process like a binomial, the variance is a function of the same parameters as the expected value. For example, the expected value of a binomial is np and its variance is $np(1 - p)$. When the observed variance exceeds this amount—after conditioning on all the predictor variables—this implies that some omitted variable is producing additional dispersion in the observed counts.

What could go wrong, if we ignore the over-dispersion? Ignoring it can lead to all of the same problems as ignoring any predictor variable. Heterogeneity in counts can be a confound, hiding effects of interest or producing spurious inferences.

So it's worth trying grappling with over-dispersion. The best solution would of course be to discover the omitted source of dispersion and include it in the model. But even when no additional variables are available, it is possible to mitigate the effects of over-dispersion. We'll consider two common and useful strategies.

The first strategy is to use a CONTINUOUS MIXTURE model in which a linear model is attached not to the observations themselves but rather to a distribution of observations. We'll spend the rest of this section outlining this kind of model, using the common beta-binomial and gamma-Poisson (negative-binomial) models of this type. These models were mentioned at the end of the previous chapter, but now we'll actually define them.

The second strategy is to employ multilevel models and estimate both the residuals of each observation and the distribution of those residuals. In practice, it is often easier to use multilevel models (GLMMs, Chapter 12) in place of beta-binomial and gamma-Poisson GLMs. The reason is that multilevel models are often easier to fit and are much more flexible. They can handle over-dispersion and other kinds of heterogeneity at the same time.

Both strategies are useful. So in the remainder of this chapter, you'll meet the commonplace mixtures that address over-dispersion. Then the next chapter introduces multilevel models.

11.3.1. Beta-binomial.

A BETA-BINOMIAL model assumes that each binomial count observation has its own probability of a success.[150] The model estimates the *distribution* of probabilities of success across cases, instead of a single probability of success. And predictor variables change the shape of this distribution, instead of directly determining the probability of each success.

This will be easier to understand in the context of an example. For example, the UCBadmit data that you met last chapter is quite over-dispersed, as long as we ignore department. This is because the departments vary a lot in baseline admission rates. You've already seen that ignoring this variation leads to an incorrect inference about applicant gender. Now let's fit a beta-binomial model, ignoring department, and see how it picks up on the variation that arises from the omitted variable.

What a beta-binomial model of these data will assume is that each observed count on each row of the data table has its own unique, unobserved probability of admission. These probabilities of admission themselves have a common distribution. This distribution is described using a beta distribution, which is a probability distribution for probabilities. Why use a beta distribution? Because it makes the mathematics easy. When we use a beta, it is mathematically possible to solve for a closed form likelihood function that averages over the unknown probabilities for each observation. See the Overthinking box at the end of this section (page 351) for details.

A beta distribution has two parameters, an average probability $\bar{p}$ and a shape parameter θ.[151] The shape parameter θ describes how spread out the distribution is. When $\theta = 2$, every probability from zero to one is equally likely. As θ increases above 2, the distribution of probabilities grows more concentrated. When $\theta < 2$, the distribution is so dispersed that extreme probabilities near zero and one are more likely than the mean. You can play around with the parameters to get a feel for the shapes this distribution can take:

R code
11.25

```
pbar <- 0.5
theta <- 5
curve( dbeta2(x,pbar,theta) , from=0 , to=1 ,
    xlab="probability" , ylab="Density" )
```

Explore different values for pbar and theta in the code above. Remember, this is a distribution for probabilities, so the horizontal axis you'll see represents different possible probability values, and the vertical axis is the density with which each probability on the horizontal is sampled from the distribution. It's weird, but you'll get used to it.

We're going to bind our linear model to $\bar{p}$, so that changes in predictor variables change the central tendency of the distribution. In mathematical form, the model is:

$$A_i \sim \text{BetaBinomial}(n_i, \bar{p}_i, \theta)$$
$$\text{logit}(\bar{p}_i) = \alpha$$
$$\alpha \sim \text{Normal}(0, 10)$$
$$\theta \sim \text{HalfCauchy}(0, 1)$$

where the outcome A is admit, and the size n is applications. This model doesn't use any predictors, but if you want to add them, just tack them onto the linear model on the second line, as usual.

The code below will load the data and then fit, using map2stan, the beta-binomial model:

R code
11.26
```
library(rethinking)
data(UCBadmit)
d <- UCBadmit
m11.5 <- map2stan(
    alist(
        admit ~ dbetabinom(applications,pbar,theta),
        logit(pbar) <- a,
        a ~ dnorm(0,2),
        theta ~ dexp(1)
    ),
    data=d,
    constraints=list(theta="lower=0"),
    start=list(theta=3),
    iter=4000 , warmup=1000 , chains=2 , cores=2 )
```

The only trick above is to use the constraints list to make sure theta is strictly positive. Also, the scale parameter theta can be rather fickle, so it's better to use dexp instead of dcauchy and to specify a reasonable start value. To see how to fit a similar model with map, inspect the examples for ?dbetabinom.

Let's take a quick look at the posterior means:

R code
11.27
```
precis(m11.5)
```

```
        Mean StdDev lower 0.89 upper 0.89 n_eff Rhat
theta   2.77   0.96       1.26       4.15  3343    1
a      -0.38   0.30      -0.86       0.10  2999    1
```

The parameter a is easy enough to interpret: It is still on the log-odds scale and defines $\bar{p}$ of the beta distribution of probabilities for each row of the data. So the implied average probability of admission, across departments, is:

R code
11.28
```
post <- extract.samples(m11.5)
quantile( logistic(post$a) , c(0.025,0.5,0.975) )
```

```
      2.5%        50%      97.5%
0.2728234  0.3988572  0.5425554
```

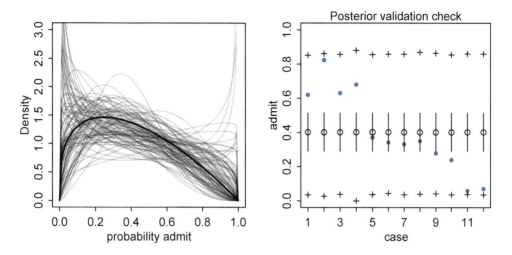

FIGURE 11.5. Left: Posterior distribution of beta distributions for m11.5. The thick curve is the posterior mean beta distribution. The lighter curves represent 100 combinations of $\bar{p}$ and θ sampled from the posterior. Right: Posterior validation check for m11.5. As a result of the widely dispersed beta distributions on the left, the raw data (blue) is contained within the prediction intervals.

The median is about 0.4, but there is a wide percentile interval.

   To really see what the model is saying about the data, we have to account for the correlation between $\bar{p}$ and θ. Together, these two parameters define a distribution of distributions. Let's plot this monster, to help in interpretation.

```
post <- extract.samples(m11.5)

# draw posterior mean beta distribution
curve( dbeta2(x,mean(logistic(post$a)),mean(post$theta)) , from=0 , to=1 ,
    ylab="Density" , xlab="probability admit", ylim=c(0,3) , lwd=2 )

# draw 100 beta distributions sampled from posterior
for ( i in 1:100 ) {
    p <- logistic( post$a[i] )
    theta <- post$theta[i]
    curve( dbeta2(x,p,theta) , add=TRUE , col=col.alpha("black",0.2) )
}
```

R code 11.29

The result is shown on the left in FIGURE 11.5. Remember that a posterior distribution simultaneously scores the plausibility of every combination of parameter values. This plot shows 100 combinations of $\bar{p}$ and θ, sampled from the posterior. The thick curve is the beta distribution corresponding to the posterior mean. The central tendency is for low probabilities of admission, less than 0.5. But the most plausible distributions allow for departments that admit most applicants.

To get a sense of how the beta distribution of probabilities of admission influences predicted counts of applications admitted, let's look at the posterior validation check:

```
postcheck(m11.5)
```

This plot is shown on the right in FIGURE 11.5. The vertical axis shows the predicted proportion admitted, for each case on the horizontal. The blue points show the empirical proportion admitted on each row of the data. The open circles are the posterior mean $\bar{p}$, with 89%
percentile interval, and the + symbols mark the 89% interval of predicted counts of admission. There is a lot of dispersion expected here. The model can't see departments, because
we didn't tell it about them. But it does see heterogeneity across rows, and it uses the beta
distribution to estimate and anticipate that heterogeneity.

While I've used a half-Cauchy prior for the scale parameter θ in this example, it is often
the case that θ is poorly identified. This is because many values of θ will produce very similar
predictions. For this reason, the half-Cauchy prior is not always efficient, but an ordinary exponential prior with appropriate scale will produce much better sampling, without changing
inference.

11.3.2. Negative-binomial or gamma-Poisson. A NEGATIVE-BINOMIAL model, more use-
fully called a GAMMA-POISSON model, assumes that each Poisson count observation has its
own rate.[152] It estimates the shape of a gamma distribution to describe the Poisson rates
across cases. Predictor variables adjust the shape of this distribution, not the expected value
of each observation. The gamma-Poisson model is very much like a beta-binomial model,
with the gamma distribution of rates (or expected values) replacing the beta distribution of
probabilities of success. Why gamma? Because it makes the mathematics easy.

Fitting a gamma-Poisson model uses the `dgampois` function. The conventional construction defines a gamma distribution described by its mean μ and scale θ. Like the beta distribution, as θ increases, the gamma distribution becomes more dispersed around its mean.
Explore the shapes this distribution can take by changing the `mu` and `theta` values in the
code below.

```
mu <- 3
theta <- 1
curve( dgamma2(x,mu,theta) , from=0 , to=10 )
```

As `theta` approaches zero, the gamma distribution approaches a Gaussian distribution with
the same mean value.

A linear model of the mean can be attached to μ by using a log link function. We won't
work an example here. But see the examples for `?dgampois` for working model fitting code.
And there are practice problems at the end of this chapter that employ gamma-Poisson models.

11.3.3. Over-dispersion, entropy, and information criteria. Both the beta-binomial and
gamma-Poisson models are maximum entropy for the same constraints as the regular binomial and Poisson. They just try to account for unobserved heterogeneity in probabilities

and rates. So while they can be a lot harder to fit to data, they can be usefully conceptualized much like ordinary binomial and Poisson GLMs. So in terms of model comparison using information criteria, a beta-binomial model is a binomial model, and a gamma-Poisson (negative-binomial) is a Poisson model.

You should not use WAIC with these models, however, unless you are very sure of what you are doing. The reason is that while ordinary binomial and Poisson models can be aggregated and disaggregated across rows in the data, without changing any causal assumptions, the same is not true of beta-binomial and gamma-Poisson models. The reason is that a beta-binomial or gamma-Poisson likelihood applies an unobserved parameter to each row in the data. When we then go to calculate log-likelihoods, how the data are structured will determine how the beta-distributed or gamma-distributed variation enters the model.

For example, a beta-binomial model like the one examined earlier in this chapter has counts on each row. The rows were departments in that case, and all of the applications for each department were assumed to have the same unknown baseline probability of acceptance. What we'd like to do is treat each application as an observation, calculating WAIC over applications. But if we do that, then we lose the fact that the beta-binomial model implies the same latent probability for all of the applicants from the same row in the data. This is a huge bother.

What to do? In most cases, you'll want to fall back on DIC, which doesn't force a decomposition of the log-likelihood. Once you see how to incorporate over-dispersion with multilevel models, in the next chapter, this obstacle will be reduced. Why? Because a multilevel model can assign heterogeneity in probabilities or rates at any level of aggregation.

Overthinking: Continuous mixtures. A distribution like the beta-binomial is called a *continuous mixture*, because every binomial count is assumed to have its own independent beta-distributed probability of success, and the beta distribution is continuous rather than discrete. So the parameters of the beta-binomial are just the number of draws in each case (the same as the "size" n of the ordinary binomial distribution) and the two parameters that describe the shape of the beta distribution. All of this implies that the probability of observing a number of successes y from a beta-binomial process is:

$$f(y|n, \alpha, \beta) = \int_0^1 g(y|n, p)h(p|\bar{p}, \theta)dp$$

where f is the beta-binomial density, g is the binomial distribution, and h is the beta density. The integral above, like most integrals in applied probability, just computes an average: the probability of y, averaged over all values of p. The p values are drawn from the beta distribution with mean $\bar{p}$ and scale θ. The probability of a success p is no longer a free parameter, as it is produced by the beta distribution. The gamma-Poisson density has a similar form, but averaging a Poisson probability over a gamma distribution of rates.

In the case of the beta-binomial, as well as the gamma-Poisson, it is possible to close the integral above. You can look up the closed-form expressions anytime you need the analytic forms. The R functions `dbetabinom` and `dgampois` provide computations from them.

11.4. Summary

This chapter introduced several new types of regression, all of which are generalizations of generalized linear models (GLMs). Ordered logistic models are useful for categorical outcomes with a strict ordering. They are built by attaching a cumulative link function to

a categorical outcome distribution. Zero-inflated models mix together two different outcome distributions, allowing us to model outcomes with an excess of zeros. Models for over-
dispersion, such as beta-binomial and gamma-Poisson, draw the expected value of each observation from a distribution that changes shape as a function of a linear model. The next
chapter further generalizes these model types by introducing multilevel models.

11.5. Practice

Easy.

11E1. What is the difference between an *ordered* categorical variable and an unordered one? Define
and then give an example of each.

11E2. What kind of link function does an ordered logistic regression employ? How does it differ
from an ordinary logit link?

11E3. When count data are zero-inflated, using a model that ignores zero-inflation will tend to induce which kind of inferential error?

11E4. Over-dispersion is common in count data. Give an example of a natural process that might
produce over-dispersed counts. Can you also give an example of a process that might produce *under*-
dispersed counts?

Medium.

11M1. At a certain university, employees are annually rated from 1 to 4 on their productivity, with
1 being least productive and 4 most productive. In a certain department at this certain university
in a certain year, the numbers of employees receiving each rating were (from 1 to 4): 12, 36, 7, 41.
Compute the log cumulative odds of each rating.

11M2. Make a version of FIGURE 11.2 for the employee ratings data given just above.

11M3. Can you modify the derivation of the zero-inflated Poisson distribution (ZIPoisson) from
the chapter to construct a zero-inflated binomial distribution?

Hard.

11H1. In 2014, a paper was published that was entitled "Female hurricanes are deadlier than male
hurricanes."[153] As the title suggests, the paper claimed that hurricanes with female names have caused
greater loss of life, and the explanation given is that people unconsciously rate female hurricanes as
less dangerous and so are less likely to evacuate.
    Statisticians severely criticized the paper after publication. Here, you'll explore the complete data
used in the paper and consider the hypothesis that hurricanes with female names are deadlier. Load
the data with:

R code
11.32
```
library(rethinking)
data(Hurricanes)
```

Acquaint yourself with the columns by inspecting the help ?Hurricanes.
    In this problem, you'll focus on predicting deaths using femininity of each hurricane's name.
Fit and interpret the simplest possible model, a Poisson model of deaths using femininity as a
predictor. You can use map or map2stan. Compare the model to an intercept-only Poisson model of
deaths. How strong is the association between femininity of name and deaths? Which storms does
the model fit (retrodict) well? Which storms does it fit poorly?

11H2. Counts are nearly always over-dispersed relative to Poisson. So fit a gamma-Poisson (aka negative-binomial) model to predict `deaths` using `femininity`. Show that the over-dispersed model no longer shows as precise a positive association between femininity and deaths, with an 89% interval that overlaps zero. Can you explain why the association diminished in strength?

11H3. In order to infer a strong association between deaths and femininity, it's necessary to include an interaction effect. In the data, there are two measures of a hurricane's potential to cause death: `damage_norm` and `min_pressure`. Consult `?Hurricanes` for their meanings. It makes some sense to imagine that femininity of a name matters more when the hurricane is itself deadly. This implies an interaction between `femininity` and either or both of `damage_norm` and `min_pressure`.

Fit a series of models evaluating these interactions. Interpret and compare the models. In interpreting the estimates, it may help to generate counterfactual predictions contrasting hurricanes with masculine and feminine names. Are the effect sizes plausible?

11H4. In the original hurricanes paper, storm damage (`damage_norm`) was used directly. This assumption implies that mortality increases exponentially with a linear increase in storm strength, because a Poisson regression uses a log link. So it's worth exploring an alternative hypothesis: that the logarithm of storm strength is what matters. Explore this by using the logarithm of `damage_norm` as a predictor. Using the best model structure from the previous problem, compare a model that uses `log(damage_norm)` to a model that uses `damage_norm` directly. Compare their DIC/WAIC values as well as their implied predictions. What do you conclude?

11H5. One hypothesis from developmental psychology, usually attributed to Carol Gilligan, proposes that women and men have different average tendencies in moral reasoning. Like most hypotheses in social psychology, it is merely descriptive. The notion is that women are more concerned with care (avoiding harm), while men are more concerned with justice and rights. Culture-bound nonsense? Yes. Descriptively accurate? Maybe.

Evaluate this hypothesis, using the `Trolley` data, supposing that `contact` provides a proxy for physical harm. Are women more or less bothered by contact than are men, in these data? Figure out the model(s) that is needed to address this question.

11H6. The data in `data(Fish)` are records of visits to a national park. See `?Fish` for details. The question of interest is how many fish an average visitor takes per hour, when fishing. The problem is that not everyone tried to fish, so the `fish_caught` numbers are zero-inflated. As with the monks example in the chapter, there is a process that determines who is fishing (working) and another process that determines fish per hour (manuscripts per day), conditional on fishing (working). We want to model both. Otherwise we'll end up with an underestimate of rate of fish extraction from the park.

You will model these data using zero-inflated Poisson GLMs. Predict `fish_caught` as a function of any of the other variables you think are relevant. One thing you must do, however, is use a proper Poisson offset/exposure in the Poisson portion of the zero-inflated model. Then use the `hours` variable to construct the offset. This will adjust the model for the differing amount of time individuals spent in the park.

12 Multilevel Models

In the year 1985, Clive Wearing lost his mind, but not his music.[154] Wearing was a musicologist and accomplished musician, but the same virus that causes cold sores, *Herpes simplex*, snuck into his brain and ate his hippocampus. The result was chronic anterograde amnesia—he cannot form new long-term memories. He remembers how to play the piano, though he cannot remember that he played it 5 minutes ago. Wearing now lives moment to moment, unaware of anything more than a few minutes into the past. Every cup of coffee is the first he has ever had.

Many statistical models also have anterograde amnesia. As the models move from one cluster—individual, group, location—in the data to another, estimating parameters for each cluster, they forget everything about the previous clusters. They behave this way, because the assumptions force them to. Any of the models from previous chapters that used dummy variables (page 152) to handle categories are programmed for amnesia. These models implicitly assume that nothing learned about any one category informs estimates for the other categories—the parameters are independent of one another and learn from completely separate portions of the data. This would be like forgetting you had ever been in a café, each time you go to a new café. Cafés do differ, but they are also alike.

Anterograde amnesia is bad for learning about the world. We want models that instead use all of the information in savvy ways. This does not mean treating all clusters as if they were the same. Instead it means learning simultaneously about each cluster while learning about the population of clusters. Doing both estimation tasks at the same time allows us to transfer information across clusters, and that transfer improves accuracy. That is the value of remembering.

Consider cafés again. Suppose we program a robot to visit two cafés, order coffee, and estimate the waiting times at each. The robot begins with a vague prior for the waiting times, say with a mean of 5 minutes and a standard deviation of 1. After ordering a cup of coffee at the first café, the robot observes a waiting time of 4 minutes. It updates its prior, using Bayes' theorem of course, with this information. This gives it a posterior distribution for the waiting time at the first café.

Now the robot moves on to a second café. When this robot arrives at the next café, what is its prior? It could just use the posterior distribution from the first café as its prior for the second café. But that implicitly assumes that the two cafés have the same average waiting time. Cafés are all pretty much the same, but they aren't identical. Likewise, it doesn't make much sense to ignore the observation from the first café. That would be anterograde amnesia.

So how can the coffee robot do better? It needs to represent the population of cafés and learn about that population. The distribution of waiting times in the population becomes the prior for each café. But unlike priors in previous chapters, this prior is actually learned from the data. This means the robot tracks a parameter for each café as well as at least two parameters to describe the population of cafés: an average and a standard deviation. As the robot observes waiting times, it updates everything: the estimates for each café as well as the estimates for the population. If the population seems highly variable, then the prior is flat and uninformative and, as a consequence, the observations at any one café do very little to the estimate at another. If instead the population seems to contain little variation, then the prior is narrow and highly informative. An observation at any one café will have a big impact on estimates at any other café.

In this chapter, you'll see the formal version of this argument and how it leads us to MUL-TILEVEL MODELS. These models remember features of each cluster in the data as they learn about all of the clusters. Depending upon the variation among clusters, which is learned from the data as well, the model pools information across clusters. This pooling tends to improve estimates about each cluster. This improved estimation leads to several, more pragmatic sounding, benefits of the multilevel approach. I mentioned them in Chapter 1. They are worth repeating.

(1) *Improved estimates for repeat sampling.* When more than one observation arises from the same individual, location, or time, then traditional, single-level models either maximally underfit or overfit the data.

(2) *Improved estimates for imbalance in sampling.* When some individuals, locations, or times are sampled more than others, multilevel models automatically cope with differing uncertainty across these clusters. This prevents over-sampled clusters from unfairly dominating inference.

(3) *Estimates of variation.* If our research questions include variation among individuals or other groups within the data, then multilevel models are a big help, because they model variation explicitly.

(4) *Avoid averaging, retain variation.* Frequently, scholars pre-average some data to construct variables. This can be dangerous, because averaging removes variation, and there are also typically several different ways to perform the averaging. Averaging therefore both manufactures false confidence and introduces arbitrary data transformations. Multilevel models allow us to preserve the uncertainty and avoid data transformations.

All of these benefits flow out of the same strategy and model structure. You learn one basic design and you get all of this for free.

When it comes to regression, multilevel regression deserves to be the default approach. There are certainly contexts in which it would be better to use an old-fashioned single-level model. But the contexts in which multilevel models are superior are much more numerous. It is better to begin to build a multilevel analysis, and then realize it's unnecessary, than to overlook it. And once you grasp the basic multilevel strategy, it becomes much easier to incorporate related tricks such as allowing for measurement error in the data and even modeling missing data itself (Chapter 14).

There are costs of the multilevel approach. The first is that we have to make some new assumptions. We have to define the distributions from which the characteristics of the clusters arise. Luckily, conservative maximum entropy distributions do an excellent job in this context. Second, there are new estimation challenges that come with the full multilevel approach. These challenges lead us headfirst into MCMC estimation. Third, multilevel models can be hard to understand, because they make predictions at different levels of the data. In many cases, we are interested in only one or a few of those levels, and as a consequence, model comparison using metrics like DIC and WAIC becomes more subtle. The basic logic remains unchanged, but now we have to make more decisions about which parameters in the model we wish to focus on.

This chapter has the following progression. First, we'll work through an extended example of building and fitting a multilevel model for clustered data. Then we'll simulate clustered data, to demonstrate the improved accuracy the approach delivers. This improved accuracy arises from the same underfitting and overfitting trade-off you met in Chapter 6. Then we'll finish by looking at contexts in which there is more than one type of clustering in the data. All of this work lays a foundation for more advanced multilevel examples in the next two chapters.

> **Rethinking: A model by any other name.** Multilevel models go by many different names, and some statisticians use the same names for different specialized variants, while others use them all interchangeably. The most common synonyms for "multilevel" are HIERARCHICAL and MIXED EFFECTS. The type of parameters that appear in multilevel models are most commonly known as RANDOM EFFECTS, which itself can mean very different things to different analysts and in different contexts.[155] And even the innocent term "level" can mean different things to different people. There's really no cure for this swamp of vocabulary aside from demanding a mathematical or algorithmic definition of the model. Otherwise, there will always be ambiguity.

12.1. Example: Multilevel tadpoles

The heartwarming focus of this example are experiments exploring Reed frog (*Hyperolius spinigularis*) tadpole mortality.[156] The natural history background to these data is very interesting. Take a look at the full paper, if amphibian life history dynamics interests you. But even if it doesn't, load the data and acquaint yourself with the variables:

```
library(rethinking)
data(reedfrogs)
d <- reedfrogs
str(d)
```

R code 12.1

```
'data.frame':	48 obs. of  5 variables:
 $ density : int  10 10 10 10 10 10 10 10 10 10 ...
 $ pred    : Factor w/ 2 levels "no","pred": 1 1 1 1 1 1 1 1 2 2 ...
 $ size    : Factor w/ 2 levels "big","small": 1 1 1 1 2 2 2 2 1 1 ...
 $ surv    : int  9 10 7 10 9 9 10 9 4 9 ...
 $ propsurv: num  0.9 1 0.7 1 0.9 0.9 1 0.9 0.4 0.9 ...
```

For now, we'll only be interested in number surviving, surv, out of an initial count, density. In the practice at the end of the chapter, you'll consider the other variables, which are experimental manipulations.

There is a lot of variation in these data. Some of the variation comes from experimental treatment. But a lot of it comes from other sources. Think of each row as a "tank," an experimental environment that contains tadpoles. There are lots of things peculiar to each tank that go unmeasured, and these unmeasured factors create variation in survival across tanks, even when all the predictor variables have the same value. These tanks are an example of a *cluster* variable. Multiple observations, the tadpoles in this case, are made within each cluster.

So we have repeat measures and heterogeneity across clusters. If we ignore the clusters, assigning the same intercept to each of them, then we risk ignoring important variation in baseline survival. This variation could mask association with other variables. If we instead estimate a unique intercept for each cluster, using a dummy variable for each tank, we instead practice anterograde amnesia. After all, tanks are different but each tank does help us estimate survival in the other tanks. So it doesn't make sense to forget entirely, moving from one tank to another.

A multilevel model, in which we simultaneously estimate both an intercept for each tank and the variation among tanks, is what we want. This will be a **VARYING INTERCEPTS** model. Varying intercepts are the simplest kind of **VARYING EFFECTS**.[157] For each cluster in the data, we use a unique intercept parameter. This is no different than the categorical variable examples from previous chapters, except now we also adaptively learn the prior that is common to all of these intercepts. This adaptive learning is the absence of amnesia discussed at the start of the chapter. When what we learn about each cluster informs all the other clusters, we learn the prior simultaneous to learning the intercepts.

Here is a model for predicting tadpole mortality in each tank, using the regularizing priors of earlier chapters:

$$s_i \sim \text{Binomial}(n_i, p_i) \qquad \text{[likelihood]}$$
$$\text{logit}(p_i) = \alpha_{\text{TANK}[i]} \qquad \text{[unique log-odds for each tank } i]$$
$$\alpha_{\text{TANK}} \sim \text{Normal}(0, 5) \qquad \text{[weakly regularizing prior]}$$

And you can fit this to the data in the standard way, using `map` or `map2stan`. We'll use `map2stan` from here onwards, because the next model will not work in `map`.

R code
12.2

```
library(rethinking)
data(reedfrogs)
d <- reedfrogs

# make the tank cluster variable
d$tank <- 1:nrow(d)

# fit
m12.1 <- map2stan(
    alist(
        surv ~ dbinom( density , p ) ,
        logit(p) <- a_tank[tank] ,
        a_tank[tank] ~ dnorm( 0 , 5 )
    ),
    data=d )
```

If you inspect the estimates, `precis(m12.1,depth=2)`, you'll see 48 different intercept offsets, one for each tank. To get each tank's expected survival probability, just take one of the `a_tank` values and then use the logistic transform. So far there is nothing new here.

Now let's fit the multilevel model, which adaptively pools information across tanks. All that is required to enable adaptive pooling is to make the prior for the `a_tank` parameters a function of its own parameters. Here is the multilevel model, in mathematical form, with the changes from the previous model highlighted in blue:

$$s_i \sim \text{Binomial}(n_i, p_i) \qquad \text{[likelihood]}$$
$$\text{logit}(p_i) = \alpha_{\text{TANK}[i]} \qquad \text{[log-odds for tank on row } i\text{]}$$
$$\alpha_{\text{TANK}} \sim \text{Normal}(\alpha, \sigma) \qquad \text{[varying intercepts prior]}$$
$$\alpha \sim \text{Normal}(0, 1) \qquad \text{[prior for average tank]}$$
$$\sigma \sim \text{HalfCauchy}(0, 1) \qquad \text{[prior for standard deviation of tanks]}$$

Notice that the prior for the α_{TANK} intercepts is now a function of two parameters, α and σ. This is where the "multi" in multilevel arises.[158] The Gaussian distribution with mean α and standard deviation σ is the prior for each tank's intercept. But that prior itself has priors for α and σ. So there are two *levels* in the model, each resembling a simpler model. In the top level, the outcome is s, the parameters are α_{TANK}, and the prior is $\alpha_{\text{TANK}} \sim \text{Normal}(\alpha, \sigma)$. In the second level, the "outcome" variable is the vector of intercept parameters, α_{TANK}. The parameters are α and σ, and their priors are $\alpha \sim \text{Normal}(0, 1)$ and $\sigma \sim \text{HalfCauchy}(0, 1)$. For more explanation of the σ prior, see the Overthinking box on the next page.

These two parameters, α and σ, are often referred to as **HYPERPARAMETERS**. They are parameters for parameters. And their priors are often called **HYPERPRIORS**. In principle, there is no limit to how many "hyper" levels you can install in a model. For example, different populations of tanks could be embedded within different regions of habitat. But in practice there are limits, both because of computation and our ability to understand the model.

Rethinking: Why Gaussian tanks? In the multilevel tadpole model, the population of tanks is assumed to be Gaussian. Why? The least satisfying answer is "convention." The Gaussian assumption is extremely common. A more satisfying answer is "pragmatism." The Gaussian assumption is easy to work with, and it generalizes easily to more than one dimension. This generalization will be important for handling varying slopes in the next chapter. But my preferred answer is instead "entropy." If all we are willing to say about a distribution is the mean and variance, then the Gaussian is the most conservative assumption (Chapter 9). There is no rule requiring the Gaussian distribution of varying effects, though. So if you have a good reason to use another distribution, then do so. The practice problems at the end of the chapter provide an example.

Fitting the model to data estimates both levels simultaneously, in the same way that our robot at the start of the chapter learned both about each café and the variation among cafés. But you cannot fit this model with map. Why? Because the likelihood must now average over the level 2 parameters α and σ. But map just hill climbs, using static values for all of the parameters. It can't see the levels. For more explanation, see the Overthinking box further down. You can however fit this model with map2stan:

R code
12.3

```
m12.2 <- map2stan(
    alist(
        surv ~ dbinom( density , p ) ,
        logit(p) <- a_tank[tank] ,
        a_tank[tank] ~ dnorm( a , sigma ) ,
        a ~ dnorm(0,1) ,
        sigma ~ dcauchy(0,1)
    ), data=d , iter=4000 , chains=4 )
```

This model fit provides estimates for 50 parameters: one overall sample intercept α, the variance among tanks σ, and then 48 per-tank intercepts. Let's check WAIC though to see the effective number of parameters. We'll compare the earlier model, m12.1, with the new multilevel model:

R code
12.4
```
compare( m12.1 , m12.2 )
```

```
          WAIC pWAIC dWAIC weight    SE  dSE
m12.2 1010.2  38.0   0.0      1 37.94   NA
m12.1 1023.3  49.4  13.1      0 43.01 6.54
```

There are two facts to note here. First, the multilevel model has only 38 effective parameters. There are 12 fewer effective parameters than actual parameters, because the prior assigned to each intercept shrinks them all towards the mean α. In this case, the prior is reasonably strong. Check the mean of sigma with precis or coef and you'll see it's around 1.6. This is a REGULARIZING PRIOR, like you've used in previous chapters, but now the amount of regularization has been learned from the data itself.[159] Second, notice that the multilevel model m12.2 has fewer effective parameters than the ordinary fixed model m12.1. This is despite the fact that the ordinary model has fewer actual parameters, only 48 instead of 50. The extra two parameters in the multilevel model allowed it to learn a more aggressive regularizing prior, to adaptively regularize. This resulted in a less flexible posterior and therefore fewer effective parameters.

Overthinking: MAP fails, MCMC wins. Why doesn't MAP estimation, using for example map, work with multilevel models? When a prior is itself a function of parameters, there are two levels of uncertainty. This means that the probability of the data, conditional on the parameters, must average over each level. Ordinary MAP estimation cannot handle the averaging in the likelihood, because in general it's not possible to derive an analytical solution. That means there is no unified function for calculating the log-posterior. So your computer cannot directly find its minimum (the maximum of the posterior).

Some other computational approach is needed. It is possible to extend the mode-finding optimization strategy to these models, but we don't want to be stuck with optimization in general. One reason is that the posterior of these models is routinely non-Gaussian. More generally, as models become more complex, a phenomenon known as *concentration of measure* guarantees that the posterior mode will be far from the posterior median. So we really need to give up optimization as a strategy. One robust solution is MCMC.

To appreciate the impact of this adaptive regularization, let's plot and compare the posterior medians from models m12.1 and m12.2. The code that follows is long, only because it decorates the plot with informative labels. The basic code is just the first part, which extracts samples and computes medians.

R code
12.5
```
# extract Stan samples
post <- extract.samples(m12.2)

# compute median intercept for each tank
# also transform to probability with logistic
d$propsurv.est <- logistic( apply( post$a_tank , 2 , median ) )
```

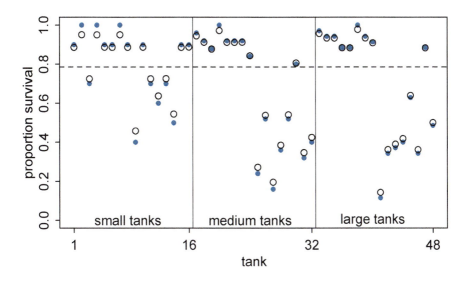

FIGURE 12.1. Empirical proportions of survivors in each tadpole tank, shown by the filled blue points, plotted with the 48 per-tank estimates from the multilevel model, shown by the black circles. The dashed line locates the overall average proportion of survivors across all tanks. The vertical lines divide tanks with different initial densities of tadpoles: small tanks (10 tadpoles), medium tanks (25), and large tanks (35). In every tank, the posterior median from the multilevel model is closer to the dashed line than the empirical proportion is. This reflects the pooling of information across tanks, to help with inference about each tank.

```
# display raw proportions surviving in each tank
plot( d$propsurv , ylim=c(0,1) , pch=16 , xaxt="n" ,
    xlab="tank" , ylab="proportion survival" , col=rangi2 )
axis( 1 , at=c(1,16,32,48) , labels=c(1,16,32,48) )

# overlay posterior medians
points( d$propsurv.est )

# mark posterior median probability across tanks
abline( h=logistic(median(post$a)) , lty=2 )

# draw vertical dividers between tank densities
abline( v=16.5 , lwd=0.5 )
abline( v=32.5 , lwd=0.5 )
text( 8 , 0 , "small tanks" )
text( 16+8 , 0 , "medium tanks" )
text( 32+8 , 0 , "large tanks" )
```

You can see the result in FIGURE 12.1. The horizontal axis is tank index, from 1 to 48. The vertical is proportion of survivors in a tank. The filled blue points show the raw proportions, computed from the observed counts. These values are already present in the data frame, in the propsurv column. The black circles are instead the varying intercept medians. The horizontal dashed line at about 0.8 is the estimated median survival proportion in the population of tanks, α. It is not the same as the empirical mean survival. The vertical gray lines divide tanks with different initial counts of tadpoles—10 (left), 25 (middle), and 35 (right).

First, notice that in every case, the multilevel estimate is closer to the dashed line than the raw empirical estimate is. It's as if the entire distribution of black circles has been shrunk towards the dashed line at the center of the data, leaving the blue points behind on the outside. This phenomenon is sometimes called SHRINKAGE, and it results from regularization (as in Chapter 6). Second, notice that the estimates for the smaller tanks have shrunk farther from the blue points. As you move from left to right in the figure, the initial densities of tadpoles increase from 10 to 25 to 35, as indicated by the vertical dividers. In the smallest tanks, it is easy to see differences between the open estimates and empirical blue points. But in the largest tanks, there is little difference between the blue points and open circles. Varying intercepts for the smaller tanks, with smaller sample sizes, shrink more. Third, note that the farther a blue point is from the dashed line, the greater the distance between it and the corresponding multilevel estimate. Shrinkage is stronger, the further a tank's empirical proportion is from the global average α.

All three of these phenomena arise from a common cause: pooling information across clusters (tanks) to improve estimates. What POOLING means here is that each tank provides information that can be used to improve the estimates for all of the other tanks. Each tank helps in this way, because we made an assumption about how the varying log-odds in each tank related to all of the others. We assumed a distribution, the normal distribution in this case. Once we have a distributional assumption, we can use Bayes' theorem to optimally (in the small world only) share information among the clusters.

What does the inferred population distribution of survival look like? We can visualize it by sampling from the posterior distribution, as usual. First we'll plot 100 Gaussian distributions, one for each of the first 100 samples from the posterior distribution of both α and σ. Then we'll sample 8000 new log-odds of survival for individual tanks. The result will be a posterior distribution of variation in survival in the population of tanks. Before we do the sampling though, remember that "sampling" from a posterior distribution is not a simulation of empirical sampling. It's just a convenient way to characterize and work with the uncertainty in the distribution. Now the sampling:

R code
12.6

```
# show first 100 populations in the posterior
plot( NULL , xlim=c(-3,4) , ylim=c(0,0.35) ,
    xlab="log-odds survive" , ylab="Density" )
for ( i in 1:100 )
    curve( dnorm(x,post$a[i],post$sigma[i]) , add=TRUE ,
    col=col.alpha("black",0.2) )

# sample 8000 imaginary tanks from the posterior distribution
sim_tanks <- rnorm( 8000 , post$a , post$sigma )

# transform to probability and visualize
```

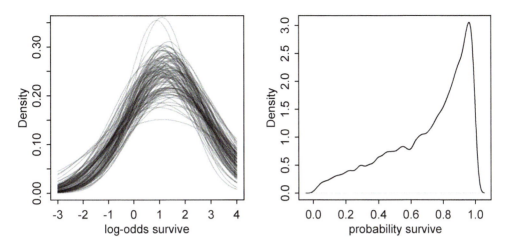

FIGURE 12.2. The inferred population of survival across tanks. Left: 100 Gaussian distributions of the log-odds of survival, sampled from the posterior of m12.2. Right: Survival probabilities for 8000 new simulated tanks, averaging over the posterior distribution on the left.

```
dens( logistic(sim_tanks) , xlab="probability survive" )
```

The results are displayed in FIGURE 12.2. Notice that there is uncertainty about both the location, α, and scale, σ, of the population distribution of log-odds of survival. All of this uncertainty is propagated into the simulated probabilities of survival.

Rethinking: Varying intercepts as over-dispersion. In the previous chapter (page 346), the beta-binomial and gamma-Poisson models were presented as ways for coping with OVER-DISPERSION of count data. Varying intercepts accomplish the same thing, allowing count outcomes to be over-dispersed. They accomplish this, because when each observed count gets its own unique intercept, but these intercepts are pooled through a common distribution, the predictions expect over-dispersion just like a beta-binomial or gamma-Poisson model would. Compared to a beta-binomial or gamma-Poisson model, a binomial or Poisson model with a varying intercept on every observed outcome will often be easier to estimate and easier to extend. There will be an example of this approach, later in this chapter.

Overthinking: Priors for variance components. The examples in this book use weakly regularizing half-Cauchy priors for variance components, the σ parameters that estimate the variation across clusters in the data. These Cauchy priors work very well in routine multilevel modeling. But there are two common contexts in which they can be problematic. First, sometimes there isn't much information in the data with which to estimate the variance. For example, if you only have 5 clusters, then that's something like trying to estimate a variance with 5 data points. Second, in non-linear models with logit and log links, floor and ceiling effects sometimes render extreme values of the variance equally plausible as more realistic values. In such cases, the trace plot for the variance parameters may swing around over very large values. It can do this, because the Cauchy prior has a very thick and long tail, extending into very large values. Such large values are typically *a priori* impossible. Often, the chain will still sample validly, but it might be highly inefficient, exhibiting small n_eff values and possibly many divergent iterations.

To improve such a model, instead of using half-Cauchy priors for the variance components, you can use exponential priors. For example:

$$s_i \sim \text{Binomial}(n_i, p_i)$$
$$\text{logit}(p_i) = \alpha_{\text{TANK}[i]}$$
$$\alpha_{\text{TANK}} \sim \text{Normal}(\alpha, \sigma)$$
$$\alpha \sim \text{Normal}(0, 1)$$
$$\sigma \sim \text{Exponential}(1)$$

The exponential prior—dexp(1) in R code—has a much thinner tail than the Cauchy does. This induces more conservatism in estimates and can help your Markov chain converge correctly. The exponential is also the maximum entropy prior for the standard deviation, provided all we want to say *a priori* is the expected value. That is to say that the only information contained in an exponential prior is the mean value and the positive constraint.

Again, using the exponential instead of the Cauchy isn't usually necessary. But there are cases, especially with non-linear models with ceiling or floor effects, in which the variance components can be only weakly identified. In those cases, you are going to have to add more strongly regularizing priors in order to make any inference at all. And of course, it is typically useful to try different priors to ensure that inference either is insensitive to them or rather to measure how inference is altered.

12.2. Varying effects and the underfitting/overfitting trade-off

Varying intercepts are just regularized estimates, but adaptively regularized by estimating how diverse the clusters are while estimating the features of each cluster. This fact is not easy to grasp, so if it still seems mysterious, this section aims to further relate the properties of multilevel estimates to the foundational underfitting/overfitting dilemma from Chapter 6.

A major benefit of using varying effects estimates, instead of the empirical raw estimates, is that they provide more accurate estimates of the individual cluster (tank) intercepts.[160] On average, the varying effects actually provide a better estimate of the individual tank (cluster) means. The reason that the varying intercepts provide better estimates is that they do a better job of trading off underfitting and overfitting.

To understand this in the context of the reed frog example, suppose that instead of experimental tanks we had natural ponds, so that we might be concerned with making predictions for the same clusters in the future. We'll approach the problem of predicting future survival in these ponds, from three perspectives:

(1) Complete pooling. This means we assume that the population of ponds is invariant, the same as estimating a common intercept for all ponds.
(2) No pooling. This means we assume that each pond tells us nothing about any other pond. This is the model with amnesia.
(3) Partial pooling. This means using an adaptive regularizing prior, as in the previous section.

First, suppose you ignore the varying intercepts and just use the overall mean across all ponds, α, to make your predictions for each pond. A lot of data contributes to your estimate of α, and so it can be quite precise. However, your estimate of α is unlikely to exactly match the mean of any particular pond. As a result, the total sample mean underfits the data. This is the COMPLETE POOLING approach, pooling the data from all ponds to produce a single estimate that is applied to every pond. This sort of model is equivalent to assuming that the variation among ponds is zero—all ponds are identical.

Second, suppose you use the survival proportions for each pond to make predictions. This means using a separate intercept for each pond. The blue points in FIGURE 12.1 are this same kind of estimate. In each particular pond, quite little data contributes to each estimate, and so these estimates are rather imprecise. This is particularly true of the smaller ponds, where less data goes into producing the estimates. As a consequence, the error of these estimates is high, and they are rather overfit to the data. Standard errors for each intercept can be very large, and in extreme cases, even infinite. These are sometimes called the NO POOL-ING estimates. No information is shared across ponds. It's like assuming that the variation among ponds is infinite, so nothing you learn from one pond helps you predict another.

Third, when you estimate varying intercepts, you use PARTIAL POOLING of information to produce estimates for each cluster that are less underfit than the grand mean and less overfit than the no-pooling estimates. As a consequence, they tend to be better estimates of the true per-cluster (per-pond) means. This will be especially true when ponds have few tadpoles in them, because then the no pooling estimates will be especially overfit. When a lot of data goes into each pond, then there will be less difference between the varying effect estimates and the no pooling estimates.

To demonstrate this fact, we'll simulate some tadpole data. That way, we'll know the true per-pond survival probabilities. Then we can compare the no-pooling estimates to the partial pooling estimates, by computing how close each gets to the true values they are trying to estimate. The rest of this section shows how to do such a simulation.

Learning to simulate and validate models and model fitting in this way is extremely valuable. Once you start using more complex models, you will want to ensure that your code is working and that you understand the model. You can help in this project by simulating data from the model, with specified parameter values, and then making sure that your method of estimation can recover the parameters within tolerable ranges of precision. Even just simulating data from a model structure has a huge impact on understanding.

12.2.1. The model. The first step is to define the model we'll be using. I'll use the same basic multilevel binomial model as before, but now with "ponds" instead of "tanks":

$$s_i \sim \text{Binomial}(n_i, p_i)$$
$$\text{logit}(p_i) = \alpha_{\text{POND}[i]}$$
$$\alpha_{\text{POND}} \sim \text{Normal}(\alpha, \sigma)$$
$$\alpha \sim \text{Normal}(0, 1)$$
$$\sigma \sim \text{HalfCauchy}(0, 1)$$

So to simulate data from this process, we need to assign values to:

- α, the average log-odds of survival in the entire population of ponds
- σ, the standard deviation of the distribution of log-odds of survival among ponds
- α_{POND}, a vector of individual pond intercepts, one for each pond

We'll also need to assign sample sizes, n_i, to each pond. But once we've made all of those choices, we can easily simulate counts of surviving tadpoles, straight from the top-level binomial process, using rbinom. We'll do it all one step at a time.

Note that the priors are part of the model when we estimate, but not when we simulate. Why? Because priors are epistemology, not ontology. They represent the initial state of information of our robot, not a statement about how nature chooses parameter values.

12.2.2. Assign values to the parameters. I'm going to assign specific values representative of the actual tadpole data, to make the upcoming plot that demonstrates the increased accuracy of the varying effects estimates. But you can come back to this step later and change them to whatever you want.

Here's the code to initialize the values of α, σ, the number of ponds, and the sample size n_i in each pond.

R code
12.7
```
a <- 1.4
sigma <- 1.5
nponds <- 60
ni <- as.integer( rep( c(5,10,25,35) , each=15 ) )
```

I've chosen 60 ponds, with 15 each of initial tadpole density 5, 10, 25, and 35. I've chosen these densities to illustrate how the error in prediction varies with sample size. The use of `as.integer` in the last line arises from a subtle issue with how Stan, and therefore `map2stan`, works. See the Overthinking box at the bottom of the page for an explanation.

The values $\alpha = 1.4$ and $\sigma = 1.5$ define a Gaussian distribution of individual pond log-odds of survival. So now we need to simulate all 60 of these intercept values from the implied Gaussian distribution with mean α and standard deviation σ:

R code
12.8
```
a_pond <- rnorm( nponds , mean=a , sd=sigma )
```

Go ahead and inspect the contents of `a_pond`. It should contain 60 log-odds values, one for each simulated pond.

Finally, let's bundle some of this information in a data frame, just to keep it organized.

R code
12.9
```
dsim <- data.frame( pond=1:nponds , ni=ni , true_a=a_pond )
```

Go ahead and inspect the contents of `dsim`, the simulated data. The first column is the pond index, 1 through 60. The second column is the initial tadpole count in each pond. The third column is the true log-odds survival for each pond.

Overthinking: Data types and Stan models. There are two basic types of numerical data in R, integers and real values. A number like "3" could be either. Inside your computer, integers and real ("numeric") values are represented differently. For example, here is the same vector of values generated as both:

R code
12.10
```
class(1:3)
class(c(1,2,3))
```

```
[1] "integer"
[1] "numeric"
```

Usually, you don't have to manage these types, because R manages them for you. But when you pass values to Stan, or another external program, often the internal representation does matter. In particular, Stan and `map2stan` sometimes require explicit integers. For example, in a binomial model, the "size" variable that specifies the number of trials must be of integer type. Stan may provide a mysterious warning message about a function not being found, when the size variable is instead of "real" type, or what R calls `numeric`. Using `as.integer` before passing the data to Stan or `map2stan` will resolve the issue.

12.2.3. Simulate survivors. Now we're ready to simulate the binomial survival process. Each pond i has n_i potential survivors, and nature flips each tadpole's coin, so to speak, with probability of survival p_i. This probability p_i is implied by the model definition, and is equal to:

$$p_i = \frac{\exp(\alpha_i)}{1 + \exp(\alpha_i)}$$

The model uses a logit link, and so the probability is defined by the logistic function.

Putting the logistic into the random binomial function, we can generate a simulated survivor count for each pond:

```
dsim$si <- rbinom( nponds , prob=logistic(dsim$true_a) , size=dsim$ni )
```
R code
12.11

As usual with R, if you give it a list of values, it returns a new list of the same length. In the above, each paired α_i (`dsim$true_a`) and n_i (`dsim$ni`) is used to generate a random survivor count with the appropriate probability of survival and maximum count. These counts are stored in a new column in `dsim`.

12.2.4. Compute the no-pooling estimates. We're ready to start analyzing the simulated data now. The easiest task is to just compute the no-pooling estimates. We can accomplish this straight from the empirical data, just by calculating the proportion of survivors in each pond. I'll keep these estimates on the probability scale, instead of translating them to the log-odds scale, because we'll want to compare the quality of the estimates on the probability scale later.

```
dsim$p_nopool <- dsim$si / dsim$ni
```
R code
12.12

Now there's another column in `dsim`, containing the empirical proportions of survivors in each pond. These are the same no-pooling estimates you'd get by fitting a model with a dummy variable for each pond and flat priors that induce no regularization.

12.2.5. Compute the partial-pooling estimates. Now to fit the model to the simulated data, using `map2stan`. I'll use a single long chain in this example, but keep in mind that you need to use multiple chains to check convergence to the right posterior distribution. In this case, it's safe. But don't get cocky.

```
m12.3 <- map2stan(
    alist(
        si ~ dbinom( ni , p ),
        logit(p) <- a_pond[pond],
        a_pond[pond] ~ dnorm( a , sigma ),
        a ~ dnorm(0,1),
        sigma ~ dcauchy(0,1)
    ),
```
R code
12.13

```
     data=dsim , iter=1e4 , warmup=1000 )
```

We've fit the basic varying intercept model above. You can take a look at the estimates for α and σ with the usual precis approach:

R code
12.14

```
precis(m12.3,depth=2)
```

```
              Mean StdDev lower 0.89 upper 0.89 n_eff Rhat
a_pond[1]     1.45   0.95      -0.11       2.89  9000    1
a_pond[2]     1.47   0.95      -0.02       2.96  9000    1
...
a_pond[59]    1.81   0.47       1.02       2.52  7314    1
a_pond[60]    2.03   0.50       1.24       2.82  9000    1
a             1.13   0.23       0.78       1.50  5848    1
sigma         1.59   0.22       1.25       1.93  2705    1
```

I've abbreviated the output, since there are 60 intercept parameters, one for each pond.

Now that we've found these estimates, let's compute the predicted survival proportions and add those proportions to our growing simulation data frame. To indicate that it contains the partial pooling estimates, I'll call the column p.partpool.

R code
12.15

```
estimated.a_pond <- as.numeric( coef(m12.3)[1:60] )
dsim$p_partpool <- logistic( estimated.a_pond )
```

If we want to compare to the true per-pond survival probabilities used to generate the data, then we'll also need to compute those, using the true_a column:

R code
12.16

```
dsim$p_true <- logistic( dsim$true_a )
```

The last thing we need to do, before we can plot the results and realize the point of this lesson, is to compute the absolute error between the estimates and the true varying effects. This is easy enough, using the existing columns:

R code
12.17

```
nopool_error <- abs( dsim$p_nopool - dsim$p_true )
partpool_error <- abs( dsim$p_partpool - dsim$p_true )
```

Now we're ready to plot. This is enough to get the basic display:

R code
12.18

```
plot( 1:60 , nopool_error , xlab="pond" , ylab="absolute error" ,
    col=rangi2 , pch=16 )
points( 1:60 , partpool_error )
```

I've decorated this plot with some additional information, displayed in FIGURE 12.3. Your own plot will look different, because of simulation variance. The pattern displayed in the figure is the central tendency. To see how to quickly re-run the model on newly simulated data, without re-compiling, see the Overthinking box at the end of this section.

The filled blue points in FIGURE 12.3 display the no-pooling estimates. The black circles show the varying effect estimates. The horizontal axis is the pond index, from 1 through

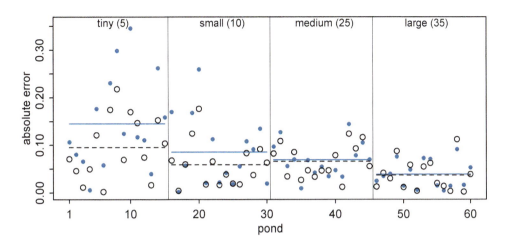

FIGURE 12.3. Error of no-pooling and partial pooling estimates, for the simulated tadpole ponds. The horizontal axis displays pond number. The vertical axis measures the absolute error in the predicted proportion of survivors,
compared to the true value used in the simulation. The higher the point,
the worse the estimate. No-pooling shown in blue. Partial pooling shown
in black. The blue and dashed black lines show the average error for each
kind of estimate, across each initial density of tadpoles (pond size). Smaller
ponds produce more error, but the partial pooling estimates are better on
average, especially in smaller ponds.

60. The vertical axis is the distance between the mean estimated probability of survival and
the actual probability of survival. So points close to the bottom had low error, while those
near the top had a large error, more than 20% off in some cases. The vertical lines divide
the groups of ponds with different initial densities of tadpoles. And finally, the horizontal
blue and black line segments show the average error of the no-pooling and partial pooling
estimates, respectively, for each group of ponds with the same initial size.

The first thing to notice about this plot is that both kinds of estimates are much more
accurate for larger ponds, on the right side. This arises because more data means better
estimates, usually. In the small ponds, sample size is small, and neither kind of estimate can
work magic. Therefore, prediction suffers on the left side of the plot. Second, note that the
blue line is always above the black dashed line. This indicates that the no-pool estimates,
shown by the blue points, have higher average error in each group of ponds. So even though
both kinds of estimates get worse as sample size decreases, the varying effect estimates have
the advantage, on average. Third, the distance between the blue line and the black dashed
line grows as ponds get smaller. So while both kinds of estimates suffer from reduced sample
size, the partial pooling estimates suffer less.

Okay, so what are we to make of all of this? Remember, back in FIGURE 12.1 (page 361),
the smaller tanks demonstrated more shrinkage towards the mean. Here, the ponds with
the smallest sample size show the greatest improvement over the naive no-pooling estimates.
This is no coincidence. Shrinkage towards the mean results from trying to negotiate the
underfitting and overfitting risks of the grand mean on one end and the individual means

of each pond on the other. The smaller tanks/ponds contain less information, and so their varying estimates are influenced more by the pooled information from the other ponds. In other words, small ponds are prone to overfitting, and so they receive a bigger dose of the underfit grand mean. Likewise, the larger ponds shrink much less, because they contain more information and are prone to less overfitting. Therefore they need less correcting. When individual ponds are very large, pooling in this way does hardly anything to improve estimates, because the estimates don't have far to go. But in that case, they also don't do any harm, and the information pooled from them can substantially help prediction in smaller ponds.

The partially pooled estimates are better on average. They adjust individual cluster (pond) estimates to negotiate the trade-off between underfitting and overfitting. This is a form of regularization, just like in Chapter 6, but now with an amount of regularization that is learned from the data itself.

But there are some cases in which the no-pooling estimates are better. These exceptions often result from ponds with extreme probabilities of survival. The partial pooling estimates shrink such extreme ponds towards the mean, because few ponds exhibit such extreme behavior. But sometimes outliers really are outliers.

Overthinking: Repeating the pond simulation. This model samples pretty quickly. Compiling the model takes up most of the execution time. Luckily the compilation only has to be done once. Then you can pass new data to the compiled model and get new estimates. Once you've compiled m12.3 once, you can use this code to re-simulate ponds and sample from the new posterior, without waiting for the model to compile again:

R code
12.19
```
a <- 1.4
sigma <- 1.5
nponds <- 60
ni <- as.integer( rep( c(5,10,25,35) , each=15 ) )
a_pond <- rnorm( nponds , mean=a , sd=sigma )
dsim <- data.frame( pond=1:nponds , ni=ni , true_a=a_pond )
dsim$si <- rbinom( nponds,prob=logistic( dsim$true_a ),size=dsim$ni )
dsim$p_nopool <- dsim$si / dsim$ni
newdat <- list(si=dsim$si,ni=dsim$ni,pond=1:nponds)
m12.3new <- map2stan( m12.3 , data=newdat , iter=1e4 , warmup=1000 )
```

The map2stan function reuses the compiled model in m12.3, passes it the new data, and returns the new samples in m12.3new. This is a useful trick, in case you want to perform a simulation study of a particular model structure. And if you ever want to extract the actual compiled Stan model, it is held in m12.3@stanfit, and you can always view its code with stancode(m12.3) and the input data (which is augmented a bit) with m12.3@data.

12.3. More than one type of cluster

We can use and often should use more than one type of cluster in the same model. For example, the observations in data(chimpanzees), which you met back in Chapter 10, are lever pulls. Each pull is within a cluster of pulls belonging to an individual chimpanzee. But each pull is also within an experimental block, which represents a collection of observations that happened on the same day. So each observed pull belongs to both an actor (1 to 7) and a block (1 to 6). There may be unique intercepts for each actor as well as for each block.

So in this section we'll reconsider the chimpanzees data, using both types of clusters simultaneously. This will allow us to use partial pooling on both categorical variables, `actor` and `block`, at the same time. We'll also get estimates of the variation among actors and among blocks.

> **Rethinking: Cross-classification and hierarchy.** The kind of data structure in `data(chimpanzees)` is usually called a CROSS-CLASSIFIED multilevel model. It is cross-classified, because actors are not nested within unique blocks. If each chimpanzee had instead done all of his or her pulls on a single day, within a single block, then the data structure would instead be *hierarchical*. However, the model specification would typically be the same. So the model structure and code you'll see below will apply both to cross-classified designs and hierarchical designs. Other software sometimes forces you to treat these differently, on account of using a conditioning engine substantially less capable than MCMC. There are other types of "hierarchical" multilevel models, types that make adaptive priors for adaptive priors. It's turtles all the way down, recall (page 13). You'll see an example in the next chapter. But for the most part, people (or their software) nearly always use the same kind of model in both cases.

12.3.1. Multilevel chimpanzees.

Let's proceed by taking the full chimpanzees model from Chapter 10 (`m10.4`, page 299) and first adding varying intercepts on actor. To add varying intercepts to this model, we just replace the fixed regularizing prior with an adaptive prior. But this time, I'll put the mean α up in the linear model, rather than down in the prior. Why? Because it will pave the way to adding more varying effects later. You'll see why, once we've pushed forward a little.

Here is the multilevel chimpanzees model in mathematical form, with the varying intercept components highlighted in blue:

$$L_i \sim \text{Binomial}(1, p_i)$$
$$\text{logit}(p_i) = \alpha + \alpha_{\text{ACTOR}[i]} + (\beta_P + \beta_{PC}C_i)P_i$$
$$\alpha_{\text{ACTOR}} \sim \text{Normal}(0, \sigma_{\text{ACTOR}})$$
$$\alpha \sim \text{Normal}(0, 10)$$
$$\beta_P \sim \text{Normal}(0, 10)$$
$$\beta_{PC} \sim \text{Normal}(0, 10)$$
$$\sigma_{\text{ACTOR}} \sim \text{HalfCauchy}(0, 1)$$

Notice that α is inside the linear model, not inside the Gaussian prior for α_{ACTOR}. This is mathematically equivalent to what you did with the tadpoles earlier in the chapter. You can always take the mean out of a Gaussian distribution and treat the distribution as a constant plus a Gaussian distribution centered on zero.

This might seem a little weird at first, so it might help train your intuition by experimenting in R. These two lines of code sample values from two identical Gaussian distributions, with mean 10 and standard deviation 1:

R code
12.20

```
y1 <- rnorm( 1e4 , 10 , 1 )
y2 <- 10 + rnorm( 1e4 , 0 , 1 )
```

Inspect the distributions of values in y1 and y2. You'll see they are the same. This feature of the Gaussian distribution arises from the independence of the mean and standard deviation. Most distributions do not have this property. But we'll exploit it here. And sometimes a given combination of model and data is more efficiently fit using one form or the other. I'll say more about this in the next chapter.

Here's the corresponding map2stan code for the model with varying intercepts on actor, but not yet on block. Note that the linear model contains α, the varying intercepts mean. The adaptive prior for the intercepts themselves has a mean of zero.

R code
12.21
```
library(rethinking)
data(chimpanzees)
d <- chimpanzees
d$recipient <- NULL       # get rid of NAs

m12.4 <- map2stan(
    alist(
        pulled_left ~ dbinom( 1 , p ) ,
        logit(p) <- a + a_actor[actor] + (bp + bpC*condition)*prosoc_left ,
        a_actor[actor] ~ dnorm( 0 , sigma_actor ),
        a ~ dnorm(0,10),
        bp ~ dnorm(0,10),
        bpC ~ dnorm(0,10),
        sigma_actor ~ dcauchy(0,1)
    ) ,
    data=d , warmup=1000 , iter=5000 , chains=4 , cores=3 )
```

Inspect the trace plot, plot(m12.4), and the posterior distribution for sigma_actor. Make sure the effective numbers of samples and Rhat values look alright. If you need to review these MCMC diagnostics, glance back at Chapter 8.

Now that the mean of the population of actors, α (a), is in the linear model, it's important to notice now that the a_actor parameters are *deviations* from a. So for any given row i, the total intercept is $\alpha + \alpha_{\mathrm{ACTOR}[i]}$. The part that varies across actors is just the deviation from the grand mean α. To compute the total intercept for each actor, you need to add samples of a to samples of a_actor:

R code
12.22
```
post <- extract.samples(m12.4)
total_a_actor <- sapply( 1:7 , function(actor) post$a + post$a_actor[,actor] )
round( apply(total_a_actor,2,mean) , 2 )
```

```
[1] -0.71  4.59 -1.02 -1.02 -0.71  0.23  1.76
```

12.3.2. Two types of cluster. To add the second cluster type, block, we merely replicate the structure for the actor cluster. This means the linear model gets yet another varying intercept, $\alpha_{\mathrm{BLOCK}[i]}$, and the model gets another adaptive prior and yet another standard deviation parameter. Here is the mathematical form of the model, with the new pieces of the machine highlighted in blue:

$$L_i \sim \text{Binomial}(1, p_i)$$
$$\text{logit}(p_i) = \alpha + \alpha_{\text{ACTOR}[i]} + \alpha_{\text{BLOCK}[i]} + (\beta_P + \beta_{PC}C_i)P_i$$
$$\alpha_{\text{ACTOR}} \sim \text{Normal}(0, \sigma_{\text{ACTOR}})$$
$$\alpha_{\text{BLOCK}} \sim \text{Normal}(0, \sigma_{\text{BLOCK}})$$
$$\alpha \sim \text{Normal}(0, 10)$$
$$\beta_P \sim \text{Normal}(0, 10)$$
$$\beta_{PC} \sim \text{Normal}(0, 10)$$
$$\sigma_{\text{ACTOR}} \sim \text{HalfCauchy}(0, 1)$$
$$\sigma_{\text{BLOCK}} \sim \text{HalfCauchy}(0, 1)$$

Each cluster variable needs its own standard deviation parameter that adapts the amount of pooling across units, be they actors or blocks. These are σ_{ACTOR} and σ_{BLOCK}, respectively. Finally, note that there is only one global mean parameter α, and both of the varying intercept parameters are centered at zero. We can't identify a separate mean for each varying intercept type, because both intercepts are added to the same linear prediction. So it is conventional to define varying intercepts with a mean of zero, so there's no risk of accidentally creating hard-to-identify parameters. There's a practice problem at the end of the chapter that leads you to explore what happens when you forget and instead include two grand mean α parameters.

Now to fit the model that uses both `actor` and `block`:

```
# prep data
d$block_id <- d$block  # name 'block' is reserved by Stan

m12.5 <- map2stan(
    alist(
        pulled_left ~ dbinom( 1 , p ),
        logit(p) <- a + a_actor[actor] + a_block[block_id] +
                    (bp + bpc*condition)*prosoc_left,
        a_actor[actor] ~ dnorm( 0 , sigma_actor ),
        a_block[block_id] ~ dnorm( 0 , sigma_block ),
        c(a,bp,bpc) ~ dnorm(0,10),
        sigma_actor ~ dcauchy(0,1),
        sigma_block ~ dcauchy(0,1)
    ) ,
    data=d, warmup=1000 , iter=6000 , chains=4 , cores=3 )
```

R code
12.23

If all goes well, you'll end up with 20,000 samples from 4 independent chains. As always, be sure to inspect the trace plots and the diagnostics. As soon as you start trusting the machine, the machine will betray your trust. In this case, you might see for the first time a warning about *divergent iterations*:

```
Warning message:
In map2stan(alist(pulled_left ~ dbinom(1, p), logit(p) <- a + a_actor[actor] +  :
  There were 3 divergent iterations during sampling.
Check the chains (trace plots, n_eff, Rhat) carefully to ensure they are valid.
```

We'll have a lot more to say about these in the next chapter. For now, they are safe to ignore. Just do as stated and inspect n_eff and Rhat.

This is easily the most complicated model we've fit in the book so far. So let's look at the estimates and take note of a few important features:

R code
12.24

```
precis(m12.5,depth=2) # depth=2 displays varying effects
plot(precis(m12.5,depth=2)) # also plot
```

```
             Mean StdDev lower 0.89 upper 0.89 n_eff Rhat
a_actor[1]  -1.17   0.93      -2.57       0.29  2333    1
a_actor[2]   4.14   1.59       1.92       6.33  4543    1
a_actor[3]  -1.48   0.94      -2.91      -0.02  2310    1
a_actor[4]  -1.48   0.94      -2.92      -0.01  2360    1
a_actor[5]  -1.17   0.94      -2.63       0.27  2354    1
a_actor[6]  -0.22   0.93      -1.65       1.25  2359    1
a_actor[7]   1.32   0.96      -0.15       2.84  2496    1
a_block[1]  -0.18   0.22      -0.53       0.11  3848    1
a_block[2]   0.04   0.18      -0.23       0.34  8595    1
a_block[3]   0.05   0.19      -0.23       0.35  7237    1
a_block[4]   0.00   0.18      -0.30       0.28  9532    1
a_block[5]  -0.04   0.18      -0.34       0.25  8327    1
a_block[6]   0.11   0.20      -0.17       0.43  5420    1
a            0.46   0.92      -0.98       1.88  2263    1
bp           0.82   0.26       0.41       1.25  6890    1
bpc         -0.13   0.30      -0.62       0.33  8360    1
sigma_actor  2.25   0.90       1.02       3.33  4892    1
sigma_block  0.22   0.18       0.01       0.43  2079    1
```

The precis plot is shown in the left-hand part of FIGURE 12.4.

First, notice that the number of effective samples, n_eff, varies quite a lot across parameters. This is common in complex models. Why? There are many reasons for this. But in this sort of model the most common reason is that some parameter spends a lot of time near a boundary. Here, that parameter is sigma_block. It spends a lot of time near its minimum of zero. As a consequence, you may also see a warning about "divergent iterations." You can wait until the next chapter to explore what these mean and what to do about them. For now, you can trust the Rhat values above, but later you'll see how to make sampling more efficient for models like these.

Second, compare sigma_actor to sigma_block and notice that the estimated variation among actors is a lot larger than the estimated variation among blocks. This is easy to appreciate, if we plot the marginal posterior distributions of these two parameters:

R code
12.25

```
post <- extract.samples(m12.5)
dens( post$sigma_block , xlab="sigma" , xlim=c(0,4) )
dens( post$sigma_actor , col=rangi2 , lwd=2 , add=TRUE )
text( 2 , 0.85 , "actor" , col=rangi2 )
text( 0.75 , 2 , "block" )
```

And this plot appears on the right in FIGURE 12.4. While there's uncertainty about the variation among actors, this model is confident that actors vary more than blocks. You can easily

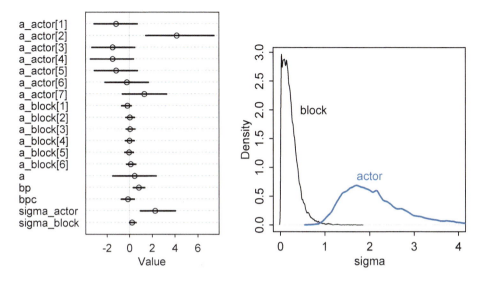

FIGURE 12.4. Left: Posterior means and 89% highest density intervals for m12.5. The greater variation across actors than blocks can be seen immediately in the a_actor and a_block distributions. Right: Posterior distributions of the standard deviations of varying intercepts by actor (blue) and experimental block (black).

see this variation in the varying intercept estimates: the a_actor distributions are much more scattered than are the a_block distributions.

As a consequence, adding block to this model hasn't added a lot of overfitting risk. Let's compare the model with only varying intercepts on actor to the model with both kinds of varying intercepts:

```
compare(m12.4,m12.5)
```

R code
12.26

```
      WAIC pWAIC dWAIC weight    SE  dSE
m12.4 531.5   8.1   0.0   0.65 19.50   NA
m12.5 532.7  10.5   1.2   0.35 19.74 1.94
```

Look at the pWAIC column, which reports the "effective number of parameters." While m12.5 has 7 more parameters than m12.4 does, it has only about 2.5 more effective parameters. Why? Because the posterior distribution for sigma_block ended up close to zero. This means each of the 6 a_block parameters is strongly shrunk towards zero—they are relatively inflexible. In contrast, the a_actor parameters are shrunk towards zero much less, because the estimated variation across actors is much larger, resulting in less shrinkage. But as a consequence, each of the a_actor parameters contributes much more to the pWAIC value.

You might also notice that the difference in WAIC between these models is small, only 1.2. This is especially small compared the standard deviation of the difference, 1.94. These two models imply nearly identical predictions, and so their expected out-of-sample accuracy is nearly identical. The block parameters have been shrunk so much towards zero that they do very little work in the model.

If you are feeling the urge to "select" `m12.4` as the best model, pause for a moment. There is nothing to gain here by selecting either model. The comparison of the two models tells a richer story—whether we include block or not hardly matters, and the `a_block` and `sigma_block` estimates tell us why. Furthermore, the standard error of the difference in WAIC between the models is twice as large as the difference itself. By retaining and reporting both models, we and our readers learn more about the experiment.

12.3.3. Even more clusters.

Adding more types of clusters proceeds the same way. At some point the model may become too complex to reliably fit to data. But Hamiltonian Monte Carlo is very capable with varying effects. It can easily handle tens of thousands of varying effect parameters. Sampling will be slow in such cases, but it will work.

So don't be shy—if you have a good theoretical reason to include a cluster variable, then you also have good theoretical reason to partially pool its parameters. As you've seen, the overfitting risk induced by including varying intercepts can be quite small, when there is little variation among the clusters. The multilevel model adaptively regularizes, helping us discover the relevance of different kinds of clusters within the data. In this way, you can think of the `sigma` parameter for each cluster as a crude measure of the cluster's relevance for explaining variation in the outcome.

12.4. Multilevel posterior predictions

Way back in Chapter 3 (page 64), I commented on the importance of MODEL CHECKING. Software does not always work as expected, and one robust way to discover mistakes is to compare the sample to the posterior predictions of a fit model. The same procedure, producing implied predictions from a fit model, is very helpful for understanding what the model means. Every model is a merger of sense and nonsense. When we understand a model, we can find its sense and control its nonsense. But as models get more complex, it is very difficult to impossible to understand them just by inspecting tables of posterior means and intervals. Exploring implied posterior predictions helps much more.

Another role for constructing implied predictions is in computing INFORMATION CRITERIA, like DIC and WAIC. These criteria provide simple estimates of out-of-sample model accuracy, the KL divergence. In practical terms, information criteria provide a rough measure of a model's flexibility and therefore overfitting risk. This was the big conceptual mission of Chapter 6.

All of this advice applies to multilevel models as well. We still often need model checks, counterfactual predictions for understanding, and information criteria. The introduction of varying effects does introduce nuance, however.

First, we should no longer expect the model to exactly retrodict the sample, because adaptive regularization has as its goal to trade off poorer fit in sample for better inference and hopefully better fit out of sample. That is what shrinkage does for us. Of course, we should never be trying to really retrodict the sample. But now you have to expect that even a perfectly good model fit will differ from the raw data in a systematic way that reflects shrinkage.

Second, "prediction" in the context of a multilevel model requires additional choices. If we wish to validate a model against the specific clusters used to fit the model, that is one thing. But if we instead wish to compute predictions for new clusters, other than the ones observed in the sample, that is quite another. We'll consider each of these in turn, continuing to use the chimpanzees model from the previous section.

12.4.1. Posterior prediction for same clusters. When working with the same clusters as you used to fit a model, varying intercepts are just parameters. The only trick is to ensure that you use the right intercept for each case in the data. If you use `link` and `sim` to do your work for you, this is handled automatically. But otherwise, there are no tricks.

For example, in `data(chimpanzees)`, there are 7 unique actors. These are the clusters. The varying intercepts model, `m12.4`, estimated an intercept for each, in addition to two parameters to describe the mean and standard deviation of the population of actors. We'll construct posterior predictions (retrodictions), using both the automated `link` approach and doing it from scratch, so there is no confusion.

Before computing predictions, note that we should no longer expect the posterior predictive distribution to match the raw data, even when the model worked correctly. Why? The whole point of partial pooling is to shrink estimates towards the grand mean. So the estimates should not necessarily match up with the raw data, once you use pooling.

The code needed to compute posterior predictions is just like the code from Chapter 10. Here it is again, computing and plotting posterior predictions for actor number 2:

```
chimp <- 2
d.pred <- list(
    prosoc_left = c(0,1,0,1),    # right/left/right/left
    condition = c(0,0,1,1),      # control/control/partner/partner
    actor = rep(chimp,4)
)
link.m12.4 <- link( m12.4 , data=d.pred )
pred.p <- apply( link.m12.4 , 2 , mean )
pred.p.PI <- apply( link.m12.4 , 2 , PI )
```
R code
12.27

And the plotting code is exactly the same as before (page 297).

To construct the same calculations without using `link`, we just have to remember the model. The only difficulty is that when we work with the samples from the posterior, the varying intercepts will be a matrix of samples. Let's take a look:

```
post <- extract.samples(m12.4)
str(post)
```
R code
12.28

```
List of 5
 $ a_actor     : num [1:8000, 1:7] -1.842 -0.225 -1.811 -0.759 -1.882 ...
 $ a           : num [1:8000(1d)] 1.291 -0.632 0.285 -0.109 1.229 ...
 $ bp          : num [1:8000(1d)] 1 1.064 1.087 0.254 0.908 ...
 $ bpC         : num [1:8000(1d)] -0.272 -0.539 -0.295 0.375 -0.218 ...
 $ sigma_actor: num [1:8000(1d)] 2.13 2.49 2.32 1.51 4.12 ...
```

The `a_actor` matrix has samples on the rows and actors on the columns. So to plot, for example, the density for actor 5:

```
dens( post$a_actor[,5] )
```
R code
12.29

The `[,5]` means "all samples for actor 5."

To construct posterior predictions, we build our own link function. I'll use the `with` function here, so we don't have to keep typing `post$` before every parameter name:

R code
12.30
```
p.link <- function( prosoc_left , condition , actor ) {
    logodds <- with( post ,
        a + a_actor[,actor] + (bp + bpC * condition) * prosoc_left
    )
    return( logistic(logodds) )
}
```

The linear model is identical to the one used to define the model, but with a single comma added inside the brackets after a_actor. Now to compute predictions:

R code
12.31
```
prosoc_left <- c(0,1,0,1)
condition <- c(0,0,1,1)
pred.raw <- sapply( 1:4 , function(i) p.link(prosoc_left[i],condition[i],2) )
pred.p <- apply( pred.raw , 2 , mean )
pred.p.PI <- apply( pred.raw , 2 , PI )
```

At some point, you will have to work with a model that link will mangle. At that time, you can return to this section and peer hard at the code above and still make progress. No matter what the model is, if it is a Bayesian model, then it is *generative*. This means that predictions are made by pushing samples up through the model to get distributions of predictions. Then you summarize the distributions to summarize the predictions.

12.4.2. Posterior prediction for new clusters. Often, the particular clusters in the sample are not of any enduring interest. In the chimpanzees data, for example, these particular 7 chimpanzees are just seven individuals. We'd like to make inferences about the whole species, not just those seven individuals. So the individual actor intercepts aren't of interest, but the distribution of them definitely is.

One way to grasp the task of construction posterior predictions for new clusters is to imagine leaving out one of the clusters when you fit the model to the data. For example, suppose we leave out actor number 7 when we fit the chimpanzees model. Now how can we assess the model's accuracy for predicting actor number 7's behavior? We can't use any of the a_actor parameter estimates, because those apply to other individuals. But we can make good use of the a and sigma_actor parameters, because those describe the population of actors.

First, let's see how to construct posterior predictions for a now, previously unobserved *average* actor. By "average," I mean an individual chimpanzee with an intercept exactly at a (α), the population mean. This simultaneously implies a varying intercept of zero. Since there is uncertainty about the population mean, there is still uncertainty about this average individual's intercept. But as you'll see, the uncertainty is much smaller than it really should be, if we wish to honestly represent the problem of what to expect from a new individual.

The first step is to make a new data list to compute predictions over. You've done this in previous chapters. Here is our new list, representing the four different treatments:

R code
12.32
```
d.pred <- list(
    prosoc_left = c(0,1,0,1),    # right/left/right/left
    condition = c(0,0,1,1),      # control/control/partner/partner
    actor = rep(2,4) )           # placeholder
```

Next, we're going to make a matrix of zeros, to replace the varying intercept samples. It's easiest to just keep the same dimension as the original matrix. In this case that means using 1000 samples for each of 7 actors. But all of the samples will be set to zero:

```
# replace varying intercept samples with zeros
# 1000 samples by 7 actors
a_actor_zeros <- matrix(0,1000,7)
```

R code
12.33

That's the only new trick. Now we just pass this new matrix to link using the optional replace argument. Make sure the new matrix is named the same as the varying intercept matrix, a_actor. Otherwise it won't replace anything that appears in the model.

```
# fire up link
# note use of replace list
link.m12.4 <- link( m12.4 , n=1000 , data=d.pred ,
    replace=list(a_actor=a_actor_zeros) )

# summarize and plot
pred.p.mean <- apply( link.m12.4 , 2 , mean )
pred.p.PI <- apply( link.m12.4 , 2 , PI , prob=0.8 )
plot( 0 , 0 , type="n" , xlab="prosoc_left/condition" ,
    ylab="proportion pulled left" , ylim=c(0,1) , xaxt="n" ,
    xlim=c(1,4) )
axis( 1 , at=1:4 , labels=c("0/0","1/0","0/1","1/1") )
lines( 1:4 , pred.p.mean )
shade( pred.p.PI , 1:4 )
```

R code
12.34

The result is displayed in FIGURE 12.5, on the left. The gray region shows the 80% interval for an actor with an average intercept. This kind of calculation makes it easy to see the impact of prosoc_left, as well as uncertainty about where the average is, but it doesn't show the variation among actors.

To show the variation among actors, we'll need to use sigma_actor in the calculation. We can again smuggle this into link by using the replace argument. This time however, we'll simulate a matrix of new varying intercepts from a Gaussian distribution defined by the adaptive prior in the model itself:

$$\alpha_{\text{ACTOR}} \sim \text{Normal}(0, \sigma_{\text{ACTOR}})$$

This implies that once we have samples for σ_{ACTOR}, we can simulate new actor intercepts from this distribution. Here's the code to do just that, using rnorm:

```
# replace varying intercept samples with simulations
post <- extract.samples(m12.4)
a_actor_sims <- rnorm(7000,0,post$sigma_actor)
a_actor_sims <- matrix(a_actor_sims,1000,7)
```

R code
12.35

Now pass the simulated intercepts into link. Note the replace list, which inserts the simulations into the posterior.

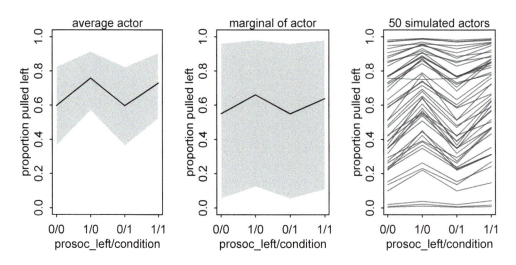

FIGURE 12.5. Posterior predictive distributions for the chimpanzees varying intercept model, m12.4. The solid lines are posterior means and the
shaded regions are 80% percentile intervals. Left: Setting the varying intercept a_actor to zero produces predictions for an *average* actor. These
predictions ignore uncertainty arising from variation among actors. Middle: Simulating varying intercepts using the posterior standard deviation
among actors, sigma_actor, produces predictions that account for variation among actors. Right: 50 simulated actors with unique intercepts sampled from the posterior. Each simulation maintains the same parameter
values across all four treatments.

R code
12.36
```
link.m12.4 <- link( m12.4 , n=1000 , data=d.pred ,
    replace=list(a_actor=a_actor_sims) )
```

Summarizing and plotting is exactly as before, and the result is displayed in the middle of
FIGURE 12.5. These posterior predictions are *marginal* of actor, which means that they average over the uncertainty among actors. In contrast, the predictions on the left just set the
actor to the average, ignoring variation among actors.

At this point, students usually ask, "So which one should I use?" The answer is, "It depends." Both are useful, depending upon the question. The predictions for an average actor
help to visualize the impact of treatment. The predictions that are marginal of actor illustrate how variable different chimpanzees are, according to the model. You probably want to
compute both for yourself, when trying to understand a model. But which you include in a
report will depend upon context.

In this case, we can do better by making a plot that displays both the treatment effect and
the variation among actors. We can do this by forgetting about intervals and instead simulating a series of new actors in each of the four treatments. By drawing a line for each actor
across all four treatments, we'll be able to visualize both the zig-zag impact of prosoc_left
as well as the variation among individuals.

What we'll do now is write a new function that simulates a new actor from the estimated population of actors and then computes probabilities of pulling the left lever for each of the four treatments. These simulations will not average over uncertainty in the posterior. We'll get that uncertainty into the plot by using multiple simulations, each with a different sample from the posterior. Here's the function:

R code
12.37

```
post <- extract.samples(m12.4)
sim.actor <- function(i) {
    sim_a_actor <- rnorm( 1 , 0 , post$sigma_actor[i] )
    P <- c(0,1,0,1)
    C <- c(0,0,1,1)
    p <- logistic(
        post$a[i] +
        sim_a_actor +
        (post$bp[i] + post$bpC[i]*C)*P
    )
    return(p)
}
```

This function takes a single argument, i, which is just the index of a sample from the posterior distribution. It then draws a random intercept for the actor, using rnorm and a particular value of sigma_actor. Then it computes probabilities p for each of the four treatments, using the same linear model, but with different predictor values inside the P and C vectors. Because these vectors are of length 4, the code spits out 4 values for p.

Now to use this function to plot 50 simulations:

R code
12.38

```
# empty plot
plot( 0 , 0 , type="n" , xlab="prosoc_left/condition" ,
    ylab="proportion pulled left" , ylim=c(0,1) , xaxt="n" , xlim=c(1,4) )
axis( 1 , at=1:4 , labels=c("0/0","1/0","0/1","1/1") )

# plot 50 simulated actors
for ( i in 1:50 ) lines( 1:4 , sim.actor(i) , col=col.alpha("black",0.5) )
```

The result is shown in the right-hand plot of FIGURE 12.5. Each trend is a simulated actor, across all four treatments on the horizontal axis. It is much easier in this plot to see both the zig-zag impact of treatment and the variation among actors that is induced by the posterior distribution of sigma_actor.

Also note the interaction of treatment and the variation among actors. Because this is a binomial model, in principle all parameters interact, due to ceiling and floor effects. For actors with very large intercepts, near the top of the plot, treatment has very little effect. These actors have strong handedness preferences. But actors with intercepts nearer the mean are influenced by treatment.

12.4.3. Focus and multilevel prediction. All of this is confusing at first. There is no uniquely correct way to always construct the predictions, and the calculations themselves probably seem a little magical. In time, it makes a lot more sense. The fact is that multilevel models contain parameters with different FOCUS. Focus here means which level of the model the parameter makes direct predictions for. It helps to organize the issue into three common cases.

First, when retrodicting the sample, the parameters that describe the population of clusters, such as α and σ_{ACTOR} in m12.4, do not influence prediction directly. Recall that these population parameters are often called **HYPERPARAMETERS**, as they are parameters for parameters. These hyperparameters had their effects during estimation, by shrinking the varying effect parameters towards a common mean. The prediction focus here is on the top level of parameters, not the deeper hyperparameters.

Second, the same is true when forecasting a new observation for a cluster that was present in the sample. For example, if we want to predict what chimpanzee number 2 will do in the next experiment, we should probably bet she'll pull the left lever, because her varying intercept was very large. The focus is again on the top level.

The third case is different. When instead we wish to forecast for some new cluster that was not present in the sample, such as a new individual or school or year or location, then we need the hyper-parameters. The hyper-parameters tell us how to forecast a new cluster, by generating a distribution of new per-cluster intercepts. This is what we did in the previous section, simulating new chimpanzees.

This is also the right thing to do whenever varying effects are used to model **OVER-DISPERSION** (page 363). In that case, we need to simulate intercepts in order to account for the over-dispersion. Here's a quick example, using the Oceanic societies example from Chapter 10, but now adding a varying intercept to each society. Here's the mathematical form of the model, with the varying intercept pieces highlighted in blue:

$$T_i \sim \text{Poisson}(\mu_i)$$
$$\log(\mu_i) = \alpha + \alpha_{\text{SOCIETY}[i]} + \beta_P \log P_i$$
$$\alpha \sim \text{Normal}(0, 10)$$
$$\beta_P \sim \text{Normal}(0, 1)$$
$$\alpha_{\text{SOCIETY}} \sim \text{Normal}(0, \sigma_{\text{SOCIETY}})$$
$$\sigma_{\text{SOCIETY}} \sim \text{HalfCauchy}(0, 1)$$

T is total_tools, P is population, and i indexes each society. The above is just a varying intercept model, but with a varying intercept for every observation. As a result, σ_{SOCIETY} ends up being an estimate of the over-dispersion among societies. Another way to think of this is that the varying intercepts α_{SOCIETY} are residuals for each society. By also estimating the distribution of these residuals, we get an estimate of the excess variation, relative to the Poisson expectation.

And here is the code to fit the over-dispersed Poisson model:

R code
12.39

```
# prep data
library(rethinking)
data(Kline)
d <- Kline
d$logpop <- log(d$population)
d$society <- 1:10

# fit model
m12.6 <- map2stan(
```

```
alist(
    total_tools ~ dpois(mu),
    log(mu) <- a + a_society[society] + bp*logpop,
    a ~ dnorm(0,10),
    bp ~ dnorm(0,1),
    a_society[society] ~ dnorm(0,sigma_society),
    sigma_society ~ dcauchy(0,1)
),
data=d ,
iter=4000 , chains=3 )
```

This model samples very efficiently, despite using 13 parameters to describe 10 observations. Remember: Varying effect parameters are adaptively regularized. So they are not completely flexible and induce much less overfitting risk. In this case, WAIC should tell you that the effective number of parameters is about 5, not 13. If you have hundreds or thousands of observations in the data, this approach still works fine. You just end up with hundreds or thousands of varying intercept estimates. You won't care about the estimates themselves. But you will care about the hyperparameters that describe the population of varying intercepts.

Now to generate posterior predictions that visualize the over-dispersion. You can display posterior predictions (retrodictions) by using postcheck(m12.6). But those predictions just use the varying intercepts, a_society, directly. They do not use the hyper-parameters. To instead see the general trend that the model expects, we'll need to simulate counterfactual societies, using the hyper-parameters α and σ_{SOCIETY}. This is the same procedure that we used for new chimpanzee actors earlier.

R code
12.40

```
post <- extract.samples(m12.6)
d.pred <- list(
    logpop = seq(from=6,to=14,length.out=30),
    society = rep(1,30)
)
a_society_sims <- rnorm(20000,0,post$sigma_society)
a_society_sims <- matrix(a_society_sims,2000,10)
link.m12.6 <- link( m12.6 , n=2000 , data=d.pred ,
    replace=list(a_society=a_society_sims) )
```

And this code will display the raw data and the new prediction envelope:

R code
12.41

```
# plot raw data
plot( d$logpop , d$total_tools , col=rangi2 , pch=16 ,
    xlab="log population" , ylab="total tools" )

# plot posterior median
mu.median <- apply( link.m12.6 , 2 , median )
lines( d.pred$logpop , mu.median )

# plot 97%, 89%, and 67% intervals (all prime numbers)
mu.PI <- apply( link.m12.6 , 2 , PI , prob=0.97 )
shade( mu.PI , d.pred$logpop )
```

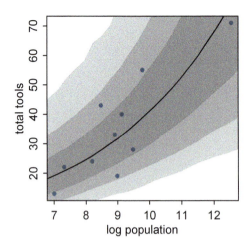

FIGURE 12.6. Posterior predictions for the over-dispersed Poisson island model, m12.6. The shaded regions are, inside to out: 67%, 89%, and 97% intervals of the expected mean. Marginalizing over the varying intercepts results in a much wider prediction region than we'd expect under a pure Poisson process.

```
mu.PI <- apply( link.m12.6 , 2 , PI , prob=0.89 )
shade( mu.PI , d.pred$logpop )
mu.PI <- apply( link.m12.6 , 2 , PI , prob=0.67 )
shade( mu.PI , d.pred$logpop )
```

The result is displayed in FIGURE 12.6. The envelope of predictions is a lot wider here than it was back in Chapter 10. This is a consequence of the varying intercepts, combined with the fact that there is much more variation in the data than a pure-Poisson model anticipates.

12.5. Summary

This chapter has been an introduction to the motivation, implementation, and interpretation of basic multilevel models. It focused on varying intercepts, which achieve better estimates of baseline differences among clusters in the data. They achieve better estimates, because they simultaneously model the population of clusters and use inferences about the population to pool information among parameters. From another perspective, varying intercepts are adaptively regularized parameters, relying upon a prior that is itself learned from the data. All of this is a foundation for the next chapter, which extends these concepts to additional types of parameters and models.

12.6. Practice

Easy.

12E1. Which of the following priors will produce more *shrinkage* in the estimates? (a) $\alpha_{\text{TANK}} \sim$ Normal$(0, 1)$; (b) $\alpha_{\text{TANK}} \sim$ Normal$(0, 2)$.

12E2. Make the following model into a multilevel model.

$$y_i \sim \text{Binomial}(1, p_i)$$
$$\text{logit}(p_i) = \alpha_{\text{GROUP}[i]} + \beta x_i$$
$$\alpha_{\text{GROUP}} \sim \text{Normal}(0, 10)$$
$$\beta \sim \text{Normal}(0, 1)$$

12E3. Make the following model into a multilevel model.

$$y_i \sim \text{Normal}(\mu_i, \sigma)$$
$$\mu_i = \alpha_{\text{GROUP}[i]} + \beta x_i$$
$$\alpha_{\text{GROUP}} \sim \text{Normal}(0, 10)$$
$$\beta \sim \text{Normal}(0, 1)$$
$$\sigma \sim \text{HalfCauchy}(0, 2)$$

12E4. Write an example mathematical model formula for a Poisson regression with varying intercepts.

12E5. Write an example mathematical model formula for a Poisson regression with two different kinds of varying intercepts, a cross-classified model.

Medium.

12M1. Revisit the Reed frog survival data, `data(reedfrogs)`, and add the `predation` and `size` treatment variables to the varying intercepts model. Consider models with either main effect alone, both main effects, as well as a model including both and their interaction. Instead of focusing on inferences about these two predictor variables, focus on the inferred variation across tanks. Explain why it changes as it does across models.

12M2. Compare the models you fit just above, using WAIC. Can you reconcile the differences in WAIC with the posterior distributions of the models?

12M3. Re-estimate the basic Reed frog varying intercept model, but now using a Cauchy distribution in place of the Gaussian distribution for the varying intercepts. That is, fit this model:

$$s_i \sim \text{Binomial}(n_i, p_i)$$
$$\text{logit}(p_i) = \alpha_{\text{TANK}[i]}$$
$$\alpha_{\text{TANK}} \sim \text{Cauchy}(\alpha, \sigma)$$
$$\alpha \sim \text{Normal}(0, 1)$$
$$\sigma \sim \text{HalfCauchy}(0, 1)$$

Compare the posterior means of the intercepts, α_{TANK}, to the posterior means produced in the chapter, using the customary Gaussian prior. Can you explain the pattern of differences?

12M4. Fit the following cross-classified multilevel model to the `chimpanzees` data:

$$L_i \sim \text{Binomial}(1, p_i)$$
$$\text{logit}(p_i) = \alpha_{\text{ACTOR}[i]} + \alpha_{\text{BLOCK}[i]} + (\beta_P + \beta_{PC} C_i) P_i$$
$$\alpha_{\text{ACTOR}} \sim \text{Normal}(\alpha, \sigma_{\text{ACTOR}})$$
$$\alpha_{\text{BLOCK}} \sim \text{Normal}(\gamma, \sigma_{\text{BLOCK}})$$
$$\alpha, \gamma, \beta_P, \beta_{PC} \sim \text{Normal}(0, 10)$$
$$\sigma_{\text{ACTOR}}, \sigma_{\text{BLOCK}} \sim \text{HalfCauchy}(0, 1)$$

Each of the parameters in those comma-separated lists gets the same independent prior. Compare the posterior distribution to that produced by the similar cross-classified model from the chapter. Also compare the number of effective samples. Can you explain the differences?

Hard.

12H1. In 1980, a typical Bengali woman could have 5 or more children in her lifetime. By the year 200, a typical Bengali woman had only 2 or 3. You're going to look at a historical set of data, when contraception was widely available but many families chose not to use it. These data reside in data(bangladesh) and come from the 1988 Bangladesh Fertility Survey. Each row is one of 1934 women. There are six variables, but you can focus on three of them for this practice problem:

(1) district: ID number of administrative district each woman resided in
(2) use.contraception: An indicator (0/1) of whether the woman was using contraception
(3) urban: An indicator (0/1) of whether the woman lived in a city, as opposed to living in a rural area

The first thing to do is ensure that the cluster variable, district, is a contiguous set of integers. Recall that these values will be index values inside the model. If there are gaps, you'll have parameters for which there is no data to inform them. Worse, the model probably won't run. Look at the unique values of the district variable:

R code
12.42
```
sort(unique(d$district))
```

```
[1]   1  2  3  4  5  6  7  8  9 10 11 12 13 14 15 16 17 18 19 20 21 22 23 24 25
[26] 26 27 28 29 30 31 32 33 34 35 36 37 38 39 40 41 42 43 44 45 46 47 48 49 50
[51] 51 52 53 55 56 57 58 59 60 61
```

District 54 is absent. So district isn't yet a good index variable, because it's not contiguous. This is easy to fix. Just make a new variable that is contiguous. This is enough to do it:

R code
12.43
```
d$district_id <- as.integer(as.factor(d$district))
sort(unique(d$district_id))
```

```
[1]   1  2  3  4  5  6  7  8  9 10 11 12 13 14 15 16 17 18 19 20 21 22 23 24 25
[26] 26 27 28 29 30 31 32 33 34 35 36 37 38 39 40 41 42 43 44 45 46 47 48 49 50
[51] 51 52 53 54 55 56 57 58 59 60
```

Now there are 60 values, contiguous integers 1 to 60.

Now, focus on predicting use.contraception, clustered by district_id. Do not include urban just yet. Fit both (1) a traditional fixed-effects model that uses dummy variables for district and (2) a multilevel model with varying intercepts for district. Plot the predicted proportions of women in each district using contraception, for both the fixed-effects model and the varying-effects model. That is, make a plot in which district ID is on the horizontal axis and expected proportion using contraception is on the vertical. Make one plot for each model, or layer them on the same plot, as you prefer. How do the models disagree? Can you explain the pattern of disagreement? In particular, can you explain the most extreme cases of disagreement, both why they happen where they do and why the models reach different inferences?

12H2. Return to the Trolley data, data(Trolley), from Chapter 11. Define and fit a varying intercepts model for these data. Cluster intercepts on individual participants, as indicated by the unique values in the id variable. Include action, intention, and contact as ordinary terms. Compare the varying intercepts model and a model that ignores individuals, using both WAIC and posterior predictions. What is the impact of individual variation in these data?

12H3. The Trolley data are also clustered by story, which indicates a unique narrative for each vignette. Define and fit a cross-classified varying intercepts model with both id and story. Use the same ordinary terms as in the previous problem. Compare this model to the previous models. What do you infer about the impact of different stories on responses?

13 Adventures in Covariance

Recall the coffee robot from the introduction to the previous chapter (page 355). This robot is programmed to move among cafés, order coffee, and record the waiting time. The previous chapter focused on the fact that the robot learns more efficiently when it pools information among the cafés. Varying intercepts are a mechanism for achieving that pooling.

Now suppose that the robot also records the time of day. The average wait time in the morning tends to be longer than the average wait time in the afternoon. This is because cafés are busier in the morning. But just like cafés vary in their *average* wait times, they also vary in their *differences* between morning and afternoon. In conventional regression, these differences in wait time between morning and afternoon are slopes, since they express the change in expectation when an indictor (or *dummy*, page 153) variable for time of day changes value. The linear model might look like this:

$$\mu_i = \alpha_{\text{CAFÉ}[i]} + \beta_{\text{CAFÉ}[i]} A_i$$

where A_i is a 0/1 indicator for *afternoon* and $\beta_{\text{CAFÉ}[i]}$ is a parameter for the expected difference between afternoon and morning for each café.

Since the robot more efficiently learns about the intercepts, $\alpha_{\text{CAFÉ}[i]}$ above, when it pools information about intercepts, it likewise learns more efficiently about the slopes when it also pools information about slopes. And the pooling is achieved in the same way, by estimating the population distribution of slopes at the same time the robot estimates each slope. The distributions assigned to both intercepts and slopes enable pooling for both, as the model (robot) learns the prior from the data.

This is the essence of the general VARYING EFFECTS strategy: Any batch of parameters with *exchangeable* index values can and probably should be pooled. Exchangeable just means the index values have no true ordering, because they are arbitrary labels. There's nothing special about intercepts; slopes can also vary by unit in the data, and pooling information among them makes better use of the data. So our coffee robot should be programmed to model both the population of intercepts and the population of slopes. Then it can use pooling for both and squeeze more information out of the data.

But here's a fact that will help us to squeeze even more information out of the data: Cafés covary in their intercepts and slopes. Why? At a popular café, wait times are on average long in the morning, because staff are very busy (FIGURE 13.1). But the same café will be much less busy in the afternoon, leading to a large difference between morning and afternoon wait times. At such a popular café, the intercept is high and the slope is far from zero, because the difference between morning and afternoon waits is large. But at a less popular café, the difference will be small. Such an unpopular café makes you wait less in the morning—because

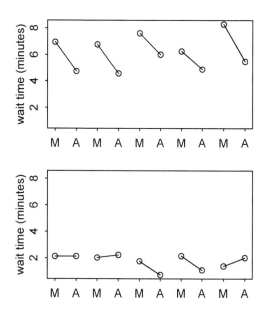

FIGURE 13.1. Waiting times at two cafés. Top: A busy café at which wait times nearly always improve in the afternoon. Bottom: An unpopular café where wait times are nearly always short. In a population of cafés like these, long morning waits (intercepts) covary with larger differences between morning and afternoon (slopes).

it's not busy—but there isn't much improvement in the afternoon. In the entire population of cafés, including both the popular and the unpopular, intercepts and slopes covary.

This covariation is information that the robot can use. If we can figure out a way to pool information *across* parameter types—intercepts and slopes—what the robot learns in the morning can improve learning about afternoons, and vice versa. Suppose for example that the robot arrives at a new café in the morning. It observes a long wait for its coffee. Even before it orders a coffee at the same café in the afternoon, it can update its expectation for how long it will wait. In the population of cafés, a long wait in the morning is associated with a shorter wait in the afternoon.

In this chapter, you'll see how to really do this, to specify VARYING SLOPES in combination with the varying intercepts of the previous chapter. This will enable pooling that will improve estimates of how different units respond to or are influenced by predictor variables. It will also improve estimates of intercepts, by borrowing information across parameter types. Essentially, varying slopes models are massive interaction machines. They allow every unit in the data to have its own unique response to any treatment or exposure or event, while also improving estimates via pooling. When the variation in slopes is large, the average slope is of less interest. Sometimes, the pattern of variation in slopes provides hints about omitted variables that explain why some units respond more or less. We'll see an example in this chapter.

The machinery that makes such complex varying effects possible will be used later in the chapter to extend the varying effects strategy to more subtle model types, including the use of continuous categories, using GAUSSIAN PROCESSES. Ordinary varying effects work only with discrete, unordered categories, such as individuals, countries, or ponds. In these cases, each category is equally different from all of the others. But it is possible to use pooling with categories such as age or location. In these cases, some ages and some locations are more similar than others. You'll see how to model covariation among continuous categories of this kind, as well as how to generalize the strategy to seemingly unrelated types of models such as phylogenetic and network regressions.

The material in this chapter is difficult. So if it suddenly seems both conceptually and computationally much more difficult, that only means you are paying attention. Material like this requires repetition, discussion, and learning from mistakes. The struggle is definitely worth it.

13.1. Varying slopes by construction

How should the robot pool information across intercepts and slopes? By modeling the joint population of intercepts and slopes, which means by modeling their covariance. In conventional multilevel models, the device that makes this possible is a joint multivariate Gaussian distribution for all of the varying effects, both intercepts and slopes. So instead of having two independent Gaussian distributions of intercepts and of slopes, the robot can do better by assigning a two-dimensional Gaussian distribution to both the intercepts (first dimension) and the slopes (second dimension).

You've been working with multivariate Gaussian distributions ever since Chapter 4, when you began using the quadratic approximation for the posterior distribution. The variance-covariance matrix, vcov, for a fit model describes how each parameter's posterior probability is associated with each other parameter's posterior probability. Now we'll use the same kind of distribution to describe the variation within and covariation among different kinds of varying effects. Varying intercepts have variation, and varying slopes have variation. Intercepts and slopes covary.

In order to see how this works and how varying slopes are specified and interpreted, let's simulate the coffee robot from the introduction. Like previous simulation exercises, this will simultaneously help you see how to conduct your own prospective power analyses, in addition to reemphasizing the generative nature of Bayesian statistical models.

> **Rethinking: Why Gaussian?** There is no reason the multivariate distribution of intercepts and slopes must be Gaussian. But there are both practical and epistemological justifications. On the practical side, there aren't many multivariate distributions that are easy to work with. The only common ones are multivariate Gaussian and multivariate Student (or "t") distributions. On the epistemological side, if all we want to say about these intercepts and slopes is their means, variances, and covariances, then the maximum entropy distribution is multivariate Gaussian.

13.1.1. Simulate the population. Begin by defining the population of cafés that the robot might visit. This means we'll define the average wait time in the morning and the afternoon, as well as the correlation between them. These numbers are sufficient to define the *average* properties of the cafés. Let's define these properties, then we'll sample cafés from them.

```
a <- 3.5            # average morning wait time
b <- (-1)           # average difference afternoon wait time
sigma_a <- 1        # std dev in intercepts
sigma_b <- 0.5      # std dev in slopes
rho <- (-0.7)       # correlation between intercepts and slopes
```
R code
13.1

These values define the entire population of cafés. To use these values to simulate a sample of cafés for the robot, we'll need to build them into a 2-dimensional multivariate Gaussian distribution. This means we need a vector of two means and 2-by-2 matrix of variances and covariances. The means are easiest. The vector we need is just:

R code
13.2
```
Mu <- c( a , b )
```

That's it. The value in a is the mean intercept, the wait in the morning. And the value in b is the mean slope, the difference in wait between afternoon and morning.

The matrix of variances and covariances is arranged like this:

$$\begin{pmatrix} \sigma_\alpha^2 & \sigma_\alpha\sigma_\beta\rho \\ \sigma_\alpha\sigma_\beta\rho & \sigma_\beta^2 \end{pmatrix}$$

The variance in intercepts is σ_α^2, and the variance in slopes is σ_β^2. These are found along the *diagonal* of the matrix. The other two elements of the matrix are the same, $\sigma_\alpha\sigma_\beta\rho$. This is the covariance between intercepts and slopes. It's just the product of the two standard deviations and the correlation. It might help to imagine an ordinary variance as the covariance of a variable with itself.

To build this matrix with R code, there are several options. I'll show you two of them, both very common. The first is to just use `matrix` to build the entire covariance matrix directly:

R code
13.3
```
cov_ab <- sigma_a*sigma_b*rho
Sigma <- matrix( c(sigma_a^2,cov_ab,cov_ab,sigma_b^2) , ncol=2 )
```

The awkward thing is that R matrices defined this way fill down each column before moving to the next row over. So the order inside the code above looks odd, but works. To see what I mean by "fill down each column," try this:

R code
13.4
```
matrix( c(1,2,3,4) , nrow=2 , ncol=2 )
```

```
     [,1] [,2]
[1,]    1    3
[2,]    2    4
```

The first column filled, and then R started over at the top of the second column.

The other common way to build the covariance matrix is conceptually very useful, because it treats the standard deviations and correlations separately. Then it matrix multiplies them to produce the covariance matrix. We're going to use this approach later on, to define priors, so it's worth seeing it now. Here's how it's done:

R code
13.5
```
sigmas <- c(sigma_a,sigma_b) # standard deviations
Rho <- matrix( c(1,rho,rho,1) , nrow=2 ) # correlation matrix

# now matrix multiply to get covariance matrix
Sigma <- diag(sigmas) %*% Rho %*% diag(sigmas)
```

If you are not sure what `diag(sigmas)` accomplishes, then try typing just `diag(sigmas)` at the R prompt.

Now we're ready to simulate some cafés, each with its own intercept and slope. Let's define the number of cafés:

```
N_cafes <- 20
```

And to simulate their properties, we just sample randomly from the multivariate Gaussian distribution defined by Mu and Sigma:

```
library(MASS)
set.seed(5) # used to replicate example
vary_effects <- mvrnorm( N_cafes , Mu , Sigma )
```

Note the set.seed(5) line above. That's there so you can replicate the precise results in the example figures. The particular number, 5, produces a particular sequence of random numbers. Each unique number generates a unique sequence. Including a set.seed line like this in your code allows others to exactly replicate your analyses. Later you'll want to repeat the example without repeating the set.seed call, or with a different number, so you can appreciate the variation across simulations.

Look at the contents of vary_effects now. It should be a matrix with 20 rows and 2 columns. Each row is a café. The first column contains intercepts. The second column contains slopes. For transparency, let's split these columns apart into nicely named vectors:

```
a_cafe <- vary_effects[,1]
b_cafe <- vary_effects[,2]
```

To visualize these intercepts and slopes, go ahead and plot them against one another.

```
plot( a_cafe , b_cafe , col=rangi2 ,
    xlab="intercepts (a_cafe)" , ylab="slopes (b_cafe)" )

# overlay population distribution
library(ellipse)
for ( l in c(0.1,0.3,0.5,0.8,0.99) )
    lines(ellipse(Sigma,centre=Mu,level=l),col=col.alpha("black",0.2))
```

FIGURE 13.2 displays a typical result. In any particular simulation, the correlation may not be as obvious. But on average, the intercepts in a_cafe and the slopes in b_cafe will have a correlation of -0.7, and you'll be able to see this in the scatterplot. The contour lines in the plot, produced by the ellipse package (make sure you install it), display the multivariate Gaussian population of intercepts and slopes that the 20 cafés were sampled from.

13.1.2. Simulate observations. We're almost done simulating. What we did above was simulate individual cafés and their average properties. Now all that remains is to simulate our robot visiting these cafés and collecting data. The code below simulates 10 visits to each café, 5 in the morning and 5 in the afternoon. The robot records the wait time during each visit. Then it combines all of the visits into a common data frame.

```
N_visits <- 10
afternoon <- rep(0:1,N_visits*N_cafes/2)
cafe_id <- rep( 1:N_cafes , each=N_visits )
```

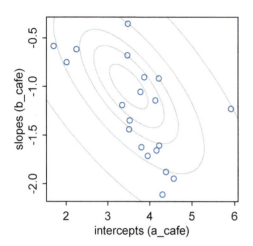

FIGURE 13.2. 20 cafés sampled from a statistical population. The horizontal axis is the intercept (average morning wait) for each cafe. The vertical axis is the slope (average difference between afternoon and morning wait) for each café. The gray ellipses illustrate the multivariate Gaussian population of intercepts and slopes.

```
mu <- a_cafe[cafe_id] + b_cafe[cafe_id]*afternoon
sigma <- 0.5  # std dev within cafes
wait <- rnorm( N_visits*N_cafes , mu , sigma )
d <- data.frame( cafe=cafe_id , afternoon=afternoon , wait=wait )
```

Go ahead and look inside the data frame d now. You'll find exactly the sort of data that is well-suited to a varying slopes model. There are multiple *clusters* in the data. These are the cafés. And each cluster is observed under different conditions. So it's possible to estimate both an individual intercept for each cluster, as well as an individual slope.

In this example, everything is *balanced*: Each café has been observed exactly 10 times, and the time of day is always balanced as well, with 5 morning and 5 afternoon observations for each café. But in general the data do not need to be balanced. Just like the tadpoles example from the previous chapter, lack of balance can really favor the varying effects analysis, because partial pooling uses information about the population where it is needed most.

Rethinking: Simulation and misspecification. In this exercise, we are simulating data from a generative process and then analyzing that data with a model that reflects exactly the correct structure of that process. But in the real world, we're never so lucky. Instead we are always forced to analyze data with a model that is MISSPECIFIED: The true data-generating process is different than the model. Simulation can be used however to explore misspecification. Just simulate data from a process and then see how a number of models, none of which match exactly the data-generating process, perform. And always remember that Bayesian inference does not depend upon data-generating assumptions, such as the likelihood, being true. Non-Bayesian approaches may depend upon sampling distributions for their inferences, but this is not the case for a Bayesian model. In a Bayesian model, a likelihood is merely a prior for the data, and inference about parameters can be shockingly insensitive to its details.

13.1.3. The varying slopes model. Now we're ready to play the process in reverse. We just generated data from a set of 20 cafés, and those cafés were themselves generated from a statistical population of cafés. Now we'll use that data to learn about the data-generating process, through a model.

The model is much like the varying intercepts models from the previous chapter. But now the joint population of intercepts and slopes appears, instead of just a distribution of varying intercepts. This is the varying slopes model, with explanation to follow:

$$W_i \sim \text{Normal}(\mu_i, \sigma) \qquad \text{[likelihood]}$$

$$\mu_i = \alpha_{\text{CAFÉ}[i]} + \beta_{\text{CAFÉ}[i]} A_i \qquad \text{[linear model]}$$

$$\begin{bmatrix} \alpha_{\text{CAFÉ}} \\ \beta_{\text{CAFÉ}} \end{bmatrix} \sim \text{MVNormal}\left(\begin{bmatrix} \alpha \\ \beta \end{bmatrix}, S \right) \qquad \text{[population of varying effects]}$$

$$S = \begin{pmatrix} \sigma_\alpha & 0 \\ 0 & \sigma_\beta \end{pmatrix} R \begin{pmatrix} \sigma_\alpha & 0 \\ 0 & \sigma_\beta \end{pmatrix} \qquad \text{[construct covariance matrix]}$$

$$\alpha \sim \text{Normal}(0, 10) \qquad \text{[prior for average intercept]}$$

$$\beta \sim \text{Normal}(0, 10) \qquad \text{[prior for average slope]}$$

$$\sigma \sim \text{HalfCauchy}(0, 1) \qquad \text{[prior stddev within cafés]}$$

$$\sigma_\alpha \sim \text{HalfCauchy}(0, 1) \qquad \text{[prior stddev among intercepts]}$$

$$\sigma_\beta \sim \text{HalfCauchy}(0, 1) \qquad \text{[prior stddev among slopes]}$$

$$R \sim \text{LKJcorr}(2) \qquad \text{[prior for correlation matrix]}$$

The likelihood and linear model need no explanation, at this point in the book. But the third line, which defines the population of varying intercepts and slopes, deserves attention.

$$\begin{bmatrix} \alpha_{\text{CAFÉ}} \\ \beta_{\text{CAFÉ}} \end{bmatrix} \sim \text{MVNormal}\left(\begin{bmatrix} \alpha \\ \beta \end{bmatrix}, S \right) \qquad \text{[population of varying effects]}$$

This line states that each café has an intercept $\alpha_{\text{CAFÉ}}$ and slope $\beta_{\text{CAFÉ}}$ with a prior distribution defined by the two-dimensional Gaussian distribution with means α and β and covariance matrix S. This statement of prior will adaptively regularize the individual intercepts, slopes, and the correlation among them.

The next line:

$$S = \begin{pmatrix} \sigma_\alpha & 0 \\ 0 & \sigma_\beta \end{pmatrix} R \begin{pmatrix} \sigma_\alpha & 0 \\ 0 & \sigma_\beta \end{pmatrix} \qquad \text{[construct covariance matrix]}$$

just states how we're constructing the covariance matrix S, by factoring it into separate standard deviations, σ_α and σ_β, and a correlation matrix R. There are other ways to go about this, but by splitting the covariance up into standard deviations and correlations, it'll be easier to later understand the inferred structure of the varying effects.

The rest of the model just defines fixed priors. The final line probably looks unfamiliar, though.

$$R \sim \text{LKJcorr}(2) \qquad \text{[prior for correlation matrix]}$$

The correlation matrix R needs a prior. It isn't easy to conceptualize what a distribution of matrices means. But in this introductory case, it isn't so hard. This particular correlation matrix is only 2-by-2 in size. So it looks like this:

$$R = \begin{pmatrix} 1 & \rho \\ \rho & 1 \end{pmatrix}$$

where ρ is the correlation between intercepts and slopes. So there's just one parameter to define a prior for. In larger matrices, with additional varying slopes, it gets more complicated. But even then the same LKJcorr prior will work.

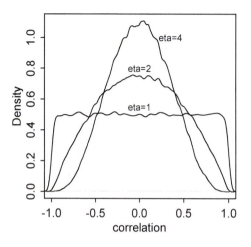

FIGURE 13.3. LKJcorr(η) probability density. The plot shows the distribution of correlation coefficients extracted from random 2-by-2 correlation matrices, for three values of η. When $\eta = 1$, all correlations are equally plausible. As η increases, extreme correlations become less plausible.

So whatever is the LKJcorr distribution? What LKJcorr(2) does is define a weakly informative prior on ρ that is skeptical of extreme correlations near -1 or 1.[161] You can think of it as a regularizing prior for correlations. This distribution has a single parameter, η, that controls how skeptical the prior is of large correlations in the matrix. When we use LKJcorr(1), the prior is flat over all valid correlation matrices. When the value is greater than 1, such as the 2 we used above, then extreme correlations are less likely. To visualize this family of priors, it will help to sample random matrices from it and plot the distribution of correlations. For example:

R code
13.11
```
R <- rlkjcorr( 1e4 , K=2 , eta=2 )
dens( R[,1,2] , xlab="correlation" )
```

This is shown in FIGURE 13.3, along with two other η values. When the matrix is larger, there are more correlations inside it, but the nature of the distribution remains the same. There is an example density for a 3-by-3 matrix in the help page examples, ?rlkjcorr.

To fit the model, we use a list of formulas that closely mirrors the model definition above. Note the use of c() to combine parameters into a vector.

R code
13.12
```
m13.1 <- map2stan(
    alist(
        wait ~ dnorm( mu , sigma ),
        mu <- a_cafe[cafe] + b_cafe[cafe]*afternoon,
        c(a_cafe,b_cafe)[cafe] ~ dmvnorm2(c(a,b),sigma_cafe,Rho),
        a ~ dnorm(0,10),
        b ~ dnorm(0,10),
        sigma_cafe ~ dcauchy(0,2),
        sigma ~ dcauchy(0,2),
        Rho ~ dlkjcorr(2)
    ) ,
    data=d ,
    iter=5000 , warmup=2000 , chains=2 )
```

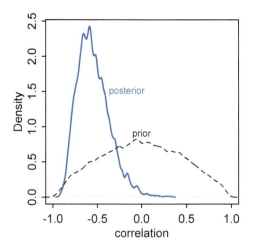

FIGURE 13.4. Posterior distribution of the correlation between intercepts and slopes. Blue: Posterior distribution of the correlation, reliably below zero. Dashed: Prior distribution, the LKJcorr(2) density.

The distribution dmvnorm2 is a multivariate Gaussian notation that takes a vector of means, c(a,b), a vector of standard deviations, sigma_cafe, and a correlation matrix, Rho. It constructs the covariance matrix internally. If you are interested in the details, you can peek at the raw Stan code with stancode(m13.1).

The first thing to notice when sampling from the model is that you'll get some of those (harmless, in this case) "severely ill-conditioned or misspecified" warnings. Once the chains get purring, those warnings will cease. You'll also see a couple of warnings stating "Warning (non-fatal): Left-hand side of sampling statement (~) contains a non-linear transform of a parameter or local variable." These are also harmless, and happen whenever you first compile a varying slopes model with map2stan. But they do deserve some explanation. See the Overthinking box at the end of this section for details.

Now instead of looking at the marginal estimates in the precis output, let's go straight to inspecting the posterior distribution of varying effects. First, let's examine the posterior correlation between intercepts and slopes.

```
post <- extract.samples(m13.1)
dens( post$Rho[,1,2] )
```

R code
13.13

The result is shown in FIGURE 13.4, with some additional decoration and the addition of the prior for comparison. The blue density is the posterior distribution of the correlation between intercepts and slopes. The posterior is concentrated on negative values, because the model has learned the negative correlation you can see in FIGURE 13.2. Keep in mind that the model did not get to see the true intercepts and slopes. All it had to work from was the observed wait times in morning and afternoon.

If you are curious about the impact of the prior, then you should change the prior and repeat the analysis. I suggest trying a flat prior, LKJcorr(1), and then a more strongly regularizing prior like LKJcorr(4) or LKJcorr(5).

Next, consider the shrinkage. The multilevel model estimates posterior distributions for intercepts and slopes of each café. The inferred correlation between these varying effects was used to pool information across them. This is just as the inferred variation among intercepts pools information among them, as well as how the inferred variation among slopes pools information among them. All together, the variances and correlation define an inferred multivariate Gaussian prior for the varying effects. And this prior, learned from the data, adaptively regularizes both the intercepts and slopes.

To see the consequence of this adaptive regularization, shrinkage, let's plot the posterior mean varying effects. Then we can compare them to raw, unpooled estimates. We'll also show the contours of the inferred prior—the population of intercepts and slopes—and this will help us visualize the shrinkage. Here's code to plot the unpooled estimates and posterior means.

R code
13.14

```
# compute unpooled estimates directly from data
a1 <- sapply( 1:N_cafes ,
        function(i) mean(wait[cafe_id==i & afternoon==0]) )
b1 <- sapply( 1:N_cafes ,
        function(i) mean(wait[cafe_id==i & afternoon==1]) ) - a1

# extract posterior means of partially pooled estimates
post <- extract.samples(m13.1)
a2 <- apply( post$a_cafe , 2 , mean )
b2 <- apply( post$b_cafe , 2 , mean )

# plot both and connect with lines
plot( a1 , b1 , xlab="intercept" , ylab="slope" ,
    pch=16 , col=rangi2 , ylim=c( min(b1)-0.1 , max(b1)+0.1 ) ,
    xlim=c( min(a1)-0.1 , max(a1)+0.1 ) )
points( a2 , b2 , pch=1 )
for ( i in 1:N_cafes ) lines( c(a1[i],a2[i]) , c(b1[i],b2[i]) )
```

And to superimpose the contours of the population:

R code
13.15

```
# compute posterior mean bivariate Gaussian
Mu_est <- c( mean(post$a) , mean(post$b) )
rho_est <- mean( post$Rho[,1,2] )
sa_est <- mean( post$sigma_cafe[,1] )
sb_est <- mean( post$sigma_cafe[,2] )
cov_ab <- sa_est*sb_est*rho_est
Sigma_est <- matrix( c(sa_est^2,cov_ab,cov_ab,sb_est^2) , ncol=2 )

# draw contours
library(ellipse)
for ( l in c(0.1,0.3,0.5,0.8,0.99) )
    lines(ellipse(Sigma_est,centre=Mu_est,level=l),
        col=col.alpha("black",0.2))
```

The result appears on the left in FIGURE 13.5. The blue points are the unpooled estimates for each café. The open points are the posterior means from the varying effects model. A line connects the points that belong to the same café. Each open point is displaced from the blue towards the center of the contours, as a result of shrinkage in both dimensions. Blue points farther from the center experience more shrinkage, because they are less plausible, given the inferred population.

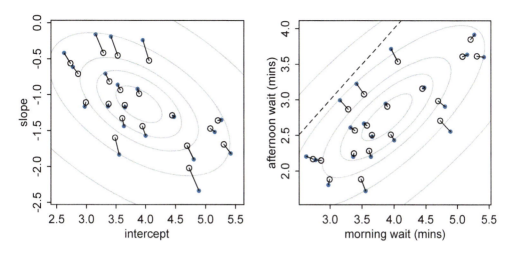

FIGURE 13.5. Shrinkage in two dimensions. Left: Raw unpooled intercepts and slopes (filled blue) compared to partially pooled posterior means (open circles). The gray contours show the inferred population of varying effects. Right: The same estimates on the outcome scale.

But notice too that shrinkage is not in direct lines towards the center. This is most obvious for the café that appears in the top-middle of the plot. That particular café had an average intercept, so it lies in the middle of the horizontal axis. But it also had an unusually high slope, so it lies at the top of the vertical axis. Pooled information from the other cafés results in skepticism about the slope. But since intercepts and slopes are correlated in the population as a whole, shrinking the slope down also shrinks the intercept. So all those angled shrinkage lines reflect the negative correlation between intercepts and slopes.

The right-hand plot in FIGURE 13.5 displays the same information, but now on the outcome scale. You can compute these average outcomes directly from your knowledge of the linear model:

```
# convert varying effects to waiting times
wait_morning_1 <- (a1)
wait_afternoon_1 <- (a1 + b1)
wait_morning_2 <- (a2)
wait_afternoon_2 <- (a2 + b2)
```

R code
13.16

The horizontal axis in the plot shows the expected morning wait, in minutes, for each café. The vertical axis shows the expected afternoon wait. Again the blue points are unpooled empirical estimates from the data. The open points are posterior predictions, using the pooled estimates. The diagonal dashed line shows where morning wait is equal to afternoon wait. What I want you to appreciate in this plot is that shrinkage on the parameter scale naturally produces shrinkage where we actually care about it: on the outcome scale. And it also implies a population of wait times, shown by the gray contours.

Overthinking: Jacobians and transforms. When you compile a `map2stan` model that contains vary-ing slopes, and therefore a multivariate prior, you will see a warning like this:

```
Warning (non-fatal): Left-hand side of sampling statement (~) contains a
non-linear transform of a parameter or local variable.
 You must call increment_log_prob() with the log absolute determinant of the
 Jacobian of the transform.
  Sampling Statement left-hand-side expression:
    v_a_blockbp_blockbpc_block ~ multi_normal_log(...)
```

This is quite confusing. What is going on here is that Stan is being cautious and warning the user about a potential problem with the model definition. The multivariate prior for the varying effects transforms the parameters, because it constructs a vector parameter with the multivariate prior. Then the individually named varying effect parameters in the `map2stan` model are transformed into that common vector. In general, parameter transformations require taking account of any change in the geometry of the posterior distribution that results from the transform. This is done by calculating the rates of change of each dimension (parameter), and the matrix that holds these rates of change is called the *Jacobian*, a matrix of partial derivatives. Multiplying the posterior probability by the absolute value of the determinate of that matrix provides the right adjustment to account for the transform.

Stan detects the presence of a transformation in the varying slopes model, and so it warns you to be sure you've taken account of any change in geometry. In this case, everything is fine, because there is no change in geometry: Each parameter is one-to-one inserted in a fixed position in the vector parameter. So the absolute Jacobian is exactly 1—no adjustment required. But it's nice that Stan is cautious—there's no way for it to verify that the transform has been properly accounted for. So instead of being alarmed, feel the warm glow that comes from knowing that Stan is looking out for you.

13.2. Example: Admission decisions and gender

Now let's return to a previous data example and incorporate varying slopes. This will allow you to appreciate how variation in slopes arises in a natural context, as well as how the correlation between intercepts and slopes can provide hints about process.

Recall the `UCBadmit` data from Chapter 10. In those data, failing to model the varying means across departments led to exactly the opposite inference of the truth. But we left some information on the floor, so to speak, by not using varying effects to pool information across departments. As a consequence, we probably overfit the smaller departments. We also ignored variation across departments in how they treated male and female applicants. Varying slopes will provide a direct way to model such variation.

Here's the data again, also constructing the dummy variable for male applications and the index variable for departments:

R code
13.17
```
library(rethinking)
data(UCBadmit)
d <- UCBadmit
d$male <- ifelse( d$applicant.gender=="male" , 1 , 0 )
d$dept_id <- coerce_index( d$dept )
```

Now we're ready to fit some models.

13.2.1. Varying intercepts. We'll begin slowly, by presenting just the varying intercept model for these data. Here's the model, with the varying intercept components in blue:

$$A_i \sim \text{Binomial}(n_i, p_i) \hspace{3cm} \text{[likelihood]}$$
$$\text{logit}(p_i) = \alpha_{\text{DEPT}[i]} + \beta m_i \hspace{2.5cm} \text{[linear model]}$$
$$\alpha_{\text{DEPT}} \sim \text{Normal}(\alpha, \sigma) \hspace{2.5cm} \text{[prior for varying intercepts]}$$
$$\alpha \sim \text{Normal}(0, 10) \hspace{3cm} \text{[prior for } \alpha\text{]}$$
$$\beta \sim \text{Normal}(0, 1) \hspace{3.2cm} \text{[prior for } \beta\text{]}$$
$$\sigma \sim \text{HalfCauchy}(0, 2) \hspace{2.7cm} \text{[prior for } \sigma\text{]}$$

The outcome variable A_i is the number of admit decisions, admit, and the sample size in each case is n_i, applications. Notice that I have placed the average intercept, α, in the linear model rather than inside the varying intercepts prior. This form is perfectly equivalent, recall, to placing α inside the prior. But it means the α_{DEPT} parameters are now displacements from the average department.

Here's the code to fit the varying intercepts model:

R code
13.18

```
m13.2 <- map2stan(
    alist(
        admit ~ dbinom( applications , p ),
        logit(p) <- a_dept[dept_id] + bm*male,
        a_dept[dept_id] ~ dnorm( a , sigma_dept ),
        a ~ dnorm(0,10),
        bm ~ dnorm(0,1),
        sigma_dept ~ dcauchy(0,2)
    ) ,
    data=d , warmup=500 , iter=4500 , chains=3 )
precis( m13.2 , depth=2 ) # depth=2 to display vector parameters
```

	Mean	StdDev	lower 0.89	upper 0.89	n_eff	Rhat
a_dept[1]	0.67	0.10	0.51	0.83	4456	1
a_dept[2]	0.63	0.12	0.44	0.81	4758	1
a_dept[3]	-0.59	0.08	-0.70	-0.46	6831	1
a_dept[4]	-0.62	0.09	-0.76	-0.48	5258	1
a_dept[5]	-1.06	0.10	-1.22	-0.90	9368	1
a_dept[6]	-2.61	0.16	-2.87	-2.36	6969	1
a	-0.59	0.64	-1.61	0.36	5115	1
bm	-0.09	0.08	-0.22	0.04	3480	1
sigma_dept	1.48	0.60	0.71	2.16	4633	1

The estimated effect of male is very similar to what we got in Chapter 10. But now we also have better estimates of the individual department average acceptance rates. You'll see that the departments are ordered from those with the highest proportions accepted to the lowest. Remember, the values above are the α_{DEPT} estimates, and so they are deviations from the global mean α, which in this case has posterior mean -0.58. So department A, "[1]" in the table, has the highest average admission rate. Department F, "[6]" in the table, has the lowest.

13.2.2. Varying effects of being male. Now let's consider the variation in gender bias among departments. Sure, overall there isn't much evidence of gender bias in the previous model. But what if we allow the effect of an applicant's being male to vary in the same way we already allowed the overall rate of admission to vary? This will constitute the varying slopes model in this context.

One extra feature of varying slopes that will arise here is that since there is substantial *imbalance* in sample size across departments and the numbers of male and female applications they received, pooling will be stronger for those cases with fewer applications. Department B, for example, received only 25 applications from females. So any estimate of how that department differently treats males and females will shrink towards the population average. In contrast, department F received hundreds of applications from both males and females. So pooling will do very little to the estimates for that department.

This is what the varying slopes model looks like, with the varying effects components in blue:

$$A_i \sim \text{Binomial}(n_i, p_i) \qquad \text{[likelihood]}$$

$$\text{logit}(p_i) = \alpha_{\text{DEPT}[i]} + \beta_{\text{DEPT}[i]} m_i \qquad \text{[linear model]}$$

$$\begin{bmatrix} \alpha_{\text{DEPT}} \\ \beta_{\text{DEPT}} \end{bmatrix} \sim \text{MVNormal}\left(\begin{bmatrix} \alpha \\ \beta \end{bmatrix}, \mathbf{S} \right) \qquad \text{[joint prior for varying effects]}$$

$$\mathbf{S} = \begin{pmatrix} \sigma_\alpha & 0 \\ 0 & \sigma_\beta \end{pmatrix} \mathbf{R} \begin{pmatrix} \sigma_\alpha & 0 \\ 0 & \sigma_\beta \end{pmatrix}$$

$$\alpha \sim \text{Normal}(0, 10) \qquad \text{[prior for } \alpha]$$

$$\beta \sim \text{Normal}(0, 1) \qquad \text{[prior for } \beta]$$

$$(\sigma_\alpha, \sigma_\beta) \sim \text{HalfCauchy}(0, 2) \qquad \text{[prior for each } \sigma]$$

$$\mathbf{R} \sim \text{LKJcorr}(2) \qquad \text{[prior for correlation matrix]}$$

The symbol m_i indicates the value of `male` for the i-th row. It is multiplied by the sum $\beta + \beta_{\text{DEPT}[i]}$, which is a total slope defined by both a value common to all departments, β, and a value unique to the department for row i, $\beta_{\text{DEPT}[i]}$.

To fit this model:

R code
13.19
```
m13.3 <- map2stan(
    alist(
        admit ~ dbinom( applications , p ),
        logit(p) <- a_dept[dept_id] +
                    bm_dept[dept_id]*male,
        c(a_dept,bm_dept)[dept_id] ~ dmvnorm2( c(a,bm) , sigma_dept , Rho ),
        a ~ dnorm(0,10),
        bm ~ dnorm(0,1),
        sigma_dept ~ dcauchy(0,2),
        Rho ~ dlkjcorr(2)
    ) ,
    data=d , warmup=1000 , iter=5000 , chains=4 , cores=3 )
```

Check for yourself that the chains mixed and converged excellently. You might get a warning about a few "divergent iterations." We'll focus on those in the next section.

We're interested in what adding varying slopes has revealed. So let's look at the marginal posterior distributions for the varying effects only:

R code
13.20

```
plot( precis(m13.3,pars=c("a_dept","bm_dept"),depth=2) )
```

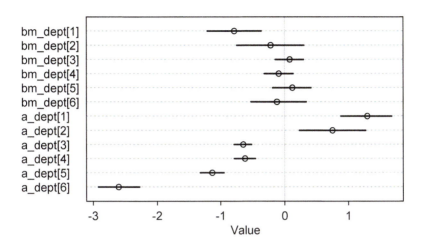

Notice that the intercepts range all over the place, while the slopes all cling close to zero. This reflects the fact that departments varied a lot in overall admission rates, but they neither discriminated much between male and female applicants nor varied much in how much they discriminated.

But there are a few departments with slopes consistent with noticeable bias, departments 1 and 2 in particular. Department 1 has a slope estimate centered almost 1 log-odds below the mean, and it doesn't have any mass on the other side of zero. Department 2 has a highly uncertain slope, but that means it also includes plausibly large effects. Just because a marginal posterior overlaps zero does not mean we should think of it as zero.

Notice also that these two departments have the largest intercepts. So let's look at the estimated correlation between intercepts and slopes next, as well as the two-dimensional shrinkage it induces.

13.2.3. Shrinkage. The posterior correlation between intercepts and slopes is shown in the left-hand plot of FIGURE 13.6. The majority of the probability mass is below zero, indicating a negative correlation. This corresponds to the fact from just above: The departments with the highest admissions rates also have the smallest slopes.

The right-hand plot shows the shrinkage in both intercepts and slopes. This plot is analogous to the shrinkage plot from the previous section, FIGURE 13.5 (page 397). The blue points are again the raw empirical (unpooled) estimates. The open points are the varying effect (adaptively pooled) estimates. The lines connect points from the same departments, and the text labels correspond to the department labels. The gray contours show the inferred population of intercepts and slopes.

Again, shrinkage follows a negative correlation, although a weaker one in this case. The department with the most shrinkage is A, which has the most extreme intercept and slope. More interesting and instructive is the shrinkage for department F. That department had an average raw slope, but an unusually low intercept. As a result, the intercept moves a little towards the mean, as a result of pooling, and the slope moves down, as a result of the negative correlation in the population. That is to say that the model thinks that, since high intercepts

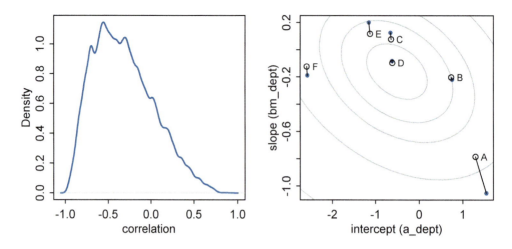

FIGURE 13.6. Left: Posterior distribution of the correlation between intercepts and slopes for the UCB admissions model, m13.3. Right: Two-
dimensional shrinkage of unpooled (blue) and adaptively pooled (open) estimates.

and associated with low slopes overall, if department F's intercepts are too small, then its
slope is probably also too big.

Overall, there isn't much shrinkage going on in this model. This is because there are a
lot of applications on each row. But even though the absolute scale of shrinkage is small here,
the nature of the shrinkage again illustrates the properties of varying effects.

13.2.4. Model comparison. To make this more interesting, we can also fit the model that
ignores gender. Then we can compare all of these models, using WAIC.

R code
13.21

```
m13.4 <- map2stan(
    alist(
        admit ~ dbinom( applications , p ),
        logit(p) <- a_dept[dept_id],
        a_dept[dept_id] ~ dnorm( a , sigma_dept ),
        a ~ dnorm(0,10),
        sigma_dept ~ dcauchy(0,2)
    ) ,
    data=d , warmup=500 , iter=4500 , chains=3 )

compare( m13.2 , m13.3 , m13.4 )
```

```
      WAIC pWAIC dWAIC weight    SE  dSE
m13.3 5191.4  11.3   0.0   0.99 57.30   NA
m13.4 5201.2   6.0   9.8   0.01 56.84 6.84
m13.2 5201.9   7.2  10.5   0.01 56.94 6.50
```

The model that ignores gender, m13.4, earns the same expected out-of-sample performance
as the model that includes a *constant* effect of gender, m13.2. The varying slopes model,
m13.3, dominates both. This is despite the fact that the *average* slope in m13.3 is nearly zero.

The average isn't what matters, however. It is the individual slopes, one for each department, that matter. If we wish to generalize to new departments, the variation in slopes suggests that it'll be worth paying attention to gender, even if the average slope is nearly zero in the population.

13.2.5. More slopes. The varying slopes strategy generalizes to as many slopes as you like, within practical limits. All that happens is that each new predictor you want to construct varying slopes for adds one more dimension to the covariance matrix of the varying effects prior. So this means one more standard deviation parameter and one more dimension to the correlation matrix.

For example, suppose the UCB admissions data also recorded the test scores of each applicant. Then we could also include test score as a predictor. But we don't have another predictor for these data. So instead we'll turn to another data set, in the next section, to go deeper into varying slopes.

13.3. Example: Cross-classified chimpanzees with varying slopes

To see how to construct a model with more than two varying effects—varying intercepts plus more than one varying slope—as well as with more than one type of cluster, we'll return to the chimpanzee experiment data that was introduced in Chapter 10. In these data, there are two types of clusters: actors and blocks. We explored *cross-classification* with two kinds of varying intercepts back on page 370. We also modeled the experiment with two different slopes: one for the effect of the prosocial option (the side of the table with two pieces of food) and one for the interaction between the prosocial option and the presence of another chimpanzee. So now we'll model both types of clusters and place varying effects on the intercepts and both slopes.

I'll also use this example to emphasize the importance of NON-CENTERED PARAMETERI-ZATION for some multilevel models. For any given multilevel model, there are several different ways to write it down. These ways are called "parameterizations." Mathematically, these alternative parameterizations are equivalent, but inside the MCMC engine they are not. Remember, how you fit the model is part of the model. Choosing a better parameterization is an awesome way to improve sampling for your MCMC model fit, and the non-centered parameterization tends to help a lot with complex varying effect models like the one you'll work with in this section. I'll hide the details of the technique in the main text. But as usual, there is an Overthinking box at the end that provides some detail.

Okay, let's construct a cross-classified varying slopes model. To maintain some sanity with this complicated model, we'll use more than one linear model in the formulas. This will allow us to compartmentalize sub-models for the intercepts and each slope. Here's what the likelihood and its linear models look like:

$$L_i \sim \text{Binomial}(1, p_i)$$
$$\text{logit}(p_i) = \mathcal{A}_i + (\mathcal{B}_{P,i} + \mathcal{B}_{PC,i}C_i)P_i \qquad \text{[linear model skeleton]}$$
$$\mathcal{A}_i = \alpha + \alpha_{\text{ACTOR}[i]} + \alpha_{\text{BLOCK}[i]} \qquad \text{[intercept model]}$$
$$\mathcal{B}_{P,i} = \beta_P + \beta_{P,\text{ACTOR}[i]} + \beta_{P,\text{BLOCK}[i]} \qquad \text{[P slope model]}$$
$$\mathcal{B}_{PC,i} = \beta_P + \beta_{PC,\text{ACTOR}[i]} + \beta_{PC,\text{BLOCK}[i]} \qquad \text{[}P \times C \text{ interaction model]}$$

The linear model for logit(p_i) defines three sub-models, one for each "effect" in the model. $\mathcal{A}_i$ is the sub-model for the intercept for observation i. Similarly, $\mathcal{B}_{P,i}$ and $\mathcal{B}_{PC,i}$ are the sub-models for the slopes. All that these sub-models do is present the model differently. If you substitute each into the first linear model, you'll get the more standard form we've been using.

But having the separate sub-models helps with explanation. The sub-models make clear that varying effects decompose ordinary intercepts and slopes into means and offsets. For each case i, the intercept is a constant mean, α, plus an offset for the actor, $\alpha_{\text{ACTOR}[i]}$, plus an offset for the block, $\alpha_{\text{BLOCK}[i]}$. The two slopes are constructed analogously.

The next part of the model are the multivariate priors. Since there are two cluster types, actors and blocks, there are two multivariate Gaussian priors. The multivariate Gaussian priors are both 3-dimensional, in this example. But in general, you can choose to have different varying effects in different cluster types. Here are the two priors in this case:

$$\begin{bmatrix} \alpha_{\text{ACTOR}} \\ \beta_{P,\text{ACTOR}} \\ \beta_{PC,\text{ACTOR}} \end{bmatrix} \sim \text{MVNormal}\left(\begin{bmatrix} 0 \\ 0 \\ 0 \end{bmatrix}, \mathbf{S}_{\text{ACTOR}} \right)$$

$$\begin{bmatrix} \alpha_{\text{BLOCK}} \\ \beta_{P,\text{BLOCK}} \\ \beta_{PC,\text{BLOCK}} \end{bmatrix} \sim \text{MVNormal}\left(\begin{bmatrix} 0 \\ 0 \\ 0 \end{bmatrix}, \mathbf{S}_{\text{BLOCK}} \right)$$

What these priors state is that actors and blocks come from two different statistical populations. Within each, the three features of each actor or block are related through a covariance matrix specific to that population. There are no means in these priors, just because we already placed the average effects—α, β_P, and β_{PC}—in the linear models.

And the map2stan code for this model looks as you'd expect, given previous examples. To define the multiple linear models, just write each into the formula list in order. I'll add some white space and comments to this formula list, to make it easier to read.

R code
13.22

```
library(rethinking)
data(chimpanzees)
d <- chimpanzees
d$recipient <- NULL
d$block_id <- d$block

m13.6 <- map2stan(
    alist(
        # likeliood
        pulled_left ~ dbinom(1,p),

        # linear models
        logit(p) <- A + (BP + BPC*condition)*prosoc_left,
        A <- a + a_actor[actor] + a_block[block_id],
        BP <- bp + bp_actor[actor] + bp_block[block_id],
        BPC <- bpc + bpc_actor[actor] + bpc_block[block_id],

        # adaptive priors
        c(a_actor,bp_actor,bpc_actor)[actor] ~
                            dmvnorm2(0,sigma_actor,Rho_actor),
        c(a_block,bp_block,bpc_block)[block_id] ~
```

```
                        dmvnorm2(0,sigma_block,Rho_block),

    # fixed priors
    c(a,bp,bpc) ~ dnorm(0,1),
    sigma_actor ~ dcauchy(0,2),
    sigma_block ~ dcauchy(0,2),
    Rho_actor ~ dlkjcorr(4),
    Rho_block ~ dlkjcorr(4)
) , data=d , iter=5000 , warmup=1000 , chains=3 , cores=3 )
```

When sampling from this model, you should notice several of those "severely ill-conditioned or misspecified" warnings. Once warmup completes, those should stop. More importantly, you should get a scary warning about "divergent iterations":

```
Warning message:
In map2stan(alist(pulled_left ~ dbinom(1, p), logit(p) <- A + (BP +  :
  There were 559 divergent iterations during sampling.
Check the chains (trace plots, n_eff, Rhat) carefully to ensure they are valid.
```

This is a very important issue with fitting varying effects models with Hamiltonian Monte Carlo. In this case, as in many, you can just use brute force: Sample very thoroughly in order to get eventual convergence. But sometimes brute force is no solution. The chains may never converge. But even when they do converge, the chains are inefficient, and these "divergent" iterations are a highly technical indicator of that fact.

This is where using the NON-CENTERED PARAMETERIZATION will help. The Overthinking box at the end of the section explains what this means in more detail. For the moment, let's just re-express the model with the alternative parameterization. The dmvnormNC density in map2stan does all the work for you, hiding the change in parameterization and making it easy to switch back and forth. Here's the same model, but now using dvnormNC for the varying effect priors, instead of the usual dmvnorm2:

R code
13.23

```
m13.6NC <- map2stan(
    alist(
        pulled_left ~ dbinom(1,p),
        logit(p) <- A + (BP + BPC*condition)*prosoc_left,
        A <- a + a_actor[actor] + a_block[block_id],
        BP <- bp + bp_actor[actor] + bp_block[block_id],
        BPC <- bpc + bpc_actor[actor] + bpc_block[block_id],
        # adaptive NON-CENTERED priors
        c(a_actor,bp_actor,bpc_actor)[actor] ~
                            dmvnormNC(sigma_actor,Rho_actor),
        c(a_block,bp_block,bpc_block)[block_id] ~
                            dmvnormNC(sigma_block,Rho_block),
        c(a,bp,bpc) ~ dnorm(0,1),
        sigma_actor ~ dcauchy(0,2),
        sigma_block ~ dcauchy(0,2),
        Rho_actor ~ dlkjcorr(4),
        Rho_block ~ dlkjcorr(4)
    ) , data=d , iter=5000 , warmup=1000 , chains=3 , cores=3 )
```

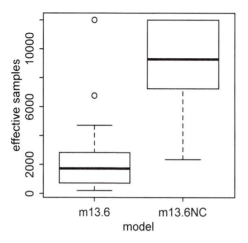

FIGURE 13.7. Distributions of effective samples, n_eff, for the ordinary and non-centered parameterizations of the cross-classified varying slopes model, m13.6 and m13.6NC, respectively. Both models arrive at equivalent inferences, but the non-centered version samples much more efficiently.

Note that there is no zero, 0, in the adaptive priors. With the non-centered parameterization, the means (a, bp, and bpc in this case) are always in the linear model, never inside the prior. In fact, as the Overthinking box at the end of the section explains, *all* of the parameters of the prior, including the correlation matrix Rho, are smuggled out and into the linear model to construct a non-centered parameterization.

How has the non-centered parameterization helped here? First, notice that m13.6NC produces no divergent iterations. It also sampled noticeably faster than m13.6. If you compare the precis output of the two models, you'll see that they arrive at the same inferences. But the n_eff values for m13.6NC are much larger. Let's show the difference in effective samples visually, using a simple boxplot:

R code
13.24

```
# extract n_eff values for each model
neff_c <- precis(m13.6,2)@output$n_eff
neff_nc <- precis(m13.6NC,2)@output$n_eff
# plot distributions
boxplot( list( 'm13.6'=neff_c , 'm13.6NC'=neff_nc ) ,
    ylab="effective samples" , xlab="model" )
```

FIGURE 13.7 displays the result. The non-centered version of the model samples much more efficiently, producing more effective samples per parameter. In practice, this means you don't need as many actual iterations, iter, to arrive at an equally good portrait of the posterior distribution. For larger data sets, the savings can mean hours of time.

This model has 54 parameters: 3 average effects, 3×7 varying effects on actor, 3×6 varying effects on block, 6 standard deviations, and 6 free correlation parameters. You can check them all for yourself with precis(m13.6NC,depth=2). But effectively the model has only about 18 parameters—check WAIC(m13.6NC). The two varying effects populations, one for actors and one for blocks, regularize the varying effects themselves. So as usual, each varying intercept or slope counts less than one effective parameter.

We can inspect the standard deviation parameters to get a sense of how aggressively the varying effects are being regularized:

R code
13.25

```
precis( m13.6NC , depth=2 , pars=c("sigma_actor","sigma_block") )
```

```
              Mean StdDev lower 0.89 upper 0.89 n_eff Rhat
sigma_actor[1] 2.33   0.90       1.12       3.46  3296    1
sigma_actor[2] 0.46   0.36       0.00       0.88  5677    1
sigma_actor[3] 0.52   0.49       0.00       1.08  5868    1
sigma_block[1] 0.22   0.20       0.00       0.46  5809    1
sigma_block[2] 0.57   0.40       0.00       1.03  3931    1
sigma_block[3] 0.51   0.42       0.00       1.01  5834    1
```

The [1] index in each vector is the varying intercept standard deviation, while the [2] and [3] are the slopes. While these are just posterior means, and the amount of shrinkage averages over the entire posterior, you can get a sense from the small values that shrinkage is pretty aggressive here. This is what takes the model from 56 actual parameters to 18 effective parameters, as measured by WAIC.

This is a good example of how varying effects adapt to the data. The overfitting risk is much milder here than it would be with ordinary fixed effects. It can of course be challenging to define and fit these models. But if you don't check for variation in slopes, you may never notice it. And even if the average slope is almost zero, there might still be substantial variation in slopes across clusters.

When you fit a model like this, with more than one linear model, link will work a little differently. It will return a list of matrices, one matrix for each linear model. So in this case:

R code
13.26

```
p <- link(m13.6NC)
str(p)
```

```
List of 4
 $ p  : num [1:1000, 1:504] 0.304 0.251 0.34 0.285 0.316 ...
 $ A  : num [1:1000, 1:504] -0.829 -1.095 -0.662 -0.92 -0.774 ...
 $ BP : num [1:1000, 1:504] 0.534 1.141 0.319 0.371 0.892 ...
 $ BPC: num [1:1000, 1:504] 0.4237 -1.1239 0.1357 0.0185 0.1143 ...
```

The only one that matters here is p, the probability at the top of the model. The others—A, BP, and BPC—just feed into it. But if you are going to use link to construct predictions, you'll need to extract the right matrix.

Finally, let's compare the varying slopes model to the simpler varying intercepts model from the previous chapter, model m12.5 (page 373). Jump back and fit that model again, if necessary. Then compare the two models using WAIC:

R code
13.27

```
compare( m13.6NC , m12.5 )
```

```
         WAIC pWAIC dWAIC weight    SE  dSE
m12.5   532.7  10.4     0   0.73 19.66   NA
m13.6NC 534.7  18.3     2   0.27 19.89  4.1
```

This is the kind of result you get when there is hardly any predictive difference between the models. The varying slopes model makes very similar predictions, because there isn't much variation across actors or blocks in the slopes. You could see that in the precis output at the top of the page. And the WAIC values reflect it.

In the end, you might simplify your life by reducing the model down to only vary the slopes with important variation. In this case, effective inference doesn't depend upon including varying effects on block at all. There are no "correct" models in this business. It's often more useful to understand the data through the consilience of multiple models than to try find any one true model.

In this example, no matter which varying effect structure you use, you'll find that actors vary a lot in their baseline preference for the left-hand lever. Everything else is much less important. But using the most complex model, m13.6NC, tells the correct story. Because the varying slopes are adaptively regularized, the model hasn't overfit much, relative to the simpler model that contains only the important intercept variation.

Overthinking: Non-centered parameterization of the multilevel model. Stan sometimes has trouble efficiently sampling a multilevel model of the typical form. This isn't just an issue with Stan. All MCMC algorithms have analogous problems. Signs of the problem include low `n_eff` values, high `Rhat` values, and divergent iteration warnings. Often running the chains long enough will produce reliable samples from the posterior, but this can be very inefficient. This was the case with m13.6 in the main text.

A better idea is to re-parameterize the model to use a *non-centered parameterization* for the multivariate Gaussian varying effects prior. The word "non-centered" here is entirely unhelpful, but it is the standard terminology, unfortunately.[162] Better to say that we're going to use a *standardized adaptive prior* for the varying effects. This means the means will be zero (you've done this already) and the standard deviations will all be one. Just like you can move the means to the linear model, you can also move the standard deviations to the linear model. How? For any given Gaussian distribution, we can always subtract out the mean and factor out the standard deviation. So for example, this:

$$y \sim \text{Normal}(\mu, \sigma)$$

is equivalent to this:

$$y = \mu + z\sigma$$
$$z \sim \text{Normal}(0, 1)$$

This fact allows us to take the means and standard deviations out of a Gaussian distribution and place them in the linear model. This leaves only a standardized multivariate Gaussian prior, with all means at zero and standard deviations at one. Technically this is the same model. But for some models and data, this form of the model will sample much more efficiently.

For the example of m13.6, here's what the non-centered linear models look like, with the standard deviation parameters highlighted in blue:

$$L_i \sim \text{Binomial}(1, p_i)$$
$$\text{logit}(p_i) = \mathcal{A}_i + (\mathcal{B}_{P,i} + \mathcal{B}_{PC,i}C_i)P_i$$
$$\mathcal{A}_i = \alpha + \alpha_{\text{ACTOR}[i]}\sigma_{\text{ACTOR},1} + \alpha_{\text{BLOCK}[i]}\sigma_{\text{BLOCK},1}$$
$$\mathcal{B}_{P,i} = \beta_P + \beta_{P,\text{ACTOR}[i]}\sigma_{\text{ACTOR},2} + \beta_{P,\text{BLOCK}[i]}\sigma_{\text{BLOCK},2}$$
$$\mathcal{B}_{PC,i} = \beta_P + \beta_{PC,\text{ACTOR}[i]}\sigma_{\text{ACTOR},3} + \beta_{PC,\text{BLOCK}[i]}\sigma_{\text{BLOCK},3}$$

Notice that each varying intercept and slope is multiplied by its corresponding scale parameter, one of the standard deviations that ordinarily appears inside the multivariate Gaussian prior for these effects. This just undoes the standardization that is imposed in the prior. Inside the priors, there are no covariance matrices now. There are just correlation matrices, since the scale of each dimension has been migrated to the linear model. (We'll take the correlation matrix out of the prior, as well. But let's leave that aside for the moment.) So these adaptive priors are essentially distributions of correlated z-scores. Here's what all of this looks like in the modified map2stan code:

```
m13.6nc1 <- map2stan(
    alist(
        pulled_left ~ dbinom(1,p),

        # linear models
        logit(p) <- A + (BP + BPC*condition)*prosoc_left,
        A <- a + za_actor[actor]*sigma_actor[1] +
                za_block[block_id]*sigma_block[1],
        BP <- bp + zbp_actor[actor]*sigma_actor[2] +
                zbp_block[block_id]*sigma_block[2],
        BPC <- bpc + zbpc_actor[actor]*sigma_actor[3] +
                zbpc_block[block_id]*sigma_block[3],

        # adaptive priors
        c(za_actor,zbp_actor,zbpc_actor)[actor] ~ dmvnorm(0,Rho_actor),
        c(za_block,zbp_block,zbpc_block)[block_id] ~ dmvnorm(0,Rho_block),

        # fixed priors
        c(a,bp,bpc) ~ dnorm(0,1),
        sigma_actor ~ dcauchy(0,2),
        sigma_block ~ dcauchy(0,2),
        Rho_actor ~ dlkjcorr(4),
        Rho_block ~ dlkjcorr(4)
    ) ,
    data=d ,
    start=list( sigma_actor=c(1,1,1), sigma_block=c(1,1,1) ),
    constraints=list( sigma_actor="lower=0", sigma_block="lower=0" ),
    types=list( Rho_actor="corr_matrix", Rho_block="corr_matrix" ),
    iter=5000 , warmup=1000 , chains=3 , cores=3 )
```

The start, constraints, and types lists at the bottom are needed to inform Stan of the proper dimension and constraints on the parameters. This model produces the same effective inference as the original version of the model, m13.6. But it samples much more efficiently, and it produces zero of those divergent iterations. To see how much more efficiently, check the n_eff from the model form above and compare it to that of the ordinary version, m13.6.

It is possible to further extend this non-centering strategy of taking parameters out of the prior. All that is left in the priors above are the correlation matrices, Rho_actor and Rho_block. These too can be extracted by using a CHOLESKY DECOMPOSITION of the correlation matrix. A Cholesky decomposition L is a way to represent a square, symmetric matrix like a correlation matrix R such that $R = LL^T$. It is a marvelous fact that you can multiply L by a vector of uncorrelated samples and end up with a vector of correlated samples. This is the trick that lets us take even the correlations out of the prior. We just sample a vector of uncorrelated z-scores and then multiply those by the Cholesky factor and the standard deviations to get the varying effects with the correct scale and correlation.

The details of the implementation are a bit technical. But there is a special density, dmvnormNC, that you can use with map2stan that will automate it. That's what we did in the main text, when we defined m13.6NC. The automation even hides the z-scores from you, but they are still there. If you check the precis(m13.6NC) output, you'll see them, as well as the Cholesky factors L_Rho_actor and L_Rho_block that are actually used in sampling. In this particular example, this form of the model samples most efficiently. Look at stancode(m13.6NC), if you are curious about the details of this golem. It's messy in there, but in the transformed parameters block you'll see the formulas that construct the ordinary varying effects from the z-scores, Cholesky factors, and standard deviation parameters.

The cost of these non-centered forms is twofold. First, it looks a lot more confusing. There are standard deviations in the linear models, so this is a model that only a mother could love. Taking the correlation out too, the Cholesky parameterization, is even more opaque. Hard-to-read models and model code limit our ability to share implementations with our colleagues, and sharing is a principle goal of scientific computation. Second, the varying effects parameters emerge as z-scores, so they have to be processed further to be interpreted.

Finally, not all combinations of model structure and data benefit from the non-centered parameterization. Sometimes the centered version—putting the means and standard deviations in the prior—is better. So you might try the form that is most natural for you personally. If it gives you trouble, try an alternative form. With some experience, different forms of the same model become familiar. There is a practice problem at the end of this chapter that may help.

13.4. Continuous categories and the Gaussian process

All of the varying effects so far, whether they were intercepts or slopes, have been defined over discrete, unordered categories. For example, cafés are unique places, and there is no sense in which café 1 comes before café 2. The "1" and "2" are just labels for unique things. The same goes for tadpole ponds, academic departments, or individual chimpanzees. By estimating unique parameters for each cluster of this kind, we can quantify some of the unique features that generate variation across clusters and covariation among the observations within each cluster. Pooling across the clusters improves accuracy and simultaneously provides a picture of the variation.

But what about continuous dimensions of variation like age or income or stature? Individuals of the same age share some of the same exposures. They listened to some of the same music, heard about the same politicians, and experienced the same weather events. And individuals of *similar* ages also experienced some of these same exposures, but to a lesser extent than individuals of the same age. The covariation falls off as any two individuals become increasingly dissimilar in age or income or stature or any other dimension that indexes background similarity. It doesn't make sense to estimate a unique varying intercept for all individuals of the same age, ignoring the fact that individuals of similar ages should have more similar intercepts. And of course, it's likely that every individual in your sample has a unique age. So then continuous differences in similarity are all you have to work with.

Luckily, there is a way to apply the varying effects approach to continuous categories of this kind. This will allow us to estimate a unique intercept (or slope) for any age, while still regarding age as a continuous dimension in which similar ages have more similar intercepts (or slopes). The general approach is known as GAUSSIAN PROCESS REGRESSION.[163] This name is unfortunately wholly uninformative about what it is for and how it works.

We'll proceed to work through a basic example that demonstrates both what it is for and how it works. The general purpose is to define some dimension along which cases differ. This might be individual differences in age. Or it could be differences in location. Then we measure the distance between each pair of cases. What the model then does is estimate a function for the covariance between pairs of cases at different distances. This covariance function provides one continuous category generalization of the varying effects approach.

13.4.1. Example: Spatial autocorrelation in Oceanic tools. When we looked at the complexity of tool kits among historic Oceanic societies, back in Chapter 10 (page 313), we used a crude binary contact predictor as a proxy for possible exchange among societies. But that variable is pretty unsatisfying. First, it takes no note of which other societies each had contact (or not) with. If all of your neighbors are small islands, then high rate of contact with them may not do much at all to tool complexity. Second, if indeed tools were exchanged among societies—and we know they were—then the total number of tools for each are truly not independent of one another, even after we condition on all of the predictors. Instead we expect close geographic neighbors to have more similar tool counts, because of exchange. Third, closer islands may share unmeasured geographic features like sources of stone or shell that lead to similar technological industries. So space could matter in multiple ways.

This is a classic setting in which to use Gaussian process regression. We'll define a distance matrix among the societies. Then we can estimate how similarity in tool counts depends upon geographic distance. You'll see how to simultaneously incorporate ordinary predictors, so that the covariation among societies with distance will both control for and be controlled by other factors that influence technology.

Let's begin by loading the data and inspecting the geographic distance matrix. I've already gone ahead and looked up the as-the-crow-flies navigation distance between each pair of societies. These distances are measured in thousands of kilometers, and the matrix of them is in the rethinking package:

R code
13.29

```
# load the distance matrix
library(rethinking)
data(islandsDistMatrix)

# display short column names, so fits on screen
Dmat <- islandsDistMatrix
colnames(Dmat) <- c("Ml","Ti","SC","Ya","Fi","Tr","Ch","Mn","To","Ha")
round(Dmat,1)
```

```
             Ml  Ti  SC  Ya  Fi  Tr  Ch  Mn  To  Ha
Malekula    0.0 0.5 0.6 4.4 1.2 2.0 3.2 2.8 1.9 5.7
Tikopia     0.5 0.0 0.3 4.2 1.2 2.0 2.9 2.7 2.0 5.3
Santa Cruz  0.6 0.3 0.0 3.9 1.6 1.7 2.6 2.4 2.3 5.4
Yap         4.4 4.2 3.9 0.0 5.4 2.5 1.6 1.6 6.1 7.2
Lau Fiji    1.2 1.2 1.6 5.4 0.0 3.2 4.0 3.9 0.8 4.9
Trobriand   2.0 2.0 1.7 2.5 3.2 0.0 1.8 0.8 3.9 6.7
Chuuk       3.2 2.9 2.6 1.6 4.0 1.8 0.0 1.2 4.8 5.8
Manus       2.8 2.7 2.4 1.6 3.9 0.8 1.2 0.0 4.6 6.7
Tonga       1.9 2.0 2.3 6.1 0.8 3.9 4.8 4.6 0.0 5.0
Hawaii      5.7 5.3 5.4 7.2 4.9 6.7 5.8 6.7 5.0 0.0
```

Notice that the diagonal is all zeros, because each society is zero kilometers from itself. Also notice that the matrix is symmetric around the diagonal, because the distance between two societies is the same whichever society we measure from.

We'll use these distances as a measure of similarity in technology exposure. This will allow us to estimate varying intercepts for each society that account for non-independence in tools as a function of their geographical similarly. The approach would be very similar for phylogenetic distance or distance in age or any other continuous dimension of similarity we think influences observations.

The first part of the model is a familiar Poisson likelihood and a varying intercept linear model, with a log link:

$$T_i \sim \text{Poisson}(\lambda_i)$$
$$\log \lambda_i = \alpha + \gamma_{\text{SOCIETY}[i]} + \beta_P \log P_i$$

The γ_{SOCIETY} parameters will be the varying intercepts in this case. But unlike typical varying intercepts, they will be estimated in light of geographic distance, not distinct category membership. I've also included an ordinary coefficient for log population. We'll be concerned with whether including spatial similarity washes out the association between log population and the total tools.

The heart of the Gaussian process is the multivariate prior for these intercepts:

$$\gamma \sim \text{MVNormal}\big([0, \ldots, 0], \mathbf{K}\big) \qquad \text{[prior for intercepts]}$$
$$\mathbf{K}_{ij} = \eta^2 \exp(-\rho^2 D_{ij}^2) + \delta_{ij}\sigma^2 \qquad \text{[define covariance matrix]}$$

The first line is the 10-dimensional Gaussian prior for the intercepts. It has 10 dimensions, because there are 10 societies in the distance matrix. The vector of means is all zeros, because we've put the grand mean α in the linear model, which makes these intercepts deviations from the expectation.

The covariance matrix for these intercepts is named $\mathbf{K}$, and the covariance between any pair of societies i and j is $\mathbf{K}_{ij}$. This covariance is defined by the formula on the second line above. This formula uses three parameters—η, ρ, and σ—to model how covariance among societies changes with distances among them. It probably looks very unfamiliar. I'll walk you through it in pieces.

The part of the formula for $\mathbf{K}$ that gives the covariance model its shape is $\exp(-\rho^2 D_{ij}^2)$. D_{ij} is the distance between the i-th and j-th societies. So what this function says is that the covariance between any two societies i and j declines exponentially with the squared distance between them. The parameter ρ determines the rate of decline. If it is large, then covariance declines rapidly with squared distance.

Why square the distance? You don't have to. This is just a model. But the squared distance is the most common assumption, both because it is easy to fit to data and has the often-realistic property of allowing covariance to decline more quickly as distance grows. This will be easy to appreciate, if we plot this function under the linear-decline alternative, $\exp(-\rho^2 D_{ij})$, and compare. We'll use a value $\rho^2 = 1$, just for the example.

R code
13.30

```
# linear
curve( exp(-1*x) , from=0 , to=4 , lty=2 ,
    xlab="distance" , ylab="correlation" )

# squared
curve( exp(-1*x^2) , add=TRUE )
```

The result is shown in FIGURE 13.8. The vertical axis here is just part of the total covariance function. You can think of it as the proportion of the maximum correlation between two societies i and j. The dashed curve is the linear distance function. It produces an exact exponential shape. The solid curve is the squared distance function. It produces a half-Gaussian decline that is initially slower than the exponential but rapidly accelerates and then becomes faster than exponential.

The last two pieces of $\mathbf{K}_{ij}$ are simpler. η^2 is the maximum covariance between any two societies i and j. The term on the end, $\delta_{ij}\sigma^2$, provides for extra covariance beyond η^2 when $i = j$. It does this because the function δ_{ij} is equal to 1 when $i = j$ but is zero otherwise. In the Oceanic societies data, this term will not matter, because we only have one observation

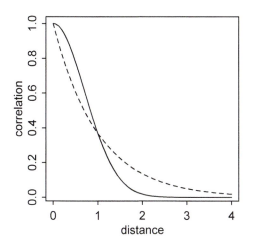

FIGURE 13.8. Shape of the function relating distance to the covariance $\mathbf{K}_{ij}$. The horizontal axis is distance. The vertical is the correlation, relative to maximum, between any two societies i and j. The dashed curve is the linear distance function. The solid curve is the squared distance function.

for each society. But if we had more than one observation per society, σ here describes how these observations covary.

The model computes the posterior distribution of ρ, η, and σ. But it also needs priors for them. We'll define priors for the square of each, and estimate them on the same scale, because that's computationally easier. We don't need σ in this model, so we'll instead just fix it at an irrelevant constant.

Now here's the full model, with the fixed priors for each parameter added at the bottom:

$$T_i \sim \text{Poisson}(\lambda_i)$$
$$\log \lambda_i = \alpha + \gamma_{\text{SOCIETY}[i]} + \beta_P \log P_i$$
$$\gamma \sim \text{MVNormal}\big((0, \dots, 0), \mathbf{K}\big)$$
$$\mathbf{K}_{ij} = \eta^2 \exp(-\rho^2 D_{ij}^2) + \delta_{ij}(0.01)$$
$$\alpha \sim \text{Normal}(0, 10)$$
$$\beta_P \sim \text{Normal}(0, 1)$$
$$\eta^2 \sim \text{HalfCauchy}(0, 1)$$
$$\rho^2 \sim \text{HalfCauchy}(0, 1)$$

Note that ρ^2 and η^2 must be positive, so we place half-Cauchy priors on them. There's nothing special about the Cauchy here. It's just a useful weakly informative prior for scale parameters like these. If you are concerned about the impact of the priors, you should repeat the sampling with different priors. A little knowledge of Pacific navigation would probably allow us a smart, informative prior on ρ^2 at least.

We're finally ready to fit the model. The distribution to use, so to signal to map2stan that you want to the squared distance Gaussian process prior, is GPL2. The rest of the code should be familiar.

R code
13.31

```
data(Kline2) # load the ordinary data, now with coordinates
d <- Kline2
d$society <- 1:10 # index observations
```

```
m13.7 <- map2stan(
    alist(
        total_tools ~ dpois(lambda),
        log(lambda) <- a + g[society] + bp*logpop,
        g[society] ~ GPL2( Dmat , etasq , rhosq , 0.01 ),
        a ~ dnorm(0,10),
        bp ~ dnorm(0,1),
        etasq ~ dcauchy(0,1),
        rhosq ~ dcauchy(0,1)
    ),
    data=list(
        total_tools=d$total_tools,
        logpop=d$logpop,
        society=d$society,
        Dmat=islandsDistMatrix),
    warmup=2000 , iter=1e4 , chains=4 )
```

Be sure to check the chains. They should sample very well. Let's check the estimates, just to check the convergence diagnostics and to verify that, as usual, the parameters themselves are hard to interpret:

R code
13.32

```
precis(m13.7,depth=2)
```

	Mean	StdDev	lower 0.89	upper 0.89	n_eff	Rhat
g[1]	-0.27	0.45	-0.94	0.42	3094	1
g[2]	-0.12	0.44	-0.76	0.55	2934	1
g[3]	-0.16	0.42	-0.79	0.47	2887	1
g[4]	0.30	0.38	-0.25	0.85	2973	1
g[5]	0.03	0.38	-0.50	0.59	2958	1
g[6]	-0.45	0.38	-1.01	0.10	3162	1
g[7]	0.10	0.37	-0.43	0.63	3018	1
g[8]	-0.26	0.37	-0.79	0.28	3160	1
g[9]	0.24	0.35	-0.25	0.76	3076	1
g[10]	-0.11	0.46	-0.84	0.58	4695	1
a	1.31	1.18	-0.57	3.14	3995	1
bp	0.24	0.12	0.05	0.42	4962	1
etasq	0.35	0.55	0.00	0.73	4238	1
rhosq	2.67	51.60	0.01	2.21	9319	1

First, note that the coefficient for log population, bp, is very much as it was before we added all this Gaussian process stuff. This suggests that it's hard to explain all of the association between tool counts and population as a side effect of geographic contact. Second, those g parameters are the Gaussian process varying intercepts for each society. Like a and bp, they are on the log-count scale, so they are hard to interpret raw.

In order to understand the parameters that describe the covariance with distance, rhosq and etasq, we'll want to plot the function they imply. Actually the joint posterior distribution of these two parameters defines a posterior distribution of covariance functions. We can get a sense of this distribution of functions—I know, this is rather meta—by plotting a bunch of them. Here we'll sample 100 from the posterior and display them along with the posterior

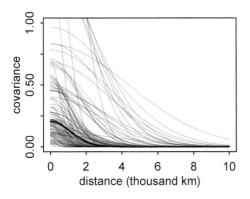

FIGURE 13.9. Posterior distribution of the spatial covariance between pairs of societies. The dark curve displays the posterior median. The thin curves show 100 functions sampled from the joint posterior distribution of ρ^2 and η^2.

median. Why use the median? Because the densities for rhosq and etasq are skewed. You can detect this in the precis output above: the mean for rhosq isn't even inside the 89% HPDI. So the median is a better measure of the center of mass than the mean. But as always, it is the entire distribution that matters. No single point within it is special.

R code
13.33

```
post <- extract.samples(m13.7)

# plot the posterior median covariance function
curve( median(post$etasq)*exp(-median(post$rhosq)*x^2) , from=0 , to=10 ,
    xlab="distance (thousand km)" , ylab="covariance" , ylim=c(0,1) ,
    yaxp=c(0,1,4) , lwd=2 )

# plot 100 functions sampled from posterior
for ( i in 1:100 )
    curve( post$etasq[i]*exp(-post$rhosq[i]*x^2) , add=TRUE ,
        col=col.alpha("black",0.2) )
```

FIGURE 13.9 shows the result. Each combination of values for ρ^2 and η^2 produces a relationship between covariance and distance. The posterior median function, shown by the thick curve, represents a center of plausibility. But the other curves show that there's a lot of uncertainty about the spatial covariance. Curves that peak at twice the posterior median peak, around 0.2, are commonplace. And curves that peak at half the median are very common, as well. There's a lot of uncertainty about how strong the spatial effect is, but the majority of posterior curves decline to zero covariance before 4000 kilometers.

It's hard to interpret these covariances directly, because they are on the log-count scale, just like everything else in a Poisson GLM. So let's consider the correlations among societies that are implied by the posterior median. First, we push the parameters back through the function for **K**, the covariance matrix:

R code
13.34

```
# compute posterior median covariance among societies
K <- matrix(0,nrow=10,ncol=10)
for ( i in 1:10 )
    for ( j in 1:10 )
        K[i,j] <- median(post$etasq) *
                exp( -median(post$rhosq) * islandsDistMatrix[i,j]^2 )
```

```
diag(K) <- median(post$etasq) + 0.01
```

Second, we convert K to a correlation matrix:

R code
13.35

```
# convert to correlation matrix
Rho <- round( cov2cor(K) , 2 )
# add row/col names for convenience
colnames(Rho) <- c("Ml","Ti","SC","Ya","Fi","Tr","Ch","Mn","To","Ha")
rownames(Rho) <- colnames(Rho)
Rho
```

```
     Ml   Ti   SC   Ya   Fi   Tr   Ch   Mn   To Ha
Ml 1.00 0.87 0.82 0.00 0.52 0.19 0.02 0.04 0.24  0
Ti 0.87 1.00 0.92 0.00 0.52 0.19 0.04 0.06 0.21  0
SC 0.82 0.92 1.00 0.00 0.37 0.30 0.07 0.11 0.12  0
Ya 0.00 0.00 0.00 1.00 0.00 0.09 0.37 0.34 0.00  0
Fi 0.52 0.52 0.37 0.00 1.00 0.02 0.00 0.00 0.76  0
Tr 0.19 0.19 0.30 0.09 0.02 1.00 0.26 0.72 0.00  0
Ch 0.02 0.04 0.07 0.37 0.00 0.26 1.00 0.53 0.00  0
Mn 0.04 0.06 0.11 0.34 0.00 0.72 0.53 1.00 0.00  0
To 0.24 0.21 0.12 0.00 0.76 0.00 0.00 0.00 1.00  0
Ha 0.00 0.00 0.00 0.00 0.00 0.00 0.00 0.00 0.00  1
```

The cluster of small societies in the upper-left of the matrix—Malekula (Ml), Tikopia (Ti), and Santa Cruz (SC)—are highly correlated, all above 0.8 with one another. As you'll see in a moment, these societies are very close together, and they also have similar tool totals. These correlations were estimating with log population in the model, remember, and so suggest some additional resemblance even accounting for the average association between population and tools. On the other end of spectrum is Hawaii (Ha), which is so far from all of the other societies that the correlation decays to zero everyplace. Other societies display a range of correlations.

To make some sense of the variation in these correlations, let's plot them on a crude map of the Pacific Ocean. The Kline2 data frame provides latitude and longitude for each society, to make this easy. I'll also scale the size of each society on the map in proportion to its log population.

R code
13.36

```
# scale point size to logpop
psize <- d$logpop / max(d$logpop)
psize <- exp(psize*1.5)-2

# plot raw data and labels
plot( d$lon2 , d$lat , xlab="longitude" , ylab="latitude" ,
    col=rangi2 , cex=psize , pch=16 , xlim=c(-50,30) )
labels <- as.character(d$culture)
text( d$lon2 , d$lat , labels=labels , cex=0.7 , pos=c(2,4,3,3,4,1,3,2,4,2) )

# overlay lines shaded by Rho
for( i in 1:10 )
    for ( j in 1:10 )
```

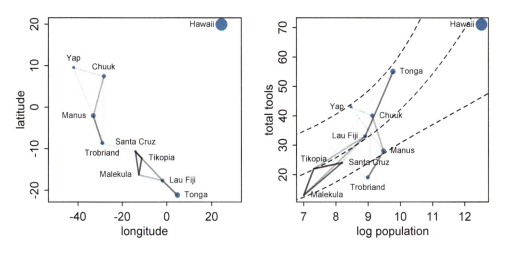

FIGURE 13.10. Left: Posterior median correlations among societies in geographic space. Right: Same posterior median correlations, now shown against relationship between total tools and log population.

```
if ( i < j )
    lines( c( d$lon2[i],d$lon2[j] ) , c( d$lat[i],d$lat[j] ) ,
        lwd=2 , col=col.alpha("black",Rho[i,j]^2) )
```

The result appears on the left side of FIGURE 13.10. Darker lines indicate stronger correlations, with pure white being zero correlation and pure black 100% correlation. The cluster of three close societies—Malekula, Tikopia, and Santa Cruz—stand out. Close societies have stronger correlations. But since we can't see total tools on this map, it's hard to see what the consequence of these correlations is supposed to be.

More sense can be made of these correlations, if we also compare against the simultaneous relationship between tools and log population. Here's a plot that combines the average posterior predictive relationship between log population and total tools with the shaded correlation lines for each pair of societies:

R code
13.37

```
# compute posterior median relationship, ignoring distance
logpop.seq <- seq( from=6 , to=14 , length.out=30 )
lambda <- sapply( logpop.seq , function(lp) exp( post$a + post$bp*lp ) )
lambda.median <- apply( lambda , 2 , median )
lambda.PI80 <- apply( lambda , 2 , PI , prob=0.8 )

# plot raw data and labels
plot( d$logpop , d$total_tools , col=rangi2 , cex=psize , pch=16 ,
    xlab="log population" , ylab="total tools" )
text( d$logpop , d$total_tools , labels=labels , cex=0.7 ,
    pos=c(4,3,4,2,2,1,4,4,4,2) )

# display posterior predictions
```

```
lines( logpop.seq , lambda.median , lty=2 )
lines( logpop.seq , lambda.PI80[1,] , lty=2 )
lines( logpop.seq , lambda.PI80[2,] , lty=2 )

# overlay correlations
for( i in 1:10 )
    for ( j in 1:10 )
        if ( i < j )
            lines( c( d$logpop[i],d$logpop[j] ) ,
                   c( d$total_tools[i],d$total_tools[j] ) ,
                   lwd=2 , col=col.alpha("black",Rho[i,j]^2) )
```

This plot appears in the right-hand side of FIGURE 13.10. Now it's easier to appreciate that the correlations among Malekula, Tikopia, and Santa Cruz describe the fact that they are below the expected number of tools for their populations. All three societies lying below the expectation, and being so close, is consistent with spatial covariance. The posterior correlations merely describe this feature of the data. Similarly, Manus and the Trobriands are geographically close, have a substantial posterior correlation, and fewer tools than expected for their population sizes. Tonga has more tools than expected for its population, and its proximity to Fiji counteracts some of the tug Fiji's smaller neighbors—Malekula, Tikopia, and Santa Cruz—exert on it. So the model seems to think Fiji would have fewer tools, if it weren't for Tonga.

Of course the correlations that this model describes by geographic distance may be the result of other, unmeasured commonalities between geographically close societies. For example, Manus and the Trobriands are geologically and ecologically quite different from Fiji and Tonga. So it could be availability of, for example, tool stone that explains some of the correlations. The Gaussian process regression is a grand and powerful descriptive model. As a result, its output is always compatible with many different causal explanations.

13.4.2. Other kinds of "distance". Gaussian process regression is routinely used with a huge variety of "distance" measures, both physical and abstract. For example, phylogenetic regressions can be constructed as a special case of Gaussian process regression in which a kind of phylogenetic distance, *patristic distance*, is used. Combined with a model of how differences between lineages accumulate with patristic distance, a covariance matrix is built to handle the potential non-independence of species. For those interested in social networks, network distance is another type of abstract distance that can be plugged into these models. Another common use for Gaussian process regression is to model cyclical covariation with time. In those cases, the covariance matrix $\mathbf{K}$ is modeled using periodic functions—sine and cosine are the easiest to use—of distance in time. This helps model seasonal influences, without imposing any hard cutoffs for seasons.

So the definition of $\mathbf{K}$ isn't the same in all Gaussian process models, but the basic strategy of modeling covariance as a function of distance is present in all such models. It is also possible to use more than one dimension of distance at the same time. This corresponds to the varying slopes strategy, in which variation within and between categories depends upon several features. But the Gaussian process merges all of these influences into a common covariance matrix and so a common intercept. It would be possible, for example, to remove the influence of population size in the Oceanic data from the linear model and merge it instead into the Gaussian process. In that case, a common approach is to define the covariance as:

$$\mathbf{K}_{ij} = \eta^2 \exp\left(-\left(\rho_D^2 D_{ij}^2 + \rho_P^2 (\log P_i - \log P_j)^2 \right) \right) + \delta_{ij}\sigma^2$$

The parameters ρ_D and ρ_P are separate *relevance* parameters for each dimension of difference in the total covariance. This approach is sometimes called AUTOMATIC RELEVANCE DETERMINATION.[164] But it's no more or less "automatic" than any other kind of coefficient estimation we've done so far in this book.

Building fancier Gaussian process models like the ones described just above is best done directly in Stan's modeling language, or some other tool like the excellent GPstuff, than with map2stan. You can get a good idea how this is done by looking at the Stan code for m13.7 with stancode(m13.7). Then read the Stan reference manual's section on Gaussian processes. It's very well explained.

13.5. Summary

This chapter extended the basic multilevel strategy of partial pooling to slopes as well as intercepts. Accomplishing this meant modeling covariation in the statistical population of parameters. The LKJcorr prior was introduced as a convenient family of priors for correlation matrices. Finally, Gaussian processes represent a practical method of extending the varying effects strategy to continuous dimensions of similarity, such as spatial, network, phylogenetic, or any other abstract distance between entities in the data. The next chapter continues to develop the broader multilevel approach by applying it to commonplace problems in statistical inference: measurement error and missing data.

13.6. Practice

Easy.

13E1. Add to the following model varying slopes on the predictor x.

$$y_i \sim \text{Normal}(\mu_i, \sigma)$$
$$\mu_i = \alpha_{\text{GROUP}[i]} + \beta x_i$$
$$\alpha_{\text{GROUP}} \sim \text{Normal}(\alpha, \sigma_\alpha)$$
$$\alpha \sim \text{Normal}(0, 10)$$
$$\beta \sim \text{Normal}(0, 1)$$
$$\sigma \sim \text{HalfCauchy}(0, 2)$$
$$\sigma_\alpha \sim \text{HalfCauchy}(0, 2)$$

13E2. Think up a context in which varying intercepts will be positively correlated with varying slopes. Provide a mechanistic explanation for the correlation.

13E3. When is it possible for a varying slopes model to have fewer effective parameters (as estimated by WAIC or DIC) than the corresponding model with fixed (unpooled) slopes? Explain.

Medium.

13M1. Repeat the café robot simulation from the beginning of the chapter. This time, set rho to zero, so that there is no correlation between intercepts and slopes. How does the posterior distribution of the correlation reflect this change in the underlying simulation?

13M2. Fit this multilevel model to the simulated café data:

$$W_i \sim \text{Normal}(\mu_i, \sigma)$$
$$\mu_i = \alpha_{\text{CAFÉ}[i]} + \beta_{\text{CAFÉ}[i]} A_i$$
$$\alpha_{\text{CAFÉ}} \sim \text{Normal}(\alpha, \sigma_\alpha)$$
$$\beta_{\text{CAFÉ}} \sim \text{Normal}(\beta, \sigma_\beta)$$
$$\alpha \sim \text{Normal}(0, 10)$$
$$\beta \sim \text{Normal}(0, 10)$$
$$\sigma \sim \text{HalfCauchy}(0, 1)$$
$$\sigma_\alpha \sim \text{HalfCauchy}(0, 1)$$
$$\sigma_\beta \sim \text{HalfCauchy}(0, 1)$$

Use WAIC to compare this model to the model from the chapter, the one that uses a multi-variate Gaussian prior. Explain the result.

13M3. Re-estimate the varying slopes model for the UCBadmit data, now using a non-centered parameterization. Compare the efficiency of the forms of the model, using n_eff. Which is better? Which chain sampled faster?

13M4. Use WAIC to compare the Gaussian process model of Oceanic tools to the models fit to the same data in Chapter 10. Pay special attention to the effective numbers of parameters, as estimated by WAIC.

Hard.

13H1. Let's revisit the Bangladesh fertility data, data(bangladesh), from the practice problems for Chapter 12. Fit a model with both varying intercepts by district_id and varying slopes of urban by district_id. You are still predicting use.contraception. Inspect the correlation between the intercepts and slopes. Can you interpret this correlation, in terms of what it tells you about the pattern of contraceptive use in the sample? It might help to plot the mean (or median) varying effect estimates for both the intercepts and slopes, by district. Then you can visualize the correlation and maybe more easily think through what it means to have a particular correlation. Plotting predicted proportion of women using contraception, with urban women on one axis and rural on the other, might also help.

13H2. Varying effects models are useful for modeling time series, as well as spatial clustering. In a time series, the observations cluster by entities that have continuity through time, such as individuals. Since observations within individuals are likely highly correlated, the multilevel structure can help quite a lot. You'll use the data in data(Oxboys), which is 234 height measurements on 26 boys from an Oxford Boys Club (I think these were like youth athletic leagues?), at 9 different ages (centered and standardized) per boy. You'll be interested in predicting height, using age, clustered by Subject (individual boy).

Fit a model with varying intercepts and slopes (on age), clustered by Subject. Present and interpret the parameter estimates. Which varying effect contributes more variation to the heights, the intercept or the slope?

13H3. Now consider the correlation between the varying intercepts and slopes. Can you explain its value? How would this estimated correlation influence your predictions about a new sample of boys?

13H4. Use `mvrnorm` (in `library(MASS)`) or `rmvnorm` (in `library(mvtnorm)`) to simulate a new sample of boys, based upon the posterior mean values of the parameters. That is, try to simulate varying intercepts and slopes, using the relevant parameter estimates, and then plot the predicted trends of height on age, one trend for each simulated boy you produce. A sample of 10 simulated boys is plenty, to illustrate the lesson. You can ignore uncertainty in the posterior, just to make the problem a little easier. But if you want to include the uncertainty about the parameters, go for it.

Note that you can construct an arbitrary variance-covariance matrix to pass to either `mvrnorm` or `rmvnorm` with something like:

R code
13.38

```
S <- matrix( c( sa^2 , sa*sb*rho , sa*sb*rho , sb^2 ) , nrow=2 )
```

where `sa` is the standard deviation of the first variable, `sb` is the standard deviation of the second variable, and `rho` is the correlation between them.

14 Missing Data and Other Opportunities

A big advantage of Bayesian inference is that it obviates the need to be clever. For example, there's a classic probability puzzle known as *Bertrand's box paradox*.[165] The version that I prefer involves pancakes. Suppose I cook three pancakes. The first pancake is burnt on both sides (BB). The second pancake is burnt on only one side (BU). The third pancake is not burnt at all (UU). Now I serve you—at random—one of these pancakes, and the side facing up on your plate is burnt. What is the probability that the other side is also burnt?

This is a hard problem, if we rely upon intuition. Most people say "one-half," but that is quite wrong. And with no false modesty, my intuition is no better. But I have learned to solve these problems by cold, hard, ruthless application of conditional probability. There's no need to be clever when you can be ruthless.

So let's get ruthless. Applying conditional probability means using what we do know to refine our knowledge about what we wish to know. In other words:

$$\Pr(\text{want to know}|\text{already know})$$

In this case, we know the up side is burnt. We want to know whether or not the down side is burnt. The definition of conditional probability tells us:

$$\Pr(\text{burnt down}|\text{burnt up}) = \frac{\Pr(\text{burnt up, burnt down})}{\Pr(\text{burnt up})}$$

This is just the definition of conditional probability, labeled with our pancake problem. We want to know if the down side is burnt, and the information we have is that the up side is burnt. We *condition* on the information, so we update our state of information in light of it. The definition tells us that the probability we want is just the probability of the burnt/burnt pancake divided by the probability of seeing a burnt side up. The probability of the burnt/burnt pancake is 1/3, because a pancake was selected at random. The probability the up side is burnt must average over each way we can get dealt a burnt top side of the pancake. This is:

$$\Pr(\text{burnt up}) = \Pr(\text{BB})(1) + \Pr(\text{BU})(0.5) + \Pr(\text{UU})(0) = (1/3) + (1/3)(1/2) = 0.5$$

So all together:

$$\Pr(\text{burnt down}|\text{burnt up}) = \frac{1/3}{1/2} = \frac{2}{3}$$

If you don't quite believe this answer, you can do a quick simulation to confirm it.

R code
14.1

```
# simulate a pancake and return randomly ordered sides
sim_pancake <- function() {
    pancake <- sample(1:3,1)
```

```
    sides <- matrix(c(1,1,1,0,0,0),2,3)[,pancake]
    sample(sides)
}

# sim 10,000 pancakes
pancakes <- replicate( 1e4 , sim_pancake() )
up <- pancakes[1,]
down <- pancakes[2,]

# compute proportion 1/1 (BB) out of all 1/1 and 1/0
num_11_10 <- sum( up==1 )
num_11 <- sum( up==1 & down==1 )
num_11/num_11_10
```

[1] 0.6777889

Two-thirds.

If you want to derive some intuition now at the end, having seen the right answer, the trick is to count *sides* of the pancakes, not the pancakes themselves. Yes, there are 2 pancakes that have at least one burnt side. And only one of those has 2 burnt sides. But it is the sides, not the pancakes, that matter. Conditional on the up side being burnt, there are three *sides* that could be down. Two of those sides are burnt. So the probability is 2 out of 3.

Probability theory is not difficult mathematically. It is just counting. But it is hard to interpret and apply. Doing so often seems to require some cleverness, and authors have an incentive to solve problems in clever ways, just to show off. But we don't need that cleverness, if we ruthlessly apply conditional probability. And that's the real trick of the Bayesian approach: to apply conditional probability in all places, for data and parameters. The benefit is that once we define our information state—our assumptions—we can let the rules of probability do the rest. The work that gets done is the revelation of the implications of our assumptions. Model fitting, as we've been practicing it, is the same un-clever approach. We define the model and introduce the data, and conditional probability does the rest, revealing the implications of our assumptions, in light of the evidence.

In this chapter, you'll meet two commonplace applications of this assume-and-deduce strategy. The first is the incorporation of MEASUREMENT ERROR into our models. The second is the estimation of MISSING DATA through BAYESIAN IMPUTATION. You'll see a fully worked, introductory example of each.

In neither application do you have to intuit the consequences of measurement errors nor the implications of missing values in order to design the models. All you have to do is state your information about the error or about the variables with missing values. Logic does the rest. Well, your computer does the rest. But it's just using fancy algorithms to perform Bayesian updating. It's not at all clever. But the implications it reveals are both counterintuitive and valuable.

14.1. Measurement error

Back in Chapter 5, you met the divorce and marriage data for the United States. Those data demonstrated a simple spurious association among the predictors, as well as how multiple regression can sort it out. What we ignored at the time is that both the divorce rate variable and the marriage rate variable are measured with substantial error, and that error is reported in the form of standard errors. Importantly, the amount of error varies a lot across States. Here, you'll see a simple and useful way to incorporate that information into the model. Then we'll let logic reveal the implications.

Begin by loading the data and plotting to display the measurement error of the outcome as an error bar:

R code
14.2

```
library(rethinking)
data(WaffleDivorce)
d <- WaffleDivorce

# points
plot( d$Divorce ~ d$MedianAgeMarriage , ylim=c(4,15) ,
    xlab="Median age marriage" , ylab="Divorce rate" )

# standard errors
for ( i in 1:nrow(d) ) {
    ci <- d$Divorce[i] + c(-1,1)*d$Divorce.SE[i]
    x <- d$MedianAgeMarriage[i]
    lines( c(x,x) , ci )
}
```

The plot is shown on the left in FIGURE 14.1. Notice that there is a lot of variation in how uncertain the observed divorce rate is, as reflected in varying lengths of the vertical line segments. Why does the error vary so much? Large States provide better samples, so their measurement error is smaller. The data are displayed this way, to show the association between the population size of each State and its measurement error, in the right-hand plot in FIGURE 14.1.

Since the values in same States are more certain than in others, it makes sense for the more certain estimates to influence the regression more. There are all manner of *ad hoc* procedures for weighting some points more than others, and these can help. But they leave a lot of information on the table. And they prevent a helpful phenomenon that arises automatically in the fully Bayesian approach: Information flows among the measurements to provide improved estimates of the data itself. So let's see how to do all of that, just by stating the information as a model.

Rethinking: Generative thinking, Bayesian inference. Bayesian models are *generative*, meaning they can be used to simulate observations just as well as they can be used to estimate parameters. One benefit of this fact is that a statistical model can be developed by thinking hard about how the data might have arisen. This includes sampling and measurement, as well as the nature of the process we are studying. Then let Bayesian updating discover the implications.

14.1.1. Error on the outcome. To incorporate measurement error, recognize that we can replace the observed data for divorce rate with a distribution.[166] In other words, typical "data" is just a special case of a probability distribution for which all of the mass is piled up on a single value. When there is instead uncertainty about the true value, that uncertainty can be replaced by a distribution that represents the information we have.

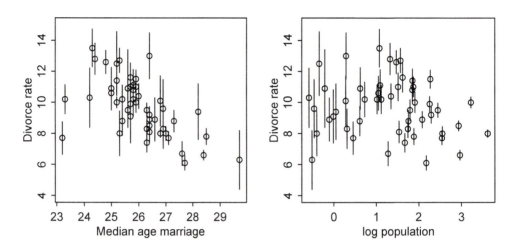

FIGURE 14.1. Left: Divorce rate by median age of marriage, States of the United States. Vertical bars show plus and minus one standard deviation of the Gaussian uncertainty in measured divorce rate. Right: Divorce rate, again with standard deviations, against log population of each State. Smaller States produce more uncertain estimates.

If this isn't intuitive to you—because you are normal—it might help to consider the problem generatively: If you wanted to simulate measurement error, you would assign a distribution to each observation and sample from it. For example, suppose the true value of a measurement is 10 meters. If it is measured with Gaussian error with standard deviation of 2 meters, this implies a probability distribution for any realized measurement y:

$$y \sim \text{Normal}(10, 2)$$

As the measurement error here shrinks, all the probability piles up on 10. But when there is error, many measurements are more and less plausible. This is what I mean by saying that ordinary data are a special case of a distribution. And here is the key insight: If we don't know the true value (10 in this example), then we can just put a parameter there and let Bayes do the rest.

To see how this works in practice, let's work with some actual data. In this example we'll use a Gaussian distribution with mean equal to the observed value and standard deviation equal to the measurement's standard error. This is the logical choice, because if all we know about the error is its standard deviation, then the maximum entropy distribution for it will be Gaussian. If we choose any other distribution, that implies we have additional information. But we don't have any additional information here, so Gaussian it is. As always, the Gaussian choice is not equivalent to assuming that the measurement error is actually Gaussian. It's just the most conservative assumption, given only a mean and variance.

Here's how to define the distribution for each divorce rate. For each observed value $D_{\text{OBS},i}$, there will be one parameter, $D_{\text{EST},i}$, defined by:

$$D_{\text{OBS},i} \sim \text{Normal}(D_{\text{EST},i}, D_{\text{SE},i})$$

All this does is define the measurement $D_{\text{OBS},i}$ as having the specified Gaussian distribution centered on the unknown parameter $D_{\text{EST},i}$. So the above defines a probability for each State i's observed divorce rate, given a known measurement error.

In combination with the rest of the model, this will allow us to estimate the plausible true values consistent with the observation. To accomplish this, we also use these D_{EST} values as the outcome variable in the regression equation. Since ordinary data are just a special case, it is fine to replace a predictor variable with a probability distribution. This will not only allow us to estimate coefficients for predictions that take into account the uncertainty in the outcome, but it will also update the estimate for divorce rate in each State.

This is a lot to take in. But we'll go one step at a time. Recall that the goal is to model divorce rate D as a linear function of age at marriage A and marriage rate R. Here's what the model looks like, with the measurement errors highlighted in blue:

$$D_{\text{EST},i} \sim \text{Normal}(\mu_i, \sigma) \qquad [\text{"likelihood" for estimates}]$$
$$\mu_i = \alpha + \beta_A A_i + \beta_R R_i \qquad [\text{linear model}]$$
$$D_{\text{OBS},i} \sim \text{Normal}(D_{\text{EST},i}, D_{\text{SE},i}) \qquad [\text{prior for estimates}]$$
$$\alpha \sim \text{Normal}(0, 10)$$
$$\beta_A \sim \text{Normal}(0, 10)$$
$$\beta_R \sim \text{Normal}(0, 10)$$
$$\sigma \sim \text{Cauchy}(0, 2.5)$$

So really the only difference between this model and a typical linear regression is replacing the outcome with a vector of parameters. Each outcome parameter also gets a second role as the unknown mean of another distribution, one that "predicts" the observed measurement. A cool implication that will arise here is that information flows in both directions—the uncertainty in measurement influences the regression parameters in the linear model, and the regression parameters in the linear model also influence the uncertainty in the measurements.

Here is the map2stan version of the model:

R code
14.3

```
dlist <- list(
    div_obs=d$Divorce,
    div_sd=d$Divorce.SE,
    R=d$Marriage,
    A=d$MedianAgeMarriage
)

m14.1 <- map2stan(
    alist(
        div_est ~ dnorm(mu,sigma),
        mu <- a + bA*A + bR*R,
        div_obs ~ dnorm(div_est,div_sd),
        a ~ dnorm(0,10),
        bA ~ dnorm(0,10),
        bR ~ dnorm(0,10),
        sigma ~ dcauchy(0,2.5)
    ) ,
```

```
    data=dlist ,
    start=list(div_est=dlist$div_obs) ,
    WAIC=FALSE , iter=5000 , warmup=1000 , chains=2 , cores=2 ,
    control=list(adapt_delta=0.95) )
```

There are three things to note in this code. First, I've turned off WAIC calculation, because the default code in WAIC will not compute the likelihood correctly, by integrating over the uncertainty in each div_est distribution. Second, I've provided a start list for the div_est values. This tells map2stan how many parameters it needs. It won't be sensitive to the exact starting values, but it makes sense to start each at the observed value for each State. Third, I've added a control list at the end. This allows us to tune the HMC algorithm. In this case, I've increased something known as the target acceptance rate, adapt_delta, to 0.95 from the default of 0.8. This means that Stan will work harder during warmup and potentially sample more efficiently. If you try this model without the control argument, you'll find that you get a few (harmless, in this case) divergent iterations.

Consider the posterior means (abbreviating the precis output below):

R code
14.4
```
precis( m14.1 , depth=2 )
```

```
              Mean StdDev lower 0.89 upper 0.89 n_eff Rhat
div_est[1]   11.77   0.68      10.68      12.88  8000    1
div_est[2]   11.19   1.06       9.54      12.92  6397    1
div_est[3]   10.47   0.62       9.47      11.44  8000    1
...
div_est[48]  10.61   0.87       9.23      11.96  7333    1
div_est[49]   8.47   0.51       7.66       9.27  7288    1
div_est[50]  11.52   1.11       9.78      13.27  5389    1
a            21.30   6.60      11.45      32.29  2465    1
bA           -0.55   0.21      -0.88      -0.20  2568    1
bR            0.13   0.08       0.01       0.25  2746    1
sigma         1.12   0.21       0.78       1.43  2215    1
```

If you look back at Chapter 5, you'll see that the former estimate for bA was about −1. Now it's almost half that, but still reliably negative. So compared to the original regression that ignores measurement error, the association between divorce and age at marriage has been reduced. This isn't true only of this example. Ignoring measurement error tends to exaggerate associations between outcomes and predictors. But it might also mask an association. It all depends upon which cases have how much error.

If you look again at FIGURE 14.1, you can see a hint of why this has happened. States with extremely low and high ages at marriage tend to also have more uncertain divorce rates. As a result those rates have been shrunk towards the expected mean defined by the regression line. FIGURE 14.2 displays this shrinkage phenomenon. On the left of the figure, the difference between the observed and estimated divorce rates is shown on the vertical axis, while the standard error of the observed is shown on the horizontal. The dashed line at zero indicates no change from observed to estimated. Notice that States with more uncertain divorce rates— farther right on the plot—have estimates more different from observed. This is your friend SHRINKAGE from the previous two chapters. Less certain estimates are improved by pooling information from more certain estimates.

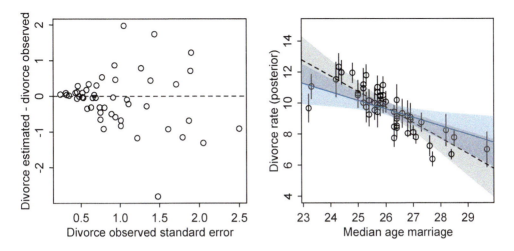

FIGURE 14.2. Left: Shrinkage resulting from modeling the measurement error. The less error in the original measurement, the less shrinkage in the posterior estimate. Right: Comparison of regression that ignores measurement error (dashed line and gray shading) with regression that incorporates measurement error (blue line and shading). The points and line segments show the posterior means and standard deviations for each posterior divorce rate, $D_{EST,i}$.

This shrinkage results in pulling divorce rates towards the regression line, as seen in the right-hand plot in the same figure. This plot shows the posterior mean divorce rate for each State against its observed median age at marriage. The vertical line segments show the posterior standard deviations of each divorce rate—the estimates have moved, but they are still uncertain.

As a result of their movement, however, the regression trend has moved. The old no-error regression is shown in gray. The fancy new with-error regression is shown in blue. Well, really both the estimates and the trend have moved one another at the same time. For a State with an uncertain divorce rate, the trend has strongly influenced the new estimate of divorce rate. For a State with a fairly certain divorce rate—a small standard error—the State has instead strongly influenced the trend. The balance of all of this information is the shift in both the estimated divorce rates and the regression relationship.

14.1.2. Error on both outcome and predictor. What happens when there is measurement error on predictor variables as well? The approach is the same. Again, consider the problem generatively: Each observed predictor value is a draw from a distribution with an unknown mean, the true value, but known standard deviation. So we define a vector of parameters, one for each unknown true value, and then make those parameters the means of a family of Gaussian distributions with known standard deviations.

In the divorce data, the measurement error for the marriage rate predictor variable also comes as standard error. So let's incorporate that information, to include measurement error for marriage rate R. Here's the updated model, with the new bits in blue:

$$D_{\mathrm{EST},i} \sim \mathrm{Normal}(\mu_i, \sigma) \qquad \text{[likelihood for outcome estimates]}$$
$$\mu_i = \alpha + \beta_A A_i + \beta_R R_{\mathrm{EST},i} \qquad \text{[linear model using predictor estimates]}$$
$$D_{\mathrm{OBS},i} \sim \mathrm{Normal}(D_{\mathrm{EST},i}, D_{\mathrm{SE},i}) \qquad \text{[prior for outcome estimates]}$$
$$R_{\mathrm{OBS},i} \sim \mathrm{Normal}(R_{\mathrm{EST},i}, R_{\mathrm{SE},i}) \qquad \text{[prior for predictor estimates]}$$
$$\alpha \sim \mathrm{Normal}(0, 10)$$
$$\beta_A \sim \mathrm{Normal}(0, 10)$$
$$\beta_R \sim \mathrm{Normal}(0, 10)$$
$$\sigma \sim \mathrm{Cauchy}(0, 2.5)$$

The R_{EST} parameters will hold the posterior distributions of the true marriage rates. And fitting the model is much like before:

R code
14.5

```
dlist <- list(
    div_obs=d$Divorce,
    div_sd=d$Divorce.SE,
    mar_obs=d$Marriage,
    mar_sd=d$Marriage.SE,
    A=d$MedianAgeMarriage )

m14.2 <- map2stan(
    alist(
        div_est ~ dnorm(mu,sigma),
        mu <- a + bA*A + bR*mar_est[i],
        div_obs ~ dnorm(div_est,div_sd),
        mar_obs ~ dnorm(mar_est,mar_sd),
        a ~ dnorm(0,10),
        bA ~ dnorm(0,10),
        bR ~ dnorm(0,10),
        sigma ~ dcauchy(0,2.5)
    ) ,
    data=dlist ,
    start=list(div_est=dlist$div_obs,mar_est=dlist$mar_obs) ,
    WAIC=FALSE , iter=5000 , warmup=1000 , chains=3 , cores=3 ,
    control=list(adapt_delta=0.95) )
```

If you inspect the precis output, you'll see that the coefficients for age at marriage and marriage rate are essentially unchanged from the previous model. So adding error on the predictor didn't change the major inference. But it did provide updated estimates of marriage rate itself (FIGURE 14.3). What has happened is that since the States with highly uncertain marriage rates tend to be small States with high marriage rates, pooling has resulted in smaller estimates for those States.

Also note that since there isn't much association between divorce and marriage rates, there is less movement of the marriage rate estimates. That is to say that there isn't much information in divorce rate to help us improve estimates of marriage rate. In contrast, since

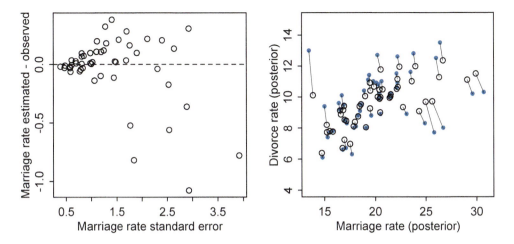

FIGURE 14.3. Left: Shrinkage for the predictor variable marriage rate. Notice that shrinkage is not balanced, but rather that the model believes the observed values tended to be overestimates. Right: Shrinkage of both divorce rate and marriage rate. Solid points are the observed values. Open points are posterior means. Lines connect pairs of points for the same State.

the relationship between divorce and median age at marriage is strong, there's a lot of information in age at marriage to help us improve estimates of divorce rate. That's why divorce estimates shrink more than marriage rate estimates do.

Of course, that information in age at marriage may be illusory, as we haven't incorporated uncertainty about age at marriage. If we had those data, in the form of standard errors or better yet a distribution of ages at marriage in each State, then we could blend that information into the model using the same approach as above.

The big take home point for this section is that when you have a distribution of values, don't reduce it down to a single value to use in a regression. Instead, use the entire distribution. Anytime we use an average value, discarding the uncertainty around that average, we risk overconfidence and spurious inference. This doesn't only apply to measurement error, but also to cases in which data are averaged before analysis.

Do not average. Instead, model.

14.2. Missing data

With measurement error, the insight is to realize that any datum can be replaced by a distribution that reflects uncertainty. This approach improves estimates of data while simultaneously honestly estimating coefficients. Information flows among the cases, through the regression model, allowing each measurement to borrow strength from the others.

Sometimes however data are simply missing—no measurement is available at all. At first, this seems like a lost cause. What can be done when there is no measurement at all, not even one with error? A lot can be done, because just like information flows among cases in the measurement error example, it can also flow from present data to missing data, as long as we are willing to make a model of the whole variable.[167]

14.2.1. Imputing neocortex. Let's return to the primate milk example, from Chapter 5. We used `data(milk)` to illustrate masking, using both neocortex percent and body mass to predict milk energy. One aspect of those data are 12 missing values in the `neocortex.perc` column. We used a COMPLETE-CASE analysis back then, which means we dropped those 12 cases from the analysis. That means we also dropped 12 perfectly good body mass and milk energy values. That left us with only 17 cases to work with. We can do better.

Now we'll see how to use Bayesian IMPUTATION to conserve and use the information in the other columns, while producing estimates for the missing values. The estimates we derive for the missing values won't be particularly crisp—they will have wide posteriors. But they will be honest, given the assumptions built into the model.

What we're going to do is build an example of **MCAR** (Missing Completely At Random) imputation.[168] With MCAR, we assume that the location of the missing values is completely random with respect to those values and all other values in the data. This is the simplest and easiest to understand form of imputation. So it makes sense to start here. It is also the assumption that is required for dropping the cases with missing values to make sense. Otherwise, dropped cases would bias the results. So really no additional assumption is being made here. We're just following through with the assumption and adding it to the model, instead of tossing out useful data.

The trick to imputation is to simultaneously model the predictor variable that has missing values together with the outcome variable. The present values will produce estimates that comprise a prior for each missing value. These priors will then be updated by the relationship between the predictor and outcome. So there will be a posterior distribution for each missing value.

The obstacle in practice is that we have to conceive of the predictor now as a mixed vector of data and parameters. In our case, the variable with missing values is neocortex percent. Call it N:

$$N = [0.55, N_2, N_3, N_4, 0.65, 0.65, ..., 0.76, 0.75]$$

For every index i at which there is a missing value, there is also a parameter N_i that will form a posterior distribution for it.

This is the model we need, with the neocortex pieces in blue:

$$
\begin{aligned}
k_i &\sim \text{Normal}(\mu_i, \sigma) && \text{[likelihood for outcome } k\text{]} \\
\mu_i &= \alpha + \beta_N N_i + \beta_M \log M_i && \text{[linear model]} \\
N_i &\sim \text{Normal}(\nu, \sigma_N) && \text{[likelihood/prior for obs/missing } N\text{]} \\
\alpha &\sim \text{Normal}(0, 10) \\
\beta_N &\sim \text{Normal}(0, 10) \\
\beta_M &\sim \text{Normal}(0, 10) \\
\sigma &\sim \text{Cauchy}(0, 1) \\
\nu &\sim \text{Normal}(0.5, 1) \\
\sigma_N &\sim \text{Cauchy}(0, 1)
\end{aligned}
$$

Note that when N_i is observed, then the third line above is a likelihood, just like any old linear regression. The model learns the distributions of ν and σ_N that are consistent with the data. But when N_i is missing and therefore a parameter, that same line is interpreted as a prior. Since the parameters ν and σ_N are also estimated, the prior is learned from the data.

One issue with this model is that it assumes each *N* value has Gaussian uncertainty. But we know that these values are bounded between zero and one, because they are proportions. So it is possible to do a little better. In the practice problems at the end of the chapter, you'll see how.

Implementing the model can be done several different ways. All of the ways are a little awkward, because the locations of missing values have to respected. The approach I'll use here hews closely to the discussion just above: We'll merge the observed values and parameters into a vector that we'll use as "data" in the regression. For convenience, map2stan automates this merging. But if you ever need to do it yourself, you can inspect the raw Stan code to see how it's done, as well as see the Overthinking box at the end of this section for a full implementation in raw Stan code.

To fit the model with map2stan, first get the data loaded and transform the predictors:

```
library(rethinking)
data(milk)
d <- milk
d$neocortex.prop <- d$neocortex.perc / 100
d$logmass <- log(d$mass)
```
R code
14.6

The formula list looks much as you'd expect:

```
# prep data
data_list <- list(
    kcal = d$kcal.per.g,
    neocortex = d$neocortex.prop,
    logmass = d$logmass )

# fit model
m14.3 <- map2stan(
    alist(
        kcal ~ dnorm(mu,sigma),
        mu <- a + bN*neocortex + bM*logmass,
        neocortex ~ dnorm(nu,sigma_N),
        a ~ dnorm(0,100),
        c(bN,bM) ~ dnorm(0,10),
        nu ~ dnorm(0.5,1),
        sigma_N ~ dcauchy(0,1),
        sigma ~ dcauchy(0,1)
    ) ,
    data=data_list , iter=1e4 , chains=2 )
```
R code
14.7

Take a look at the estimates:

```
precis(m14.3,depth=2)
```
R code
14.8

	Mean	StdDev	lower 0.89	upper 0.89	n_eff	Rhat
neocortex_impute[1]	0.63	0.05	0.55	0.71	7520	1
neocortex_impute[2]	0.63	0.05	0.54	0.70	7965	1
neocortex_impute[3]	0.62	0.05	0.54	0.70	6319	1

```
neocortex_impute[4]     0.65    0.05        0.57        0.72 10000      1
neocortex_impute[5]     0.70    0.05        0.62        0.78 10000      1
neocortex_impute[6]     0.66    0.05        0.58        0.73  8443      1
neocortex_impute[7]     0.69    0.05        0.61        0.77 10000      1
neocortex_impute[8]     0.70    0.05        0.62        0.77 10000      1
neocortex_impute[9]     0.71    0.05        0.64        0.79 10000      1
neocortex_impute[10]    0.65    0.05        0.57        0.72 10000      1
neocortex_impute[11]    0.66    0.05        0.58        0.74 10000      1
neocortex_impute[12]    0.70    0.05        0.62        0.78 10000      1
a                      -0.55    0.47       -1.24        0.24  2375      1
bN                      1.93    0.73        0.77        3.09  2336      1
bM                     -0.07    0.02       -0.11       -0.04  3264      1
nu                      0.67    0.01        0.65        0.69  7336      1
sigma_N                 0.06    0.01        0.04        0.08  4892      1
sigma                   0.13    0.02        0.09        0.17  4036      1
```

Each of the 12 imputed distributions for missing values is shown here, along with the ordinary regression parameters below them. To see how including all cases has impacted inference, let's do a quick comparison to the estimates that drop missing cases.

R code
14.9

```
# prep data
dcc <- d[ complete.cases(d$neocortex.prop) , ]
data_list_cc <- list(
    kcal = dcc$kcal.per.g,
    neocortex = dcc$neocortex.prop,
    logmass = dcc$logmass )

# fit model
m14.3cc <- map2stan(
    alist(
        kcal ~ dnorm(mu,sigma),
        mu <- a + bN*neocortex + bM*logmass,
        a ~ dnorm(0,100),
        c(bN,bM) ~ dnorm(0,10),
        sigma ~ dcauchy(0,1)
    ) ,
    data=data_list_cc , iter=1e4 , chains=2 )
precis(m14.3cc)
```

```
        Mean StdDev lower 0.89 upper 0.89 n_eff Rhat
a      -1.07   0.56      -1.96      -0.20  2112    1
bN      2.77   0.87       1.40       4.13  2070    1
bM     -0.10   0.03      -0.14      -0.05  2251    1
sigma   0.14   0.03       0.10       0.18  2430    1
```

By including the incomplete cases, the posterior mean for neocortex has gone from 2.8 to 1.9, and the mean for body mass has diminished from -0.1 to -0.07. So by using all the cases, the strength of the inferred relationships has diminished. This might make you sad, but ask yourself whether you would want your colleagues to use all the data, even if it meant inferring a weaker relationship. Then apply that same standard to yourself. But wait until the next model, before getting too depressed.

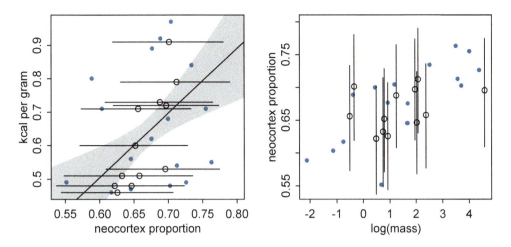

FIGURE 14.4. Left: Inferred relationship between milk energy (vertical) and neocortex proportion (horizontal), with imputed values shown by open points. The line segments are 89% posterior intervals. Right: Inferred relationship between the two predictors, neocortex proportion and log mass. Imputed values again shown by open points.

Let's do some plotting to visualize what's happened here. FIGURE 14.4 displays both the inferred relationship between milk energy and neocortex (left) and the relationship between the two predictors (right). Both plots show imputed neocortex values in blue, with 90% confidence intervals shown by the blue line segments. Although there's a lot of uncertainty in the imputed values—hey, Bayesian inference isn't magic, just logic—they do show a gentle tilt towards the regression line. This has happened because the observed values provide information that guides the estimation of the missing values.

The right-hand plot shows the inferred relationship between the predictors. We already know that these two predictors are positively associated—that's what creates the masking problem. But notice here that the imputed values do not show an upward slope. They do not, because the imputation model—the first regression with neocortex (observed and missing) as the outcome—assumed no relationship. So odds are we can improve this model by changing the imputation model to estimate the relationship between the two predictors.

14.2.2. Improving the imputation model. So let's do that now, make a better model of the predictor with missing values, a model that accounts for the association among the predictors themselves. This will let us exploit more information in the data.

The notion is to change the imputation line of the model from the simple:

$$N_i \sim \text{Normal}(\nu, \sigma_N)$$

to the slightly more complicated:

$$N_i \sim \text{Normal}(\nu_i, \sigma_N)$$
$$\nu_i = \alpha_N + \gamma_M \log M_i$$

where α_N and γ_M now describe the linear relationship between neocortex and log mass. The objective is to extract information from the observed cases and exploit it to improve the estimates of the missing values inside N.

Here's the map2stan implementation:

```
m14.4 <- map2stan(
    alist(
        kcal ~ dnorm(mu,sigma),
        mu <- a + bN*neocortex + bM*logmass,
        neocortex ~ dnorm(nu,sigma_N),
        nu <- a_N + gM*logmass,
        a ~ dnorm(0,100),
        c(bN,bM,gM) ~ dnorm(0,10),
        a_N ~ dnorm(0.5,1),
        sigma_N ~ dcauchy(0,1),
        sigma ~ dcauchy(0,1)
    ) ,
    data=data_list , iter=1e4 , chains=2 )
precis(m14.4,depth=2)
```

	Mean	StdDev	lower 0.89	upper 0.89	n_eff	Rhat
neocortex_impute[1]	0.63	0.03	0.58	0.69	10000	1
neocortex_impute[2]	0.63	0.04	0.57	0.69	7574	1
neocortex_impute[3]	0.62	0.04	0.56	0.67	10000	1
neocortex_impute[4]	0.65	0.03	0.59	0.70	10000	1
neocortex_impute[5]	0.66	0.04	0.61	0.72	8867	1
neocortex_impute[6]	0.63	0.04	0.57	0.69	10000	1
neocortex_impute[7]	0.68	0.03	0.63	0.74	10000	1
neocortex_impute[8]	0.70	0.03	0.65	0.75	10000	1
neocortex_impute[9]	0.71	0.04	0.66	0.77	10000	1
neocortex_impute[10]	0.66	0.03	0.61	0.72	8116	1
neocortex_impute[11]	0.68	0.03	0.62	0.73	10000	1
neocortex_impute[12]	0.74	0.04	0.69	0.80	10000	1
a	-0.87	0.48	-1.61	-0.11	3053	1
bN	2.44	0.75	1.22	3.57	2967	1
bM	-0.09	0.02	-0.12	-0.05	3288	1
gM	0.02	0.01	0.01	0.03	6887	1
a_N	0.64	0.01	0.62	0.66	5646	1
sigma_N	0.04	0.01	0.03	0.05	4737	1
sigma	0.13	0.02	0.09	0.16	4964	1

The marginal posterior for gM confirms that the two predictors are positively associated, as you already knew. The model uses that positive association to improve the imputation now.

FIGURE 14.5 displays the same kind of plots as before, but now for the new imputation model. On the left, the slope has slightly increased (mean 1.5 now instead of 1.2) and the posterior interval for each imputed value has narrowed. On the right, you can see now that the model has imputed in a way to preserve the positive association between neocortex and log mass.

In general, the goal is to model the joint distribution of all variables, both predictors and outcomes. The joint information then makes the intuitively impossible—inferring missing

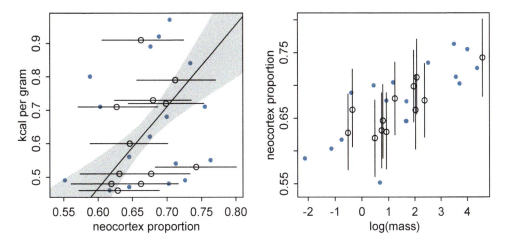

FIGURE 14.5. Same relationships as shown in FIGURE 14.4, but now for the imputation model that estimates the association between the predictors. The information in the association between predictors has been used to infer a stronger relationship between milk energy and the imputed values.

data—a matter of deduction. Of course there is no escape from assumption: We must be willing to make assumptions to arrive at deductions. But assumptions are not oaths. You can try different assumptions to explore the sensitivity of inference to them. And often, as in this case, the most conservative assumptions are hard to object to and make better use of the information in the data. The only courage required is to jointly model all of the data.

14.2.3. Not at random. In many cases, it is more plausible that missing values are not randomly distributed across cases. Instead, certain values of outcomes or predictors are more likely to induce missingness. For example, suppose species with small brains are harder to measure. Then missingness might actually be correlated with neocortex percent. If you can make a model of how missingness arises in the data, you can use Bayesian inference to figure out what the data say about the model. The same principle applies to more difficult situations, such as when there is sampling bias, such that certain species or individuals or events are more likely to be included in the data.

Overthinking: Imputation algorithm. To implement the imputation model directly in Stan, and therefore to understand how it actually works, the only tricky part is to make a vector that simultaneously tells us both which values inside `neocortex.prop` are missing and also the location of the corresponding parameter in the vector of missing values parameters. This code will construct a list of indexes to help with this problem:

R code
14.11

```
nc_missing <- ifelse( is.na(d$neocortex.prop) , 1 , 0 )
nc_missing <- sapply( 1:length(nc_missing) ,
    function(n) nc_missing[n]*sum(nc_missing[1:n]) )
nc_missing
```

```
 [1]  0  1  2  3  4  0  0  0  5  0  0  0  0  6  7  0  8  0  9  0 10  0 11  0  0
[26] 12  0  0  0
```

Everyplace there's a zero, the value has been observed. Otherwise, the number is the index of the corresponding parameter that estimates the value. Okay, now we just need to remove the NA values from the neocortex variable, so Stan doesn't complain about them.

R code
14.12
```
nc <- ifelse( is.na(d$neocortex.prop) , -1 , d$neocortex.prop )
```

Now we're ready for the Stan model code. I'll define this directly in an R variable, instead of using an external file. But you can do it either way. See the Stan reference manual for more details.

R code
14.13
```
model_code <- '
data{
    int N;
    int nc_num_missing;
    vector[N] kcal;
    real neocortex[N];
    vector[N] logmass;
    int nc_missing[N];
}
parameters{
    real alpha;
    real<lower=0> sigma;
    real bN;
    real bM;
    vector[nc_num_missing] nc_impute;
    real mu_nc;
    real<lower=0> sigma_nc;
}
model{
    vector[N] mu;
    vector[N] nc_merged;
    alpha ~ normal(0,10);
    bN ~ normal(0,10);
    bM ~ normal(0,10);
    mu_nc ~ normal(0.5,1);
    sigma ~ cauchy(0,1);
    sigma_nc ~ cauchy(0,1);
    // merge missing and observed
    for ( i in 1:N ) {
        nc_merged[i] <- neocortex[i];
        if ( nc_missing[i] > 0 ) nc_merged[i] <- nc_impute[nc_missing[i]];
    }
    // imputation
    nc_merged ~ normal( mu_nc , sigma_nc );
    // regression
    mu <- alpha + bN*nc_merged + bM*logmass;
    kcal ~ normal( mu , sigma );
}'
```

Aside from the loop in the model block that merges the observed and parameter values, it looks much like the mathematical model stated earlier. It's just two regressions, really. But since they're fit simultaneously and share parameters, they share information.

Finally, we can build the data and start lists and fire this puppy up:

R code
14.14
```
data_list <- list(
    N = nrow(d),
    kcal = d$kcal.per.g,
```

```
    neocortex = nc,
    logmass = d$logmass,
    nc_missing = nc_missing,
    nc_num_missing = max(nc_missing)
)
start <- list(
    alpha=mean(d$kcal.per.g), sigma=sd(d$kcal.per.g),
    bN=0, bM=0, mu_nc=0.68, sigma_nc=0.06,
    nc_impute=rep( 0.5 , max(nc_missing) )
)
library(rstan)
m14.3stan <- stan( model_code=model_code , data=data_list , init=list(start) ,
    iter=1e4 , chains=1 )
```

You'll find this produces the same results as before, but now with much less mystery.

14.3. Summary

This chapter has been a quick introduction to the design and implementation of measurement error and missing data models. In both cases, we must make assumptions about the distributions of variables. These assumptions then cascade through the model, changing not only the inferences from the data, but also our inferences about the data itself. This is an example of the invaluable capability of logic, as instantiated in Bayesian inference, to automatically discover counterintuitive implications of information. Design of the model is always up to you, however.

14.4. Practice

Easy.

14E1. Rewrite the Oceanic tools model (from Chapter 10) below so that it assumes measured error on the log population sizes of each society.

$$T_i \sim \text{Poisson}(\mu_i)$$
$$\log \mu_i = \alpha + \beta \log P_i$$
$$\alpha \sim \text{Normal}(0, 10)$$
$$\beta \sim \text{Normal}(0, 1)$$

14E2. Rewrite the same model so that it allows imputation of missing values for log population. There aren't any missing values in the variable, but you can still write down a model formula that would imply imputation, if any values were missing.

Medium.

14M1. Using the mathematical form of the imputation model in the chapter, explain what is being assumed about how the missing values were generated.

14M2. In earlier chapters, we threw away cases from the primate milk data, so we could use the neocortex variable. Now repeat the WAIC model comparison example from Chapter 6, but use imputation on the neocortex variable so that you can include all of the cases in the original data. The simplest form of imputation is acceptable. How are the model comparison results affected by being able to include all of the cases?

14M3. Repeat the divorce data measurement error models, but this time double the standard errors. Can you explain how doubling the standard errors impacts inference?

Hard.

14H1. The data in data(elephants) are counts of matings observed for bull elephants of differing ages. There is a strong positive relationship between age and matings. However, age is not always assessed accurately. First, fit a Poisson model predicting MATINGS with AGE as a predictor. Second, assume that the observed AGE values are uncertain and have a standard error of ± 5 years. Re-estimate the relationship between MATINGS and AGE, incorporating this measurement error. Compare the inferences of the two models.

14H2. Repeat the model fitting problem above, now increasing the assumed standard error on AGE. How large does the standard error have to get before the posterior mean for the coefficient on AGE reaches zero?

14H3. The fact that information flows in all directions among parameters sometimes leads to rather unintuitive conclusions. Here's an example from missing data imputation, in which imputation of a single datum reverses the direction of an inferred relationship. Use these data:

R code
14.15
```
set.seed(100)
x <- c( rnorm(10) , NA )
y <- c( rnorm(10,x) , 100 )
d <- list(x=x,y=y)
```

These data comprise 11 cases, one of which has a missing predictor value. You can quickly confirm that a regression of y on x for only the complete cases indicates a strong positive relationship between the two variables. But now fit this model, imputing the one missing value for x:

$$y_i \sim \text{Normal}(\mu_i, \sigma)$$
$$\mu_i = \alpha + \beta x_i$$
$$x_i \sim \text{Normal}(0, 1)$$
$$\alpha \sim \text{Normal}(0, 100)$$
$$\beta \sim \text{Normal}(0, 100)$$
$$\sigma \sim \text{HalfCauchy}(0, 1)$$

What has happened to the posterior distribution of β? Be sure to inspect the full density. Can you explain the change in inference?

15 Horoscopes

Statistics courses and books—this one included—tend to resemble horoscopes. There are two senses to this resemblance.

First, in order to remain plausibly correct, they must remain tremendously vague. This is because the targets of the advice, for both horoscopes and statistical advice, are diverse. But only the most general advice applies to all cases. A horoscope uses only the basic facts of birth to forecast life events, and a textbook statistical guide uses only the basic facts of measurement and design to dictate a model. It is easy to do better, once more detail is available. In the case of statistical analysis, it is typically only the scientist who can provide that detail, not the statistician.[169]

Second, there are strong incentives for both astrologers and statisticians to exaggerate the power and importance of their advice. No one likes an astrologer who forecasts doom, and few want a statistician who admits the answers as desired are not in the data as collected. Scientists desire results, and they will buy and attend to statisticians and statistical procedures that promise them. What we end up with is too often *horoscopic*: vague and optimistic, but still claiming critical importance.[170]

Statistical inference is indeed critically important. But only as much as every other part of research. Scientific discovery is not an additive process, in which sin in one part can be atoned by virtue in another. Everything interacts.[171] So equally when science works as intended as when it does not, every part of the process deserves our attention. Statistical analysis can neither be uniquely credited with science's success, nor can it be uniquely blamed for its failures and follies.

And there are plenty of failures and follies. Science, you may have heard, is not perfect. *The Lancet* is one of the oldest and most prestigious medical journals in the world. This is what its editor-in-chief, Richard Horton, wrote in its pages in 2015:[172]

> The case against science is straightforward: much of the scientific literature, perhaps half, may simply be untrue. Afflicted by studies with small sample sizes, tiny effects, invalid exploratory analyses, and flagrant conflicts of interest, together with an obsession for pursuing fashionable trends of dubious importance, science has taken a turn towards darkness.

How do we know that much of the published scientific literature is untrue? There are two major methods.

First, it is hard to repeat many published findings, even those in the best journals.[173] Some of this lack of repeatability arises from methodological subtleties, not because the findings are false. But many famous findings cannot be repeated, no matter who tries. There is a sense in which this should be unsurprising, given the nature of statistical testing. See the

Rethinking box on page 51. But the high false-discovery rate has become a great concern, partly because many placed unrealistic faith in significance testing and partly because it is hugely expensive to try to develop drugs and therapies from unrepeatable medical findings. It is even more expensive to design policy around false nutritional, psychological, economic, or ecological discoveries.[174] But the basic reputation of science is also at stake, all material costs aside. Why pay attention to breathlessly announced new discoveries, when as many as half of them will turn out to be false?

Second, the history of the sciences is equal parts wonder and blunder. The periodic table of the elements looks impressive now, but its story is unglamorous. There were more false elemental discoveries than there are current elements in the periodic table.[175] Don't think that all these false discoveries were performed by frauds and cranks. Enrico Fermi (1901–1954) was one of the greatest physicists of the 20th century. He discovered two heavy elements, ausonium (Ao, atomic number 93) and hesperium (Es, atomic number 94). These atomic numbers are now assigned to neptunium and plutonium, because Fermi had not actually discovered either. He mistook a mix of lighter already-discovered elements. These sorts of errors, and many other sorts of errors, were routine on the path to the current periodic table. Its story is one of error, ego, fraud, and correction. Other sciences look similar. Philosophers of science actually have a term, *the pessimistic induction*, for the observation that because most science has been wrong, most science is wrong.[176]

How can we reconcile such messy history, and widespread contemporary failure, with obvious successes like General Relativity? Science is a population-level process of variation and selective retention. It does not operate on individual hypotheses, but rather on populations of hypotheses. It comprises a mix of dynamics that may, over long periods of time, reveal the clockwork of nature.[177] But these same dynamics generate error. So it's entirely possible for most findings at any one point in time to be false but for science in the long term to still function. This is analogous to how natural selection can adapt a biological population to its environment, even though most individual variation in any one generation is maladaptive.

What is included in these dynamics? Here's a list of some salient pieces of the dynamic of scientific discovery, in no particular order. You might make your own list here, as there's nothing special about mine.

(1) Quality of theory and predictions: If most theories are wrong, most findings will be false positives. Karl Popper argued that all that matters for a theory to be scientific is that it be falsifiable. But for science to be effective, we must require more of theory. There was a brief quantitative version of this argument on page 51. A good theory specifies precise predictions that provide precise tests, and more than one model is usually necessary.

(2) Dynamics of research funding: Who gets funded, and how does the process select for particular forms of research? If there are no sources of long-term funding, then necessary long-term research will not be done. If people who already have funding judge who gets new funding, research may become overly conservative and possibly corrupt.

(3) Quality of measurement: Research design matters, all agree; but often this is forgotten when interpreting statistical analyses. A persistent problem is designs with low signal-to-noise ratios.[178] Poor signal will not mean no findings, just unreliable ones.

(4) Quality of data analysis: The topic of this book, but still much broader than it has indicated. Many common practices in the sciences exacerbate false discovery.[179] If you are not designing your analysis before you see the data, then your analysis may overfit the data in ways that regularization cannot reliably address.

(5) Quality of peer review: Good pre-publication peer review is invaluable. But much of it is not so good. Many mistakes get through, and many brilliant papers do not. Peer review selects for hyperbole, since honestly admitting limitations of work only hurts a paper's chances. Is this nevertheless the best system we can devise? Let's hope not.

(6) Publication: We agonize over bias in measurement and statistical analysis, but then allow it all back in during publication.[180] Incentives for positive findings and newsworthy results distort the design of research and how it is summarized.[181]

(7) Post-publication peer review: What happens to a finding after publication is just as important as what happens before. It is common for invalid analyses to be published in top-tier journals, only to be torn apart on blogs.[182] But there is no system for linking published papers to later peer criticism, and there are few formal incentives to conduct it. Even retracted papers continue to be cited.

(8) Replication and meta-analysis: The most important aspects of science are repetition and synthesis.[183] No single study is definitive, but incentives to replicate and summarize are weaker than incentives to produce novel findings. Top-tier journals prioritize news.

We tend to focus on the statistical analysis, perhaps because it is the only piece for which we have formulas and theorems. But every piece deserves attention and improvement. Sadly, many pieces are not under individual control, so social solutions are needed.

But there is an aspect of science that you do personally control: openness. Pre-plan your research together with the statistical analysis. Doing so will improve both the research design and the statistics. Document it in the form of a mock analysis that you would not be ashamed to share with a colleague. Register it publicly, perhaps in a simple repository, like Github or any other. But your webpage will do just fine, as well. Then collect the data. Then analyze the data as planned. If you must change the plan, that's fine. But document the changes and justify them. Provide all of the data and scripts necessary to repeat your analysis. Do not provide scripts and data "on request," but rather put them online so reviewers of your paper can access them without your interaction. There are of course cases in which full data cannot be released, due to privacy concerns. But the bulk of science is not of that sort.

The data and its analysis are the scientific product. The paper is just an advertisement. If you do your honest best to design, conduct, and document your research, so that others can build directly upon it, you can make a difference.

Rethinking: Statistics is to math as cooking is to chemistry. In a uniquely valuable essay,[184] Terry Speed stated that "statistics is no closer to mathematics than cooking is to chemistry." What this means is that while each field has a basis in another, each is sufficiently abstracted from its base to make its practice mostly depend upon other factors. In cooking, abstract heuristics are more useful than the chemical laws that explain them, and human psychology and culture can dominate. In statistics, context is king. General mathematical considerations always matter, but mathematical foundations solve few, if any, of the contingent problems that we confront in the context of a study.

Endnotes

Chapter 1

1. I draw this metaphor from Collins and Pinch (1998), *The Golem: What You Should Know about Science*. It is very similar to E. T. Jaynes' 2003 metaphor of statistical models as robots, although with a less precise and more monstrous implication. [1]

2. There are probably no algorithms nor machines that never break, bend, or malfunction. A common citation for this observation is Wittgenstein (1953), *Philosophical Investigations*, section 193. Malfunction will interest us, later in the book, when we consider more complex models and the procedures needed to fit them to data. [2]

3. See Mulkay and Gilbert (1981). I sometimes teach a PhD core course that includes some philosophy of science, and PhD students are nearly all shocked by how little their casual philosophy resembles that of Popper or any other philosopher of science. The first half of Ian Hacking's *Representing and Intervening* (1983) is probably the quickest way into the history of the philosophy of science. It's getting out of date, but remains readable and broad minded. [4]

4. Maybe best to begin with Popper's last book, *The Myth of the Framework* (1996). I also recommend interested readers to go straight to a modern translation of Popper's earlier *Logic of Scientific Discovery*. Chapters 6, 8, 9 and 10 in particular demonstrate that Popper appreciated the difficulties with describing science as an exercise in falsification. Other later writings, many collected in *Objective knowledge: An evolutionary approach*, show that Popper viewed the generation of scientific knowledge as an evolutionary process that admits many different methods. [4]

5. Meehl (1967) observed that this leads to a methodological paradox, as improvements in measurement make it easier to reject the null. But since the research hypothesis has not made any specific quantitative prediction, more accurate measurement doesn't lead to stronger corroboration. See also Andrew Gelman's comments in a September 5, 2014 blog post: http://andrewgelman.com/2014/09/05/confirmationist-falsificationist-paradigms-science/. [5]

6. George E. P. Box is famous for this dictum. As far as I can tell, his first published use of it was as a section heading in a 1979 paper (Box, 1979). Population biologists like myself are more familiar with a philosophically similar essay about modeling in general by Richard Levins, "The Strategy of Model Building in Population Biology" (Levins, 1966). [5]

7. Ohta and Gillespie (1996). [5]

8. Hubbell (2001). The theory has been productive in that it has forced greater clarity of modeling and understanding of relations between theory and data. But the theory has had its difficulties. See Clark (2012). For a more general skeptical attitude towards "neutrality," see Proulx and Adler (2010). [5]

9. For direct application of Kimura's model to cultural variation, see for example Hahn and Bentley (2003). All of the same epistemic problems reemerge here, but in a context with much less precision of theory. Hahn and Bentley have since adopted a more nuanced view of the issue. See their comment to Lansing and Cox (2011), as well as the similar comment by Feldman. [5]

10. Gillespie (1977). [5]

11. Lansing and Cox (2011). See objections by Hahn, Bentley, and Feldman in the peer commentary to the article. [7]

12. See Cho (2011) for a December 2011 summary focusing on debates about measurement. [8]

13. For an autopsy of the experiment, see http://profmattstrassler.com/articles-and-posts/particle-physics-basics/neutrinos/neutrinos-faster-than-light/opera-what-went-wrong/. [9]

14. See Mulkay and Gilbert (1981) for many examples of "Popperism" from practicing scientists, including famous ones. [9]

15. For an accessible history of some measurement issues in the development of physics and biology, including early experiments on relativity and abiogenesis, I recommend Collins and Pinch (1998). Some scientists have read this book as an attack on science. However, as the authors clarify in the second edition, this was not their intention. Science makes myths, like all cultures do. That doesn't necessarily imply that science does not work. See also Daston and Galison (2007), which tours concepts of objective measurement, spanning several centuries. [9]

16. The first chapter of Sober (2008) contains a similar discussion of *modus tollens*. Note that the statistical philosophy of Sober's book is quite different from that of the book you are holding. In particular, Sober is weakly anti-Bayesian. This is important, because it emphasizes that rejecting *modus tollens* as a model of statistical inference has nothing to do with any debates about Bayesian versus non-Bayesian tools. [9]

17. Popper himself had to deal with this kind of theory, because the rise of quantum mechanics in his lifetime presented rather serious challenges to the notion that measurement was unproblematic. See Chapter 9 in his *Logic of Scientific Discovery*, for example. [9]

18. See the Afterword to the 2nd edition of Collins and Pinch (1998) for examples of textbooks getting it wrong by presenting tidy fables about the definitiveness of evidence. [10]

19. A great deal has been written about the sociology of science and the interface of science and public interest. Interested novices might begin with Kitcher (2011), *Science in a Democratic Society*, which has a very broad topical scope and so can serve as an introduction to many dilemmas. [10]

20. Yes, even procedures that claim to be free of assumptions do have assumptions and are a kind of model. All systems of formal representation, including numbers, do not directly reference reality. For example, there is more than one way to construct "real" numbers in mathematics, and there are important consequences in some applications. In application, all formal systems are like models. See http://plato.stanford.edu/entries/philosophy-mathematics/ for a short overview of some different stances that can be sustained towards reasoning in mathematical systems. [10]

21. Saint Augustine, in *City of God*, famously admonished against trusting in luck, as personified by Fortuna: "How, therefore, is she good, who without discernment comes to both the good and to the bad?" See also the introduction to Gigerenzer et al. (1990). Rao (1997) presents a page from an old book of random numbers, commenting upon how seemingly useless such a thing would have been in previous eras. [10]

22. Most scholars trace frequentism to British logician John Venn (1834–1923), as for example presented in his 1876 book. Speaking of the proportion of male births in all births, Venn said, "probability *is* nothing but that proportion" (page 84). Venn taught Fisher some of his maths, so this may be where Fisher acquired his opposition to Bayesian probability. [11]

23. Fisher (1956). See also Fisher (1955), the first major section of which discusses the same point. Some people would dispute that Fisher was a "frequentist," because he championed his own likelihood methods over the methods of Neyman and Pearson. But Fisher definitely rejected the broader Bayesian approach to probability theory. See Endnote 28. [11]

24. This last sentence is a rephrasing from Lindley (1971): "A statistician faced with some data often embeds it in a family of possible data that is just as much a product of his fantasy as is a prior distribution." Dennis V. Lindley (1923–2013) was a prominent defender of Bayesian data analysis when it had very few defenders. [11]

25. It's hard to find an accessible introduction to image analysis, because it's a very computational subject. At the intermediate level, see Marin and Robert (2007), Chapter 8. You can hum over their mathematics and still

acquaint yourself with the different goals and procedures. See also Jaynes (1984) for spirited comments on the history of Bayesian image analysis and his pessimistic assessment of non-Bayesian approaches. There are better non-Bayesian approaches since. [12]

26. Binmore (2009) describes the history within economics and related fields and provides a critique that I am sympathetic to. [12]

27. See Gigerenzer et al. (2004). [13]

28. Fisher (1925), page 9. See Gelman and Robert (2013) for reflection on intemperate anti-Bayesian attitudes from the middle of last century. [13]

29. See McGrayne (2011) for a non-technical history of Bayesian data analysis. See also Fienberg (2006), which describes (among many other things) applied use of Bayesian multilevel models in election prediction, beginning in the early 1960s. [13]

30. See Fienberg (2006), page 24. [15]

31. See Wang et al. (2015) for a vivid example. [15]

32. I borrow this phrasing from Silver (2012). Silver's book is a well-written, non-technical survey of modeling and prediction in a range of domains. [15]

33. See Theobald (2010) for a fascinating example in which multiple non-null phylogenetic models are contrasted. [16]

34. See Sankararaman et al. (2012) for a thorough explanation, including why current evidence suggests that there really was interbreeding. [16]

Chapter 2

35. Morison (1942). Globe illustration modified from public domain illustration at the Wikipedia entry for Martin Behaim. In addition to underestimating the circumference, Colombo also overestimated the size of Asia and the distance between mainland China and Japan. [19]

36. This distinction and vocabulary derive from Savage (1962). [19]

37. See Robert (2007) for thorough coverage of the decision-theoretic optimality of Bayesian inference. [19]

38. See Simon (1969) and chapters in Gigerenzer et al. (2000). [20]

39. See Cox (1946). Jaynes (2003) and Van Horn (2003) explain the Cox theorem and its role in inference. [24]

40. See Gelman and Robert (2013) for examples. [24]

41. I first encountered this globe tossing strategy in Gelman and Nolan (2002). Since I've been using it in classrooms, several people have told me that they have seen it in other places, but I've been unable to find a primeval citation, if there is one. [28]

42. There is actually a set of theorems, the *No Free Lunch* theorems. These theorems—and others which are similar but named and derived separately—effectively state that there is no optimal way to pick priors (for Bayesians) or select estimators or procedures (for non-Bayesians). See Wolpert and Macready (1997) for example. [31]

43. This is a subtle point that will be expanded in other places. On the topic of accuracy of assumptions versus information processing, see e.g. Appendix A of Jaynes (1985): The Gaussian, or normal, error distribution needn't be physically correct in order to be the most useful assumption. [32]

44. Kronecker (1823–1891), an important number theorist, was quoted as stating "God made the integers, all else is the work of man" (*Die ganzen Zahlen hat der liebe Gott gemacht, alles andere ist Menschenwerk*). There appears to be no consensus among mathematicians about which parts of mathematics are discovered rather than invented. But all admit that applied mathematical models are "work of man." [32]

45. This approach is usually identified with Bruno de Finetti and L. J. Savage. See Kadane (2011) for review and explanation. [35]

46. See Berger and Berry (1988), for example, for further exploration of these ideas. [35]

Chapter 3

47. Gigerenzer and Hoffrage (1995). There is a large empirical literature, which you can find by searching forward on the Gigerenzer and Hoffrage paper. [50]

48. Feynman (1967) provides a good defense of this device in scientific discovery. [50]

49. For a binary outcome problem of this kind, the posterior density is given by `dbeta(p,w+1,n-w+1)`, where p is the proportion of interest, w is the observed count of water, and n is the number of tosses. If you're curious about how to prove this fact, look up "beta-binomial conjugate prior." I avoid discussing the analytical approach in this book, because very few problems are so simple that they have exact analytical solutions like this. [51]

50. See Ioannidis (2005) for another narrative of the same idea. The problem is possibly worse than the simple calculation suggests. On the other hand, real scientific inference is usually more subtle than mere truth or falsehood of an hypothesis. But many, if not most, scientists tend to think in such binary terms, so this calculation should be disturbing. [52]

51. See Box and Tiao (1973), page 84 and then page 122 for a general discussion. [56]

52. Gelman et al. (2013a), page 33, comment on differences between percentile intervals and HPDIs. [57]

53. Fisher (1925), in Chapter III within section 12 on the normal distribution. There are a couple of other places in the book in which the same resort to convenience or convention is used. Fisher seems to indicate that the 5% mark was already widely practiced by 1925 and already without clear justification. [58]

54. Fisher (1956). [58]

55. See Henrion and Fischoff (1986) for examples from the estimation of physical constants, such as the speed of light. [58]

56. Robert (2007) provides concise proofs of optimal estimators under several standard loss functions, like this one. It also covers the history of the topic, as well as many related issues in deriving good decisions from statistical procedures. [59]

57. Rice (2010) presents an interesting construction of classical Fisherian testing through the adoption of loss functions. [61]

58. See Hauer (2004) for three tales from transportation safety in which testing resulted in premature incorrect decisions and a demonstrable and continuing loss of human life. [61]

59. It is poorly appreciated that coin tosses are very hard to bias, as long as you catch them in the air. Once they land and bounce and spin, however, it is very easy to bias them. [67]

60. E. T. Jaynes (1922–1998) said all of this much more succinctly: Jaynes (1985), page 351, "It would be very nice to have a formal apparatus that gives us some 'optimal' way of recognizing unusual phenomena and inventing new classes of hypotheses that are most likely to contain the true one; but this remains an art for the creative human mind." See also Box (1980) for a similar perspective. [68]

Chapter 4

61. Leo Breiman, at the start of Chapter 9 of his classic book on probability theory (Breiman, 1968), says "there is really no completely satisfying answer" to the question "why normal?" Many mathematical results remain mysterious, even after we prove them. So if you don't quite get why the normal distribution is the limiting distribution, you are in good company. [73]

62. For the reader hungry for mathematical details, see Frank (2009) for a nicely illustrated explanation of this, using Fourier transforms. [74]

63. Technically, the distribution of sums converges to normal only when the original distribution has finite variance. What this means practically is that the magnitude of any newly sampled value cannot be so big as to overwhelm all of the previous values. There are natural phenomena with effectively infinite variance, but we won't be working with any. Or rather, when we do, I won't comment on it. [74]

64. Howell (2010) and Howell (2000). See also Lee and DeVore (1976). Much more raw data is available for download from https://tspace.library.utoronto.ca/handle/1807/10395. [79]

65. Jaynes (2003), page 21–22. See that book's index for other mentions in various statistical arguments. [81]

66. See Jaynes (1986) for an entertaining example concerning the beer preferences of left-handed kangaroos. There is an updated 1996 version of this paper available online. [81]

67. The strategy is the same grid approximation strategy as before (page 39). But now there are two dimensions, and so there is a geometric (literally) increase in bother. The algorithm is mercifully short, however, if not transparent. Think of the code as being six distinct commands. The first two lines of code just establish the range of μ and σ values, respectively, to calculate over, as well as how many points to calculate in-between. The third line of code expands those chosen μ and σ values into a matrix of all of the combinations of μ and σ. This matrix is stored in a data frame, `post`. In the monstrous fourth line of code, shown in expanded form to make it easier to read, the log-likelihood at each combination of μ and σ is computed. This line looks so awful, because we have to be careful here to do everything on the log scale. Otherwise rounding error will quickly make all of the posterior probabilities zero. So what `sapply` does is pass the unique combination of μ and σ on each row of `post` to a function that computes the log-likelihood of each observed height, and adds all of these log-likelihoods together (`sum`). In the fifth line, we multiply the prior by the likelihood to get the product that is proportional to the posterior density. The priors are also on the log scale, and so we add them to the log-likelihood, which is equivalent to multiplying the raw densities by the likelihood. Finally, the obstacle for getting back on the probability scale is that rounding error is always a threat when moving from log-probability to probability. If you use the obvious approach, like `exp( post$prod )`, you'll get a vector full of zeros, which isn't very helpful. This is a result of R's rounding very small probabilities to zero. Remember, in large samples, all unique samples are unlikely. This is why you have to work with log-probability. The code in the box dodges this problem by scaling all of the log-products by the maximum log-product. As a result, the values in `post$prob` are not all zero, but they also aren't exactly probabilities. Instead they are relative posterior probabilities. But that's good enough for what we wish to do with these values. [84]

68. The most accessible of Galton's writings on the topic has been reprinted as Galton (1989). [92]

69. See Reilly and Zeringue (2004) for an example using predator-prey dynamics. [94]

70. The implied definition of α in a parabolic model is $\alpha = \mathrm{E}\, y_i - \beta_1\, \mathrm{E}\, x_i - \beta_2\, \mathrm{E}\, x_i^2$. Now even when the average x_i is zero, $\mathrm{E}\, x_i = 0$, the average square will likely not be zero. So α becomes hard to directly interpret again. [112]

Chapter 5

71. "How to Measure a Storm's Fury One Breakfast at a Time." *The Wall Street Journal*: September 1, 2011. [119]

72. See Meehl (1990), in particular the "crud factor" described on page 204. [119]

73. Simpson (1951). Simpson's paradox is very famous in statistics, probably because recognizing it increases the apparent usefulness of statistical modeling. It's a lot less known outside of statistics. [119]

74. Debates about causal inference go back a long time. David Hume is key citation. One curious obstacle in modern statistics is that classic causal reasoning requires that if A causes B, then B will always appear when A appears. But with probabilistic relationships, like those described in most contemporary scientific models, it is unsurprising to talk about probabilistic causes, in which B only sometimes follows A. See http://plato.stanford.edu/entries/causation-probabilistic/. [120]

75. See Pearl (2014) for an accessible introduction, with discussion. See also Rubin (2005) for a related approach. [120]

76. Data from Table 2 of Hinde and Milligan (2011). [135]

77. Rosenbaum (1984) calls it *concomitant variable bias*. See also Chapter 9 in Gelman and Hill (2007). There isn't really any standard terminology for this issue. It is a component of generalized mediation analysis, and some fields discuss it under that banner. [150]

78. Provided the posterior distributions are Gaussian, you could, however, get the variance of their sum by adding their variances and twice their covariance. The variance of the sum (or difference) of two normal distributions a and b is given by $\sigma_a^2 + \sigma_b^2 + 2\rho\sigma_a\sigma_b$, where ρ is the correlation between the two. [154]

79. See Gelman and Stern (2006) for further explanation, and see Nieuwenhuis et al. (2011) for some evidence of how commonly this mistake occurs. [157]

80. See Stigler (1981) for historical context. There are a number of legitimate ways to derive the method of least squares estimation. Gauss' approach was Bayesian, but a probability interpretation isn't always necessary. [159]

81. These data are modified from an example in Grafen and Hails (2002). [163]

Chapter 6

82. *De Revolutionibus*, Book 1, Chapter 10. [165]

83. See e.g. Akaike (1978), as well as discussion in Burnham and Anderson (2002). [167]

84. When priors are flat and models are simple, this will always be true. But later in the book, you'll work with other types of models, like multilevel regressions, for which adding parameters does not necessarily lead to better fit to sample. [167]

85. Data from Table 1 of McHenry and Coffing (2000). [168]

86. See Grünwald (2007) for a book-length treatment of these ideas. [172]

87. There are many discussions of bias and variance in the literature, some much more mathematical than others. For a broad treatment, I recommend Chapter 7 of Hastie, Tibshirani and Friedman's 2009 book, which explores BIC, AIC, cross-validation and other measures, all in the context of the bias-variance trade-off. [174]

88. I first encountered this kind of example in Jaynes (1976), page 246. Jaynes himself credits G. David Forney's 1972 information theory course notes. Forney is an important figure in information theory, having won several awards for his contributions. [175]

89. Shannon (1948). For a more accessible introduction, see the venerable textbook *Elements of Information Theory*, by Cover and Thomas. Slightly more advanced, but having lots of added value, is Jaynes' (2003, Chapter 11) presentation. A foundational book in applying information theory to statistical inference is Kullback (1959), but it's not easy reading. [177]

90. See two famous editorials on the topic: Shannon (1956) and Elias (1958). Elias' editorial is a clever work of satire and remains as current today as it was in 1958. Both of these one-page editorials are readily available online. [177]

91. I really wish I could say there is an accessible introduction to maximum entropy, at the level of most natural and social scientists' math training. If there is, I haven't found it yet. Jaynes (2003) is an essential source, but if your integral calculus is rusty, progress will be very slow. Better might be Steven Frank's papers (2009; 2011) that explain the approach and relate it to common distributions in nature. You can mainly hum over the maths in these and still get the major concepts. See also Harte (2011), for a textbook presentation of applications in ecology. [179]

92. Kullback and Leibler (1951). Note however that Kullback and Leibler did not name this measure after themselves. See Kullback (1987) for Solomon Kullback's reflections on the nomenclature. For what it's worth, Kullback and Leibler make it clear in their 1951 paper that Harold Jeffreys had used this measure already in the development of Bayesian statistics. [179]

93. Under somewhat general conditions, for many common model types, a difference between two deviances has a chi-squared distribution. The factor of 2 is there to scale it that way. [182]

94. Akaike (1973). See also Akaike (1974, 1978, 1981), where AIC was further developed and related to Bayesian approaches. Population biologists tend to know about AIC from Burnham and Anderson (2002), which strongly advocates its use. [189]

95. A common approximation in the case of small N is AICc $= D_{\text{train}} + \frac{2k}{1-(k+1)/N}$. As N grows, this expression approaches AIC. See Burnham and Anderson (2002). [189]

96. Lunn et al. (2013) contains a fairly understandable presentation of DIC, including a number of different ways to compute it. [190]

97. Watanabe (2010). Gelman et al. (2013b) re-dub WAIC the "Watanabe-Akaike Information Criterion" to give explicit credit to Watanabe, in the same way people renamed AIC after Akaike. Gelman et al. (2013b) is worthwhile also for the broad perspective it takes on the inference problem. [190]

98. Schwarz (1978). [192]

99. Gelman and Rubin (1995). See also section 7.4, page 182, of Gelman et al. (2013a). [192]

100. Stone (1977). [194]

101. This is closely related to minimum description length. See Grünwald (2007). [195]

102. See Vehtari and Gelman (2014) for definition and discussion. [198]

103. See Burnham and Anderson (2002) for a thorough argument in favor of this interpretation. The authors note that some aspects of the procedure remain heuristic, because priors within models are left behind when using information criteria, unlike when using Bayes factors. For some, that's a virtue. For others, a sin. [199]

104. See both Burnham and Anderson (2002) and Claeskens and Hjort (2008). [200]

105. William Henry Harrison's military history earned him the nickname "Old Tippecanoe." Tippecanoe was the sight of a large battle between Native Americans and Harrison, in 1811. In popular opinion, he was a war hero. But in popular imagination, Harrison was cursed by the Native Americans in the aftermath of the battle. [205]

Chapter 7

106. All manatee facts here taken from Lightsey et al. (2006); Rommel et al. (2007). Scarchart in figure from the free educational materials at http://www.learner.org/jnorth/tm/manatee/RollCall.html. [209]

107. Wald (1943). See Mangel and Samaniego (1984) for a more accessible presentation and historical context. [209]

108. Wald (1950). Wald's foundational paper is Wald (1939). Fienberg (2006) is a highly recommended read for historical context. For more technical discussions, see Berger (1985), Robert (2007), and Jaynes (2003) page 406. [211]

109. GDP is Gross Domestic Product. It's the most common measure of economic performance, but also one of the silliest. Using GDP to measure the health of an economy is like using heat to measure the quality of a chemical reaction. [211]

110. From Nunn and Puga (2011). [212]

111. Riley et al. (1999). [212]

112. See Ioannidis (2005) for the claim that false positives are common. See also Simmons et al. (2011). In principle, it is possible to be too conservative for any particular circumstance, so different subfields need different attitudes towards risks taken in statistical inference. [218]

113. A good example is the extensive modern tunnel system in the Faroe Islands. The natural geology of the islands is very rugged, such that it has historically been much easier to travel by water than by land. But in the late 20th century, the Danish government invested heavily in tunnel construction, greatly reducing the effective ruggedness of the islands. [224]

114. Modified example from Grafen and Hails (2002), which is a great non-Bayesian applied statistics book you might also enjoy. It has a rather unique geometric presentation of some of the standard linear models. [226]

115. Data from Nettle (1998). [238]

Chapter 8

116. See the introduction of Gigerenzer et al. (1990) for more on this history. See also Rao (1997) for an example page from a book of random numbers, with similar commentary on the cultural shift. [241]

117. The traveling individual metaphor is one of two common metaphors. The other is of a mountain climber who maps a mountain range by random jumps. See Kruscke (2011) for a very similar story-based explanation about a politician who raises funds at different locations. Kruschke's book is excellent. It has a rather different style and coverage than this one, so may bring a lot of added value to the reader, in terms of getting a different perspective and a different set of examples. [242]

118. Metropolis et al. (1953). The algorithm has been named after the first author of this paper, however it's not clear how each co-author participated in discovery and implementation of the algorithm. Among the other authors were Edward Teller, most famous as the father of the hydrogen bomb, and Marshall Rosenbluth, a renown physicist in his own right, as well as their wives Augusta and Arianna (respectively), who did much of the computer programming. Nicholas Metropolis lead the research group. Their work was in turn based on earlier work with Stanislaw Ulam: Metropolis and Ulam (1949). [245]

119. Hastings (1970). [245]

120. Geman and Geman (1984) is the original. See Casella and George (1992) as well. Note that Gibbs sampling is named after physicist and mathematician J. W. Gibbs, one of the founders of statistical physics. However, Gibbs died in the year 1903, long before even the Metropolis algorithm was invented. Instead the strategy is named after Gibbs, both to honor him and in light of the algorithm's connections to statistical physics. [245]

121. Chapter 16 of Jaynes (2003). [246]

122. See Hoffman and Gelman (2011). [247]

123. See Robert and Casella (2011) for a concise history of MCMC that covers both computation and mathematical foundations. [247]

124. Gelman (2006) recommends the half-Cauchy because it is approximately uniform in the tail and still weak near zero, without the odd behavior of traditional priors like the inverse-gamma. Polson and Scott (2012) examine this prior in more detail. Simpson et al. (2014) also note that the half-Cauchy prior has useful features, but recommend instead an exponential prior. Either is equally useful in all the examples in this book. [249]

125. For some more detail and background citations, see Chapter 6 in Brooks et al. (2011). [256]

126. Gelman and Rubin (1992). [257]

Chapter 9

127. Grosberg (1998). [267]

128. Williams (1980). See also Caticha and Griffin (2007); Griffin (2008) for a clearer argument with some worked examples. See Jaynes (1988) for historical context. [268]

129. Jaynes (2003), page 351. [271]

130. Williams (1980). [272]

131. Williams (1980). See also Caticha and Griffin (2007); Griffin (2008) for a clearer argument with some worked examples. See Jaynes (1988) for historical context. [272]

132. E. T. Jaynes called this phenomenon "entropy concentration." See Jaynes (2003), pages 365–370. [273]

133. A generalized normal distribution has variance $\alpha^2 \Gamma(3/\beta)/\Gamma(1/\beta)$. We can define a family of such distributions with equal variance by choosing the shape β and solving for the α that makes the variance expression equal to any chosen σ^2. The solution is $\alpha = \sigma\sqrt{\frac{\Gamma(1/\beta)}{\Gamma(3/\beta)}}$. This density is provided by `rethinking` as `dgnorm`, in case you want to play around with it. [273]

134. I learned this proof from Keith Conrad's "Probability distributions and maximum entropy" notes, found online. [274]

135. The first line of the function just samples 3 uniform random numbers, with no joint constraint. The second line then solves for the relative value of the 4th value, by using the stated expected value G. The rest of the function just normalizes to a probability distribution and computes entropy. [277]

136. McCullagh and Nelder (1989) is the central citation for the conventional generalized linear models. The term "generalized linear model" is due to Nelder and Wedderburn (1972). The terminology can be confusing, because there is also the "general linear model." Nelder later regretted the choice. See Senn (2003), page 127. [281]

137. Frank (2007). [283]

138. Not a real distribution. [284]

139. Nuzzo (2014). See also Simmons et al. (2011). [287]

Chapter 10

140. Silk et al. (2005). [292]

141. Bickel et al. (1975). [304]

142. Simpson (1951). [309]

143. See Pearl (2014), for example. So much has been written about Simpson's paradox that you can find it explained in seemingly contradictory ways. [309]

144. Kline and Boyd (2010). [313]

145. There seems to be no primordial citation for this transformation. A common citation is Baker (1994), who cites a lot of prior *ad hoc* use. McCullagh and Nelder (1989) explain the transformation beginning on page 209. [325]

146. Welsh and Lind (1995). [330]

Chapter 11

147. McCullagh (1980) is credited with introducing and popularizing this approach. See also Fullerton (2009) for an overview with comparison of different model types. [332]

148. Cushman et al. (2006). [332]

149. See Lambert (1992) for the first presentation of this type of model. The basic zero-inflated approach is older, but Lambert presented the version we use here, with log and logit links to two separate linear models. [343]

150. Williams (1975, 1982). Bolker (2008) contains a clear presentation in the context of ecological data. [347]

151. Another very common parameterization is $\alpha = \bar{p}\theta$ and $\beta = (1 - \bar{p})\theta$. The $\bar{p}$ and θ version is more useful for modeling, because we typically want to attach a linear model to the beta distribution's central tendency, one measure of which is $\bar{p}$. [347]

152. Hilbe (2011) is an entire book devoted to gamma-Poisson regression. [350]

153. Jung et al. (2014). [352]

Chapter 12

154. Wearing's wife Deborah has written a book about their life after the illness (Wearing, 2005). His story has also appeared in a number of documentaries. A quick internet search will turn up a number of news articles, as well. [355]

155. See section 6, page 20, of Gelman (2005) for an entertaining list of wildly different definitions of "random effect." [357]

156. Vonesh and Bolker (2005). [357]

157. I adopt the terminology of Gelman (2005), who argues that the common term *random effects* hardly aids with understanding, for most people. Indeed, it seems to encourage misunderstanding, partly because the terms *fixed* and *random* mean different things to different statisticians. See pages 20–21 of Gelman's paper. I fully realize, however, that by trying to spread Gelman's alternative jargon, I am essentially spitting into a very strong wind. [358]

158. It's also common for the "multi" to refer to multiple linear models. This is especially true in the literature on "hierarchical linear models." Regardless, we're talking about the same kind of robot here. [359]

159. Note that there is still uncertainty about the regularization. So this model isn't exactly the same as just assuming a regularizing prior with a constant standard deviation 1.6. Instead the intercepts for each tank average over the uncertainty in σ (and α). [360]

160. This fact has been understood much longer than multilevel models have been practical to use. See Stein (1955) for an influential non-Bayesian paper. [364]

Chapter 13

161. Lewandowski et al. (2009). The "LKJ" part of the name comes from the first letters of the last names of the authors, who themselves called the approach the "onion method." For use in Bayesian models, see the explanation in the latest version of the Stan reference manual. [394]

162. See Gelfand et al. (1995), as well as Roberts and Sahu (1997). See also Papaspiliopoulos et al. (2007) for a more recent overview. See Betancourt and Girolami (2013) for a discussion focusing of Hamiltonian Monte Carlo. [408]

163. See Neal (1998) for a highly cited overview, with notes on implementation. [410]

164. See MacKay and Neal (1994); Neal (1996). [419]

Chapter 14

165. Joseph Bertrand, 1889, *Calcul des probabilités*. [423]

166. See Carroll et al. (2012) for an overview of both Bayesian and non-Bayesian approaches to measurement error. The topic of measurement error is often very specific to different disciplines and contexts, because the nature of error can be very specific. [425]

167. See Molenberghs et al. (2014) for an overview of contemporary approaches, Bayesian and otherwise. [431]

168. See Rubin (1976); Rubin and Little (2002) for background and additional terminology. Section 4 of Rubin's 1976 article is valuable for the clear definitions of causes of missing data. [432]

Chapter 15

169. See Speed (1986) for extended comments like this, aimed at statisticians. You can find a copy of this essay online with a quick internet search. [441]

170. A related phenomenon in popular culture and in science is the *Forer effect* or *Barnum effect*. See Forer (1949) and Meehl (1956). [441]

171. There have been a few attempts to model these mutual interactions. See McElreath and Smaldino (2015). [441]

172. Horton (2015). [441]

173. Perhaps it is better to say "especially those in the best journals." See Ioannidis (2005) for a very highly cited review and argument. See also Ioannidis (2012) and citations therein. There is a lot of recent work in this area, including the Many Labs Replication Projects for social psychology, which have both confirmed and rejected famous textbook findings. [441]

174. A particularly infamous example of an un-replicable economic finding that had a big impact on policy is Reinhart and Rogoff (2010). Although apparently, if not actually, influential in national and international budget debates, the finding was based on odd inclusion criteria and an Excel spreadsheet error. See Herndon et al. (2014). Many other false findings result from no error at all, just misleading samples. The answer is not always in the data, remember. But if you torture the data long enough, it will confess. [442]

175. Fontani et al. (2014). This is a fantastic book which catalogs and explains hundreds of false discoveries in elemental chemistry and physics. [442]

176. Laudan (1981). To be fair, there are several ways to interpret the pessimistic induction. Newtonian mechanics, for example, is strictly wrong. But it's an amazingly successful theory nevertheless. I made a similar point about the geocentric model of the solar system, back in Chapter 4. But there are plenty of less successful theories that have also turned out to be false, despite being held to be true for decades or generations. [442]

177. This is the standard view in history and philosophy of science. See for introduction Campbell (1985); Hull (1988); Kitcher (2000); Popper (1963, 1996). [442]

178. See Sedlemeier and Gigerenzer (1989) and more recent publications on the same topic. [442]

179. See for examples relevant to the process of discovery: Gelman and Loken (2013, 2014); Simmons et al. (2011, 2013). [443]

180. See Fanelli (2012); Franco et al. (2014); Rosenthal (1979). This one has the best title of the genre: Ferguson and Heene (2012). [443]

181. Ecologist Art Shapiro published his satirical "Laws of Field Ecology Research" in *Bulletin of the Entomological Society of Canada* in the early 1980s. I can't find the original citation, but a copy provided by Art reads: "Law #4: Never state explicitly the limits on generalizing from your results. The referees will take you at your word and recommend rejection." Sadly that has always been my experience as well. [443]

182. Two excellent examples of this phenomenon occurred in 2014 and 2015. First, Lin et al. (2014) published an analysis of gene expression that was terribly confounded by batch effects. Basically, they ran a bad experiment. Yoav Gilad discovered this and released a reanalysis on Twitter, later published as Gilad and Mizrahi-Man (2015). The original authors continue to deny the results were in error, and the saga continues. The second involves a competition held by Lior Pachter on his blog: https://liorpachter.wordpress.com/2015/05/26/pachters-p-value-prize/. I recommend reading the whole thing, including the comments, which is where the action is. [443]

183. Replication and meta-analysis obviously interact strongly with all the other forces. For a unique article addressing replication and meta-analysis for the incentives they provide in the quality of research, see O'Rourke and Detsky (1989). [443]

184. Speed (1986), "Questions, answers and statistics." You can find a copy of this essay online with a quick internet search. [443]

Bibliography

Akaike, H. (1973). Information theory and an extension of the maximum likelihood principle. In Petrov, B. N. and Csaki, F., editors, *Second International Symposium on Information Theory*, pages 267–281.

Akaike, H. (1974). A new look at the statistical model identification. *IEEE Transactions on Automatic Control*, 19(6):716–723.

Akaike, H. (1978). A Bayesian analysis of the minimum AIC procedure. *Ann. Inst. Statist. Math.*, 30:9–14.

Akaike, H. (1981). Likelihood of a model and information criteria. *Journal of Econometrics*, 16:3–14.

Baker, S. G. (1994). The multinomial-Poisson transformation. *Journal of the Royal Statistical Society, Series D*, 43(4):495–504.

Berger, J. O. (1985). *Statistical decision theory and Bayesian Analysis*. Springer-Verlag, New York, 2nd edition.

Berger, J. O. and Berry, D. A. (1988). Statistical analysis and the illusion of objectivity. *American Scientist*, pages 159–165.

Betancourt, M. J. and Girolami, M. (2013). Hamiltonian Monte Carlo for hierarchical models. arXiv:1312.0906.

Bickel, P. J., Hammel, E. A., and O'Connell, J. W. (1975). Sex bias in graduate admission: Data from Berkeley. *Science*, 187(4175):398–404.

Binmore, K. (2009). *Rational Decisions*. Princeton University Press.

Bolker, B. (2008). *Ecological Models and Data in R*. Princeton University Press.

Box, G. E. P. (1979). Robustness in the strategy of scientific model building. In Launer, R. and Wilkinson, G., editors, *Robustness in Statistics*. Academic Press, New York.

Box, G. E. P. (1980). Sampling and Bayes' inference in scientific modelling and robustness. *Journal of the Royal Statistical Society A*, 143:383–430.

Box, G. E. P. and Tiao, G. C. (1973). *Bayesian Inference in Statistical Analysis*. Addison-Wesley Pub. Co., Reading, Mass.

Breiman, L. (1968). *Probability*. Addison-Wesley Pub. Co.

Brooks, S., Gelman, A., Jones, G. L., and Meng, X., editors (2011). *Handbook of Markov Chain Monte Carlo*. Handbooks of Modern Statistical Methods. Chapman & Hall/CRC.

Burnham, K. and Anderson, D. (2002). *Model Selection and Multimodel Inference: A Practical Information-Theoretic Approach*. Springer-Verlag, 2nd edition.

Campbell, D. T. (1985). Toward an epistemologically-relevant sociology of science. *Science, Technology, & Human Values*, 10(1):38–48.

Carroll, R. J., Ruppert, D., Stefanski, L. A., and Crainiceanu, C. M. (2012). *Measurement Error in Nonlinear Models: A Modern Perspective*. CRC Press, 2nd edition.

Casella, G. and George, E. I. (1992). Explaining the Gibbs sampler. *The American Statistician*, 46(3):167–174.

Caticha, A. and Griffin, A. (2007). Updating probabilities. In Mohammad-Djafari, A., editor, *Bayesian Inference and Maximum Entropy Methods in Science and Engineering*, volume 872 of *AIP Conf. Proc.*

Cho, A. (2011). Superluminal neutrinos: Where does the time go? *Science*, 334(6060):1200–1201.

Claeskens, G. and Hjort, N. (2008). *Model Selection and Model Averaging.* Cambridge University Press.

Clark, J. S. (2012). The coherence problem with the unified neutral theory of biodiversity. *Trends in Ecology and Evolution*, 27:198–2002.

Collins, H. M. and Pinch, T. (1998). *The Golem: What You Should Know about Science.* Cambridge University Press, 2nd edition.

Cox, R. T. (1946). Probability, frequency and reasonable expectation. *American Journal of Physics*, 14:1–10.

Cushman, F., Young, L., and Hauser, M. (2006). The role of conscious reasoning and intuition in moral judgment: Testing three principles of harm. *Psychological Science*, 17(12):1082–1089.

Daston, L. J. and Galison, P. (2007). *Objectivity.* MIT Press, Cambridge, MA.

Elias, P. (1958). Two famous papers. *IRE Transactions: on Information Theory*, 4:99.

Fanelli, D. (2012). Negative results are disappearing from most disciplines and countries. *Scientometrics*, 90(3):891–904.

Ferguson, C. J. and Heene, M. (2012). A vast graveyard of undead theories: Publication bias and psychological science's aversion to the null. *Perspectives on Psychological Science*, 7(6):555–561.

Feynman, R. (1967). *The character of physical law.* MIT Press.

Fienberg, S. E. (2006). When did Bayesian inference become "Bayesian"? *Bayesian Analysis*, 1(1):1–40.

Fisher, R. A. (1925). *Statistical Methods for Research Workers.* Oliver and Boyd, Edinburgh.

Fisher, R. A. (1955). Statistical methods and scientific induction. *Journal of the Royal Statistical Society B*, 17(1):69–78.

Fisher, R. A. (1956). *Statistical methods and scientific inference.* Hafner, New York, NY.

Fontani, M., Costa, M., and Orna, M. V. (2014). *The Lost Elements: The Periodic Table's Shadow Side.* Oxford University Press, Oxford.

Forer, B. (1949). The fallacy of personal validation: A classroom demonstration of gullibility. *Journal of Abnormal and Social Psychology*, 44:118–123.

Franco, A., Malhotra, N., and Simonovits, G. (2014). Publication bias in the social sciences: Unlocking the file drawer. *Science*, 345:1502–1505.

Frank, S. (2007). *Dynamics of Cancer: Incidence, Inheritance, and Evolution.* Princeton University Press, Princeton, NJ.

Frank, S. A. (2009). The common patterns of nature. *Journal of Evolutionary Biology*, 22:1563–1585.

Frank, S. A. (2011). Measurement scale in maximum entropy models of species abundance. *Journal of Evolutionary Biology*, 24:485–496.

Fullerton, A. S. (2009). A conceptual framework for ordered logistic regression models. *Sociological Methods & Research*, 38(2):306–347.

Galton, F. (1989). Kinship and correlation. *Statistical Science*, 4(2):81–86.

Gelfand, A. E., Sahu, S. K., and Carlin, B. P. (1995). Efficient parameterisations for normal linear mixed models. *Biometrika*, (82):479–488.

Gelman, A. (2005). Analysis of variance: Why it is more important than ever. *The Annals of Statistics*, 33(1):1–53.

Gelman, A. (2006). Prior distributions for variance parameters in hierarchical models. *Bayesian Analysis*, 1:515–534.

Gelman, A., Carlin, J. C., Stern, H. S., Dunson, D. B., Vehtari, A., and Rubin, D. B. (2013a). *Bayesian Data Analysis.* Chapman & Hall/CRC, 3rd edition.

Gelman, A. and Hill, J. (2007). *Data Analysis Using Regression and Multilevel/Hierarchical Models.* Cambridge University Press.

Gelman, A., Hwang, J., and Vehtari, A. (2013b). Understanding predictive information criteria for Bayesian models.

Gelman, A. and Loken, E. (2013). The garden of forking paths: Why multiple comparisons can be a problem, even when there is no 'fishing expedition' or 'p-hacking' and the research hypothesis was posited ahead of time. Technical report, Department of Statistics, Columbia University.

Gelman, A. and Loken, E. (2014). Ethics and statistics: The AAA tranche of subprime science. *CHANCE*, 27(1):51–56.

Gelman, A. and Nolan, D. (2002). *Teaching Statistics: A Bag of Tricks*. Oxford University Press.

Gelman, A. and Robert, C. P. (2013). "Not only defended but also applied": The perceived absurdity of Bayesian inference. *The American Statistician*, 67(1):1–5.

Gelman, A. and Rubin, D. (1992). Inference from iterative simulation using multiple sequences. *Statistical Science*, 7:457–511.

Gelman, A. and Rubin, D. B. (1995). Avoiding model selection in Bayesian social research. *Sociological Methodology*, 25:165–173.

Gelman, A. and Stern, H. (2006). The difference between "significant" and "not significant" is not itself statistically significant. *The American Statistician*, 60(4):328–331.

Geman, S. and Geman, D. (1984). Stochastic relaxation, Gibbs distributions, and the Bayesian restoration of images. *IEEE Transactions on Pattern Analysis and Machine Intelligence*, 6(6):721–741.

Gigerenzer, G. and Hoffrage, U. (1995). How to improve Bayesian reasoning without instruction: Frequency formats. *Psychological Review*, 102:684–704.

Gigerenzer, G., Krauss, S., and Vitouch, O. (2004). The null ritual: What you always wanted to know about significance testing but were afraid to ask. In Kaplan, D., editor, *The Sage handbook of quantitative methodology for the social sciences*, pages 391–408. Sage Publications, Inc., Thousand Oaks.

Gigerenzer, G., Swijtink, Z., Porter, T., Daston, L., Beatty, J., and Kruger, L. (1990). *The Empire of Chance: How Probability Changed Science and Everyday Life*. Cambridge University Press.

Gigerenzer, G., Todd, P., and The ABC Research Group (2000). *Simple Heuristics That Make Us Smart*. Oxford University Press, Oxford.

Gilad, Y. and Mizrahi-Man, O. (2015). A reanalysis of mouse encode comparative gene expression data. *F1000Research*, 4(121).

Gillespie, J. H. (1977). Sampling theory for alleles in a random environment. *Nature*, 266:443–445.

Grafen, A. and Hails, R. (2002). *Modern Statistics for the Life Sciences*. Oxford University Press, Oxford.

Griffin, A. (2008). *Maximum Entropy: The Universal Method for Inference*. PhD thesis, University of Albany, State University of New York, Department of Physics.

Grosberg, A. (1998). Entropy of a knot: Simple arguments about difficult problem. In Stasiak, A., Katrich, V., and Kauffman, L. H., editors, *Ideal Knots*, pages 129–142. World Scientific.

Grünwald, P. D. (2007). *The Minimum Description Length Principle*. MIT Press, Cambridge MA.

Hacking, I. (1983). *Representing and Intervening: Introductory Topics in the Philosophy of Natural Science*. Cambridge University Press, Cambridge.

Hahn, M. W. and Bentley, R. A. (2003). Drift as a mechanism for cultural change: an example from baby names. *Proceedings of the Royal Society B*, 270:S120–S123.

Harte, J. (2011). *Maximum Entropy and Ecology: A Theory of Abundance, Distribution, and Energetics*. Oxford Series in Ecology and Evolution. Oxford University Press, Oxford.

Hastie, T., Tibshirani, R., and Friedman, J. (2009). *The Elements of Statistical Learning: Data Mining, Inference, and Prediction*. Springer, 2nd edition.

Hastings, W. (1970). Monte Carlo sampling methods using Markov chains and their applications. *Biometrika*, 57(1):97–109.

Hauer, E. (2004). The harm done by tests of significance. *Accident Analysis & Prevention*, 36:495–500.

Henrion, M. and Fischoff, B. (1986). Assessing uncertainty in physcial constants. *American Journal of Physics*, 54:791–798.

Herndon, T., Ash, M., and Pollin, R. (2014). Does high public debt consistently stifle economic growth? A critique of Reinhart and Rogoff. *Cambridge Journal of Economics*, 38(2):257–279.

Hilbe, J. M. (2011). *Negative Binomial Regression*. Cambridge University Press, Cambridge, 2nd edition.

Hinde, K. and Milligan, L. M. (2011). Primate milk synthesis: Proximate mechanisms and ultimate perspectives. *Evolutionary Anthropology*, 20:9–23.

Hoffman and Gelman (2011). The No-U-Turn Sampler: Adaptively setting path lengths in Hamiltonian Monte Carlo.

Horton, R. (2015). What is medicine's 5 sigma? *The Lancet*, 385(April 11):1380.

Howell, N. (2000). *Demography of the Dobe !Kung*. Aldine de Gruyter, New York.

Howell, N. (2010). *Life Histories of the Dobe !Kung: Food, Fatness, and Well-being over the Life-span*. Origins of Human Behavior and Culture. University of California Press.

Hubbell, S. P. (2001). *The Unified Neutral Theory of Biodiversity and Biogeography*. Princeton University Press, Princeton.

Hull, D. L. (1988). *Science as a Process: An Evolutionary Account of the Social and Conceptual Development of Science*. University of Chicago Press, Chicago, IL.

Ioannidis, J. P. A. (2005). Why most published research findings are false. *PLoS Medicine*, 2(8):0696–0701.

Ioannidis, J. P. A. (2012). Why science is not necessarily self-correction. *Perspectives on Psychological Science*, 7(6):645–654.

Jaynes, E. T. (1976). Confidence intervals vs Bayesian intervals. In Harper, W. L. and Hooker, C. A., editors, *Foundations of Probability Theory, Statistical Inference, and Statistical Theories of Science*, page 175.

Jaynes, E. T. (1984). The intutive inadequancy of classical statistics. *Epistemologia*, 7:43–74.

Jaynes, E. T. (1985). Highly informative priors. *Bayesian Statistics*, 2:329–360.

Jaynes, E. T. (1986). Monkeys, kangaroos and *N*. In Justice, J. H., editor, *Maximum-Entropy and Bayesian Methods in Applied Statistics*, page 26. Cambridge University Press, Cambridge.

Jaynes, E. T. (1988). The relation of Bayesian and maximum entropy methods. In Erickson, G. J. and Smith, C. R., editors, *Maximum Entropy and Bayesian Methods in Science and Engineering*, volume 1, pages 25–29. Kluwer Academic Publishers.

Jaynes, E. T. (2003). *Probability Theory: The Logic of Science*. Cambridge University Press.

Jung, K., Shavitt, S., Viswanathan, M., and Hilbe, J. M. (2014). Female hurricanes are deadlier than male hurricanes. *Proceedings of the National Academy of Sciences USA*, 111(24):8782–8787.

Kadane, J. B. (2011). *Principles of Uncertainty*. Chapman & Hall/CRC.

Kitcher, P. (2000). Reviving the sociology of science. *Philosophy of Science*, 67:S33–S44.

Kitcher, P. (2011). *Science in a Democratic Society*. Prometheus Books, Amherst, New York.

Kline, M. A. and Boyd, R. (2010). Population size predicts technological complexity in Oceania. *Proc. R. Soc. B*, 277:2559–2564.

Kruscke, J. K. (2011). *Doing Bayesian Data Analysis*. Academic Press, Burlington, MA.

Kullback, S. (1959). *Information theory and statistics*. John Wiley and Sons, NY.

Kullback, S. (1987). The Kullback-Leibler distance. *The American Statistician*, 41(4):340.

Kullback, S. and Leibler, R. A. (1951). On information and sufficiency. *Annals of Mathematical Statistics*, 22(1):79–86.

Lambert, D. (1992). Zero-inflated Poisson regression, with an application to defects in manufacturing. *Technometrics*, 34:1–14.

Lansing, J. S. and Cox, M. P. (2011). The domain of the replicators: Selection, neutrality, and cultural evolution (with commentary). *Current Anthropology*, 52:105–125.

Laudan, L. (1981). A confutation of convergent realism. *Philosophy of Science*, 48(1):19–49.

Lee, R. B. and DeVore, I., editors (1976). *Kalahari Hunter-Gatherers: Studies of the !Kung San and Their Neighbors*. Harvard University Press, Cambridge.

Levins, R. (1966). The strategy of model building in population biology. *American Scientist*, 54.

Lewandowski, D., Kurowicka, D., and Joe, H. (2009). Generating random correlation matrices based on vines and extended onion method. *Journal of Multivariate Analysis*, 100:1989–2001.

Lightsey, J. D., Rommel, S. A., Costidis, A. M., and Pitchford, T. D. (2006). Methods used during gross necropsy to determine watercraft-related mortality in the Florida manatee (*Trichechus manatus latirostris*). *Journal of Zoo and Wildlife Medicine*, 37(3):262–275.

Lin, S., Lin, Y., Nery, J. R., Urich, M. A., Breschi, A., Davis, C. A., Dobin, A., Zaleski, C., Beer, M. A., Chapman, W. C., Gingeras, T. R., Ecker, J. R., and Snyder, M. P. (2014). Comparison of the transcriptional landscapes between human and mouse tissues. *Proc. Natl. Acad. Sci. U.S.A.*, 111(48):17224–17229.

Lindley, D. V. (1971). Estimation of many parameters. In Godambe, V. P. and Sprott, D. A., editors, *Foundations of Statistical Inference*. Holt, Rinehart and Winston, Toronto.

Lunn, D., Jackson, C., Best, N., Thomas, A., and Spiegelhalter, D. (2013). *The BUGS Book*. CRC Press.

MacKay, D. J. C. and Neal, R. M. (1994). Automatic relevance determination for neural networks. Technical report, Cambridge University.

Mangel, M. and Samaniego, F. (1984). Abraham Wald's work on aircraft survivability. *Journal of the American Statistical Association*, 79:259–267.

Marin, J.-M. and Robert, C. (2007). *Bayesian Core: A Practical Approach to Computational Bayesian Statistics*. Springer.

McCullagh, P. (1980). Regression models for ordinal data. *Journal of the Royal Statistical Society, Series B*, 42:109–142.

McCullagh, P. and Nelder, J. A. (1989). *Generalized Linear Models*. Chapman & Hall/CRC, Boca Raton, Florida, 2nd edition.

McElreath, R. and Smaldino, P. (2015). Replication, communication, and the population dynamics of scientific discovery. *PLoS One*, 10(8):e0136088. doi:10.1371/journal.pone.0136088.

McGrayne, S. B. (2011). *The Theory That Would Not Die: How Bayes' Rule Cracked the Enigma Code, Hunted Down Russian Submarines, and Emerged Triumphant from Two Centuries of Controversy*. Yale University Press.

McHenry, H. M. and Coffing, K. (2000). *Australopithecus* to *Homo*: Transformations in body and mind. *Annual Review of Anthropology*, 29:125–146.

Meehl, P. E. (1956). Wanted—a good cookbook. *The American Psychologist*, 11:263–272.

Meehl, P. E. (1967). Theory-testing in psychology and physics: A methodological paradox. *Philosophy of Science*, 34:103–115.

Meehl, P. E. (1990). Why summaries of research on psychological theories are often uninterpretable. *Psychological Reports*, 66:195–244.

Metropolis, N., Rosenbluth, A., Rosenbluth, M., Teller, A., and Teller, E. (1953). Equations of state calculations by fast computing machines. *Journal of Chemical Physics*, 21(6):1087–1092.

Metropolis, N. and Ulam, S. (1949). The Monte Carlo method. *Journal of the American Statistical Association*, 44(247):335–341.

Molenberghs, G., Fitzmaurice, G., Kenward, M. G., Tsiatis, A., and Verbeke, G. (2014). *Handbook of Missing Data Methodology*. CRC Press.

Morison, S. E. (1942). *Admiral of the Ocean Sea: A Life of Christopher Columbus*. Little, Brown and Company, Boston.

Mulkay, M. and Gilbert, G. N. (1981). Putting philosophy to work: Karl Popper's influence on scientific practice. *Philosophy of the Social Sciences*, 11:389–407.

Neal, R. (1996). *Bayesian Learning for Neural Networks*. Springer-Verlag, New York.

Neal, R. M. (1998). Regression and classification using Gaussian process priors. In Bernardo, J. M., editor, *Bayesian Statistics*, volume 6, pages 475–501. Oxford University Press.

Nelder, J. and Wedderburn, R. (1972). Generalized linear models. *Journal of the Royal Statistical Society, Series A*, 135:370–384.

Nettle, D. (1998). Explaining global patterns of language diversity. *Journal of Anthropological Archaeology*, 17:354–74.

Nieuwenhuis, S., Forstmann, B. U., and Wagenmakers, E.-J. (2011). Erroneous analyses of interactions in neuroscience: a problem of significance. *Nature Neuroscience*, 14(9):1105–1107.

Nunn, N. and Puga, D. (2011). Ruggedness: The blessing of bad geography in Africa. *Review of Economics and Statistics*.

Nuzzo, R. (2014). Statistical errors. *Nature*, 506:150–152.

Ohta, T. and Gillespie, J. H. (1996). Development of neutral and nearly neutral theories. *Theoretical Population Biology*, 49:128–142.

O'Rourke, K. and Detsky, A. S. (1989). Meta-analysis in medical research: Strong encouragement for higher quality in individual research efforts. *Journal of Clinical Epidemiology*, 42(10):1021–1024.

Papaspiliopoulos, O., Roberts, G. O., and Skold, M. (2007). A general framework for the parametrization of hierarchical models. *Statistical Science*, (22):59–73.

Pearl, J. (2014). Understanding Simpson's paradox. *The American Statistician*, 68:8–13.

Polson, N. G. and Scott, J. G. (2012). On the half-Cauchy prior for a global scale parameter. *Bayesian Analysis*, 7:887–902.

Popper, K. (1963). *Conjectures and Refutations: The Growth of Scientific Knowledge*. Routledge, New York.

Popper, K. (1996). *The Myth of the Framework: In Defence of Science and Rationality*. Routledge.

Proulx, S. R. and Adler, F. R. (2010). The standard of neutrality: still flapping in the breeze? *Journal of Evolutionary Biology*, 23:1339–1350.

Rao, C. R. (1997). *Statistics and Truth: Putting Chance To Work*. World Scientific Publishing.

Reilly, C. and Zeringue, A. (2004). Improved predictions of lynx trappings using a biologial model. In Gelman, A. and Meng, X., editors, *Applied Bayesian Modeling and Causal Inference from Incomplete-Data Perspectives*, pages 297–308. John Wiley and Sons.

Reinhart, C. and Rogoff, K. (2010). Growth in a time of debt. *American Economic Review*, 100(2):573–578.

Rice, K. (2010). A decision-theoretic formulation of Fisher's approach to testing. *The American Statistician*, 64(4):345–349.

Riley, S. J., DeGloria, S. D., and Elliot, R. (1999). A terrain ruggedness index that quantifies topographic heterogeneity. *Intermountain Journal of Sciences*, 5:23–27.

Robert, C. and Casella, G. (2011). A short history of Markov chain Monte Carlo: Subjective recollections from incomplete data. In Brooks, S., Gelman, A., Jones, G., and Meng, X.-L., editors, *Handbook of Markov Chain Monte Carlo*, chapter 2. CRC Press.

Robert, C. P. (2007). *The Bayesian Choice: from decision-theoretic foundations to computational implementation*. Springer Texts in Statistics. Springer, 2nd edition.

Roberts, G. O. and Sahu, S. K. (1997). Updating schemes, correlation structure, blocking and parameterisation for the Gibbs sampler. *Journal of the Royal Statistical Society, Series B*, (59):291–317.

Rommel, S. A., Costidis, A. M., Pitchford, T. D., Lightsey, J. D., Snyder, R. H., and Haubold, E. M. (2007). Forensic methods for characterizing watercraft from watercraft-induced wounds on the Florida manatee (*Trichechus manatus latirostris*). *Marine Mammal Science*, 23(1):110–132.

Rosenbaum, P. R. (1984). The consequences of adjustment for a concomitant variable that has been affected by the treatment. *Journal of the Royal Statistical Society A*, 147(5):656–666.

Rosenthal, R. (1979). The file drawer problem and tolerance for null results. *Psychological Bulletin*, 86(3):638–641.

Rubin, D. B. (1976). Inference and missing data. *Biometrika*, 63:581–592.

Rubin, D. B. (2005). Causal inference using potential outcomes: Design, modeling, decisions. *Journal of the American Statistical Association*, 100(469):322–331.

Rubin, D. B. and Little, R. J. A. (2002). *Statistical analysis with missing data*. Wiley, New York, 2nd edition.

Sankararaman, S., Patterson, N., Li, H., Pääbo, S., and Reich, D. (2012). The date of interbreeding between Neandertals and modern humans. *PLoS Genetics*, 8(10):e1002947.

Savage, L. J. (1962). *The Foundations of Statistical Inference*. Methuen.

Schwarz, G. E. (1978). Estimating the dimension of a model. *Annals of Statistics*, 6:461–464.

Sedlemeier, P. and Gigerenzer, G. (1989). Do studies of statistical power have an effect on the power of studies? *Psychological Bulletin*, 105(2):309–316.

Senn, S. (2003). A conversation with John Nelder. *Statistical Science*, 18:118–131.

Shannon, C. E. (1948). A mathematical theory of communication. *The Bell System Technical Journal*, 27:379–423.

Shannon, C. E. (1956). The bandwagon. *IRE Transactions: on Information Theory*, 2:3.

Silk, J. B., Brosnan, S. F., Vonk, J., Henrich, J., Povinelli, D. J., Richardson, A. S., Lambeth, S. P., Mascaro, J., and Schapiro, S. J. (2005). Chimpanzees are indifferent to the welfare of unrelated group members. *Nature*, 437:1357–1359.

Silver, N. (2012). *The Signal and the Noise: Why So Many Predictions Fail—but Some Don't*. Penguin Press, New York.

Simmons, J. P., Nelson, L. D., and Simonsohn, U. (2011). False-positive psychology: Undisclosed flexibility in data collection and analysis allows presenting anything as significant. *Psychological Science*, 22:1359–1366.

Simmons, J. P., Nelson, L. D., and Simonsohn, U. (2013). Life after p-hacking. SSRN Scholarly Paper ID 2205186, Social Science Research Network, Rochester, NY.

Simon, H. (1969). *The Sciences of the Artificial*. MIT Press, Cambridge, Mass.

Simpson, D. P., Martins, T. G., Riebler, A., Fuglstad, G.-A., Rue, H., and Sørbye, S. H. (2014). Penalising model component complexity: A principled, practical approach to constructing priors. *arXiv:1403.4630v3*.

Simpson, E. H. (1951). The interpretation of interaction in contingency tables. *Journal of the Royal Statistical Society, Series B*, 13:238–241.

Sober, E. (2008). *Evidence and Evolution: The logic behind the science*. Cambridge University Press, Cambridge.

Speed, T. (1986). Questions, answers and statistics. In *International Conference on Teaching Statistics 2*. International Association for Statistical Education.

Stein, C. (1955). Inadmissibility of the usual estimator for the mean of a multivatiate normal distribution. In *Proceedings of the Third Berkeley Symposium of Mathematical Statistics and Probability*, volume 1, pages 197–206, Berkeley. University of California Press.

Stigler, S. M. (1981). Gauss and the invention of least squares. *The Annals of Statistics*, 9(3):465–474.

Stone, M. (1977). An asymptotic equivalence of choice of model by cross-validation and Akaike's criterion. *Journal of the Royal Statistical Society B*, 39(1):44–47.

Theobald, D. L. (2010). A formal test of the theory of universal common ancestry. *Nature*, 465:219–222.

Van Horn, K. S. (2003). Constructing a logic of plausible inference: A guide to Cox's theorem guide to Cox's theorem. *International Journal of Approximate Reasoning*, 34:3–24.

Vehtari, A. and Gelman, A. (2014). WAIC and cross-validation in Stan. Technical report, Aalto University.

Venn, J. (1876). *The Logic of Chance*. Macmillan and co, New York, 2nd edition.

Vonesh, J. R. and Bolker, B. M. (2005). Compensatory larval responses shift trade-offs associated with predator-induced hatching plasticity. *Ecology*, 86:1580–1591.

Wald, A. (1939). Contributions to the theory of statistical estimation and testing hypotheses. *Annals of Mathematical Statistics*, 10(4):299–326.

Wald, A. (1943). A method of estimating plane vulnerability based on damage of survivors. Technical report, Statistical Research Group, Columbia University.

Wald, A. (1950). *Statistical Decision Functions*. J. Wiley, New York.

Wang, W., Rothschild, D., Goel, S., and Gelman, A. (2015). Forecasting elections with non-representative polls. *International Journal of Forecasting*, 31(3):980–991.

Watanabe, S. (2010). Asymptotic equivalence of Bayes cross validation and Widely Applicable Information Criterion in singular learning theory. *Journal of Machine Learning Research*, 11:3571–3594.

Wearing, D. (2005). *Forever Today: A True Story of Lost Memory and Never-Ending Love*. Doubleday.

Welsh, Jr., H. H. and Lind, A. (1995). Habitat correlates of the Del Norte salamander, Plethodon elongatus (Caudata: Plethodontidae) in northwestern California. *Journal of Herpetology*, 29:198–210.

Williams, D. A. (1975). The analysis of binary responses from toxicological experiments involving reproduction and teratogenicity. *Biometrics*, 31:949–952.

Williams, D. A. (1982). Extra-binomial variation in logistic linear models. *Journal of the Royal Statistical Society, Series C*, 31(2):144–148.

Williams, P. M. (1980). Bayesian conditionalisation and the principle of minimum information. *British Journal for the Philosophy of Science*, 31:131–144.

Wittgenstein, L. (1953). *Philosophical Investigations*.

Wolpert, D. and Macready, W. (1997). No free lunch theorems for optimization. *IEEE Transactions on Evolutionary Computation*, page 67.

Citation index

Topic index